DRUG DELIVERY

Longqin Hu, Department of Pharmaceutical Chemistry, Ernest Mario School of Pharmacy, Rutgers, The State University of New Jersey, Piscataway, NJ 08854

W. Griffith Humphreys, Department of Metabolism and Pharmacokinetics, Bristol-Myers Squibb Pharmaceutical Research Institute, P.O. Box 5400, Princeton, NJ 08543

D. Nedra Karunaratne, Department of Pharmaceutical Chemistry, School of Pharmacy, The University of Kansas, 2095 Constant Avenue, Lawrence, KS 66047

Naoki Kobayashi, Department of Biopharmaceutics and Drug Metabolism, Graduate School of Pharmaceutical Sciences, Kyoto University, Sakyo-ku, Kyoto 606-8501, Japan

Lisa A. Kueltzo, Department of Pharmaceutical Sciences, University of Colorado Health Sciences Center, Denver, CO 80262

Christopher P. Leamon, Endocyte, Inc., 1205 Kent Avenue, West Lafayette, IN 47906

Philip S. Low, Purdue University, West Lafayette, IN 47907

Terry O. Matsunaga, Imarx Therapeutics Inc., Tucson, AZ 85719

C. Russell Middaugh, Department of Pharmaceutical Chemistry, University of Kansas, Lawrence, KS 66047

Donald W. Miller, Department of Pharmaceutical Sciences, University of Nebraska Medical Center, Omaha, NE 68198-6025

Marilyn E. Morris, Department of Pharmaceutical Sciences, School of Pharmacy and Pharmaceutical Sciences, University at Buffalo, State University of New York, Amherst, NY 14260

John Mountzouris, Abgent Inc., 6310 Nancy Ridge Drive, Suite 106, San Diego, CA 92121

Eric J. Munson, Department of Pharmaceutical Chemistry, University of Kansas, 2095 Constant Avenue, Lawrence, KS 66047

Ka-yun Ng, Department of Pharmaceutical Sciences, School of Pharmacy, University of Colorado Health Sciences Center, Denver, CO 80262

Shihong Nicolaou, Technology Transfer and Intellectual Property Services, University of California, San Diego, 9500 Gilman Drive, La Jolla, CA 92093

Makiya Nishikawa, Department of Biopharmaceutics and Drug Metabolism, Graduate School of Pharmaceutical Sciences, Kyoto University, Sakyo-ku, Kyoto 606-8501, Japan

Erik Rytting, Department of Pharmaceutical Chemistry, School of Pharmacy, The University of Kansas, 2095 Constant Avenue, Lawrence, KS 66047

CONTRIBUTORS

Kenneth L. Audus, Department of Pharmaceutical Chemistry, School of Pharmacy, The University of Kansas, 2095 Constant Avenue, Lawrence, KS 66047

Dewey H. Barich, Department of Pharmaceutical Chemistry, University of Kansas, 2095 Constant Avenue, Lawrence, KS 66047

John A. Bontempo, Biopharmaceutical Product Development Consultant, 18 Benjamin Street, Somerset, NJ 08873

Anna Maria Calcagno, Department of Pharmaceutical Chemistry, The University of Kansas, Lawrence, KS 66047

Hervé Le Calvez, Abgent Inc., 6310 Nancy Ridge Drive, Suite 106, San Diego, CA 92121

William F. Elmquist, Department of Pharmaceutics, University of Minnesota, Minneapolis, MN 55455

Fang Fang, NexBio, Inc., 6330 Nancy Ridge Drive, Suite 105, San Diego, CA 92121

Kosi Gramatikoff, Abgent Inc., 6310 Nancy Ridge Drive, Suite 106, San Diego, CA 92121

Xiangming Guan, Department of Pharmaceutical Sciences, College of Pharmacy, South Dakota State University, Brookings, SD 57007

Chao Han, GlaxoSmithKline, Collegeville, PA 19426

Anthony J. Hickey, School of Pharmacy, University of North Carolina, Kerr Hall, Chapel Hill, NC 27599 important to drug delivery, and should be an excellent desk reference for medicinal chemists interested in gaining an overview of this field. In addition, there is a need to strengthen the education of our future medicinal chemists in the area of drug delivery. This book can also serve as a textbook for graduate students or advanced level undergraduate students interested in a career in the pharmaceutical industry.

The book starts with chapters that cover general drug delivery issues such as physicochemical and biological barriers, various pathways for drug delivery, formulation, pharmacokinetic and pharmacodynamic issues, metabolism, and cell culture models used in studying drug delivery. Then it moves on to cover specific drug delivery strategies. At the end, we have added one chapter on intellectual property so as to give readers a general idea of how to protect their intellectual property when doing drug delivery research. Each chapter is structured in such a way that it gives an overview of the specific subject, and also goes into details with selected examples so that there is an in depth discussion with extensive references. With this kind of structure, the book will be valuable to both novices and experts.

We would like to thank Ms. Neeta Raje for her diligent work in assisting BW in organizing and coordinating the editing and processing of the manuscripts.

BINGHE WANG
TERUNA SIAHAAN
RICHARD SOLTERO

PREFACE

The pharmaceutical industry, at least the part represented by the major pharmaceutical companies, has undergone a fundamental transformation during the last decade. This transformation has changed the "silo" structure, where different "steps" of the drug discovery process were treated independent of each other, to that of an integrated approach. This transformation was brought about by several factors. One important reason was that about 60% of new chemical entities (NCE) fail clinical trials because of poor pharmacokinetics, undesirable metabolic properties, and toxicity. NCEs do not fail due to a lack of biological activities per se. In addition, the transformation came about because of development in areas such as combinatorial chemistry, high throughput screening, genomics, and proteomics that has allowed for the rapid identification of a large number of biologically active compounds. Therefore, there is a tremendous pressure to have a mechanism to identify a "winner" among a large pool of candidates or "play out" the failures early on. An integrated approach helps to achieve this goal by evaluating various factors such as formulation, permeation, metabolism, and toxicity early in the drug discovery and development process. Such a practice also helps to bring issues that impact development to the awareness of medicinal chemists so that many structural features detrimental to clinical development would be avoided at the designing stage.

Among all the factors that effect the clinical development of a NCE, drug delivery occupies a special place. In an integrated drug discovery approach, one has to start considering the delivery properties of a NCE at the design stage. This means that medicinal chemists are the first line of researchers who have to consider this issue. However, most entry-level medicinal chemists are trained as synthetic chemists, and have very little exposure to drug delivery. Without a basic understanding of the issues that effect drug delivery properties, it is hard for the medicinal chemists to address this issue. This book systematically examines various subject areas

CONTENTS

Published by John Wiley & Sons, Inc., Hoboken, New Jersey.
Published simultaneously in Canada.

Library of Congress Cataloging-in-Publication Data:
Wang, Binghe.
Drug delivery : principles and applications / Binghe Wang, Teruna Siahaan, Richard soltero.
p. cm.
Includes index.
ISBN 0-471-47489-4 (cloth)
1. Drug delivery systems. 2. Pharmaceutical chemistry. I. Siahaan, Teruna. II. Soltero, Richard. III. Title.
RS199.5.W36 2005
615′.6–dc22 2004020585

Printed in the United States of America

10 9 8 7 6 5 4 3 2 1

DRUG DELIVERY:

Principles and Applications

BINGHE WANG
Georgia State University

TERUNA SIAHAAN
University of Kansas

RICHARD SOLTERO
DirectPharma, Inc.

WILEY-
INTERSCIENCE

A JOHN WILEY & SONS, INC. PUBLICATION

Teruna J. Siahaan, Department of Pharmaceutical Chemistry, The University of Kansas, Lawrence, KS 66047

Peter S. Silverstein, Department of Pharmaceutical Chemistry, School of Pharmacy, The University of Kansas, 2095 Constant Avenue, Lawrence, KS 66047

Rick Soltero, Pharma Directions, Inc.

Yoshinobu Takakura, Department of Biopharmaceutics and Drug Metabolism, Graduate School of Pharmaceutical Sciences, Kyoto University, Sakyo-ku, Kyoto 606-8501, Japan

Veena Vasandani, Department of Pharmaceutical Chemistry, School of Pharmacy, The University of Kansas, 2095 Constant Avenue, Lawrence, KS 66047

Binghe Wang, Department of Chemistry, Georgia State University, Atlanta, GA 30303

Guijun Wang, Department of Chemistry, University of New Orleans, New Orleans, LA 70148

Amber M. Young, Department of Pharmaceutical Chemistry, School of Pharmacy, The University of Kansas, 2095 Constant Avenue, Lawrence, KS 66047

Bradley Yops, Department of Pharmaceutical Chemistry, School of Pharmacy, The University of Kansas, 2095 Constant Avenue, Lawrence, KS 66047

Mark T. Zell, Pfizer Global Research and Development, Ann Arbor Laboratories, 2800 Plymouth Road, Ann Arbor, MI 48105

Shuzhong Zhang, Department of Pharmaceutical Sciences, School of Pharmacy and Pharmaceutical Sciences, University at Buffalo, State University of New York, Amherst, NY 14260

Yan Zhang, Department of Pharmaceutical Sciences, University of Nebraska Medical Center, Omaha, NE 68198-6025

1

FACTORS THAT IMPACT THE DEVELOPABILITY OF DRUG CANDIDATES: AN OVERVIEW

CHAO HAN
GlaxoSmithKline, Collegeville, PA 19426
BINGHE WANG
Department of Chemistry, Georgia State University, Atlanta, GA 30303

1.1. ISSUES FACING THE PHARMACEUTICAL INDUSTRY

Drug discovery is a long, arduous, and expensive process. It was estimated that the total expenditure for research and development in the U.S. pharmaceutical industries

Drug Delivery: Principles and Applications Edited by Binghe Wang, Teruna Siahaan, and Richard Soltero
ISBN 0-471-47489-4

was over $20 billion a year in the late 1990s,' and this figure has been increasing.[1] The average cost for every new drug (a new chemical entity, NCE) from research laboratory to patients is a staggering number: $400 to $650 million,[2–4] and the whole process may take up to 14 years![5] Because of the high cost, there is tremendous pressure to maximize efficiency and minimize the time it takes to discover and bring a drug to the market. In order to do this, it is necessary to analyze the entire drug discovery and development process and identify steps where changes can be made to increase efficiency and save time. Analyzing the entire drug discovery and development process will help reveal where maximal improvements can be expected with some effort.

The entire endeavor of bringing a new drug from idea to market is generally divided into several stages: target/disease identification, hit identification/discovery, hit optimization, lead selection and further optimization, candidate identification, and clinical trials.[6] Each stage has many aspects and components. A target is identified early in the discovery period, when there is sufficient evidence to validate the relationship between this target and a disease of interest. Tens of thousands of new compounds are then synthesized and screened against the target to identify a few compounds (hits) with the desired biological activity. Analogs of these selected compounds are then screened further for better activity and optimized in order to identify a small number of compounds for testing in pharmacological models. These efficacious compounds (leads) are further optimized for their biopharmaceutical properties, and the most drug-like compounds (drug candidate, only one or two) are then selected for further development. The drug discovery and development path, with emphasis on the discovery stages, is schematically illustrated in Figure 1.1.

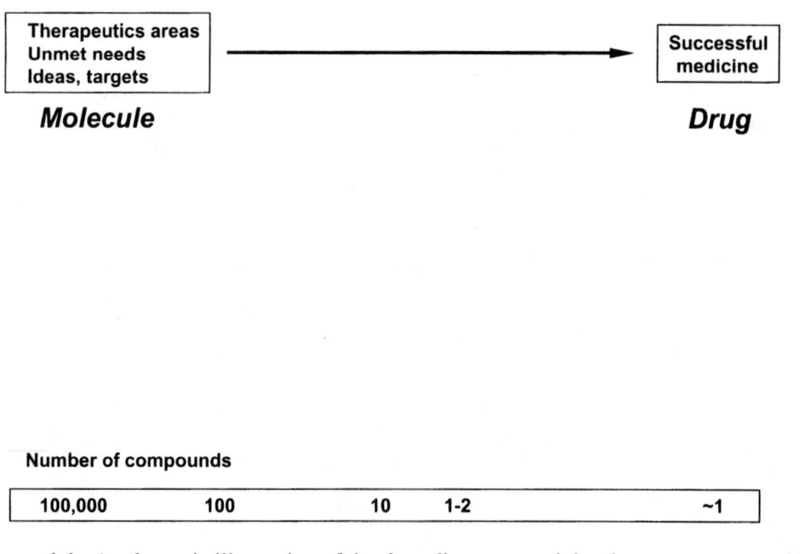

Figure 1.1. A schematic illustration of the drug discovery and development process with the estimated number of compounds shown for each step.

Of those drug candidates with most drug-like properties, only about 40% make their way to evaluation in humans (Phase I clinical trial).[7] Unfortunately, the historical average reveals an almost 90% overall attrition rate in clinical trials;[7] in another words, only 1 compound makes it to market from among 10 compounds tested in humans. Results from another statistical analysis gave a similar success rates for NCEs for which an IND (investigational new drug) was filed during 1990–1992.[8] This high attrition rate obviously does not produce the long-term success desired by both the pharmaceutical and health care industries.

In order to reduce the failure rate, it is necessary to analyze how and where failures occur. More than 10 years ago, Prentis et al.[9] analyzed the cause of the high attrition rate based on data from seven UK-based pharmaceutical companies from 1964 to 1985. The results revealed that 39% of the failure was due to poor pharmacokinetic properties in humans; 29% was due to a lack of clinical efficacy; 21% was due to toxicity and adverse effects; and about 6% was caused by commercial limitations. Although not enough detailed information was available, it is believed that some of these causes are interrelated. For instance, toxicity or lack of efficacy can be caused by poor or undesired pharmacokinetic properties. With the understanding that most failure was not due to a lack of "biological activities" per se as defined by *in vitro* testing, there is a drive to incorporate the evaluation of the other major factors that may potentially precipitate developmental failures in the early drug discovery and candidate selection processes. This is intended to reduce the rate of late-stage failures, which is most costly. This point is further substantiated by the studies indicating that the major cost in drug discovery and development occurs at late stages.[10] For example, in a $400 million total R&D cost,[4] preclinical research costs probably account for only tens of million dollars, whereas clinical studies cost hundreds of millions of dollars (Figure 1.2).

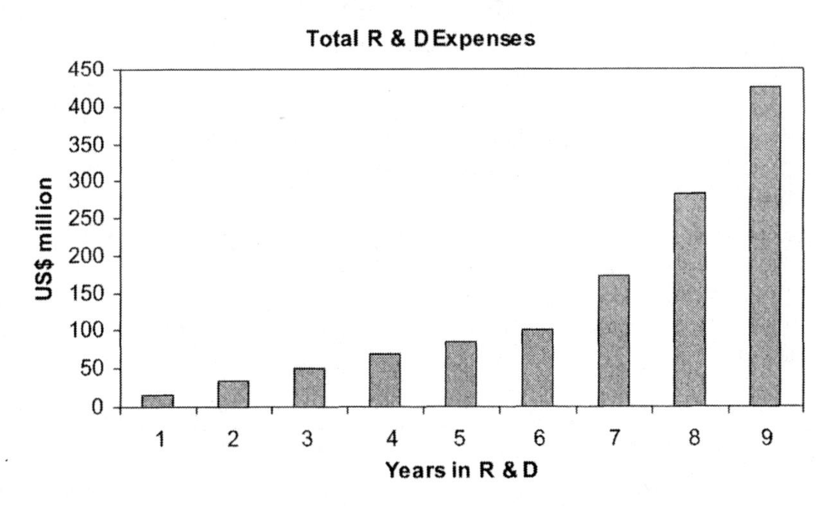

Figure 1.2. Estimated annual expenses based on a hypothetical $400 million total R&D expense for the development of a NCE within a nine-year period (launch in 10th year). Data from Drews, J. and Ryser, S. (1997).[4]

Another factor that is fueling the movement for early integration of multiple disciplines in the drug discovery and development processes is the rapid development of chemical and biological sciences. The past decade has seen tremendous advances in both areas. Advances in combinatorial chemistry, molecular and cellular biology, high-throughput screening, and genomic research have provided both great opportunities and challenges to the pharmaceutical industry. With the rapid development in biological sciences, current interests in therapeutic targets are more focused on rational targets such as receptors, enzymes, and hormones with well-characterized structures and functions. New technologies such as combinatorial chemistry, automation in high-throughput screening, and better instrumentation in bioanalysis have also significantly accelerated the lead identification and discovery process[11] for a given target. With these new technologies, large pharmaceutical research organizations are capable of synthesizing and screening several thousand compounds or more in a year or two to find potential drug candidates.[12] These efforts typically result in the discovery of many lead compounds or potential candidates for a target in the drug discovery process. Then there is the question of how to pick a winner and how to minimize failures. This requires a thorough evaluation of all the factors that are known to affect the developability of a NEC at the early stages. These factors may include efficacy, pharmacokinetics, pharmacodynamics, toxicology, and drug-drug interactions based on the metabolism and substrate properties of certain transporters and enzymes, as well as physicochemical properties, many of which are related to drug delivery issues. For this reason, a drug discovery and development program is more like a symphony (not just a cross-functional action) of multiple sciences including chemistry, biology, toxicology, clinical science, and pharmaceutical engineering.

Under the pressure to reduce the cost and shorten the time needed to bring an NCE to the market, many major pharmaceutical organizations have undergone rapid and drastic changes in the past decade, both in terms of organizational structures and fundamental approaches, in order to develop an integrated approach to drug discovery and development.[13] A conference entitled "Opportunities for Integration of Pharmacokinetics, Pharmacodynamics, and Toxicokinetics in Rational Drug Development"[14] was the landmark event in this fundamental change in the pharmaceutical industry.[15] A brand new concept, "ensuring developability," was introduced and well accepted, which employs criteria for drug development throughout the entire drug discovery and development processes. Under the guidance of such criteria, a drug discovery and development team will not only maximize the chance of success by selecting the best developable drug candidate, but will also play off the failures faster and more cheaply.

The paradigm shifts mostly involve the integration of research activities in functional areas such as pharmacokinetics and drug metabolism, pharmaceutical development, safety assessment, and process chemistry into drug discovery and development process in the very early stages of discovery. The inputs from these functional areas, as well as those from clinical, regulatory, commercial, and marketing groups in the early stages, help to minimize costly mistakes in late stages of development and have become more and more important to the success of the drug discovery and development process. Developability is an overall evaluation of the drug-like properties of a NCE. Many of the recent changes in the pharmaceutical industry have been driven by the concept of ensuring developability. These changes, that is, the integration of multifunctional areas in drug discovery and development, ensure that the NCEs of interest will be successful in every step toward the final goal.

Below is a brief introduction to the factors that impact developability and a discussion on why the examination of drug delivery issues is very important in helping to ensure the developability of a drug candidate.

1.2. FACTORS THAT IMPACT DEVELOPABILITY

In most pharmaceutical companies, many efforts have been made to create a clear framework for selecting compound(s) with minimal ambiguity for further progression. Such a framework is not a simple list of the factors that impact the quality of a drug-like molecule. This framework, which is more often referred as "developability criteria," is a comprehensive summary of the characteristics, properties, and qualities of the NCE(s) of interest, which normally consist of preferred profiles with a minimally acceptable range. The preferred profile describes the optimal goal for selection and further progression of a candidate, whereas the minimum range gives the acceptable properties for a compound that is not ideal but may succeed. Molecules that do not meet the criteria will not be considered further. Such criteria cover all the functional areas in drug development. Some of the major developability considerations are briefly described in the following subsections.

1.2.1. Commercial Goal

It does not need to be emphasized that we are in a business world. Generally speaking, a product needs to be profitable to be viable. Therefore, early inputs from commercial, marketing, and medical outcome professionals are very important for setting up a projective product profile, which profoundly affects the creation of the developability criteria for the intended therapeutics. In general, this portfolio documents the best possible properties of the product and the minimum acceptable ones that may succeed based on the studies of market desires. These studies should be based on the results of professional analyses of the medical care needs, potential market, and existing leading products for the same, similar, or related indications. The following aspects need to be well thought out and fully justified before the commencement of a project: (1) therapeutic strategy; (2) dose form and regimen; and (3) the best possible safety profile, such as the therapeutic window, potential drug interactions, and any other potentially adverse effects. Using the development of an anticancer agent as an example for therapeutic strategy selection, one may consider the choice of developing a chemotherapeutic (directly attacking the cancer cells) versus an antiangiogenic agent (depriving cancer cells of their nutrients), or combined or stand-alone therapy. In deciding the optimal dose form and regimen, one may consider whether an oral or intravenous (iv) formulation, or both, should be developed, and whether the drug should be given once daily or in multiple doses. The results of such an analysis form the framework for developing the developability criteria and become the guideline in setting up the criterion for each desired property. For example, pharmacokinetic properties such as the half-life and oral bioavailability of a drug candidate will have a direct impact on developing a drug that is to be administered orally once a day.

1.2.2. The Chemistry Efforts

Medicinal chemistry is always the starting point and driver of drug discovery programs. In a large pharmaceutical R&D organization, early discovery of bioactive compounds (hits) can be carried out either by random, high-throughput screening of compound libraries, by rational design, or both. Medicinal chemists will then use the structural information of the pharmacophore thus identified to optimize the structures. Chemical tractability needs to be examined carefully at the very beginning when a new chemical series is identified. Functional modifications around the core structure are carefully analysed. After the examination of a small number of compounds, the initial exploratory *structure–activity relationship* (SAR) or quantitative SAR (QSAR) should be developed. Blackie et al.[16] described how the establishment of exploratory SAR helped the discovery of a potent oral bioavailable phospholipase A_2 inhibitor. In this example, numerous substructural changes were made, leading to the most active compounds; this is normally done in parallel with several different chemical series. For medicinal chemists, it is important that many different SARs are considered, developed, and integrated into their efforts at the same time, providing more opportunities to avoid undesirable properties unrelated to their intended biological activities. Such factors, again, may include potential P450 inhibition, permeability, selectivity, stability, solubility, etc.

Structural novelty of the compounds (i.e., can this product be patented?), complexity of synthetic routes, scalability (can the syntheses be scaled up in an industrial way?) and the cost of starting materials (cost of goods at the end of the game), and potential environmental and toxicity issues will all need to be closely examined at early stages of the drug discovery and development processes. It is never too early to put these thoughts into action.

1.2.3. Target Validation in Animal Models

Although drug discovery efforts almost always start with *in vitro* testing, it is well recognized that promising results of such testing do not always translate into efficacy. There are numerous reasons for this to happen, some of which are well understood and others that are not. Therefore, target validation in animal models before clinical trials in humans is a critical step. Before a drug candidate is fully assessed for its safety and brought to a clinical test, demonstration of the efficacy of a biologically active compound (e.g., active in an enzyme binding assay) in pharmacological models (*in vivo*, if available) is considered a milestone in the process of discovering a drug candidate. Many cases exemplify the challenges and importance of pharmacological models. For example, inhibitors of the integrin receptor $\alpha_v\beta_3$ have been shown to inhibit endothelial cell growth, which implies their potential as clinically useful antiangiogenic agents for cancer treatment.[17] However, the proposed mechanism did not work in animal models, although compounds were found to be very active *in vitro*.[18,19] What has been recognized is that the integrin receptor $\alpha_v\beta_3$ may not be the exclusive pathway on which cell growth depends. Its inhibition may induce a compensatory pathway for angiogenesis.

Ideally, an *in vivo* model should comprise all biochemical, cellular, and physiological complexities, as in a real-life system, which may predict the behavior of a potential drug candidate in human much more accurately than an *in vitro* system. In order to have a biological hypothesis tested in the system with validity, a compound has to be evaluated in many other regards. Knowing the pharmacokinetic parameters such as absorption, distribution, and metabolism in the animal species that is used in the pharmacological model is critical. Showing successful drug delivery in an animal model serves as an important milestone.

The pharmacokinetics/pharmacodynamics relationship, systemic and tissue levels of drug exposure, frequency of dosing following which the drug may demonstrate efficacy, and the strength of efficacy are very important factors that may affect further development of an NEC. They are all directly or indirectly related to drug delivery.

1.2.4. Pharmacokinetics and Drug Metabolism

Pharmacokinetics and drug metabolism are more often abbreviated as DMPK. The importance of DMPK in drug discovery and development practices is reflected in the statistics of the attrition rate.[9] Most of the changes in the pharmaceutical industry during the past decade occurred in DMPK[15] and related fields. The overall goal of DMPK in drug discovery and development is to predict the behavior of a drug candidate in humans. Nevertheless, the focus could be different at different stage of the process. Pharmacokinetics (PK) parameters in animal species that will be used in pharmacological (as noted briefly in the previous paragraph) and safety assessment models provide very important insights (systemic and tissue exposures) for those studies. The results of PK studies in several animal species generate the data for physiologically based models or allometric scaling[20,21] to predict the basic pharmacokinetic behavior of a compound in humans. Assays using human tissues, cells, and genetically engineered cell lines provide a tremendous amount of information before the real clinical studies begin. Optimizing DMPK developability factors is immensely beneficial for finding the candidate with best potential for success.[22]

The desirable (or undesirable) biological effects of a drug *in vivo* normally are directly related to its exposure. One of these factors, namely, the total systemic exposure, maximum concentration, or duration of the concentration above a certain level, is usually used as a parameter that is correlated with the drug's efficacy and adverse effects.[23] The exposure at a given dose is governed by (1) the ability of the body to remove the drug as a xenobiotic and (2) the route by which the drug is delivered. Blood or plasma clearance is often used as a measure of the ability to eliminate a drug molecule from the systemic circulation. A low to moderate clearance molecule is desirable in most situations unless a fast-action, short-duration drug is needed.[24]

A drug can be directly introduced into the systemic circulation by several methods. However, for convenience and many other reasons, oral dosage forms are preferred in many situations. Therefore, oral bioavailability of the compound is one of the very important developability criteria for oral drug delivery. Many factors affect the oral bioavailability of a drug. These factors will be discussed in detail in several chapters. In addition to clearance and bioavailability, other major pharmacokinetic parameters also should be evaluated.

Volume of distribution is a conceptual pharmacokinetic parameter that scales the extent of a drug distributed into the tissues. A well-known parameter, elimination half-life, can be derived from clearance and volume of distribution. It is a very important developability criterion that warrants the desired dose regimen. It should be noted here that half-life must be discussed in the context of a biologically relevant concentration. A purely mathematically derived half-life is sometimes biological irrelevant. Some more definitive explanations and comprehensive discussion of the major pharmacokinetic parameters and their biological relevance have been extensively reviewed.[25,26] These parameters should be examined across several different preclinical species to predict the behavior in humans. The DMPK topics will be discussed in Chapters 5 and 6.

Inhibition and induction of drug-metabolizing enzymes,[27,28] P-glycoprotein (P-gp) substrate property,[29,30] plasma protein binding and binding kinetics,[31,32] and metabolic stability in the microsomes or hepatocytes from different species including humans,[33] as well as the metabolic pathway and the metabolite identified,[34] are all very important developability measurements in the assessment of safety, potential drug-drug interaction, and predictability. These factors need to be optimized and carefully examined against developability criteria. Drug metabolism–related issues are outlined and discussed in Chapter 5. The impact of the transporter, including the efflux transporter in drug delivery and the models used to study and address the issues, will be discussed in Chapters 18, 2, and 3.

1.2.5. Preparation for Pharmaceutical Products

Before the early 1990s, the solid state, salt form, aqueous solubility, and dosing formulation for agents used in pharmacological, pharmacokinetic, and toxicological studies were not of major concern. However, an inappropriate salt version or solid form may cause potential drug delivery and stability problems (both physicochemically and chemically) during formulation and pharmaceutical engineering. It is now understood that the investigation of the physicochemical properties of an NCE against developability criteria should start early in the R&D processes.

Chapter 4 discusses the physicochemical properties that have a major impact on drug delivery.

Aqueous solubility is one of the most important physicochemical properties. It is believed that a drug has to be in solution to be absorbed.[35] From the pharmaceutical development point of view, the solid state form is another important factor that affects solubility, the dissolution rate, and eventually developability. The solid state form is the determinant of, to some extent, physicochemical stability, intellectual property, and formulation scalability; this factor should be carefully examined and optimized. Change in crystallinerity from different chemical processes, in some cases, results in a big difference in bioavailability when the drug is delivered by a solid dosage formulation.

Many of these properties could change when the salt version and form change. The salt with the best solubility, dissolution rate (which therefore could result in the best bioavailability if given as a solid dose), stability, and other properties such as moisture absorption should be selected before a molecule enters full development.[36] *In situ* salt screening is a new technology used to select the right salt form for a drug candidate.[37] For instance, the HCl salt[38] was formerly almost the default version for a weak base; however, it has been shown in many cases not to be the best.[39] Application of these screening processes in early drug development is one of the major steps in integrating pharmaceutical development into drug discovery and development.

Preclinical safety assessment (toxicology) is another functional area, which serves as a milestone in drug discovery and development. The NCEs have to be evaluated for their potential genetic toxicity, as well as for acute, short-term, and long-term toxicity. The results are crucial for further development of the compound. Although the principle and importance of toxicology will not be discussed in this book, many efforts in DMPK and pharmaceutics are made to assure drug delivery in the animal models used in toxicological studies. Metabolic profiles of a drug candidate in the species used in the toxicology studies should be compared with those from human tissues for major differences. The profiles are also examined for potential active/toxic metabolite(s). The factors that have an impact on drug delivery will be extensively discussed in the following chapters.

Process chemistry is a large functional area that can have major impacts on a drug's developability, but it will not be covered in this book. Although the developability criteria in this area will not be discussed here, it is important to point out that quite often collaboration with process chemists is also required early on in order to find the right salt and solid state form.

1.2.6. Remarks on Developability Criteria

The concept of ensuring developability in drug discovery and development represents an integration of all functional areas that impact the efficiency, success rate, and timetable of a drug's development. Coordination of these multifunctional, interlinked, parallel, ongoing scientific and technological research activities is a new challenge to the management of a drug discovery and development enterprise.

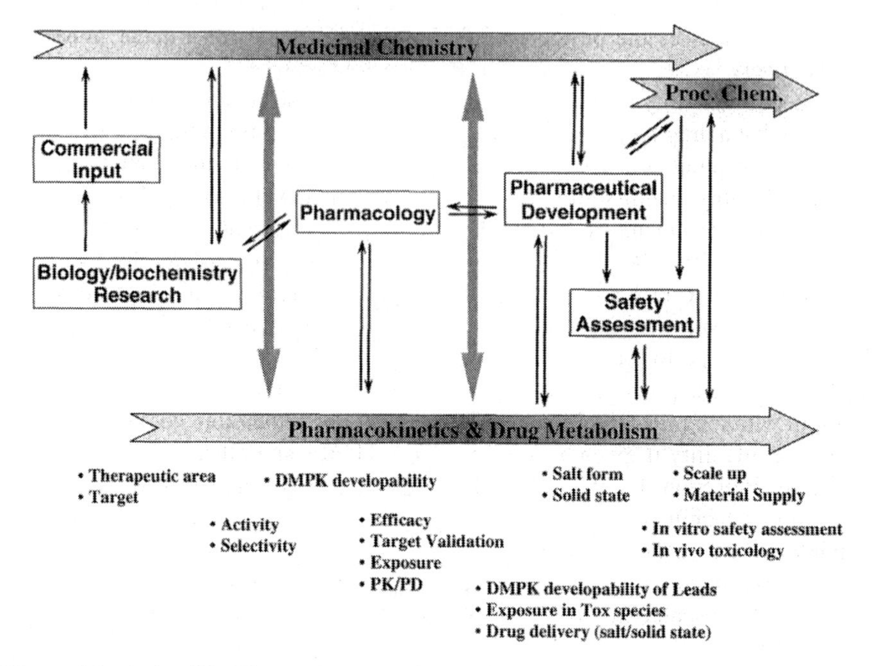

Figure 1.3. A simplified illustration of the involvement, collaboration, and interrelationship of different functional areas in a preclinical research and development organization. The bullet points summarize the major developability factors examined at different stages.

Figure 1.3 is a simplified scheme of the interrelationship of major functional areas and their roles in drug discovery and development.

1.3. DRUG DELIVERY FACTORS THAT IMPACT DEVELOPABILITY

Delivery of a pharmaceutical agent to the systemic circulation, and consequently to the site of action to produce a desired pharmacological effect, is the ultimate goal of drug delivery. The developability of a drug candidate from a drug delivery perspective has become the core of developability criteria in drug development. As discussed in the previous subsections, many other factors in developability criteria are closely related to drug delivery; this holds true from the research laboratory to clinical trails and from early discovery to postmarket development. In order to accomplish this task, one has to overcome numerous barriers that hinder drug delivery.

In a biological system, multiple mechanisms exist to protect the system from exposure to almost any foreign substance while preserving nutrient uptake. The physiological arrangement and the chemical and biochemical barriers associated with the physiological structures form the first line of defense. Any drug, delivered by any route, will almost certainly encounter some of these barriers before reaching at the site of action. These barriers, as well as their physiological and biochemical functions and their role in drug delivery, will be discussed in detail in Chapter 2. The special situations related to drug delivery to the central nervous system (CNS) is covered in Chapter 3.

How a drug molecule interacts with these barriers is very much determined by the properties of the molecule. These properties are the physicochemical and biochemical characteristics of the molecule. In Chapter 4, physicochemical properties and their implications for formulation and drug delivery will be extensively discussed.

Pharmacokinetics and pharmacodynamics provide a general approach by allowing mathematical modeling of the interaction of a drug molecule with the entire biological system to predict drug concentrations in the systemic circulation and therefore providing a prediction of pharmacological responses. Better understanding of the system will allow a pharmaceutical scientist to utilize and manipulate the system for the purpose of drug delivery. Chapter 5 discusses the basic principles and topics in pharmacokinetics and pharmacodynamics. Approaches in drug delivery based on an understanding of pharmacokinetic principles are essential in pharmaceutical development.

Developability in drug delivery is an overall assessment of all important factors. Take oral drug delivery as an example.[40] Solubility is important because a drug molecule has to be dissolved to be absorbed. Some lipophilicity is essential for the molecule to cross cell membranes by diffusion. In order to finally reach the systemic circulation, the molecule has to survive various chemical and biochemical attacks in the gastrointestinal system and the liver. A flow chart describing sequentially the factors that can impact drug delivery is illustrated in Figure 1.4. The order

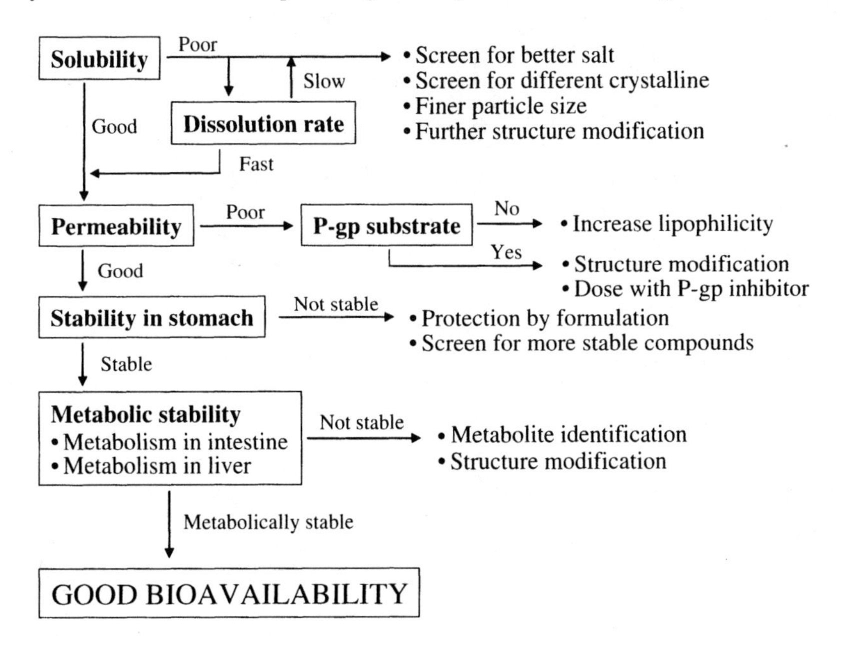

Figure 1.4. The evaluation steps of various factors that impact the oral bioavailability of a drug candidate.

in which these factors are listed could also be the order of logical thinking when one plans to tackle an oral drug delivery problem. It can also be a reference point for other routes of delivery.

It is believed that the permeability and metabolic stability of a drug molecule are two major factors in drug delivery or in the prediction of a drug's absorption[41] when the molecule is in solution. Permeability can be further detailed by passive diffusion and transporter-mediated process. Metabolism of a drug molecule in the liver and intestine can be evaluated by *in vitro* experimental methods. In many cases, *in vitro* metabolism (intrinsic clearance) can be used to predict *in vivo* metabolic clearance.[42] Drug metabolism–related issues are discussed in depth in Chapter 6. It is obvious that when efflux transporters such as P-gp are involved, the predictability of *in vivo* clearance using metabolic intrinsic clearance becomes uncertain.[43] A more in-depth understanding of drug transporters and their function in combination with our knowledge of drug metabolism will help predict oral absorption.[44,45] Transporter-related drug deliver issues, as well as *in vivo* and *in vitro* models used to address these issues, are discussed in the following chapters.

Although not discussed in detail in this book, in addition to parenteral (e.g., iv infusion) drug delivery, many other routes of drug delivery are developed for convenience, safety, specific targeting, and delivery of special agents. Most of the physiological and biochemical issues discussed in oral and CNS delivery can be extrapolated to the situations in other drug delivery routes. Knowledge of the physiological and biological barriers for each specific delivery route will help medicinal chemists to design drug candidates with optimal drug delivery properties or at least to avoid obvious problems. Prodrug approaches, utilization of metabolic activation to target a specific organ, and taking advantages of a substrate of specific transporters or carriers are some invaluable examples in modern drug delivery. Many of these issues are discussed in various chapters.

The aim of this book is to provide a basic understanding of the major issues in drug delivery. More detailed examination of various topics can be found in the references cited.

REFERENCES

1. Price, Waterhouse, Coopers. *Pharma 2005*. New York, **1998**, pp. 1–20.
2. Collins, M. A.; Shaw, I.; Billington, D. D. *Drug Des. Discov.* **1999**, *16*, 181–194.
3. Drews, J.; Ryser, S. *Nature Biotechnol.* **1997**, *15*, 1318–1319.
4. Drews, J.; Ryser, S. *Drug Disc. Today* **1997**, *2*, 365–372.
5. DiMasi, J. A. *Clin. Pharmacol. Ther.* **2001**, *69*, 286–296.
6. Kuhlmann, J. *Int. J. Clin. Pharmacol. Ther.* **1997**, *35*, 541–552.
7. Venkatesh, S.; Lipper, R. A. *J. Pharm. Sci.* **1999**, *89*, 145–154.
8. DiMasi, J. A. *Clin. Pharmacol. Ther.* **2001**, *69*, 297–307.
9. Prentis, R. A.; Lis, Y.; Walker, S. R. *Br. J. Clin. Pharmacol.* **1988**, *25*, 387–396.
10. Huff, J. L.; Barry, P. A. *Emerg. Infect. Dis.* **2003**, *9*, 246–250.

11. Drews, J. *Science* **2000**, *287*, 1960–1964.
12. Hopfinger, A. J.; Duca, J. S. *Curr. Opin. Chem. Biol.* **2000**, *11*, 97–103.
13. Railkar, A. S.; Sandhu, H. K.; Spence, E.; Margolis, R.; Tarantino, R.; Bailey, C. A. *Pharm. Res.* **1996**, *13*, S-278.
14. *Integration of Pharmacokinetics, Pharmacodynamics, and Toxicokinetics in Rational Drug Development*. Plenum Press: New York and London, **1993**.
15. Lesko, L. J.; Rowland, M.; Peck, C. C.; Blaschke, T. F. *Pharm. Res.* **2000**, *17*, 1335–1344.
16. Blackie, J. A.; Bloomer, J. C.; Brown, M. J. B.; Cheng, H. Y.; Elliott, R. L.; Hammond, B.; Hickey, D. M. B.; Ife, R. J.; Leach, C. A.; Lewis, V. A.; Macphee, C. H.; Milliner, K. J.; Moores, K. E.; Pinto, I. L.; Smith, S. A.; Stansfield, I. G.; Stanway, S. J.; Taylor, M. A.; Theobald, C. J.; Whittaker, C. M. *Bioorg. Med. Chem. Lett.* **2002**, *12*, 2603–2606.
17. Eliceiri, B. P.; Cheresh, D. A. *J. Clin. Invest.* **1999**, *103*, 1227–1230.
18. Miller, W. H.; Alberts, D. P.; Bhatnger, P. K.; et al. *J. Med. Chem.* **2000**, *43*, 22–26.
19. Carron, C. P.; Meyer, D. M.; Pegg, J. A.; Engleman, V. W.; Mickols, M. A.; Settle, S. L.; Westlin, W. F.; Ruminski, P. G.; Nickols, G. A. *Cancer Res.* **1998**, *58*, 1930–1935.
20. Mahmood, I. *Am. J. Ther.* **2002**, *9*, 35–42.
21. Mahmood, I.; Balian, J. D. *Clin. Pharmacokinet.* **1999**, *36*, 1–11.
22. Lin, J. H. *Ernst Schering Research Foundation Workshop* **2002**, *37*, 33–47.
23. Woodnutt, G. *J. Antimicrob. Chemother.* **2000**, *46*, 25–31.
24. Bodor, N.; Buchwald, P. *Med. Res. Rev.* **2000**, *20*, 58–101.
25. Benet, L. Z. *Eur. J. Respir. Dis.* **1984**, *65*, 45–61.
26. Benet, L. Z.; Zia-Amirhosseini, P. *Toxicol. Pathol.* **1995**, *23*, 115–123.
27. Rodrigues, A. D.; Lin, J. L. *Curr. Opin. Chem. Biol.* **2001**, *5*, 396–401.
28. Lin, J. H. *Curr. Drug Metab.* **2000**, *1*, 305–331.
29. van Asperen, J.; van Tellingen, O.; Beijnen, J. H. *Pharm. Res.* **1998**, *37*, 429–435.
30. Wacher, V. J.; Salphati, L.; Benet, L. Z. *Adv. Drug Delivery Rev.* **2001**, *46*, 89–102.
31. Tawara, S.; Matsumoto, S.; Kamimura, T.; Goto, S. *Antimicrob. Agents Chemother.* **1992**, *36*, 17–24.
32. Talbert, A. M.; Tranter, G. E.; Holmes, E.; Francis, P. L. *Anal. Chem.* **2002**, *74*, 446–452.
33. Rodrigues, A. D.; Wong, S. L. *Adv. Pharmacol.* **2000**, *43*, 65–101.
34. Baillie, T. A.; Cayen, M. N.; Fouda, H.; Gerson, R. J.; Green, J. D.; Grossman, S. J.; Klunk, L. J.; LeBlanc, B.; Perkins, D. G.; Shipley, L. A. *Toxicol. Appl. Pharmacol.* **2002**, *182*, 188–196.
35. Rowland, M.; Tozer, T. N. *Clinical Pharmacokinetics: Concepts and Applications*. Lippincott Williams & Wilkins: Philadelphia, **1995**, chapter 9, pp. 119–136.
36. Morris, K. R.; Fakes, M. G.; Thakur, A. B.; Newman, A. W.; Singh, A. K.; Venit, J. J.; Spagnuolo, C. J.; Serajuddin, A. T. M. *Int. J. Pharm.* **1994**, *105*, 209–217.
37. Tong, W.-Q.; Whitesell, G. *Pharm. Dev. Technol.* **1998**, *3*, 215–223.
38. Berge, S. M.; Bighley, L. D.; Monkhouse, D. C. *J. Pharm. Sci.* **1977**, *66*, 1–19.
39. Gould, P. L. *Int. J. Pharm.* **1986**, *33*, 201–217.
40. Martinez, M. N.; Amidon, G. L. *J. Clin. Pharmacol.* **2002**, *42*, 620–643.
41. Lipinski, C. A.; Lombardo, F.; Dominy, B. W.; Feeney, P. J. *Adv. Drug Delivery Rev.* **1997**, *23*, 3–25.

42. Houston, J. B. *Biochem. Pharmacol.* **1994**, *47*, 1469–1479.
43. Wu, C. Y.; Benet, L. Z.; Hebert, M. F.; Gupta, S. K.; Rowland, M.; Gomez, D. Y.; Wacher, V. J. *Clin. Pharmacol. Ther.* **1995**, *58*, 492–497.
44. Benet, L. Z.; Cummins, C. L. *Adv. Drug Delivery Rev.* **2001**, *50*, S3–S11.
45. Polli, J. W.; Wring, S. A.; Humpherys, J. E.; Huang, L.; Morgan, J. B.; Webster, L.; Serabjit-singh, C. S. *JEPT* **2001**, *299*, 620–628.

2

PHYSIOLOGICAL, BIOCHEMICAL, AND CHEMICAL BARRIERS TO ORAL DRUG DELIVERY

Anna Maria Calcagno and Teruna J. Siahaan

Department of Pharmaceutical Chemistry
The University of Kansas
Lawrence, KS 66047

Drug Delivery: Principles and Applications Edited by Binghe Wang, Teruna Siahaan, and Richard Soltero
ISBN 0-471-47489-4

2.1. INTRODUCTION

The development of various biotechnological drug products (peptidomimetics, peptides, proteins, and oligonucleotides) has brought about new challenges in drug delivery. These challenges are due to the physicochemical properties of the peptides and peptidomimetics and the presence of physiological, biochemical, and chemical barriers. The oral delivery of these molecules has been the focus of many review articles.[1–3] This chapter focuses on the various barriers that these types of drugs must overcome to reach their sites of action after oral delivery.

The body contains many biological barriers that serve to protect its interior from a variety of external invaders and toxins. The skin is the largest such obstacle, while the blood-brain barrier forms the tightest barrier to penetration of molecules from the bloodstream to the brain. Similarly, for a drug molecule to be orally bioavailable, it has to traverse the epithelial layer of the gastrointestinal tract. Thus, many factors for enhancing the delivery of molecules through this intestinal mucosal barrier must be considered. The various components of the biological barriers will be discussed in greater detail in this chapter.

Several different obstacles must be overcome for the delivery of drugs through the intestinal mucosa or the blood-brain barrier. These obstructions to drug delivery can be categorized as physiological, biochemical, and chemical barriers. The physiological barrier in the intestinal mucosa or the blood-brain barrier protects the body from various molecules such as toxins by inhibiting their passage through the barriers. Next, the drug must overcome the biochemical barrier which consists of metabolizing enzymes, that can degrade it. Finally, the drug has to have optimal physiochemical properties for its permeation across the biological barriers. Thus, these various barriers have to be taken into account when designing drugs with improved absorption characteristics.

2.2. PHYSIOLOGICAL BARRIERS TO DRUG DELIVERY

The luminal side of the gastrointestinal tract is covered with an aqueous mucus layer that is secreted by the goblet cells. Before reaching the epithelial layer of the intestinal mucosa, a drug molecule must penetrate this mucus layer, which has a thickness that varies from 100 to 150 μm. This mucus layer acts as a filter for molecules with a molecular mass of 600–800 Daltons.[3] The mucus is composed of glycoproteins, which trap water within this layer with a turnover rate of 12–24 hours. Drug penetration through this mucus and unstirred water layer is the rate-limiting step before the drug reaches the surface of the enterocytes.[3]

Immediately below this mucus layer is a single layer of columnar epithelial cells that are joined together by tight intercellular junctions to form a barrier to the systemic delivery of orally administered drugs. This layer of cells is composed of enterocytes, goblet cells, endocrine cells, and paneth cells.[4] The number of goblet cells from the small intestine to the distal colon differs; only 10% of the cells in the small intestine are goblet cells, increasing to 24% in the distal colon.[5] There are four regions of the gastric epithelium from the proximal to the distal stomach—nonglandular stratified squamous, cardiac, glandular proper gastric (fundic), and pyloric. Each region has a different physiological function. The toughness of the stratified squamous region allows it to resist food abrasion, while the cardiac region is responsible for the production of mucus and bicarbonate. Pepsinogen and hydrochloric acid are secreted from the proper gastric region. The pyloric section is associated with the release of gastrin and pepsinogen.[6] Both villi and crypts are lined by the epithelial cell layer. The microvilli amplify the surface area of the intestine,[7] and absorb nutrients and secrete mucus. The crypts are responsible for cell renewal.

The epithelial layer is in immediate contact with the lumen of the gastrointestinal tract. The lamina propria, which functions as a structural support for the epithelial layer, is situated on the basolateral side of the epithelial layer. The lamina propria contains lymph vessels, smooth muscle cells, nerves, and blood vessels, which nourish the epithelium. The muscularis mucosa makes up the deepest layer, which is thought to be involved in contractility.[8] A more detailed description of the forces that hold together the epithelial layer is provided below.

A drug can cross the intestinal mucosa via several different mechanisms, depending upon its physiochemical properties. Hydrophobic drugs that can partition through cell membranes are more likely to cross the intestinal mucosa through the transcellular pathway (Figure 2.1, Pathway A). Hydrophilic drugs cannot penetrate the cellular membranes; therefore, they must use the paracellular pathway (Figure 2.1, Pathway B). Unfortunately, this pathway is restricted by the presence of tight junctions. Only molecules that have a hydrodynamic radius (<11 Å can pass through this pathway.[9] Therefore, the transport of peptides via this pathway is very limited. Pathway C is another way in which a drug can penetrate the intestinal

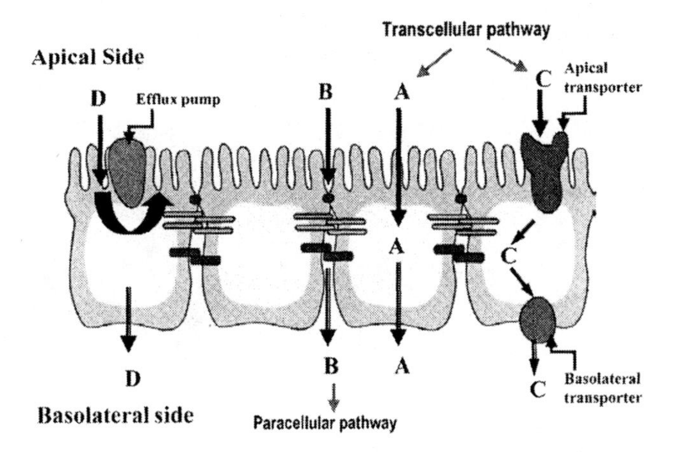

Figure 2.1. The pathways that a drug can take to cross the intestinal mucosa barrier. Pathway A is the transcellular route in which a drug passively permeates the cell membranes. Pathway B is the paracellular route; the drug passively diffuses via the intercellular junctions. Pathway C is the route of active transport of the drug by transporters. Pathway D is the route of drug permeation that is modified by efflux pumps.

mucosa. This pathway involves receptor-mediated uptake of the drug. For example, dipeptide transporters have been found to transport drugs from the lumen to the blood side of the intestinal mucosa. Finally, the intestinal mucosa has efflux pumps (Figure 2.1, Pathway D). These pumps can create an efflux of drugs, which have been partitioned through the membranes. The characteristics of these pumps will be described in a later section.

2.2.1. Paracellular Pathway

This physiological barrier exists to provide protection from the entry of toxins, bacteria, and viruses from the apical side to the basolateral side, and it allows the passage of selective molecules and cells. The intercellular junctions can be divided into three different regions: (1) tight junctions (*zonula occludens*), (2) adherens junctions (*zonula adherens*), and (3) *desmosomes*. The intercellular junctions form an 80-nm-long tortuous path between the two adjacent cells that runs the entire lateral side of the cell, as discovered by transmission electron microscopic studies.[4]

2.2.1.1. Tight Junctions At the most apical portion of the cells, the tight junctions or *zonula occludens* function to bring adjacent cells into close apposition. This is defined as the gate function of the tight junction.[10] The components of the intercellular junctions are depicted in Figure 2.2. These areas of apposition have been referred to as "kisses," and they form branching fibrils that circumscribe the cells, as seen by freeze-fracture electron microscopy.[11] Tight junctions cause

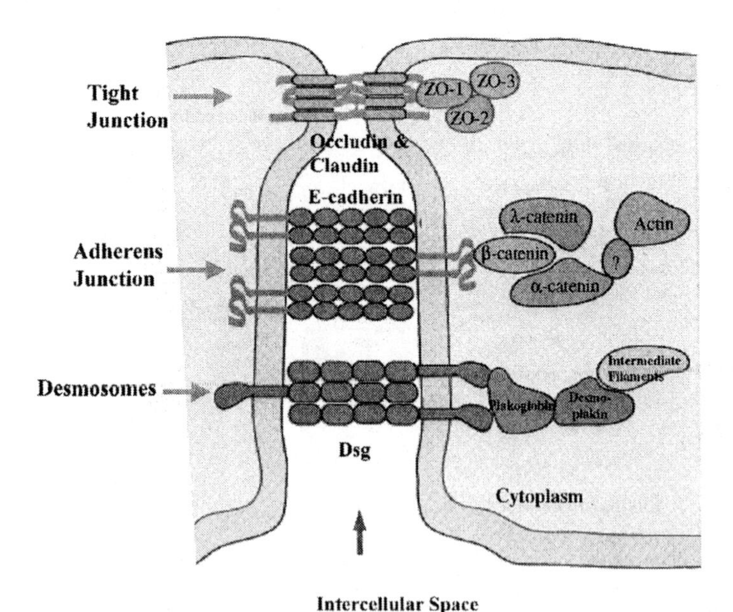

Figure 2.2. The intercellular junction is mediated by proteins at different levels: (1) tight junction (*zonula occludens*), (2) adherens junction (*zonula adherens*), and (3) desmosomes.

cell surface polarity that produces the fence function and restricts free diffusion of lipids and proteins from the apical plasma membrane to the basolateral surface.[12,13] Thus, paracellular permeation of drug through the intercellular junctions depends on the pore size of the tight junctions. The smallest pores are found on the villus tips, and the regions in the crypts contain the largest pores.[6] The integrity of the tight junctions is calcium dependent, and removal of calcium causes a rearrangement of the tight junction proteins.[14,15] It is possible that that removal of calcium disrupts the integrity of the E-cadherin–E-cadherin interactions at the adherens junctions (see below). A number of cytokines and growth factors have also been shown to decrease the barrier function of the tight junctions.[16]

Occludin and claudin proteins have been found at the tight junctions; they are involved in both gate and fence functions.[17] Both of these proteins have four transmembrane domains, two extracellular domains that form loops, and a cytoplasmic carboxyl tail.[18] The extracellular loops play a vital role in creating cell-cell contact.[13] The hyperphosphorylated form of occludin is the main form located in the tight junction.[18] There are three scaffolding proteins associated with the tight junctions: ZO-1, ZO-2, and ZO-3. These proteins belong to the membrane-associated guanine kinases (MAGUK) family. ZO-1 stabilizes the tight junction by interacting with occludin and claudin, cross-linking them to the actin cytoskeleton.

2.2.1.2. Adherens Junctions Immediately below the tight junction is an area known as the *zonula adherens* or adherens junction, which is responsible for cell-cell adhesion. Researchers have also shown that formation of the tight junctions depends on the prior formation of the cadherin-cadherin interactions within this region.[13] The paracellular path has the tight junction and intercellular junction working in series to make up the resistance across this pathway.[19] This resistance is the reciprocal of the permeability, which is dependent upon size and charge in the paracellular path.[20] It has been shown that the resistance increases as the number of tight junctional strands increases. The transepithelial electrical resistance has been reported to vary in the human intestine with the jejunum displaying 20 ohms/cm^2 and the large intestine showing 100 ohms/cm^2.[5] Within the *zonula adherens*, the perijunctional actin-myosin II ring encircles the epithelial cells, impacting solute permeation in this region.[19]

Cell-cell adhesion within the *zonula adherens* is controlled by E-cadherin.[21,22] This is a 120 kD glycoprotein consisting of three domains: an extracellular domain, a single transmembrane domain, and a cytoplasmic domain. The extracellular domain is further divided into 110 amino acid repeats known as EC1–EC5.[23] E-cadherins are calcium-dependent molecules that interact in a homotypic fashion. E-cadherin protrudes from the cell surface as a dimer (*cis*-dimer), and this *cis*-dimer interacts with another *cis*-dimer from the opposing cell to form a *trans*-dimer. The exact mechanism of the *cis*- and *trans*-dimer interaction of cadherins is not known. Several models have been suggested in the literature, including the "zipper-model."[24] The highly conserved cytoplasmic domain of cadherins has been shown to be necessary for the adhesion process; it interacts with α- and β-catenins, which link the cadherins to the actin cytoskeleton.[25]

2.2.1.3. Desmosomes The last region of the paracellular pathway is the desmosome, which is located nearest the basolateral membrane surface of the enterocyte. Studies indicate that the intermediate filaments are connected to the desmosomes by the desmoplakins.[26] The major desmosomal cadherin located in this area is desmoglein. Desmocollin, a second desmosomal cadherin, is required for the binding of desmoplakins to the intermediate filaments.[26] This region appears to be less critical for the function of the paracellular path than the two regions situated nearer the apical membrane.

2.2.2. Transcellular Pathway

A drug with the appropriate physiochemical characteristics can traverse through the cell by passive diffusion. In the case of peptides or peptidomimetics, their physicochemical properties may not be suitable for permeation through the cell membrane via the transcellular pathway. The drug molecules must travel through the lipid bilayers that make up the membranes. The bilayers consist of four regions: (1) the outermost region, which has a large number of water molecules and is accountable for the interactions with other proteins and membranes; (2) the next region, which contains the polar headgroups, causing this region to have the highest molecular density and making it the most difficult region for diffusion; (3) the third region, which contains the nonpolar tails that form the barrier to penetration based on limitations on molecular size and shape; and (4) the inner region, which is the most hydrophobic and acts as the hydrophobic barrier.[27] The resistance across the transcellular path can be visualized as resistors in a series, where the apical and basolateral membranes act as the two resistors.[19] These membranes form the rate-limiting barriers to the passive flow of molecules.

The membranes form one of the obstacles in this delivery route. The drug molecules must also traverse the cytosol before exiting through the basolateral membranes. Within the cytosol, various drug-metabolizing enzymes reside, which metabolize the drug molecules and can lower the drug transport via this route.

2.3. BIOCHEMICAL BARRIERS TO DRUG DELIVERY

Great interindividual variability can be seen in the metabolism of drugs as a result of differing enzyme activity due to inhibition, induction, genetic polymorphisms, or even disease state.[28] Enzymes found within the intestine are from two sources, mammalian and bacteria-associated. The mammalian enzymes are located within the lumen and in the enterocytes. Enzymes from the microflora within the ileum and colon have also been identified.[8] This discussion will focus on degradation by the mammalian enzymes.

2.3.1. Metabolizing Enzymes

Within the lumen of the stomach, a mixture of hydrochloric acid and proteolytic pepsins is the first metabolic barrier that a peptide drug will encounter.

Subsequently, the hydrolysis of acidic proteins occurs at pH 2–5; this is especially the case for peptides containing aspartate residues.[3] Larger proteins are quite susceptible to this gastric proteolysis, while smaller peptides are unaffected by this mixture.

Fricker and Drewe describe the luminal enzymes of the upper small intestine as the second barrier.[3] Trypsin, chymotrypsin, elastase, and carboxypeptidase A and B are positioned in the lumen of the duodenum. Their highest activity occurs at pH 8. These enzymes degrade 30–40% of large proteins within the duodenum to small peptides within 10 minutes.[3] Small peptides have been shown to be stable against these pancreatic proteases.

The major enzymatic barrier occurs within the brush border and in the cytosol of the enterocytes, both of which contain peptidases. These enzymes degrade smaller peptides ranging from di- to tetrapeptides.[3] Furthermore, there is an increase in brush border peptidase activity from the upper duodenum to the lower ileum.[3] The peptidases selective for tripeptides are located primarily within the brush border, whereas cytosolic proteases have dipeptides as their substrates. Evidence has shown that the metabolic enzyme activity decreases along the intestine to a nearly negligible rate within the colon, yet the permeability of the colon epithelium remains good.[2] This highlights the potential for targeting the colon to bypass the enzymatic barrier of the intestine for peptide delivery, preventing the degradation that occurs within the intestine. The pH at the intestinal surface on the brush border is 5.5–6.0, which is more acidic than the pH of the lumen.[29] The enterocyte has an intercellular pH of 7.0–7.2. In addition, gastrointestinal pH changes in the fasted and fed states; this topic has been reviewed elswhere.[30]

The proximal small intestine shows the greatest metabolic activity due to its large surface area and the plethora of intestinal enzymes and transporters.[8] Phase I and II enzymes have also been identified in the intestine. The most notable Phase I enzymes are those of the CYP superfamily. The P450 enzymes are also present in the intestinal walls in concentrations approximately 20 times less than those seen within the liver; however, their metabolism of drugs is comparable to the activity seen in the liver.[28,31] The activity of the enzymes varies within the area of the gastrointestinal tract. The highest activity of the P450 enzymes is displayed in the proximal part of the gastrointestinal tract, and their activity decreases distally.[28] The greatest concentration of P450 enzymes is found in the villus tips of the upper and middle third of the intestine.[27] Great variability in activity has been noted both intra- and interindividually due to exposure of the enterocytes to external stimuli such as food and drugs that can either induce or inhibit these enzymes. These intestinal P450 enzymes are more responsive to inducers than are their hepatic counterparts.[27] Although the blood flow to the intestine is lower than to the liver, the villus tip has a large surface where the enzyme can interact with its substrate, allowing extensive metabolism.[32] Metabolic activity in the intestine has been shown to be route-dependent, and the metabolism is greater for oral administration of drugs than for intravenous dosing.[28,33] In this case, intestinal metabolism occurs during the initial absorption of the drug across the intestinal barrier, and the metabolism is lower with the recirculation of the drug. The major factor that influences the route-dependent metabolism is the residence time of the drug within the enterocyte. The residence time can be lengthened by binding in the cytoplasm, the activity of efflux pumps, and limited blood flow or, conversely, shortened by basolateral clearance and basolateral transporters.[28]

The CYP1, CYP2, and CYP3 subfamilies are mostly involved in xenobiotic metabolism. Various isoenzymes that possess their own drug substrates are present for each subfamily. Within the human small intestine, CYP1A1, CYP2C, CYP2D6, and CYP3A4 have been isolated.[8] Because of the numerous polymorphs of CYP2D6, characterization of intestinal levels of this enzyme has been quite difficult. CYP3A4 is the most abundant of the intestinal P450 enzymes, making up more than 70% of the intestinal CYPs.[34] Reports indicate that there are structural similarities between the intestinal and hepatic CYP enzymes; however, they appear to be independently regulated.[8] Food interactions have been shown to affect the regulation of the intestinal CYP enzymes. Grapefruit juice inhibits CYP3A, while grilled and smoked foods induce CYP1A1 activity.[8] Variations in the population in reference to these enzymes can also confound the issue of degradation for peptide pharmaceuticals.

Conjugating enzymes, also referred to as "Phase II metabolizers," are also found in the intestine. Glucuronyltransferase, *N*-acetyltransferase, sulfotransferase, and glutathione-*S*-transferase show high activity for the intestinal Phase II enzymes.[8] Conjugates that are formed by these enzymes within the cell are reported to be substrates of the multidrug resistance–associated protein family (MRP) of transporters and are excreted into the lumen.[35] The MRP family of transporters consists of ATP-dependent transporters that excrete organic anions. At this time, additional studies are needed to understand the role of these enzymes in the degradation of peptide drugs. Metabolizing enzymes and drug transporters in the process of drug delivery will be discussed in greater detail in another chapter.

2.3.2. Transporters and Efflux Pumps

Active transporters for peptides within the intestine were first detected in experiments in the late 1960s and early 1970s. Since that time, many characteristics of these transporters have been determined. Substrates are generally peptides consisting of two or three amino acids that can be transported through the brush border membrane in a carrier-mediated, pH-dependent fashion.[3] Energy is required to move these peptides against a concentration gradient into the cell, and these carriers also display saturability.

Although most transporters are situated on the apical membrane, researchers have found some that are located only on the basolateral membrane surface. The Na^+/A amino acid transporter, Na^+/ASC amino acid transporter, GLUT2 hexose transporter, and the Na^+-independent folic acid transporter are examples of such basolateral transporters.[27] PepT1, an apical H^+/dipeptide transporter, is most abundant in the villus tip, and its concentration increases from the duodenum to the ileum. In times of starvation, there is an increase in the expression of this transporter. On the basolateral membrane, PepT2 functions as the H^+/dipeptide transporter to allow the substrate to exit the enterocyte.[27]

P-glycoprotein (P-gp) is a known MRP that serves as an efflux pump.[36,37] It is located within the brush borders of the villus tips of the intestine and has been found throughout the small and large intestines. The concentration of P-gp increases from the stomach to the colon.[6] The substrate specificity for P-gp covers a broad range of molecular structures, and the affinity varies as a function of the intestinal site.[27] A common feature of the substrates is hydrophobicity. As mentioned previously, the efflux pumps assist the intestinal metabolism by returning the drug to the lumen, allowing the metabolizing enzymes to work on the drug another time as well as preventing product inhibition by removing primary metabolites that have been formed.[31] This interaction is enhanced due to the colocalization of the CYP3A enzymes and P-gp on the apical membrane, as well as the overlap in substrate specificities and shared inducers and inhibitors.[27] Grapefruit juice also interferes with the transport mediated by P-gp;[7] however, not all substrates for the CYP3A enzyme behave as substrates for P-gp.[6] P-gp functions as a defense mechanism against xenobiotics in other biological barriers, as it is also expressed in apical surfaces of epithelial cells of the liver, kidney, pancreas, and colon as well as in the capillary endothelium of the brain.[38]

2.4. CHEMICAL BARRIERS TO DRUG DELIVERY

The chemical structure of a drug determines its solubility and permeability profiles. In turn, the concentration at the intestinal lumen and the permeation of the drug across the intestinal mucosa are responsible for the rate and extent of absorption.[39] Unfavorable physicochemical properties have been a limiting factor in the oral absorption of peptides and peptidomimetics.[40] The structural factors involved in the permeation of peptides will be described here.

2.4.1. Hydrogen-Bonding Potential

Hydrogen-bonding potential has been shown to be an important factor in the permeation of peptides. Studies *in vivo* and in various cell culture models of the blood-brain barrier and intestinal mucosa indicate that desolvation or hydrogen-bonding potential regulates the permeation of peptides.[39,41–43] The energy needed to desolvate the polar amide bonds in the peptide to allow it to enter and traverse the cell membrane is the principle behind the concept of hydrogen-bonding potential. For small organic molecules, the octanol-water partition coefficient is the best predictor of cell membrane permeation with a sigmoidal relationship.[42] However, this is not the case with peptides; the desolvation energy or hydrogen-bonding potential is a better predictor for membrane permeation of peptides. Burton et al. have reported partition coefficients of model peptides in n-octanol/Hanks' balanced salt solution (HBSS), isooctane/HBSS, and heptane/ethylene glycol systems.[41] Two experimental methods were developed to measure the desolvation energy or the hydrogen-bonding potential of peptides. In the first method, the hydrogen-bonding potential is calculated from the difference between the partition coefficients of octanol/water and isooctane/water. The second method involves measuring the partition coefficient of peptide in heptane/ethylene glycol; this method correlates well with the hydrogen-bonding potential and provides a simpler and more direct measurement.[41]

2.4.2. Other Properties

Physicochemical properties of the peptide are also important determinants in the passage of drugs via the paracellular path. Size, charge, and hydrophilicity are the factors influencing paracellular permeation.[40] A change in the hydrophilicity of a peptide may alter its route of permeation; as the hydrophilicity of a peptide decreases, its lipophilicity increases, causing a shift in permeation of the peptide from the paracellular to the transcellular route. Molecules with radii larger than 11 Å are unable to penetrate the tight junctions.[44] Studies of Caco-2 cells confirm that drug permeation via the paracellular path is size-dependent, and this highlights the sieving abilities of the intercellular junctions.[40]

Although the paracellular path is negatively charged, the effect of charge on paracellular permeation is not well understood. One study suggests that a positive net charge on a peptide produces the best paracellular permeation, but another study suggests that a -1 or -2 charge is most effective in paracellular transport.[40] It has also been suggested that the effect of charge is negligible as the molecular size of the peptide increases.[40]

2.5. DRUG MODIFICATIONS TO ENHANCE TRANSPORT ACROSS BIOLOGICAL BARRIERS

Several methods have been explored to improve drug permeation across biological barriers.[1,44–47] One method involves chemical modification of drug entities such as prodrugs and peptidomimetics. Another method is to design a formulation that enhances drug permeation through the biological barriers.

2.5.1. Prodrugs and Structural Modifications

A prodrug approach has been utilized to optimize drugs; a prodrug is defined as a chemical derivative that is inactive pharmacologically until it is converted *in vivo* to the active drug moiety. Recently, a targeted prodrug design has emerged in which prodrugs have been used to target membrane transporters or enzymes.[45] This method improves oral drug absorption or site-specific drug delivery. Extensive knowledge of the structure, the distribution within biological barriers, and substrate specificities is needed to target a desired transporter.

Prodrug strategies have been very successful with small molecules; however, the use of prodrugs for peptides has been less frequent.[44] The cyclic peptide prodrug approach has been shown to improve membrane permeation. In this method, the

N and C termini of the peptide are connected via a linker to form a cyclic peptide. The linker can be cleaved by esterase to release the peptide. The cyclic peptide prodrug formation increases the intramolecular hydrogen bonding and lowers the hydrogen-bonding potential to water molecules as solvent. In addition, the lipophilicity of the cyclic prodrug increases, which shifts its transport from paracellular to transcellular.[48] It has also been reported that cyclic peptides are less susceptible to amino- and carboxypeptidases than linear peptides because the amino and carboxy terminals are protected from these enzymes.[44]

Peptide structural modification has been applied to improve membrane permeation of peptides. Metabolism of peptide pharmaceuticals can occur in various regions along the route to oral absorption, and inhibition of this degradation is advantageous in enhancing drug delivery. To improve enzymatic stability, peptides have been converted to peptidomimetics. In this case, the peptide bond is converted to its bioisostere, which is stable to proteolytic enzymes. Other structural modification strategies to improve membrane permeation of peptides include lipidization, halogenation, glycosylation, cationization, and conjugation to polymers.[46]

2.5.2. Formulations

Peptide absorption can be improved by designing an optimal formulation.[47,49] Several methods to enhance peptide absorption have been suggested, including addition of ion-pairing and complexation molecules, nonsurfactant membrane permeation enhancers, surfactant adjuvants, or combinations of these additives.[47] Addition of perturbants of tight junctions such as cytoskeletal agents, oxidants, hormones, calcium chelators, and bacterial toxins the formulation has been investigated to improve drug permeation.[49] Another novel delivery system involves the use of mucoadhesives to enhance drug delivery because of their long retention time at the targeted mucosal membrane; lectins have been identified as potential carriers for peptides in an oral mucoadhesive system.[1] Coadministration of peptides with inhibitors of metabolizing enzymes has also been suggested to increase oral absorption.[47,50,51]

Modulation of the intercellular junctions by inhibiting the cadherin-cadherin interaction at the adherens junction has also been investigated. Peptides derived from the sequence of the extracellular domain of E-cadherin have been shown to modulate the intercellular junction of bovine brain microvessel endothelial cell (BBMEC) and Madin-Darby canine kidney (MDCK) cell monolayers. These peptides enhance the paracellular penetration of marker molecules such as ^{14}C-mannitol and lower the transepithelial resistance of the monolayers.[52–55] The use of these cadherin-derived peptides as adjuvants to enhance paracellular permeability of drugs is still under investigation.

2.6. CONCLUSIONS

The absorption of an orally administered peptide depends on the successful passage of the peptide through the several barriers to drug delivery. The peptide can pass either between or through the cells, depending on its physicochemical properties. The gastrointestinal epithelial layer is a formidable obstacle to the passage of a peptide. Recent studies have shown that metabolism within the intestine forms a major obstruction to drug absorption. The concerted activity of these drug-metabolizing enzymes and efflux systems increases the problem. Although many challenges exist in traversing the intestinal epithelial layer, pharmaceutical scientists and medicinal chemists are overcoming them with innovative peptide drug products that optimize pharmacological activity and enhance drug delivery.

REFERENCES

1. Kompella, U. B.; Lee, V. H. *Adv. Drug. Deliv. Rev.* **2001**, *46*, 211–245.
2. Lipka, E.; Crison, J.; Amidon, G. L. *J. Contr. Rel.* **1996**, *39*, 121–129.
3. Fricker, G.; Drewe, J. *J. Pept. Sci.* **1996**, *2*, 195–211.
4. Hochman, J. H.; Artursson, P. *J. Contr. Rel.* **1994**, *29*, 253–267.
5. Hilgendorf, C.; Spahn-Langguth, H.; Regardh, C. G.; Lipka, E.; Amidon, G. L.; Langguth, P. *J. Pharm. Sci.* **2000**, *89*, 63–75.
6. Martinez, M.; Amidon, G.; Clarke, L.; Jones, W. W.; Mitra, A.; Riviere, J. *Adv. Drug Deliv. Rev.* **2002**, *54*, 825–850.
7. Fricker, G.; Miller, D. S. *Pharmacol. Toxicol.* **2002**, *90*, 5–13.
8. Doherty, M. M.; Charman, W. N. *Clin. Pharmacokinet.* **2002**, *41*, 235–253.
9. Adson, A.; Raub, T. J.; Burton, P. S.; Barsuhn, C. L.; Hilgers, A. R.; Audus, K. L.; Ho, N. F. *J. Pharm. Sci.* **1994**, *83*, 1529–1536.
10. Yap, A. S.; Mullin, J. M.; Stevenson, B. R. *J. Membr. Biol.* **1998**, *163*, 159–167.
11. Anderson, J. M.; Van Itallie, C. M. *Am. J. Physiol.* **1995**, *269*, G467–G475.
12. Cereijido, M.; Valdes, J.; Shoshani, L.; Contreras, R. G. *Annu. Rev. Physiol.* **1998**, *60*, 161–177.
13. Mitic, L. L.; Anderson, J. M. *Annu. Rev. Physiol.* **1998**, *60*, 121–142.
14. Gonzalez-Mariscal, L.; Chavez de Ramirez, B.; Cereijido, M. *J. Membr. Biol.* **1985**, *86*, 113–125.
15. Rothen-Rutishauser, B.; Riesen, F. K.; Braun, A.; Gunthert, M.; Wunderli-Allenspach, H. *J. Membr. Biol.* **2002**, *188*, 151–162.
16. Walsh, S. V.; Hopkins, A. M.; Nusrat, A. *Adv. Drug Deliv. Rev.* **2000**, *41*, 303–313.
17. Tsukita, S.; Furuse, M.; Itoh, M. *Curr. Opin. Cell Biol.* **1999**, *11*, 628–633.
18. Lapierre, L. A. *Adv. Drug Deliv. Rev.* **2000**, *41*, 255–264.
19. Madara, J. L. *Annu. Rev. Physiol.* **1998**, *60*, 143–159.
20. Burton, P. S.; Goodwin, J. T.; Vidmar, T. J.; Amore, B. M. *J. Pharmacol. Exp. Ther.* **2002**, *303*, 889–895.
21. Takeichi, M. *Annu. Rev. Biochem.* **1990**, *59*, 237–252.
22. Shore, E. M.; Nelson, W. J. *J. Biol. Chem.* **1991**, *266*, 19672–19680.
23. Chothia, C.; Jones, E. Y. *Annu. Rev. Biochem.* **1997**, *66*, 823–862.
24. Pertz, O.; Bozic, D.; Koch, A. W.; Fauser, C.; Brancaccio, A.; Engel, J. *Embo J.* **1999**, *18*, 1738–1747.

25. Yap, A. S.; Brieher, W. M.; Gumbiner, B. M. *Annu. Rev. Cell. Dev. Biol.* **1997**, *13*, 119–146.
26. Cowin, P.; Burke, B. *Curr. Opin. Cell Biol.* **1996**, *8*, 56–65.
27. Martinez, M. N.; Amidon, G. L. *J. Clin. Pharmacol.* **2002**, *42*, 620–643.
28. Agoram, B.; Woltosz, W. S.; Bolger, M. B. *Adv. Drug Deliv. Rev.* **2001**, *50*, S41–S67.
29. Tsuji, A.; Tamai, I. *Pharm. Res.* **1996**, *13*, 963–977.
30. Charman, W. N.; Porter, C. J.; Mithani, S.; Dressman, J. B. *J. Pharm. Sci.* **1997**, *86*, 269–282.
31. Engman, H. A.; Lennernas, H.; Taipalensuu, J.; Otter, C.; Leidvik, B.; Artursson, P. *J. Pharm. Sci.* **2001**, *90*, 1736–1751.
32. Wacher, V. J.; Salphati, L.; Benet, L. Z. *Adv. Drug Deliv. Rev.* **2001**, *46*, 89–102.
33. Cong, D.; Doherty, M.; Pang, K. S. *Drug Metab. Dispos.* **2000**, *28*, 224–235.
34. Benet, L. Z.; Izumi, T.; Zhang, Y.; Silverman, J. A.; Wacher, V. J. *J. Contr. Rel.* **1999**, *62*, 25–31.
35. Suzuki, H.; Sugiyama, Y. *Eur. J. Pharm. Sci.* **2000**, *12*, 3–12.
36. Wagner, D.; Spahn-Langguth, H.; Hanafy, A.; Koggel, A.; Langguth, P. *Adv. Drug Deliv. Rev.* **2001**, *50*, S13–S31.
37. Benet, L. Z.; Cummins, C. L. *Adv. Drug Deliv. Rev.* **2001**, *50*, S3–S11.
38. Zhang, Y.; Benet, L. Z. *Clin. Pharmacokinet.* **2001**, *40*, 159–168.
39. Goodwin, J. T.; Conradi, R. A.; Ho, N. F.; Burton, P. S. *J. Med. Chem.* **2001**, *44*, 3721–3729.
40. Pauletti, G. M.; Okumu, F. W.; Borchardt, R. T. *Pharm. Res.* **1997**, *14*, 164–168.
41. Burton, P. S.; Conradi, R. A.; Hilgers, A. R.; Ho, N. F.; Maggiora, L. L. *J. Contr. Rel.* **1992**, *19*, 87–98.
42. Burton, P. S.; Conradi, R. A.; Ho, N. F.; Hilgers, A. R.; Borchardt, R. T. *J. Pharm. Sci.* **1996**, *85*, 1336–1340.
43. Chikhale, E. G.; Ng, K. Y.; Burton, P. S.; Borchardt, R. T. *Pharm. Res.* **1994**, *11*, 412–419.
44. Pauletti, G. M.; Gangwar, S.; Siahaan, T. J.; Aube, J.; Borchardt, R. T. *Adv. Drug Deliv. Rev.* **1997**, *27*, 235–256.
45. Han, H. K.; Amidon, G. L. *AAPS Pharm. Sci.* **2000**, *2*.
46. Witt, K. A.; Gillespie, T. J.; Huber, J. D.; Egleton, R. D.; Davis, T. P. *Peptides* **2001**, *22*, 2329–2343.
47. Aungst, B. J. *J. Pharm. Sci.* **1993**, *82*, 979–987.
48. Okumu, F. W.; Pauletti, G. M.; Vander Velde, D. G.; Siahaan, T. J.; Borchardt, R. T. *Pharm. Res.* **1997**, *14*, 169–175.
49. Lutz, K. L.; Siahaan, T. J. *J. Pharm. Sci.* **1997**, *86*, 977–984.
50. Mizuma, T.; Ohta, K.; Koyanagi, A.; Awazu, S. *J. Pharm. Sci.* **1996**, *85*, 854–857.
51. Gotoh, S.; Nakamura, R.; Nishiyama, M.; Quan, Y. S.; Fujita, T.; Yamamoto, A.; Muranishi, S. *J. Pharm. Sci.* **1996**, *85*, 858–862.
52. Pal, D.; Audus, K. L.; Siahaan, T. J. *Brain. Res.* **1997**, *747*, 103–113.
53. Lutz, K. L.; Siahaan, T. J. *Drug Deliv.* **1997**, *10*, 187–193.
54. Makagiansar, I. T.; Avery, M.; Hu, Y.; Audus, K. L.; Siahaan, T. J. *Pharm. Res.* **2001**, *18*, 446–453.
55. Sinaga, E.; Jois, S. D.; Avery, M.; Makagiansar, I. T.; Tambunan, U. S.; Audus, K. L.; Siahaan, T. J. *Pharm. Res.* **2002**, *19*, 1170–1179.

3

PATHWAYS FOR DRUG DELIVERY TO THE CENTRAL NERVOUS SYSTEM

YAN ZHANG[1] AND DONALD W. MILLER[2]

Department of Pharmaceutical Sciences, University of Nebraska Medical Center, Omaha, NE 68198-6025

[1]Current address, Pfizer Global Research and Development, PDM Department, Building 20, Room 332 E, 2800 Plymouth Road, Ann Arbor, MI 48105
[2]Corresponding author

Drug Delivery: Principles and Applications Edited by Binghe Wang, Teruna Siahaan, and Richard Soltero
ISBN 0-471-47489-4

3.1. INTRODUCTION

3.1.1. Importance of Drug Delivery to the Central Nervous System (CNS)

Most drugs produce their required pharmacological response by altering cell function or structure in a concentration-dependent and reversible manner. Unless the target tissue is the blood and the drug is given intravenously or intra-arterially, the pharmacological response is dependent on the distribution of the drug to the target tissue in therapeutically relevant concentrations for sufficient periods of time to produce the clinically desired response. Issues pertaining to drug absorption and tissue distribution are often overlooked in the high-throughput screening process used to identify compounds with pharmacological activity. However, for CNS active compounds, such considerations are essential due to the presence of the blood-brain and blood-cerebral spinal fluid barriers, which restrict the passage of most compounds into the brain.

The impact of the blood-brain barrier (BBB) in CNS drug therapy has recently been eloquently reviewed by Pardridge.[1] The problems associated with the delivery of large molecules to the brain in therapeutically relevant concentrations have long been recognized. However, misconceptions concerning the ability of small molecules to penetrate the BBB and the effectiveness of small molecules in treating CNS-related pathologies have contributed to an underappreciation of the importance of drug delivery to the brain. Thus, the search for new and effective treatments for CNS diseases must concentrate not only on the drug target within the CNS but also on the efficient delivery of the molecule to the site of action within the brain. The purpose of this chapter is to provide a basic understanding of the current obstacles in drug delivery to the brain and the various approaches that have been used to increase CNS concentrations of drugs.

3.1.2. Cellular Barriers to Drug Delivery in the CNS

There are two cellular barriers that separate the brain extracellular fluid from the blood (see Figure 3.1). The first and largest interface is the brain capillary endothelial cells that form the BBB. The brain capillaries are a continuous layer of endothelial cells connected by well-developed tight junctional complexes.[2] As a result of the tight junctions, passive diffusion of drugs and solutes between the endothelial cells is restricted. In addition, brain capillary endothelial cells lack fenestrations (water-filled pores or channels within the plasma membrane) and have reduced pinocytic activity. These characteristics further restrict the movement of compounds from the blood into the extracellular environment of the brain.[3–5]

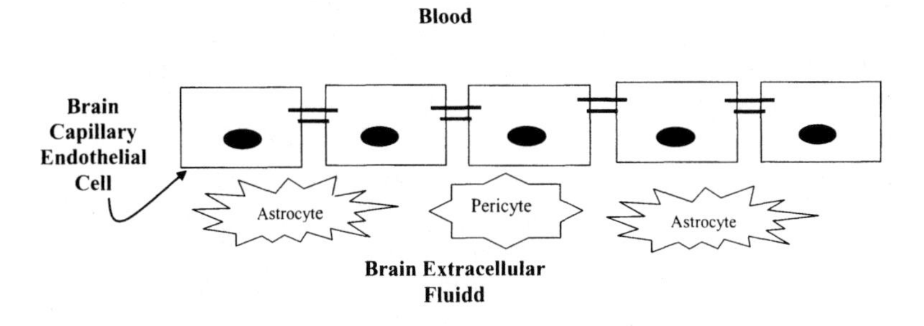

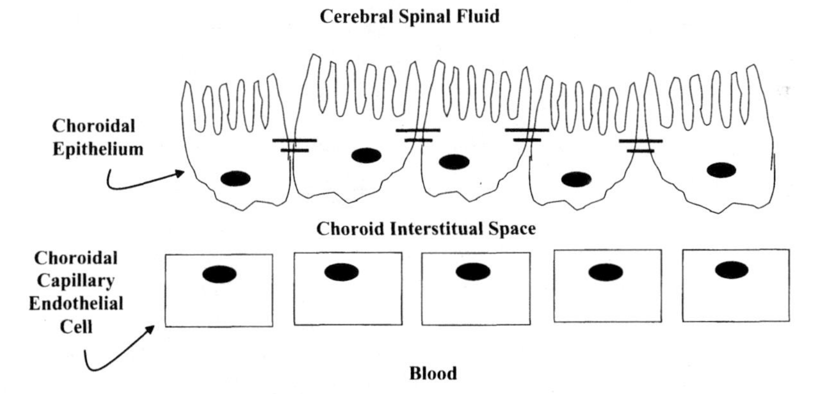

Figure 3.1. Schematic representation of the cellular components of the blood-brain (Panel A) and blood-cerebral spinal fluid (Panel B) barriers. The blood-brain barrier consists of continuous type endothelial cells with complex tight junctions to limit paracellular diffusion. The astrocytes and pericytes located in close proximity to the brain endothelial cells release various endogenous factors that modulate endothelial cell permeability. In contrast, the choroid endothelial cells are fenestrated and the blood-cerebral spinal fluid barrier properties are provided by the tight junctions formed between the choroid epithelial cells.

The presence of tight junctions between the brain capillary endothelial cells means that the paracellular pathway for drug delivery is highly restricted. Lipid-soluble drugs with a molecular mass of less than 600 can pass through the BBB via passive diffusion through the brain endothelial cells.[3] Factors that influence passive diffusion include molecular volume, charge, and the hydrogen-bonding potential of the compound.[6,7] Besides the diffusional pathway, compounds can also move across the BBB through vesicular transport and via specific transport or carrier systems within the brain endothelial cells. Furthermore, a variety of enzymes in the brain endothelial cells act as a two-way metabolic barrier capable of breaking down blood-borne substances and brain metabolites. These enzymes also play important roles in controlling the transport of a variety of compounds, both endogenous and exogenous, from the blood entering the brain.

The second barrier involved in regulation of the extracellular environment of the brain is the blood–cerebrospinal fluid barrier (BCSFB). The BCSFB is a composite barrier made up of the choroid plexuses and the arachnoid membranes of the circumventricular organs. Unlike the capillaries that form the BBB, the circumventricular organs of the brain have fenestrated "leaky" capillaries, and their barrier function is provided by the tight junctions between the epithelial cells in the choroid plexus.[8,9] The epithelial cells in the choroid plexus that form the BCSFB have complex tight junctions on the CSF (apical) side of the cells. These tight junctions formed by the epithelial cells in the choroid plexus are slightly more permeable than those found in the endothelial cells of the BBB.[9,10]

Although the apical membrane of the epithelial cells forming the BCSFB has numerous microvilli, the total surface area is still substantially smaller than the BBB.[11] It has been estimated that in the human brain there are approximately 100 billion capillaries with a total surface area of 20 m^2.[1] Given the density of the capillary network in the brain, and the close proximity of neuronal cells to these capillaries, most drug delivery approaches have focused on either circumventing the BBB or increasing drug passage through the BBB.

3.1.3. General Approaches for Increasing Brain Penetration of Drugs

There are three general routes for brain delivery of drugs. One approach is to circumvent the difficulties associated with drug permeability in the BBB and/or BCSFB by direct central administration of the drug into the brain. An alternative approach is to temporarily break down the BBB, allowing the therapeutic agents to enter the brain from the blood through a more permeable BBB. The third and final approach uses chemical modifications of the drug and/or knowledge of the biology of the brain endothelial cells forming the BBB to improve the transcellular passage of drugs into the CNS. Each of these strategies is discussed below.

3.2. DIRECT ADMINISTRATION OF DRUGS INTO THE BRAIN

3.2.1. Intracerebral Administration

One strategy for delivering drugs to the brain is to circumvent the problems associated with penetration of the BBB by direct injection of drugs into the brain. This approach is invasive, requiring a craniotomy in which a small hole is drilled in the head for intracerebroventricular (ICV) or intracerebral (IC) drug administration into the brain. An advantage of this approach is that a wide range of compounds and formulations can be considered for ICV or IC administration. Thus, both large- and small-molecule therapeutics can be delivered, either alone or in various polymer formulations, to achieve sustained release.

Aside from the invasiveness of the craniotomy procedure and the implications this may have for long-term therapy, the biggest disadvantage of direct implantation of drug into the CNS is related to the limited brain distribution of the drug. This is illustrated in the studies by Krewson and coworkers,[12] in which polymer implants containing radiolabeled nerve growth factor were placed in rat brain and diffusion of the factor monitored by autoradiography. In these studies, diffusion of nerve growth factor from the polymer implant was limited to 2–3 mm. Limited CNS distribution was also observed following one-time bolus ICV injections of brain-derived neurotrophic factor in rats.[13] This phenomenon of limited tissue distribution is not restricted to large macromolecules. Hoistad et al.[14] reported a diffusion distance of only 1 mm following striatal IC infusion of radiolabeled dopamine and mannitol in rats.

Despite these limitations, successful application of the central administration route for drug delivery to the brain has been reported in laboratory animals. Studies by Mairs et al.[15] showed significant tumor uptake and retention of radioiododeoxyuridine following IC administration in a rat glioma model. Furthermore, IC administration of sustained-release polymers containing the antiepileptic agent phenytoin was effective in preventing cobalt-induced seizures in rats.[16] However, these results should be tempered by the scaling issues present when going from small laboratory animals to humans. For small laboratory animals, a focal seizure area or tumor mass may be treatable with a 2–3 mm diffusion radius from the point of drug administration. However, from the perspective of human brain diseases there are few, if any, conditions that would be expected to respond to therapy with such a limited brain tissue distribution. At best, this approach to CNS delivery would require a highly potent therapeutic agent and a tissue target with a highly restricted area of distribution.

3.2.2. Intrathecal Administration

Recently, intrathecal administration has been reexamined as a means of circumventing the BBB delivery problems associated with large macromolecules such as proteins and peptides. Intrathecal administration involves the injection or infusion of drugs into the cerebrospinal fluid (CSF) that surrounds the spinal cord. At first glance, intrathecal administration as a method for delivering drugs to the brain would seem contradictory, as this route is commonly employed to produce localized analgesia without the CNS complications that administration of systemic anesthetics and analgesic agents produce. This lack of CNS effect observed with intrathecal administration of anesthetic and analgesic agents is due to the rapid removal of drug from the CSF to the systemic bloodstream.[17] However, it should be noted that the agents commonly administered by the intrathecal route for pain management are small lipophilic molecules. Evidence suggests that proteins administered intrathecally have a much slower clearance from the CSF.[18] Thus, the intrathecal route of administration may be a viable strategy for delivering large molecular weight macromolecules with limited lipophilic properties.

Evidence of the validity of the intrathecal route for delivery of proteins to the brain is found in studies with the centrally acting adipose regulatory protein, leptin. This 16 kDa protein is secreted by fat cells and transported into the CNS, where it acts to reduce appetite and promote weight loss. Recent studies by McCarthy et al.[19] examined the appearance of leptin in the brain following intrathecal injection in baboons using positron emission tomography. The studies were remarkable in that they indicate that leptin, administered via intrathecal injection, was able to travel from the lower lumbar region to the hypothalamus within 3 hours of injection. Even more remarkable is that the amount of leptin detected in the brain following intrathecal administration was estimated to be well above the concentrations required for biological activity.[19] In support of these studies, separate experiments by Yaksh et al.[20] reported dose-dependent suppression of body weight and food consumption in rats given 14-day continuous intrathecal infusion of leptin. Together these studies provide support for the intrathecal route for the delivery of large molecular weight molecules to the brain.

3.2.3. Nasal Administration

Another delivery option for bypassing the BBB is through intranasal (IN) administration. Compared to intracerebral and intrathecal administration, IN administration is a noninvasive means of delivery of therapeutic agents to the CNS.[21,22] Because of the unique connection between the nose and the brain, the olfactory neural pathway provides a route of delivery for various compounds into the CNS. These include toxic agents such as pathogens, viruses, and toxic metals. However, the same pathway can also be used to delivery various therapeutic agents including small molecules and proteins to the CNS (for review, see Refs. 21 and 23).

Small molecules such as cocaine and cephalexin can be transported directly to the CNS from the nasal cavity.[24,25] Sakane and colleagues[25] reported that cephalexin preferentially entered the CSF after nasal administration compared to intravenous (IV) and intraduodenal administration in rats. The levels of cephalexin in CSF were 166-fold higher 15 minutes after nasal administration than those of the other two routes. Most recently, studies by Wang and colleagues showed that the ratio of the methotrexate AUC_{CSF} value between the IN route and the IV injection was over 13-fold.[26]

In addition to small molecules, a number of protein therapeutic agents, such as neurotrophic factors[27] and insulin,[28] have been successfully delivered to the CNS using IN delivery in a variety of species. The therapeutic benefit of IN delivery of proteins has been demonstrated by Liu et al. in rat stroke models.[29] Their studies demonstrated that insulin-like growth factor I (IGF-I) could be delivered to the brain directly from the nasal cavity, even though IGF-I did not cross the BBB efficiently by itself. As a consequence, IN IGF-I markedly reduced infarct volume and improved neurological function following focal cerebral ischemia. Research in humans has also provided evidence for direct delivery of therapeutic agents to the CNS from the nasal cavity. Studies by Kern and colleagues have demonstrated CNS effects of IN insulin in humans without altering plasma glucose or insulin level.[28]

IN administration is a promising approach for rapid-onset delivery of medications to the CNS bypassing the BBB. However, there are also limitations. One of the biggest limitations is insufficient drug absorption through the nasal mucosa. Many drug candidates cannot be developed for the nasal route because they are not absorbed well enough to produce therapeutic effects.[30] Another constraint concerning nasal administration is that a small administration volume is required, beyond which the formulation will be drained out into the pharynx and swallowed.[31]

In summary, the advantages of IN delivery to the CNS are considerable. IN administration has been shown to improve the bioavailability of many presystemically metabolized drugs entering CNS, eliminating the need for systemic delivery and reducing unwanted systemic side effects. It is also rapid and noninvasive. Furthermore, IN delivery does not require any modification of the therapeutic drugs and does not require the drugs to be coupled to any carriers.

3.3. BBB DISRUPTION

Under normal conditions, the complex tight junctions that form between the brain capillary endothelial cells restrict the paracellular diffusion of molecules and solutes in the BBB. Modification of the tight junctions, causing controlled and transient increases in the permeability properties of the brain capillaries, is another strategy that has been used to increase drug delivery to the brain. Methods for disrupting BBB integrity through the breakdown of tight junctions include the systemic administration of hyperosmotic solutions,[32,33] vasoactive compounds such as bradykinin and related analogs,[34] and various alkylglycerols.[35,36] Each of these approaches is described in greater detail below.

3.3.1. Osmotic Agents

Disruption of the BBB through the use of osmotic agents has been extensively studied in both laboratory animals and clinically in the treatment of brain tumors.[37,38] In most cases, a hypertonic solution of an inert sugar, such as mannitol or arabinose, ranging from 1.4 to 1.8 M, is delivered into the cerebral circulation through bolus injection or short-term infusion into the carotid artery.[33, 39–41] The proposed cellular mechanism behind osmotic disruption of the BBB involves the physical pulling apart/breaking of tight junctions due to the shrinkage of cerebral endothelial cells and expansion of the blood volume caused by the addition of the hyperosmotic agent.[32,33,40] As the disruption of the BBB is contingent on the presence of hyperosmotic agents in the blood, the BBB resumes its normal barrier functions within hours of returning the osmolarity of the blood to normal.[42] During this period when the tight cellular junctions between the brain capillary endothelial cells have been compromised, paracellular diffusion of water-soluble drugs and solutes into the brain is enhanced.

Increased delivery of drugs to the brain following osmotic disruption of the BBB has been demonstrated in a variety of settings. Increases in both small- and large-molecule delivery to the brain have been reported.[41] The time course for disruption of BBB integrity and the subsequent return of the barrier function following osmotic disruption appear to be variable. Studies in rats suggested that the onset of BBB opening was rapid, with maximal responses observed within 5 minutes of hyperosmotic mannitol administration.[41] Likewise, return of normal BBB integrity was noted within minutes following cessation of the osmotic agent. More recent studies in humans suggest that while disruption of BBB permeability in response to hyperosmotic mannitol was rapid, with increases in BBB permeability observed within 1 minute of the osmotic agent, the barrier properties were not reestablished for several hours.[42]

Osmotic disruption of the BBB has been used to increase the delivery of chemotherapeutic agents to the brain in the treatment of CNS tumors in rats.[43–45] The clinical benefits of increased delivery of chemotherapeutic agents to the brain through osmotic disruption of the BBB have been demonstrated by the increased survival rates observed in patients with primary CNS lymphoma[46,47] and malignant gliomas.[47,48]

While osmotic BBB disruption has been used primarily for enhancing the response to small-molecule chemotherapeutic agents in brain tumors, applications in gene and protein therapy in the CNS are also being explored. Studies in rats using radiolabeled monoclonal antibodies demonstrated a 100-fold greater delivery of antibodies from the blood to the brain following osmotic BBB disruption.[49] While the actual amount of antibody delivered to the brain following osmotic BBB disruption was still relatively small, 0.72% of the total dose, the relatively long residence time observed for the monoclonal antibody in the CNS suggests that biologically relevant levels of proteins can be delivered to the brain with this approach. Osmotic disruption has also proven successful in the delivery of viral vectors to the brain.[50–52] Studies using adenoviral vector containing the *Escherichia coli* beta-galactosidase gene showed no detectable viral delivery to the brain under normal conditions following intracarotid infusion of the virus.[50] However, osmotic disruption of the BBB resulted in detectable beta-galactosidase expression in the brain. The extent of beta-galactosidase expression in the brain was directly correlated with the magnitude of BBB disruption.[50] More recent studies by Abe et al.[52] examined responses of human brain tumor xenografts to tumor suppressor gene therapy. In these studies, BBB disruption with mannitol increased the tumor response to p53 adenoviral vector. Together these studies suggest that osmotic disruption of the BBB can be used for a variety of both small- and large-molecule therapeutic agents.

3.3.2. Bradykinin Analogs

The BBB can also be disrupted by pharmacological means. Several endogenous proinflammatory vasoactive agents, such as bradykinin, histamine, nitric oxide, and various leukotrienes, are known to induce increases in BBB permeability in a concentration- and time-dependent manner. These vasoactive compounds are characteristically ultra-short-acting due to either rapid deactivation through metabolic processes or tachyphylaxis at the receptor signal transduction level. Furthermore, many of these endogenous agents have narrow therapeutic windows and dose-limiting side effects. Thus, although capable of producing increases in BBB permeability, these endogenous agents have proven difficult to apply safely for CNS drug delivery.

In an effort to circumvent the problems associated with the use of endogenous vasoactive agents, investigators have explored the use of structurally modified analogs. The best example of this is the bradykinin analog labradimil (Cereport). Bradykinin-induced increases in BBB permeability occur through a kininergic B2 receptor-mediated opening of tight junctions between the brain capillary endothelial cells.[53,54] However, with a plasma half-life on the order of seconds,[55] the effects of bradykinin on BBB permeability are short-lived, requiring carotid artery infusion. In contrast, Cereport has amino acid substitutions that result in significantly greater plasma and tissue stability than bradykinin.[56] This, coupled with the B2 receptor selectivity of Cereport, provides for a more consistent and reproducible effect on BBB permeability without many of the toxic side effects observed with bradykinin (see Ref. 57 for review). Based on studies in both rats and dogs, the effects of Cereport on BBB permeability can be produced following either intracarotid or IV administration, with both the time to onset following administration and the time to restoration of BBB permeability following cessation of Cereport being observed within 5 minutes.[58]

As with osmotic disruption of the BBB, Cereport has been used to increase delivery of chemotherapeutic agents in experimental brain tumor models.[59–61] Interestingly, as opposed to osmotic disruption, the permeability increases observed with Cereport were greater in the area in and around the brain tumor compared to nontumor regions of the brain.[62,63] The effect of Cereport on BBB and blood-tumor barrier permeability has been shown to cause significant increases in tumor responsiveness to chemotherapeutic agents in various brain tumor models.[64,65] While fewer human studies have been performed with Cereport than with osmotic disruption of the BBB, Phase II multinational clinical trials indicate improved therapeutic responses using IV or intra-arterial Cereport together with carboplatin in the treatment of gliomas.[66]

Although Cereport appears to have greater effects on the brain microvasculature in and around the tumor site, it has also been used to increase BBB permeability in conditions other than brain tumor therapy. Emerich and colleagues[67] studied the delivery of the centrally acting opiate analgesic, loperamide, and the resulting effects on pain perception in rodents. In these studies, coadministration of loperamide with Cereport produced a twofold enhancement in pain response time using the tail flick assay. As these effects were inhibited by the opiate receptor antagonist, naloxone, the investigators concluded that Cereport could be used to enhance drug delivery to the brain under normal conditions.[67] Recent studies by Bidanset and coworkers[68] described the use of Cereport to enhance the delivery and effectiveness of antiviral therapy in the CNS. In these studies, acyclovir was administered either alone or in conjunction with Cereport in herpes simplex virus–infected rats. Coadministration of acyclovir with Cereport resulted in a two- to threefold increase in acyclovir accumulation in the brain.[68] The increased delivery of acyclovir observed with Cereport also resulted in a significantly lower viral load throughout the brain. Together these studies suggest that short-term, reversible disruption of BBB permeability with Cereport can be used to impact CNS drug delivery in a variety of conditions.

3.3.3. Alkylglycerols

A relatively new approach for transient disruption of the BBB involves the systemic administration of various alkylglycerols. Erdlenbruch and colleagues[36] reported a reversible and concentration-dependent increase in BBB permeability to several anticancer and antibiotic agents. The extent of BBB disruption varied from a 2-fold to a 200-fold increase in methotrexate, depending on the length of the alkyl group and the number of glycerols present in the structure. As with the bradykinin analog Cereport, the time to onset and reversal of the BBB-disrupting actions of the alkylglycerols are very short, occurring within minutes of application and returning to normal within minutes of cessation of alkylglycerol administration.[36] Recent studies examining the effects of alkylglycerols on large-macromolecule permeability in isolated brain capillaries suggests that the increases in permeability are due to temporary breakdown of the tight junctions between the cells.[35] Although the exact mechanism(s) for the transient BBB disruption observed with the alkylglycerols are unknown, the concentration and structure dependency of the response would suggest interaction with receptor sites within the brain microvasculature.

The use of alkylglycerols to increase BBB permeability of anticancer agents has been examined in a rat glioma tumor model.[35,69,70] In these studies, several different alkylglycerols were examined for increasing the CNS delivery of methotrexate in C6 glioma-bearing rats. Studies indicated significantly greater methotrexate accumulation at both the tumor site and non-tumor-bearing sites within the brain.[35,70] The effects of the alkylglycerols were concentration-dependent. However, the magnitude of the increase in methotrexate delivery to the tumor site detected with the highest doses of alkylglycerols examined was comparable to that observed with osmotic disruption, and was significantly greater than that observed with the bradykinin analog, Cereport.[70] While human clinical trials have not been initiated, studies in the rat glioma brain tumor model have detected no long-term signs of toxicity with the alkylglycerols, suggesting that these agents may have promise in the treatment of brain tumors.

3.4. TRANSCELLULAR DELIVERY ROUTES IN THE BBB

Aside from either circumventing the BBB completely through central administration of a drug or reversibly opening the BBB, increased delivery of therapeutic agents to the brain can be accomplished through improved transcellular migration. The transcellular routes available in brain capillary endothelial cells are shown in

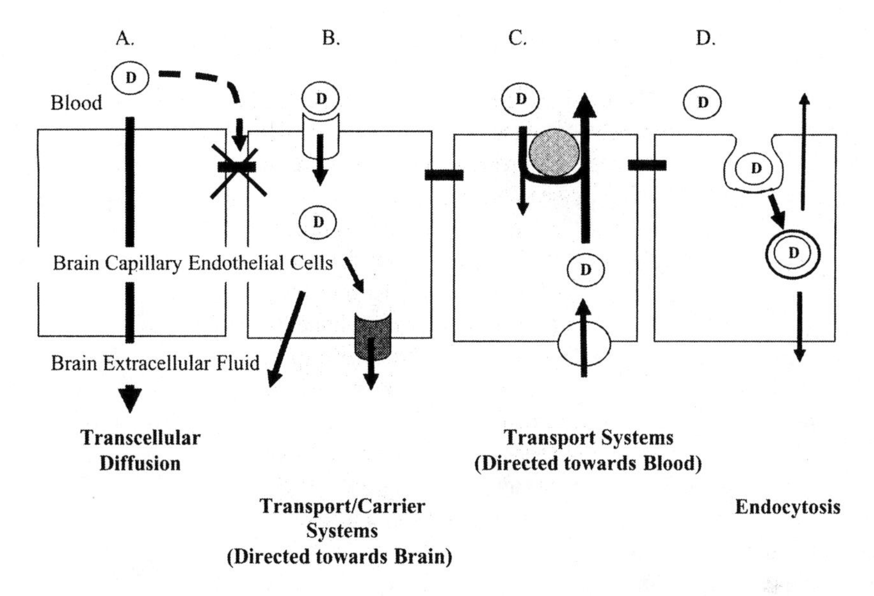

Figure 3.2. Potential mechanisms for drug movement across the blood-brain barrier. Routes of passage include passive diffusion through the brain capillary endothelial cells (A); utilization of inwardly directed (i.e. towards brain) transport or carrier systems expressed on brain capillary endothelial cells (B); utilization of outwardly directed (i.e. towards blood) efflux transport systems (C); or inclusion in various endocytic vesicular transport processes occurring within the brain capillary endothelial cells (D).

Figure 3.2 and include passive diffusion, specific transport systems, and endocytic processes present in the brain microvasculature. The design of drugs or drug delivery systems that improve CNS delivery through one or more of the transcellular pathways identified in Figure 3.2 have several advantages over the other strategies. First, since they increase the transcellular permeability of a drug across the BBB, a much greater area of the brain can be treated compared to direct administration of the drug into the brain or ventricle. Second, since they alter the transcellular permeability of a specific drug, there is less chance of neurotoxicity compared to that associated with BBB disruption techniques. And finally, by focusing on specific transcellular transport processes present in the BBB, one can achieve a targeted delivery of drugs to the brain.

3.4.1. Passive Diffusion

Key factors influencing the passive diffusion of drugs across the BBB are lipid solubility and molecular size. The relationship is described by the equation $D = \log P/MS^{1/2}$, where D is diffusion, log P is lipophilicity, and MS is molecular size. Thus, improving the passive diffusion of drugs across the BBB can be accomplished by either increasing lipophilicity or reducing molecular size. As lipophilicity is dependent on polarity and ionization, modification and/or masking of functional groups on drugs provide a method for improving passive diffusion across the BBB.

One method employed for increasing the lipophilicity of a drug is the creation of a prodrug. In this approach, water-soluble compounds with polar functional groups such as acids or amides, are chemically modified to create derivatives with increased lipid-solubility (See Chapter 15 for more information on prodrugs.) The most common prodrugs are esters, since by appropriate esterification of molecules containing carboxylic, hydroxyl, or thiol functional groups, it is feasible to obtain derivatives with almost any desired lipophilicity or hydrophilicity.[71] The prototypical example of such an approach to enhance drug delivery in the brain is heroin. Heroin, the diacetyl ester of morphine, rapidly enters the brain due to its high lipophilicity. Once in the brain, it is hydrolyzed to morphine, which is less lipophilic and has a slower diffusion rate back into the bloodstream, prolonging its time course of action in the brain.[72] This same approach has been employed with other, more clinically relevant therapeutic agents such as the anticancer agent, chlorambucil,[73] and the neurotransmitters dopamine[74] and gamma-aminobutyric acid (GABA).[75]

Chemical design approaches aimed at increasing the lipophilicity of macromolecules for improved CNS distribution have also been examined. Peptides have the potential to be potent pharmaceutical agents for the treatment of many CNS diseases. However, the delivery of conventional linear peptides or peptidomimetics to the brain presents a significant challenge to pharmaceutical scientists. This is due to the combination of high enzymatic degradation of peptides in biological fluids and limited diffusion of peptides across biological membranes.[76] One approach that has been shown to improve both the metabolic stability and membrane permeability of peptides is a process termed "cyclization."[77–80] In this approach, linear peptides are converted into a cyclic conformation via the addition of a chemical linker between selected amino acids.

It should be noted that not all cyclization reactions result in improved CNS delivery. Recent studies by Yang et al. examined the delivery of an opioid peptide, H-Tyr-D-Ala-Gly-Phe-D-Leu-OH (DADLE), using the cyclic prodrug approach.[76] In these studies, a cyclic conformation of DADLE was created using esterase-sensitive linkages. The DADLE prodrugs exhibited metabolic stability to exo- and endopeptidases, and esterase-catalyzed bioconversion of the prodrugs to DADLE was observed both *in vitro* and *in vivo*. The cyclic prodrugs displayed improved *in vivo* stability compared with the parent drug, DADLE. However, the cyclized prodrugs of DADLE were unable to deliver significant amounts of DADLE to the brain because of their rapid biliary excretion and poor BBB permeation. The poor BBB permeability of these cyclized peptides was attributed to interactions with drug efflux transporters, such as P-glycoprotein (P-gp) and multidrug resistance–associated protein 2 (MRP2) expressed in the endothelial cells forming the BBB. If these efflux transporters were inhibited, the cell permeation of the prodrugs was significantly enhanced compared to that of DADLE.[76]

Lower than expected BBB permeability has also been noted for cyclosporin A (CSA), a neutral cyclic peptide composed of 11 hydrophobic amino acids.[81] The high lipid solubility of CSA allows it to traverse the gastrointestinal barrier, making it one of the few peptide drugs that is efficacious following oral ingestion.[82] However, despite the high degree of lipid solubility of CSA, the transport of this peptide through the BBB is surprisingly slow.[83] The low transport of CSA results not only from the plasma protein binding, but also from the substrate activity for P-gp.[84]

Increasing lipophilicity through chemical modifications of the drug has to be weighed against other factors that influence the overall therapeutic efficacy of the drug. As discussed above in regard to the cyclic peptides, creating a more lipophilic compound can change not only the passive permeability into the capillary endothelial cell but also potential interactions with drug efflux transport proteins that result in the removal of drugs from the brain capillaries. This phenomenon has been observed with small-molecule ester prodrugs, where the ester derivative is removed from the brain capillary endothelial cell before its conversion to the active drug.[85] Additional considerations include the effects that chemical modifications may have on the interaction of the drug with its receptor or target site, plasma protein binding, and enzyme metabolism.[86] This is particularly the case for small molecules where the chemical modifications can have major impacts on binding to the target site. A final consideration is the impact of the chemical modification on drug accumulation at other nontarget tissue sites. This is especially true for those chemical modifications that improve lipophilicity, as such compounds will have enhanced tissue penetration outside of the CNS.

3.4.2. Inwardly Directed Transport Systems in the BBB

To meet the metabolic needs of the brain, the capillary endothelial cells that form the BBB express many selective carrier/transport systems for delivering essential nutrients from the blood to the brain. These transport systems include those for various amino acids, glucose, and assorted nucleosides.[71,87] An approach to increasing the transcellular passage of drugs across the BBB into the brain is to design drugs that structurally resemble or can be linked to endogenous compounds that are transported into brain by the carriers or transporters expressed in the brain microvessel endothelial cells.[88,89]

3.4.2.1. Amino Acid Transporters Several amino acids carrier systems are present in the BBB. These include a large neutral amino acid transporter, System L, a cationic amino acid transporter, System y^+ (both of which are Na^+-independent systems), the anionic amino acid transporter, System X^-, and neutral and/or cationic amino acid transporters, Systems A and $B^{o,+}$, that require an Na^+ gradient.[90] Of the amino acid transporters, System L has been most exploited for drug delivery purposes.[91] L-Dopa is the prototypical example of a drug that is transported by System L in the BBB and was one of the first drugs demonstrated to be taken up into the brain by a carrier mechanism.[92,93] L-Dopa is an endogenous large amino acid and is a precursor of the neurotransmitter dopamine.[86] System L is also involved in the transport of other drugs such as melphalan, baclofen, and gabapentin across the BBB [92,94–97].

3.4.2.2. Glucose Transporters The brain has a high metabolic demand for glucose. To accommodate the energy requirements of the CNS, glucose is transported from the blood to the brain through specific transport systems. The primary glucose transporter (GLUT) present in the brain capillary endothelial cells is GLUT1.[98] Compared to other nutrient transport/carrier systems in the BBB, GLUT1 has the highest capacity (more than 10–50 times greater than that of amino acid and carboxylic acid transporters) and therefore represents an attractive target for drug delivery to the CNS. Unfortunately, while various glucose-based analogs have been shown to act as ligands for the GLUT1 transporter, few have shown transporter activity.[99,100] Glycosylated analogs of various opioid compounds have shown increased CNS analgesic properties compared to the unglycosylated compounds.[101,102] However, it is unclear with the glycosylated peptide analogs whether the increased CNS response is due to enhanced BBB penetration or simply to increased stability of the peptides in blood.

3.4.2.3. Monocarboxylic Acid Transporter Systems for transporting monocarboxylic acids such as lactic acid, acetic acid, and ketone bodies both into and out of the CNS are abundant in the BBB. The best-characterized organic acid transporter in the BBB is the monocarboxylic acid transporter (MCT). The MCT has been detected on both the luminal (blood) and ablumenal (brain) plasma membranes of brain capillary endothelial cells[103] and likely participates in the bidirectional transport of organic acids into and out of the brain. An example of a drug entering the CNS through the MCT is salicylic acid.[104] More recently, the various cholesterol-lowering 3-hydroxy-3-methylglutaryl coenzyme A (HMG-CoA) reductase inhibitors have displayed carrier-mediated transport activity in brain capillary endothelial cells.[105] Given the presence of the moncarboxylic acid moiety in these agents and the MCT1-specific transport observed in intestinal epithelial cells, it has been postulated that the pH-dependent uptake of the HMG-CoA reductase inhibitors in the BBB is due, at least in part, to MCT.[91]

3.4.2.4. Nucleoside Transporters There are two general types of nucleoside transporter expressed in the brain capillary endothelial cells forming the BBB (see Ref. 91 for review): facilitative nucleoside transporters that carry selective nucleosides either into or out of the cell, depending on the presence of a concentration gradient (referred to as "equilibrative nucleoside transporters"), and active, sodium-dependent transporters that can move selective nucleosides into the cell against a concentration gradient (referred to as "concentrative nucleoside transporters"). Studies in rats indicate both equilibrative[106] and concentrative[107] nucleoside transport systems in the BBB and BCSFB, suggesting that these transporters have important roles in maintaining the appropriate nucleoside levels in the brain. The various equilibrative and concentrative nucleoside transporters display selectivity for either purine or pyrimidine nucleosides. Based on *in vivo* studies examining the BBB permeability of purine and pyrimidine nucleoside analogs, it would appear that the nucleoside transporters for purine-based nucleosides are more functionally active than the pyrimidine-selective transporters.[108]

There are several examples of drugs that are substrates for various nucleoside transport systems. The anticancer agent, gemcitabine, has high affinity for the concentrative nucleoside transporter, CNT1, and much lower affinity for an equilibrative nucleoside transporter.[109] The antiviral agents, 3′-azido-3′-azidodeoxythymidine (AZT) and 2′,3′-dideoxycytidine (ddC), are low-affinity substrates for the pyrimidine-sensitive, CNT1.[110] While the various nucleoside transporters would appear to be likely targets for CNS delivery of nucleoside-based antiviral and anticancer drugs, most studies to date have shown only limited BBB permeability with these agents.[111] This could be due to the relatively low affinity of these therapeutic agents for the nucleoside transporters present in the BBB. Alternatively, in the case of AZT and related compounds, a high-capacity efflux transport system(s) in the BBB efficiently removes these agents from the brain.[112]

3.4.2.5. Peptide Transport Systems Both inwardly and outwardly directed peptide transport systems are present in the brain capillary endothelial cells forming the BBB.[91] Unlike the intestinal epithelial barrier, where the primary peptide transporters are PepT1 and PepT2, there is little evidence for these transporters in the BBB. It should be noted that despite the absence of PepT1 and PepT2 in the brain endothelial cells of the BBB, recent studies indicate the presence of PepT2 in choroid epithelial cells forming the BCSFB.[113] The exact molecular nature of these peptide transporters remains to be determined, and to date they have been defined in more functional terms based on substrates. Nonetheless, specific saturable transport systems have been identified in the BBB for glutathione,[114] various opioid peptides,[115] peptide hormones such as arginine vasopressin, melanocyte stimulating hormone, and luteinizing hormone releasing hormone,[116–118] and various growth factors and cytokines including epithelial growth factor, interleukin 1, and tumor necrosis factor.[119–121]

3.4.2.6. Considerations for Carrier-Mediated Transport in the CNS The design of drugs that utilize one or more of the various endogenous transport systems in the BBB is an attractive approach for increasing drug delivery to the brain. Taking advantage of specific transport systems provides a more targeted approach to CNS drug delivery than physicochemical alterations aimed at creating a more lipophilic therapeutic agent or prodrug. However, there are several considerations when using carrier-mediated transport systems to increase drug permeability in the BBB. First, the transport systems prevalent in the BBB are relatively selective systems designed to aid the passage of essential nutrients and metabolites into and out of the brain. Thus, chemical modifications made to therapeutic agents to target specific transport or carrier systems in the BBB are much more restricted than those used to enhance lipophilicity. In addition, the resulting modifications are likely to result in compounds with lower affinity for the transporter than the endogenous ligand. An example of this can be seen with the nucleoside-based antiviral agents. Both AZT and ddC are transported by the concentrative nucleoside transporter, CNT1. However, the affinity of these agents for the transporter is approximately 25-fold lower than that of endogenous pyrimidine-based nucleosides.[110] An additional consideration is the potential transporter interactions of the drug with the endogenous ligand. An example of this is the well-characterized diminished therapeutic effectiveness of L-Dopa in patients with Parkinson's disease when the drug is taken following a protein-rich meal. The decrease in pharmacological activity can be explained in terms of lowered distribution of L-dopa into the brain due to saturation of the System L amino acid transporter at the BBB with high plasma concentrations of amino acids generated from the meal.[122] Thus, caution must be used in the design and administration of carrier-mediated drugs to ensure that (1) plasma fluctuations of endogenous substrates do not severely disrupt drug delivery to the CNS and (2) the drug does not reduce brain nutrient delivery below the minimum tolerable level.[93]

3.4.3. Vesicular Transport in the BBB

There are two general types of vesicular transport processes: fluid-phase endocytosis and adsorptive endocytosis. While both processes require energy and can be inhibited by metabolic inhibitors, only adsorptive endocytosis involves an initial binding or interaction of the molecule with the plasma membrane of the cell. As such, vesicular transport due to adsorptive endocytosis is a saturable, ligand-selective phenomenon. In general, vesicular transport in the brain capillary endothelial cells is reduced compared to other capillary beds. However, several large macromolecules of importance for normal brain function are transported from the blood into the brain through receptor-mediated endocytosis. Thus, the design of therapeutic agents and biomacromolecules to utilize these specific receptor-mediated transport processes in the brain capillary endothelial cells represents another approach for enhancing transcellular permeability across the BBB. The most well-characterized receptor-mediated vesicular transport processes in the BBB are discussed below.

3.4.3.1. Transferrin Receptor–Mediated Vesicular Transport Serum transferrin is a monomeric glycoprotein with a molecular weight of 80 kDa that is crucial for the transport of iron throughout the body.[123] Iron enters the cell as a complex with transferrin through an endocytic process that is initiated by the binding of transferrin to its receptor on the plasma membrane.[124] The brain capillary endothelial cells have a high density of transferrin receptors on their surface compared to other types of cells.[125] The binding of transferrin to its receptor on the brain capillary endothelial cells triggers the internalization of the transferrin-iron complex. Inside the brain endothelial cell, the iron is removed from the transferrin in the endosome, and through the vesicular cell-sorting process, iron is released into the brain extracellular fluid, and transferrin and its receptor are recycled back to the luminal (blood) plasma membrane.

The prevalence of transferrin receptors in the BBB and the resulting vesicular transport that occurs following binding to the receptor have stimulated interest in the potential use of this transport system for targeted drug delivery to the brain. However, the biggest limitation in utilizing this transport pathway is that transferrin itself undergoes cellular processing and is ultimately recycled back to the luminal surface of the brain capillary endothelial cell. Because of this, the use of the transferrin molecule as a drug carrier is not likely to enhance BBB permeability. To get around this issue, researchers have identified a murine monoclonal antibody to the transferring receptor, OX26, which appears to be suitable for use as a drug carrier for this transport system.[126,127] There are three important characteristics of OX26 that make it ideal as a drug carrier. First, unlike other antibodies to the transferrin receptor, this antibody binds to the receptor and triggers endocytosis. Second, the cellular processing of the internalized antibody is such that a significant portion of the internalized OX26 actually undergoes exocytosis (release) at the abluminal (brain side) plasma membrane.[126,128] The third important characteristic is that the OX26 antibody binds to an extracellular epitope on the transferrin receptor that is distinct from the transferrin ligand-binding site; thus the OX-26 monoclonal antibody does not interfere with transferrin binding to its receptor on the brain endothelial cells.[129] The OX-26 antibody has proven to be an effective brain delivery vector, as it has been conjugated to a variety of drugs including methotrexate,[128] nerve growth factor,[130] and brain-derived neurotrophic factor.[131,132]

3.4.3.2. Insulin Receptor–Mediated Vesicular Transport Insulin is a pancreatic peptide hormone with important functions in glucose regulation. The presence of insulin receptors in the CNS,[133] coupled with the neurotropic and neuromodulatory actions of insulin in neuronal cells, suggest that insulin may have important functions within the brain as well.[134,135] The finding that few neurons even express insulin mRNA[136] and that brain levels of insulin are directly correlated with the concentration of the peptide in the blood suggests that insulin has a non-CNS origin. Studies demonstrating the presence of high-affinity insulin receptors on the luminal plasma membrane of brain microvessel endothelial cells and their involvement in the vesicular transport of insulin indicate that the peptide penetrates the BBB through a receptor-mediated transport process.[137,138]

Several studies support the potential use of insulin as a transport vector for the delivery of therapeutic agents and macromolecules to the brain. Studies by Kabanov et al.[139] examined the use of polymer micelles for the delivery of the antipsychotic agent, haloperidol, to the brain in mice. Conjugation of insulin to the polymer micelles improved the CNS responses to haloperidol while decreasing the deposition of the micelles in peripheral organs such as the lung and liver. These studies, together with more recent investigations in cultured brain microvessel endothelial cells demonstrating that the transport of the insulin-conjugated micelles is a saturable process inhibited by excess free insulin,[140] suggest that the insulin-conjugated micelles undergo a receptor-mediated vesicular transport process in the BBB. Insulin has also been used as a BBB transport vector for proteins. Studies by Fukuta and coworkers[141] examined horseradish peroxidase (HRP) activity in the brain following IV injections of either HRP or HRP conjugated to insulin. Those mice receiving IV injections of the insulin-conjugated HRP had significantly higher peroxidase activity in the brain compared to either vehicle or HRP-treated mice.[141] Together these studies demonstrate the feasibility of the insulin receptor as a transport system into the brain.

Despite the encouraging results observed in laboratory animals with the use of insulin as a transport vector in the BBB, there are some issues that need to be resolved prior to human use. In the studies of Fukuta and colleagues described above, administration of the insulin-HRP conjugate resulted in hypoglycemia. Efforts to identify biologically inactive peptide fragments of insulin that retain receptor binding and transcellular transport properties have had only limited success.[142] Another issue is the relatively rapid metabolic clearance of insulin from the blood. In this regard, IGF-I and -II have receptors on the brain endothelial cells that appear to serve a transport function in the BBB. An advantage of developing these systems for drug transport into the brain is that it takes much higher concentrations of IGF to cause hypoglycemia.[127,142]

3.4.4. Drug Efflux Transporter Systems in the BBB

While much effort has been spent understanding the various inwardly directed (i.e., from blood to brain) carriers and transporters and their potential utilization for drug delivery, the impact of outwardly directed (i.e., from brain to blood) transport systems on CNS drug delivery is a relatively recent area of interest. Of special interest are the drug efflux transporters such as P-gp, MRP, and breast cancer resistance protein (BCRP). These transporters are part of the larger ATP binding cassette (ABC) family of proteins that remove a wide variety of compounds from the cell through an ATP-dependent active transport process.[143] Originally found to be overexpressed in various drug-resistant cancer cells, these same proteins are present in normal cells such as intestinal and renal epithelial cells, hepatocytes, and brain capillary endothelial cells, where they influence the absorption, distribution, and elimination of a variety of drugs.

3.4.4.1. P-glycoprotein (P-gp) P-gp is the best-characterized drug efflux transporter. The protein was first discovered in 1976 by Juliano and Ling[144] in drug-resistant Chinese hamster ovary (CHO) cells. As this transporter influenced the cellular levels of a variety of anticancer agents, it was termed "permeability-glycoprotein" (P-gp). The transporter was originally identified in brain capillary endothelial cells forming the BBB in the late 1980s.[145,146] However, the actual functional consequences of P-gp expression in the BBB were noted well in advance of its molecular characterization. This is shown in studies by Levin[147] in which some 25 compounds were evaluated for BBB permeability. For the most part, there was a clear correlation between BBB permeability and lipophilicity. However, there were several compounds whose BBB permeability was much lower than would be predicted based on lipophilicity.[147] While a clear explanation for these findings could not be presented at the time, it is interesting to note that all the compounds displaying lower than anticipated BBB permeability in Levin's original drug screen have since been found to be P-gp substrates.

More objective evidence of the role of P-gp in limiting BBB permeability can be found in the studies using *mdr1* knockout mice that lack the gene that encodes P-gp. The studies by Schinkel and colleagues[148] were the first to demonstrate the impact that P-gp has on the brain distribution of selected drugs using *mdr1* knockout mice. In these studies the brain levels of drugs such as vincristine and ivermectin increased by as much as 80–100-fold in the mice lacking P-gp. Since these initial studies, P-gp drug efflux in the BBB has been implicated in the reduced brain penetration of a number of structurally diverse drugs including digoxin,[149] CSA,[150] intraconazole,[151] various antiviral protease inhibitors,[152] and opioid analgesics.[153]

3.4.4.2. Multidrug Resistance–Associated Protein (MRP) There are currently nine different homologs of MRP (MRP1-MRP9) (see Table 3.1). Although there is some overlap in substrates for P-gp and MRP, the MRP proteins preferentially transport organic anions, glutathione-, glucuronide-, and sulfate conjugates, while P-gp favors more lipophilic compounds that are either neutral or positive in charge (for reviews, see Refs. 154). The ability of MRP1 to transport hydrophobic, nonanionic drugs such as vincristine and etoposide requires the cotransport of reduced glutathione.[156] While all the proteins in the MRP family examined to date can transport organic anions, the MRP4 and MRP5 homologs show selectivity for nucleoside and nucleoside-based therapeutics.[156]

TABLE 3.1 General Characteristics of MRP Homologues

MRP Homolo	Gene	Tissue Distribution	Transport Characteristics[a]			BBB Expression[b]
			OA	CNJ	NU	
MRP1	*MRP1* *ABCC1*	Ubiquitous	+	+	−	+
MRP2	*MRP2* *ABCC2*	Liver, kidney and intestine	+	+	−	+/−
MRP3	*MRP3* *ABCC3*	Liver, kidney, intestine, adrenal and pancreas	+	+	−	+/−
MRP4	*MRP4* *ABCC4*	Prostate, lung, testis, ovary pancreas, skeletal muscle, bladder, gallbladder	+	+/−	+	++
MRP5	*MRP5* *ABCC5*	Ubiquitous	+	−	+	++
MRP6	*MRP6* *ABCC6*	Liver and kidney	+	+	?	++
MRP7	*MRP7* *ABCC7*	Skin, testis, colon	?	?	?	?
MRP8	*MRP8* *ABCC8*	Ubiquitous	?	?	?	?
MRP9	*MRP9* *ABCC9*	Testis, ovary and prostate	?	?	?	?

[a]OA-organic anion; CNJ-drug conjugates (includes glucuronide and/or glutathione, sulfate conjugates); NU-nucleosides.

[b]+/− variable expression; + —— +++ low to high expression; ? expression unknown.

In contrast to P-gp, unequivocal demonstration of the importance of MRP transporters in restricting BBB permeability has been difficult. Historically, a high-capacity efflux transporter(s) for organic anions such as taurocholic acid,[157] valproic acid,[158] and the nucleoside-based antiviral drug, AZT,[159] has been widely recognized in the BBB even if the exact molecular nature of the transporter(s) was unclear. Studies using freshly isolated brain capillaries and cultured brain microvessel endothelial cells have reported the expression of several different MRP analogs in brain capillary endothelial cells.[155,160,161] These findings, together with the known substrate characteristics of the MRP transportors, suggest that at least a portion of the carrier-mediated efflux of organic anions in the BBB is attributable to MRP. However, examination of the brain accumulation of MRP-sensitive compounds in *mrp1* knockout mice has been somewhat disappointing. Studies by Wijnholds et al.[162] reported no significant increases in brain accumulation of etoposide in *mrp1* knockout mice compared to wild-type controls. Using fluorescein to access MRP transporter activity, Sun et al.[163] reported significant increases in the brain accumulation of fluorescein following treatment with probenecid. However, when brain levels of fluorescein were compared following IV administration in *mrp1* knockout mice, no significant differences were observed compared to wild-type mice. In contrast, recent studies by Sugiyama et al.[164] reported significant increases in the brain accumulation of the glucuronide conjugate of 17 beta estradiol in *mrp1* knockout mice compared to controls. Given the data supporting the expression of multiple homologs of MRP in the BBB, the apparent discrepancies observed in the *mrp1* knockout mice studies may be due to interaction of the various compounds with other MRP-related drug efflux transport systems in the BBB.

3.4.4.3. Breast Cancer Resistance Protein (BCRP) BCRP is one of the newer ABC transport proteins to be examined. The transporter was first identified in drug-resistant breast cancer cells.[165] Structurally, BCRP differs from both P-gp and MRP in that it is a "half-transporter" and requires the formation of a protein dimer for functional transporter activity.[166] There is a great deal of substrate overlap between P-gp and BCRP.[167] In addition to being overexpressed in various tumor cells, BCRP has been detected in normal tissues such as the placenta, small intestine, liver, and capillary and venous endothelial cells.[156]

The examination of BCRP in the BBB and its influence on drug permeability is in its early stages. However, preliminary studies suggest that BCRP, along with P-gp and MRP, may be an important contributor to the limited brain penetration of drugs. Recent studies have shown the expression of BCRP, or a closely related protein, in cultured porcine brain capillary endothelial cells.[168,169] Interestingly, the expression of this particular drug efflux transporter in cultured brain endothelial cells was increased in the presence of hydrocortisone,[168] suggesting that BCRP may be induced by exposure to selected agents. While the potential role of BCRP in BBB permeability remains to be determined, studies using cultured porcine brain capillary endothelial cells expressing BCRP showed bidirectional differences in daunorubicin permeability, consistent with a BCRP-mediated efflux transport system.[169]

3.4.4.4. Modulation of Drug Efflux Activity to Increase CNS Drug Delivery The most promising strategy to increase the CNS delivery of drugs that are subject to active efflux has been coadministration with another substrate, or inhibitor of the transporter, to competitively saturate the capacity of the efflux transporters.[170–172] To date, most attention has focused on identification of P-gp-modulating agents. While the first generation of P-gp-modulating agents was very effective in inhibiting P-gp function *in vitro*, the low potency of the modulators resulted in significant toxicity when the agents were introduced into animal models.[173,174] Transporter selectivity of the modulating agents was also a limitation.[175] The newer P-gp modulators have much greater potency and selectivity for the P-gp efflux transporter.[175] These newer P-gp modulators have been used to increase the brain penetration of a number of compounds with characteristically low BBB permeability such as antiviral protease inhibitors,[176] the anticancer agent, paclitaxel,[177] and the antifungal agent, itraconazole.[178] In each case, increases in the brain accumulation of drug were observed despite similar drug levels in the plasma, indicating that the increased brain penetration was due to inhibition of P-gp at the BBB.

An alternative to using pharmacological agents as selective modulators of drug efflux activity, is the use of various polymer formulation components to inhibit drug efflux transporters.[179–181] An advantage of using formulational components to inhibit drug efflux transporters over pharmacological agents is the general safety record of the various polymer formulations in humans. The mechanism by which polymers modulate drug efflux transporter activity involves alterations in membrane fluidity and/or alterations in ATP availability.[180,181] One particular polymer formulation, Pluronic block copolymer (P85), has been shown to enhance the permeability of a wide variety of drugs in an *in vitro* model of the BBB.[181] These studies suggest that P85 can be used to selectively improve drug permeability in the BBB through inhibition of drug efflux transporter activity.

3.5. SUMMARY

The delivery of both small- and large-molecule therapeutics to the brain at sufficient levels to treat CNS pathologies is a challenge for many current and emerging drugs. The challenge should not be viewed as insurmountable, as there are a number of approaches that can be used to improve brain delivery of therapeutic agents. In deciding on the delivery approach, there are several considerations. One is the nature of the disease. An approach such as BBB disruption that might be acceptable for acute or intermittent delivery of therapeutic agents may be unpractical and/or toxic for chronic brain pathologies. Furthermore, if the disease involves a small focal lesion, one might consider direct administration of drug or drug-polymer matrix at or near the site of the lesion. A second consideration is the therapeutic molecule itself. Drug delivery approaches considered for large molecules such as peptides and proteins may not be feasible for small-molecule drugs. While the intrathecal route may prove adequate for delivering highly potent peptide or protein therapeutics, such a route would be ineffective for a small molecule. Likewise, targeting of the therapeutic agent to one of the many nutrient/metabolite carriers in the BBB may be a reasonable approach for a low molecular weight drug but not for delivering macromolecules to the brain. By understanding both the disease process and the obstacles limiting drug penetration in the brain, a rational strategy for improving delivery can be selected and developed.

ACKNOWLEDGMENTS

The authors acknowledge research support grants from the National Institutes of Health (NS36831, CA93558 and AG17294), the Nebraska Research Initiative, and the Nebraska Department of Health, Cancer and Smoking Related Diseases Grant Program.

REFERENCES

1. Pardridge, W. *Mol. Intervent.* **2003**, *3*, 90–105.
2. Reese, T. S.; Karnovsky, M. J. *J. Cell Biol.* **1967**, *34*, 207–217.
3. Oldendorf, W. H. *Proc. Soc. Exp. Biol. Med.* **1974**, *147*, 813–816.
4. Rapoport, S. I.; Ohno, K.; Pettigrew, K. D. *Brain Res.* **1979**, *172*, 354.
5. Zlokovic, B. V. *Pharm. Res.* **1995**, *12*, 1395–1406.
6. Abraham, M.; Chadha, H. S.; Mitchell, R. C. *J. Pharm. Sci.* **1994**, *83*, 1257–1268.
7. Seelig, A.; Gottschlich, R.; Devant, R. M. *Proc. Natl. Acad. Sci. USA* **1994**, *91*, 68–72.
8. Davson, H.; Segal, M. B. *Physiology of the CSF and Blood-Brain Barrier.* CRC Press: Boca Raton, FL, 1996.
9. Segal, M. In *Introduction to the Blood-Brain Barrier: Methodology, Bilogy and Pathology*; W. M. Pardridge, Ed. Cambridge University Press: Cambridge, **1998**, pp. 251–258.
10. Meller, K. *Cell Tissue Res.* **1985**, *239*, 189–201.
11. Johanson, C. E. In *Implications of the Blood-Brain Barrier and Its Manipulation: Basic Science Aspects*; E. A. Neuwelt, Ed. Plenum Publishing: New York, 1989, vol. 1, pp. 223–260.
12. Krewson, C.; Klarman, M.; Saltzman, W. *Brain Res.* **1995**, *680*, 196–206.
13. Yan, Q, M. C.; Sun, J.; Radeke, M. J.; Feinstein, S. C.; Miller, J. A. *Exp. Neurol.* **1994**, *127*, 23–26.
14. Hoistad, M. K. J.; Andbjer, B.; Jansson, A.; Fuxe, K. *Eur J. Neurosci.* **2000**, *12*, 2505–2514.
15. Mairs, RJ, W. C.; Angerson, W. J.; Whateley, T. L.; Reza, M. S.; Reeves, J. R.; Robertson, L. M.; Neshasteh-Riz, A.; Rampling, R.; Owens, J.; Allan, D.; Graham, D. I. *Br. J. Cancer* **2000**, *82*, 74–80.
16. Tamargo, RJ, R. L.; Kossoff, E. H.; Tyler, B. M.; Ewend, M. G.; Aryanpur, J. J. *Epilepsy Res.* **2002**, *48*, 145–155.

17. Bernards, C. In *Spinal Drug Delivery*; TL Yaksh, Ed. Elsevier: Amsterdam, **1999**, pp. 239–252.
18. LeBel, C. In *Spinal Drug Delivery*; TL Yaksh, Ed. Elsevier: Amsterdam, **1999**, pp. 543–554.
19. McCarthy TJ, B. W.; Farrell, C. L.; Adamu, S.; Derdeyn, C. P.; Synder, A. Z.; LaForest, R.; Litzinger, D. C.; Martin, D.; LeBel, C. P.; Welch, M. J.; *J. Pharmacol. Exp. Ther.* **2002**, *301*, 878–883.
20. Yaksh, TL, S. B.; LeBel, C. L.; *Neuroscience* **2002**, *110*, 703–710.
21. Frey, W. H. *Drug Deliv. Technol.* **2002**, *2*, 46–49.
22. Illum, L. *Eur. J. Pharm. Sci.* **2000**, *11*, 1–18.
23. Mathison, S.; Nagilla, R.; Kompella, U. B. *J. Drug Target* **1998**, *5*, 415–441.
24. Chow, H. N. S.; Chen, Z.; Natsuura, G. T. *J. Pharm. Sci.* **1999**, *88*, 754–758.
25. Sakane, T.; Akizuki, M.; Yoshida, M.; Yamashita, S.; Nadai, T.; Hashida, M.; Sezaki, H. *J. Pharm. Pharmacol.* **1991**, *43*, 449–451.
26. Wang, F.; Jiang, X.; Lu, W. *Int. J. Pharm.* **2003**, *263*, 1–7.
27. Frey, W. H.; Liu, J.; Chen, X.; Thorne, R. G. *Drug Deliv.* **1997**, *4*, 87–92.
28. Kern, W.; Born, J.; Schreiber, H.; Fehm, H. L. *Diabetes* **1999**, *48*, 557–563.
29. Liu, X.-F.; Fawcett, J. R.; Thorne, R. G.; DeFor, T. A.; Frey, W. H. *J. Neurosci.* **2001**, *187*, 91–97.
30. Li, L.; Gorukanti, S.; Choi, Y. M.; Kim, K. H. *Int. J. Pharm.* **2000**, *199*, 65–76.
31. Gizurarson, S. *Adv. Drug Deliv. Rev.* **1993**, *11*, 329–347.
32. Brightman, M. W.; Hori, M.; Rapoport, S. I.; Reese, T. S.; Westergaard, E. *J. Comp. Neurol.* **1973**, *152*, 317–326.
33. Rapoport, S. I.; Robinson, P. J. *Ann. N.Y. Acad. Sci.* **1986**, *481*, 250–266.
34. Raymond, J. J.; Robertson, D. M.; Dinsdale, H. B. *Can. J. Neurol. Sci.* **1986**, *13*, 214–220.
35. Erdlenbruch, B, A. M.; Fricker, G.; Miller, D. S.; Kugler, W.; Eibl, H.; Lakomek, M. *Br. J. Pharm.* **2003**, *140*, 1201–1210.
36. Erdlenbruch, B, J. V.; Eibl, H.; Lakomek, M.; *Exp. Brain Res.* **2000**, *135*, 417–422.
37. Gumerlock, M. K.; Neuwelt, E. A. In *Physiology and Pharmacology of the Blood-Brain Barrier*, M. W. B. Bradbury, Ed. Springer-Verlag: Berlin, **1992**, vol. 103, pp. 525–542.
38. Rapoport, S. I. *Cell. Mol. Neurobiol.* **2000**, *20*, 217–230.
39. Greenwood, J. In *Physiology and Pharmacology of the Blood-Brain Barrier*, M. W. B. Bradbury, Ed. Springer-Verlag: Berlin, **1992**, vol. 103, pp. 459–486.
40. Greenwood, J.; Luthert, P. J.; Pratt, O. E.; Lantos, P. L. *J. Cereb. Blood Flow Metab.* **1988**, *8*, 9–15.
41. Cosolo, W. C.; Martinello, P.; Louis, W. S.; Christophidis, N. *Am. J. Physiol.* **1989**, *256*, R443–R447.
42. Siegal, T. R. R.; Bokstein, F.; Schwartz, A.; Lossos, A.; Shalom, E.; Chisin, R.; Gomori, J. M. *J. Neurosurg.* **2000**, *92*, 599–605.
43. Kroll, RA, P. M.; Muldoon, L. L.; Roman-Goldstein, S.; Fiamengo, S. A.; Neuwelt, E. A. *Neurosurgery* **1998**, *43*, 879–886.
44. Remsen, LG, T. P.; Hellstrom, I.; Hellstrom, K. E.; Neuwelt, E. A. *Neurosurgery* **2000**, *46*, 704–709.

45. Neuwelt EA, B. P.; McCormick, C. I.; Remsen, L. G.; Kroll, R. A.; Sexton, G. *Clin Cancer Res.* **1998**, *4*, 1549–1555.

46. Neuwelt, E. A.; Goldman, D. L. *J. Clin. Oncol.* **1991**, *9*, 1580–1590.

47. Williams, P. C.; Henner, W. D. *Neurosurgery* **1995**, *37*, 17–28.

48. Neuwelt, E. A.; Howieson, J.; Frenkel, E. P.; et al. *Neurosurgery* **1986**, *19*, 573–582.

49. Neuwelt, EA, M. J.; Frenkel, E.; Barnett, P. A.; McCormick, C. I. *Am. J. Physiol.* **1986**, *250*, R875–R883.

50. Nilaver, G, M. L.; Kroll, R. A.; Pagel, M. A.; Breakefield, X. O.; Davidson, B. L.; Neuwelt, E. A. *Proc. Natl. Acad. Sci. USA* **1995**, *92*, 9829–9833.

51. Doran, SE, R. X.; Betz, A. L.; Pagel, M. A.; Neuwelt, E. A.; Roessler, B. J.; Davidson, B. L. *Neurosurgery* **1995**, *36*, 965–970.

52. Abe, T, W. H.; Bookstein, R.; Maneval, D. C.; Chiocca, E. A.; Basilion, J. P. *Cancer Gene Ther.* **2002**, *9*, 228–235.

53. Sanovich, E.; Bartus, R. T.; Friden, P. M.; Dean, R. L.; Le, H. Q.; Brightman, M. W. *Brain Res.* **1995**, *705*, 125–135.

54. Bartus, R. T.; Elliot, P. J.; Hayward, N. J.; Dean, R. L.; McEwen, E.; Fisher, S. K. *Immunopharmacology* **1996**, *33*, 270–278.

55. McCarthy, DA, P. D.; Nicolaides, E. D. *J. Pharmacol. Exp. Ther.* **1965**, *148*, 117–122.

56. Marceau, F, K. M.; Regoli, D. *Can. J. Physiol. Pharmacol.* **1983**, *59*, 921–926.

57. Borlongan, CV, a. E. D. *Brain Res. Bull.* **2003**, *60*, 297–306.

58. Fike, JR, G. G.; Mesiwala, A. H.; Shin, H. J.; Nakagawa, M.; Lamborn, K. R.; Seilhan, T. M.; Elliott, P. J. *J. Neurooncol.* **1998**, *37*, 199–215.

59. Elliott, PJ, H. N.; Dean, R. L.; Blunt, D. G. *Cancer Res.* **1996**, *56*, 3998–4005.

60. Elliott, PJ, H. N.; Huff, M. R.; Nagle, T. L.; Black, K. L.; Bartus, R. T. *Exp. Neurol.* **1996**, *141*, 214–224.

61. Emerich D, S. P.; Dean, R.; Lafreniere, D.; Agostino, M.; Wiens, T.; Xiong, H.; Hasler, B.; March, J.; Pink, M.; Kim, B. S.; Bartus, R. *J. Pharm. Exp. Ther.* **2000**, *296*, 632–641.

62. Inamura T, a. B. K. *J. Cereb. Blood Flow Metab.* **1994**, *14*, 862–870.

63. Inamura T, N. T.; Bartus, R.; Black, K. *J. Neurosurg.* **1994**, *81*, 752–758.

64. Emerich, D, S. P.; Dean, R.; Agostino, M.; Hasler, B.; Pink, M.; Xiong, H.; Kim, B. S.; Bartus, R. *Br. J. Cancer* **1999**, *80*, 964–970.

65. Bartus, R, S. P.; Marsh, J.; Agostino, M.; Perkins, A.; Emerich, D. *J. Pharm. Exp. Ther.* **2000**, *293*, 903–911.

66. Gregor, A.; Lind, M.; Newman, H.; Grant, R.; Hadley, D.; Barton, T.; Osborn, C. *J. Neurooncol.* **1999**, *44*, 137–145.

67. Emerich D, S. P.; Pink, M.; Bloom, F.; Bartus, R. *Brain Res.* **1998**, *801*, 259–266.

68. Bidanset, DJ, P. L.; Rybak, R.; Palmer, J.; Sommadossi J.-P.; Kern, E. R. *Antimicrob. Agents Chemother.* **2001**, *45*, 2316–2323.

69. Erdlenbruch, B, J. V.; Kugler, W.; Eibl, H.; Lakomek, M. *Cancer Chemother. Pharmacol.* **2002**, *50*, 299–304.

70. Erdlenbruch, B, S. C.; Kugler, W.; Heinemann, D. E. H.; Herms, J.; Eibl, H.; Lakomek, M. *Br. J. Pharm.* **2003**, *139*, 685–694.

71. Greig, N. H. In *Physiology and Pharmacology of the Blood-Brain Barrier*, M. W. B. Bradbury, Ed. Springer-Verlag: Berlin, **1992**, vol. 103, pp. 487–523.

72. Oldendorf, W. H. In *The Ocular and Cerebrospinal Fluids* L. Z. Bito, H. Davson, J. D. Fenstermacher, Eds. Academic Press: 1977, pp. 177–190.
73. Greig, N. H.; Daly, E.; Sweeney, D. J.; Rapoport, S. I. *Cancer Chemother. Pharmacol.* **1990**, *25*, 320–325.
74. Bodor, N.; Simpkins, J. W. *Science* **1983**, *221*, 65–67.
75. Jacob, J. N.; Shashoua, V. E.; Campbell, A.; Baldessarini, R. J. *J. Med. Chem.* **1985**, *28*, 106–110.
76. Yang, J. Z.; Chen, W.; Borchardt, R. T. *J. Pharmacol. Exp. Ther.* **2002**, *303*, 840–848.
77. Weber, S. J.; Greene, D. L.; Sharma, S. D.; Yamamura, H. I.; Kramer, T. H.; Burks, T. F.; Hruby, V. J.; Hersh, L. B.; Davis, T. P. *J. Pharmacol. Exp. Ther.* **1991**, *259*, 1109–1117.
78. Weber, S. J.; Greene, D. L.; Hruby, V. J.; Yamamura, H. I.; Porreca, F.; Davis, T. P. *J. Pharmacol. Exp. Ther.* **1992**, *263*, 1308–1316.
79. Hruby, V. J. *Life Sci.* **1982**, *31*, 189–199.
80. Cardona, V. M. F.; Hartley, O.; Botti, P. *J. Peptide Res.* **2003**, *61*, 152–157.
81. Wenger, R. In *Cyclosporin A*; D. J. G. White, Ed. Elsevier Biomedical Press: New York, **1982**, pp. 19–34.
82. Pardridge, W. M. In *Peptide Drug Delivery to the Brain*; W. M. Pardridge, Ed. Raven Press: New York, **1991**, pp. 123–148.
83. Cefalu, W. T.; Pardridge, W. M. *J. Neurochem.* **1985**, *45*, 1954–1956.
84. Schinkel, AH, W. E.; van Deemter, L.; Mol, C. A.; Borst, P. *J. Clin. Invest.* **1995**, *96*, 1698–1705.
85. Cox, DS, S. K.; Gao, H.; Raje, S.; Eddington, N. D. *J. Pharm. Sci.* **2001**, *90*, 1540–1552.
86. Greig, N. H. In *Implications of the Blood-Brain Barrier and Its Manipulation*; E. A. Neuwelt, Ed. Plenum: New York, **1989**, vol. 1, pp. 311–367.
87. Smith, Q. R.; Takasato, Y. *Ann. NY Acad. Sci.* **1986**, *481*, 186–201.
88. Audus, K. L.; Chikhale, P. J.; Miller, D. W.; Thompson, S. E.; Borchardt, R. T. *Adv. Drug Res.* **1992**, *23*, 3-53.
89. Miller, D. W.; Kato, A.; Ng, K.-Y.; Chikhale, E. G.; Borchardt, R. T. In *Peptide Based Drug Design: Controlling Transport and Metabolism*; M. D. Taylor, G. L. Amidon, Eds. American Chemical Society: Washington, DC, **1995**, pp. 475–500.
90. Smith, Q. R.; Stoll, J. In *Introduction to the Blood-Brain Barrier*; M. Pardridge, Ed. Cambridge University Press: Cambridge, **1988**.
91. Tamai, I.; Tsuji, A. *J. Pharm. Sci.* **2000**, *89*, 1371–1388.
92. Wade, L. A.; Katzman, R. *J. Neurochem.* **1975**, *25*, 837.
93. Smith, Q. R. In *Frontiers in Cerebral Vascular Biology: Transport and Its Regulation*, L. R. Drewes, A. L. Betz, Eds. Plenum Press: New York, **1993**, pp. 83–93.
94. Greig, N. H.; Momma, S.; Sweeney, D. J.; Smith, Q. R.; Rapoport, S. I. *Cancer Res.* **1987**, *47*, 1571–1576.
95. van Bree, J. B. M. M.; Audus, K. L.; Borchardt, R. T. *Pharm. Res.* **1988**, *5*, 369–371.
96. van Bree, J. B. M. M.; Heijigers-Feijen, C. D.; de Boer, A. G.; Danhof, M.; Breimer, D. D. *Pharm. Res.* **1991**, *8*, 259–262.
97. Welty, D. F.; et al. *Epilepsy Res.* **1993**, *16*, 175–181.
98. Pardridge, W. *Physiol. Rev.* **1983**, *63*, 1481–1535.

99. Halmos, T, S. M.; Antonakis, K.; Scherman, D. *Eur. J. Pharmacol.* **1996**, *318*, 477–484.
100. Brunet-Desruet, MD, G. C.; Morin, C.; Comet, M.; Fagret, D. *Nucl. Med. Biol.* **1998**, *25*, 473–480.
101. Polt, R, P. F.; Szabo, L. Z.; Bilsky, E. J.; Davis, P.; Abbruscato, T. J.; Davis, T. P.; Horvath, R.; Yamamura, H. I.; Hruby, V. J. *Proc. Natl. Acad. Sci. USA* **1994**, 7114–7118.
102. Negri, L.; Lattanzi, R.; Tabacco, F.; Scolaro, B.; Rocchi, R. *Br. J. Pharmacol.* **1998**, *124*, 1516–1522.
103. Gerhart, DZ, E. B.; Zhdankina, O. Y.; Leino, R. L.; Drews, L. R. *Am. J. Physiol.* **1997**, *273*, E207–E213.
104. Terasaki, T, K. Y.; Ohnishi, T.; Tsuji, A. *J. Pharm. Pharmacol.* **1991**, *43*, 172–176.
105. Saheki, A, T. T.; Tamai, I.; Tsuji A. *Pharm. Res.* **1994**, *11*, 305–311.
106. Anderson, C.; Xiong, W.; Geiger, J.; Young, J.; Cass, C.; Baldwin, S.; Parkinson, F. *J. Neurochem.* **1999**, *73*, 867–873.
107. Anderson, C.; Xiong, W.; Young, J.; Cass, C.; Parkinson, F. *Mol. Brain Res.* **1996**, *42*, 358–361.
108. Cornford, E.; Oldendorf, W. *Biochim. Biophys. Acta* **1975**, *394*, 211–219.
109. Mackey, J.; Mani, R.; Selner, M.; Mowles, D.; Young, J.; Belt, J.; Crawford, C.; Cass, C. *Cancer Res.* **1998**, *58*, 4349–4357.
110. Yao, S.; Cass, C.; Young, J. *Mol. Pharm.* **1996**, *50*, 388–393.
111. Wu, D.; Clement, J.; Pardridge, W. *Brain Res.* **1998**, *791*, 313–316.
112. Takasawa, K.; Terasaki, T.; Suzuki, H.; Sugiyama, Y. *J. Pharmacol. Exp. Ther.* **1997**, *281*, 369–375.
113. Novotny, A.; Xiang, J.; Stummer, W.; Teuscher, N.; Smith, D.; Keep, R. *J. Neurochem.* **2000**, *75*, 321–328.
114. Kannan, R.; Kuhlenkamp, J.; Ookhtens, M.; Kaplowitz, N. *J. Pharmacol. Exp. Ther.* **1992**, *263*, 964–970.
115. Zlokovic, B.; Mackic, J.; Djunricic, B.; Davson, H. *J. Neurochem.* **1989**, *53*, 1333–1341.
116. Zlokovic, B.; Hyman, S.; McComb, J.; Lipovac, M.; Tang, G.; Davson, H. *Biochim. Biophys. Acta.* **1990**, *1025*, 191–198.
117. Wilson, J. *Psychopharmacology* **1988**, *96*, 262–266.
118. Barrera, C.; Kastin, A.; Fasold, M.; Banks, W. *Am. J. Physiol.* **1991**, *261*, E312–E318.
119. Pan, W.; Kastin, A. *Peptides* **1999**, *20*, 1091–1098.
120. Banks, W.; Kastin, A.; Durham, D. *Brain Res. Bull.* **1989**, *23*, 433–437.
121. Gutierrez, E.; Banks, W.; Kastin, A. *J. Neuroimmunol.* **1993**, *47*, 169–176.
122. Nutt, J. G.; Woodward, W. R.; Hammerstad, J. P.; Carter, J. H.; Anderson, J. L. *N. Engl. J. Med.* **1984**, *310*, 483–488.
123. Aisen, P.; Listowsky, I. *Annu. Rev. Biochem.* **1980**, *49*, 357–393.
124. McClelland, A.; Kuhn, L. C.; Ruddle, F. H. *Cell* **1984**, *39*, 267–274.
125. Jefferies, W. A.; Brandon, M. R.; Hunt, S. V.; Williams, A. F.; Gatter, K. C.; Mason, D. Y. *Nature* **1984**, *312*, 162–163.
126. Pardridge, W. M.; Buciak, J. L.; Friden, P. M. *J. Pharmacol. Exp. Ther.* **1991**, *259*, 66–70.
127. Pardridge, W. M.; *Peptide Drug Delivery to the Brain.* Raven Press: New York, **1991**.

128. Friden, P. M.; Walus, L. R.; Musso, G. F.; Taylor, M. A.; Malfroy, B.; Starzyk, R. M. *Proc. Natl. Acad. Sci. USA* **1991**, *88*, 4771–4775.

129. Tsuji, A.; Tamai, I. In *Introduction to the Blood-Brain Barrier: Methodology, Biology and Pathology*; W. M. Pardridge, Ed. Cambridge University Press: Cambridge, **1998**, pp. 238–247.

130. Friden, P. M.; Walus, L. R. *Adv. Exp. Med. Biol.* **1993**, *331*, 129–136.

131. Wu, D.; Pardridge, W. M. *Proc. Natl. Acad. Sci. USA* **1999**, *96*, 254–259.

132. Zhang, Y.; Pardridge, W. M. *Brain Res.* **2001**, *889*, 49–56.

133. Baskin, D.; Wilcox, B.; Figlewicz, D.; Dorsa, D. *Trends Neurosci.* **1988**, *11*, 107–111.

134. Knusel, B.; Michel, P.; Schaber, J.; Hefti, F. *J. Neurosci.* **1990**, *10*, 558–570.

135. Palovik, R.; Phillips, M.; Kappy, M.; Raizada, M. *Brain Res.* **1984**, *309*, 187–191.

136. Young, W. *Neuropeptides* **1986**, *8*, 93–97.

137. Pardridge, W. M.; Eisenberg, J.; Yang, J. *J. Neurochem.* **1985**, *44*, 1771–1778.

138. Miller, D. W.; Keller, B. T.; Borchardt, R. T. *J. Cell. Physiol.* **1994**, *161*, 333–341.

139. Kabanov, A.; Chekhonin, V.; Alakhov, V.; Batrakova, E.; Lebedev, A.; Melik-Nubarov, N.; Arzhakov, S.; Levashov, A.; Morozov, G.; Severin, E.; et al. *FEBS Lett.* **1989**, *258*, 343–345.

140. Batrakova, E.; Han, H.; Miller, D.; Kabanov, A. *Pharm. Res.* **1998**, *15*, 1525–1532.

141. Fukuta, M.; Okada, H.; Iinuma, S.; Yanai, S.; Toguchi, H. *Pharm. Res.* **1994**, *11*, 1681–1688.

142. Witt, K. W.; Gillespie, T. J.; Huber, J. D.; Egleton, R. D.; Davis, T. P. *Peptides* **2001**, *22*, 2329–2343.

143. Borges-Walmsley, M.; McKeegan, K.; Walmsley, A. *Biochem. J.* **2003**, *376*, 313–338.

144. Juliano, R. L.; Ling, V. *Biochim. Biophys. Acta* **1976**, *455*, 152–162.

145. Thiebaut, F.; Tsuruo, T.; Hamada, H.; Gottesman, M. M.; Pastan, I.; Willingham, M. C. *J. Histochem. Cytochem.* **1989**, *37*, 159–164.

146. Cordon-Cardo, C.; O'Brien, J. P.; Casals, D.; Rittman-Grauer, L.; Biedler, J. L.; Melamed, M. R.; Bertino, J. R. *Proc. Natl. Acad. Sci. USA* **1989**, *86*, 695–698.

147. Levin, V. *J. Med. Chem.* **1980**, *23*, 682–684.

148. Schinkel, A. H.; Smith, J. J. M.; van Tellingen, O.; Beijnen, J. H.; Wagenaar, E.; van Deemter, L.; Mol, C. A. A. M.; van der Valk, M. A.; Robanus-Mandaag, E. C.; te Riele, H. P. J.; Berns, A. J. M.; Borst, P. *Cell* **1994**, *77*, 491–502.

149. Batrakova, E.; Miller, D.; Li, S.; Alakhov, V.; Kabanov, A.; Elmquist, W. *J. Pharmacol. Exp. Ther.* **2001**, *296*, 556–562.

150. Kwei, G.; Alvaro, R.; Chen, Q.; Jenkins, H.; Hop, C.; Keohane, C.; Ly, V.; Strauss, J.; Wang, R.; Wang, Z.; Pippert, T. *Drug Metab. Dispos.* **1999**, *27*, 581–587.

151. Miyama, T.; Takanaga, H.; Matsuo, H.; Yamano, K.; Yamamoto, K.; Iga, T.; Naito, M.; Tsuruo, T.; Ishizuka, H.; Kawahara, Y.; Sawada, Y. *Antimicrob. Agents Chemother.* **1998**, *42*, 1738–1744.

152. Kim, R.; Fromm, M.; Wandel, C.; Leake, B.; Wood, A.; Roden, D.; Wilkinson, G. *J. Clin. Invest.* **1998**, *101*, 289–294.

153. Chen, C.; Pollack, G. *J. Pharmacol. Exp. Ther.* **1998**, *287*, 545–552.

154. Borst, P.; Evers, R.; Kool, M.; Wijnholds, J. *J. Natl. Cancer Inst.* **2000**, *92*, 1295–1302.

155. Zhang, Y.; Han, H.; Elmquist, W. F.; Miller, D. W. *Brain Res.* **2000**, *876*, 148–153.

156. Schinkel, A.; Jonker, J. *Adv. Drug Deliv.* **2003**, *55*, 3–29.
157. Kitazawa, T.; Terasaki, T.; Suzuki, H.; Kakee, A.; Sugiyama, Y. *J. Pharmacol. Exp. Ther.* **1998**, *286*, 890–895.
158. Adkinson, K. D. K.; Artu, A. A.; Powers, K. M.; Shen, D. D. *J. Pharmacol. Exp. Ther.* **1994**, *268*, 797–805.
159. Wang, Y.; Sawchuk, R. J. *J. Pharm. Sci.* **1995**, *84*, 871–876.
160. Huai-Yun, H.; Secrest, D. T.; Mark, K. S.; Carney, D.; Brandquist, C.; Elmquist, W. F.; Miller, D. W. *Biochem. Biophys. Res. Commun.* **1998**, *243*, 816–820.
161. Seetharaman, S.; Barrand, M. A.; Maskell, L.; Scheper, R. J. *J. Neurochem.* **1998**, *70*, 1151–1159.
162. Wijnholds, J.; Scheffer, G. L.; van der Valk, M.; Beijnen, J. H.; Scheper, R. J.; Borst, P. *J. Exp. Med.* **1998**, *188*, 797–808.
163. Sun, H.; Johnson, D. R.; Finch, R. A.; Sartorelli, A. C.; Miller, D. W.; Elmquist, W. F. *Biochem. Biophys. Res. Commun.* **2001**, *284*, 863–869.
164. Sugiyama, D.; Kusuhara, H.; Lee, Y.; Sugiyama, Y. *Pharm. Res.* **2003**, *20*, 1394–1400.
165. Doyle, L.; Yang, W.; Abruzzo, L.; Krogmann, T.; Gao, Y.; Rishi, A.; Ross, D. *Proc. Natl. Acad. Sci. USA* **1998**, *95*, 15665–15670.
166. Kage, K.; Tsukahara, S.; Sugiyama, T.; Asada, S.; Ishikawa, E.; Tsuruo, T.; Sugimoto, Y. *Int. J. Cancer* **2002**, *97*, 626–630.
167. Litman, T.; Druley, T. E.; Stein, W. D.; Bates, S. E. *Cell Mol. Life Sci.* **2001**, *58*, 931–959.
168. Eisenblatter, T.; Galla, H. *Biochem. Biophys. Res. Commun.* **2002**, *293*, 1273–1278.
169. Eisenblatter, T.; Huwel, S.; Galla, H. *Brain Res.* **2003**, *971*, 221–231.
170. Samuel, B. L.; et al.*J. Clin. Pharmacol. Ther.* **1993**, *54*, 421–429.
171. Drion, N.; Lemaire, M.; Lefauconnier, J. M.; Scherrmann, J. M. *J. Neurochem.* **1996**, *67*, 1688–1693.
172. Hughes, C. S.; Vaden, S. L.; Manaugh, C. A.; Price, G. S.; Hudson, L. C. *J. Neurooncol.* **1998**, *37*, 45–54.
173. Pennock, G. D.; Dalton, W. S.; Roeske, W. R.; Appleton, C. P.; Mosley, K.; Plezia, P.; Miller, T. P.; Salmon, S. E. *J. Natl. Cancer Inst.* **1991**, *83*, 105–110.
174. Habgood, M. D.; Begley, D. J.; Abbott, N. J. *Cell. Mol. Neurobiol.* **2000**, *20*, 231–253.
175. Dantzig, A.; Alwis, D. D.; Burgess, M. *Adv. Drug Deliv. Rev.* **2003**, *55*, 133–150.
176. Choo, E.; Leake, B.; Wandel, C.; Imamura, H.; Wood, A.; Wilkinson, G.; Kim, R. *Drug Metab. Dispos.* **2000**, *28*, 655–660.
177. Kemper, E.; Zandbergen, A. V.; Cleypool, C.; Mos, H.; Boogerd, W.; Beijnen, J.; Tellingen, O. V. *Clin. Cancer Res.* **2003**, *9*, 2849–2855.
178. Imbert, F.; Jardin, M.; Fernandez, C.; Gantier, J.; Dromer, F.; Baron, G.; Mentre, F.; Beijsterveldt, L. V.; Singlas, E.; Gimenez, F. *Drug Metab. Dispos.* **2003**, *31*, 319–325.
179. Zastre, J.; Jackson, J.; Bajwa, M.; Liggins, R.; Iqbal, F.; Burt, H. *Eur. J. Pharm. Biopharm.* **2002**, *54*, 299–309.
180. Rege, B.; Kao, J.; Polli, J. *Eur. J. Pharm. Sci.* **2002**, *16*, 237–246.
181. Kabanov, A.; Batrakova, E.; Miller, D. *Adv. Drug Deliv. Rev.* **2003**, *55*, 151–164.

4

PHYSICOCHEMICAL PROPERTIES, FORMULATION, AND DRUG DELIVERY

DEWEY H. BARICH AND ERIC J. MUNSON

Department of Pharmaceutical Chemistry, University of Kansas, 2095 Constant Avenue, Lawrence, KS 66047

MARK T. ZELL

Pfizer Global Research and Development, Ann Arbor Laboratories, 2800 Plymouth Road, Ann Arbor, MI 48105

Drug Delivery: Principles and Applications Edited by Binghe Wang, Teruna Siahaan, and Richard Soltero
ISBN 0-471-47489-4

4.1. INTRODUCTION

The goal of drug formulation and delivery is to administer a drug at a therapeutic concentration to a particular site of action for a specified period of time. The design of the final formulated product for drug delivery depends upon several factors. First, the drug must be administered using a narrow set of parameters that are defined by the therapeutic action of the drug. These parameters include the site of action (either targeted to a specific region of the body or systemic), the concentration of the drug at the time of administration, the amount of time the drug must remain at a therapeutic concentration, and the initial release rate of the drug for oral/controlled release systems. Second, the drug must remain physically and chemically stable in the formulation for at least 2 years. Third, the choice of delivery method must reflect the preferred administration route for the drug, such as oral, parenteral, or transdermal.

A complete knowledge of the relevant therapeutic and physicochemical properties of the drug is required to determine the proper formulation and delivery method of a drug. For example, the physicochemical properties of the drug strongly influence the choice of delivery methods. This creates a problem in dividing this chapter into specific sections, (See Figure 4.1 below for a pictorial representation of the interrelationships of the three main topics of this chapter) as a discussion

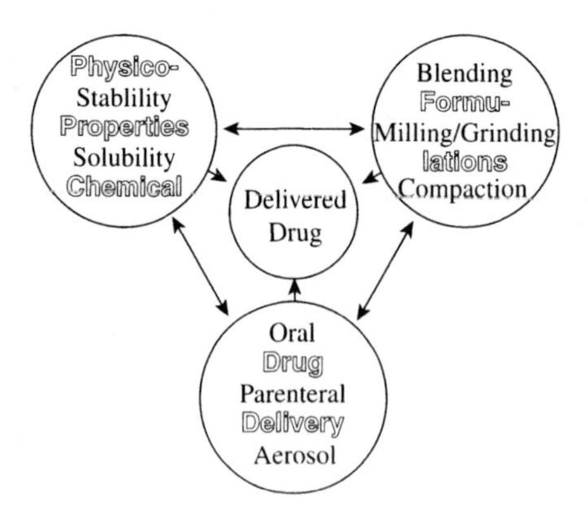

Figure 4.1. Schematic diagram showing the interdependence of physicochemical properties, formulation, and drug delivery.

of the important physicochemical properties of a drug will be different for oral administration of a solid tablet compared to parenteral administration of a drug in solution. For this reason, we have chosen to take a broad approach in the physicochemical properties section in discussing the basic physicochemical properties that are determined for almost all drugs. A similar approach has been taken in the formulation and delivery sections.

This chapter is divided into three sections. In Section 4.2, the two most relevant physicochemical properties for drug delivery, solubility and stability, are discussed. In addition to providing a basic understanding of the importance of solubility and stability to drug delivery, methods to enhance solubility and physical and chemical stability are described. Section 4.3 focuses on the processes required for the proper drug formulation. Since most drugs are administered in the solid state, the formulation process for tablets is described in detail. Finally, Section 4.4 discusses some of the basic drug delivery methods, with an emphasis on the physicochemical properties that impact those methods.

4.2. PHYSICOCHEMICAL PROPERTIES

The most important goal in the delivery of a drug is to bring the drug concentration to a specific level and maintain it at that level for a specified period of time. Stability and solubility are two key physicochemical properties that must be considered when designing a successful drug formulation. Many challenges must be overcome to formulate a product that has sufficient chemical and physical stability to not degrade during the shelf life of the product, yet has sufficient solubility (and dissolution rate) to reach the required therapeutic level.

The physicochemical properties of the drug both in solution and in the solid state play a critical role in drug formulation. The solid-state form of the drug is often preferred, because it is often more chemically stable, easier to process, and more convenient to administer than liquid formulations. However, if the drug is in the solid state, it must dissolve before it can be therapeutically active, and once it is in solution, it must be both sufficiently soluble and chemically stable. For these reasons, it is critical to determine the physicochemical properties of the drug both in solution and in the solid state.

There are several parameters that affect the solubility and chemical stability of a drug in solution. The pH of the solution can dramatically affect both the solubility and chemical stability of the drug. Buffer concentration/composition and ionic strength can also have an effect, especially on chemical stability. The hydrophobic/hydrophilic nature of the drug influences solubility. A typical characterization of a drug will start with a study of the chemical stability of the drug as a function of pH. The structure of the degradation products will be characterized to determine the mechanism of the degradation reaction.

In the solid state, the form of the drug will affect both its solubility and its physicochemical stability. A full characterization of the drug in the solid state will often include a determination of the melting point and heat of fusion using differential scanning calorimetry, loss of solvent upon heating using thermogravimetric analysis, and a characterization of the molecular state of the solid using diffraction and spectroscopic techniques.

In the following two sections, solubility and stability will be discussed as they relate to drug formulation. In the solubility section, the emphasis is on methods to increase solubility. In the stability section, the emphasis is on describing the types of reactions that lead to decreased stability.

4.2.1. Solubility

A drug must be maintained at a specific concentration to be therapeutically active. In many cases the drug's solubility is lower than the required concentration, in which case the drug is no longer effective.[1] There is a trend in new drug molecules toward larger molecular weights, which often leads to lower solubility. The ability to formulate a soluble form of a drug is becoming both more important and more challenging. This has resulted in extensive research on methods to increase drug solubility.

Solubility is affected by many factors. One of the most important factors is pH. Other factors that affect the solubility of the drug include temperature, hydrophobicity of the drug, solid form of the drug, and the presence of complexing agents in solution.

For drugs with low solubility, special efforts must be made to bring the concentration into the therapeutically active range. In this section, some of the common methods to increase solubility will be discussed: salt versus free form, inclusion compounds, prodrugs, solid form selection, and dissolution rate. It should be noted that efforts to increase solubility also have an influence (often negative) on the stability of a compound. For this reason, the most soluble form is often not the first choice when formulating the drug.

4.2.1.1. Salt versus Free Forms One of the easiest ways to increase the solubility of a therapeutic agent is to make a corresponding salt form of the drug. The salt form must be made from either the free acid or free base. Carboxylic acids are the most common acidic functional groups found in drug molecules, while amines are the most common basic groups. An important consideration in the choice of salt versus free form of a drug is that the pH changes, depending upon its location in the intestinal tract. In the stomach, the pH is typically 1–3, and changes to 6–8 in the small intestine. Since the majority of adsorption occurs in the small intestine, it is often desirable to have the maximum solubility at neutral to basic pH values. In general, the acid form of a drug will be ionized at intestinal pH values and therefore will be more soluble, whereas the basic form will be unionized and less soluble. Salts are typically more soluble than the free forms, although this often comes with increased hydrophilicity and a possible decrease in chemical stability due to increased moisture sorption.

Usually the choice of salt versus free forms is based upon the physicochemical properties of the individual compound. However, some generalizations can be made. Free acid forms of a drug usually have adequate solubility and dissolution rates at pH values found in the intestine, and salts of weak bases may be preferred to the free forms because of higher solubility and dissolution rates. It should also be noted that the counter ion can have a dramatic effect upon the solubility and/or stability of the drug. Salt form screening is routinely performed on compounds to determine the counter ion that possesses the best combination of solubility and stability.

4.2.1.2. Inclusion Compounds Another method for improving solubility is to create an inclusion compound between the drug molecule and a host molecule. To be effective, the host/guest inclusion compound must have a higher solubility than the individual drug molecule. An inclusion complex of a drug is usually not crystalline and thus should have higher solubility than a crystalline material. Cyclodextrins complexed to drugs are an example of inclusion compounds commonly used in pharmaceutics.

Cyclodextrins are nonreducing cyclic oligosaccharides made up of six to eight glucopyranose molecules. This class of molecules has a unique structure that is often represented as a tapered doughnut (with the opening at one side larger than the other). The guest molecule then fits inside this cavity and is much less likely to crystallize. Such complexes are also used to improve drug stability by reducing interactions between the drug and its environment. Chemically modified cyclodextrins, which exhibit different stabilizing effects than the natural forms, are also used. They also increase the solubility of insoluble drugs by complexing the drug with the cyclodextrin, generating a metastable form of the drug. Two examples of drugs whose solubility is enhanced by cyclodextrins are prednisolone[2] and prostaglandin E_1.[3] Figure 4.2 shows an example demonstrating the improvement in solubility provided by sulfobutyl ether-β-cyclodextrin (Captisol) for prednisolone.[4]

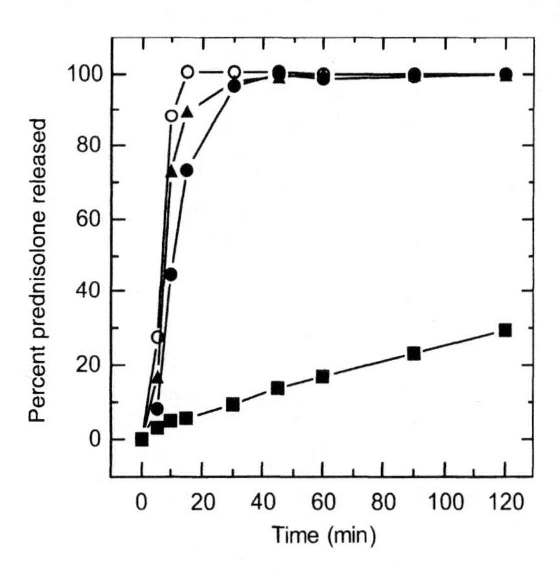

Figure 4.2. Plot of percent prednisolone released versus time for different complexes of cyclodextrin and prednisolone. Used with permission from Kluwer Academic/Plenum Publishers and the original authors.

4.2.1.3. Prodrugs Prodrugs are chemically modified forms of the drug that commonly contain an additional functional group (e.g., an ester group) designed to enhance solubility, stability, and/or transport across a biological membrane. Once the prodrug is inside the body, the additional functional group is cleaved off, either hydrolytically or enzymatically, leaving the drug so that it may fulfill its therapeutic function. Examples of prodrugs (given as prodrug [drug]) that improve solubility include fosphenytoin [phenytoin],[5–7] valacyclovir [acyclovir],[8–10] and capecitabine [5-fluorouracil].[11]

4.2.1.4. Solid Form Selection A drug can exist in multiple forms in the solid state. If the two forms have the same molecular structure but different crystal packing, then they are polymorphs. Pseudopolymorphs (or solvatomorphs) differ in the level of hydration/solvation between forms. Polymorphs and pseudopolymorphs in principle will have a different solubility, melting point, dissolution rate, etc. While less thermodynamically stable, polymorphs have higher solubilities; they also have the potential to convert to the more thermodynamically stable form. This form conversion can lead to reduced solubility for the formulated product. One example is ritonavir, a protease inhibitor compound used to treat acquired immune deficiency syndrome (AIDS). Marketed by Abbott Labs as Norvir, this compound began production in a semisolid form and an oral liquid form. In July 1998, dissolution tests of several new batches of the product failed. The problem was traced to the appearance of a previously unknown polymorph (Form II) of the compound. This form is thermodynamically more stable than Form I and therefore is less soluble. In this case, the solubility is at least a factor of 2 below that of Form I.[12] The discovery of this new polymorph ultimately led to a temporary withdrawal of the solid form of Norvir from the market and a search for a new formulation.

4.2.1.5. Dissolution Rate While not directly related to solubility, the ability to rapidly reach the therapeutic concentration may be useful for fast-acting therapeutic agents, and may compensate for drugs that may have sufficient solubility but are metabolized/excreted too quickly to reach the desired concentration. An example of a method to enhance dissolution is the WOWTAB technology developed by Yamanouchi Pharma.[13]

4.2.2. Stability

Formulation scientists must consider two types of stability: chemical and physical. Physical stability is the change in the physical form of the drug—for example, an amorphous form changing into a crystalline form. The chemical composition remains the same as it was prior to crystallization, but the drug now has different physical properties. Chemical stability is a change in the molecular structure through a chemical reaction. Hydrolysis and oxidation are two common chemical degradation pathways.

4.2.2.1. Physical Stability Physical stability can refer to molecular level changes, such as polymorphic changes, or macroscopic changes, such as dissolution rate or tablet hardness. At the molecular level, form changes include amorphous to crystalline, changes in crystalline form (polymorphism), and changes in solvation state (solvatomorphism). The impact of polymorphic changes on the solubility of ritonavir was discussed in the previous section. In general, a metastable solid form may convert to a more thermodynamically stable form, and it is usually desirable to market the most stable form if possible to avoid such transformations. The presence of seed crystals of the more stable form may initiate or accelerate the conversion from the metastable form to the more stable form. In addition, the presence of solvents, especially water, may cause formation of a solvate with significantly different physicochemical properties. Desolvation is also a possible reaction. For drug formulations, the choice of salt forms (hydrates, solvates, polymorphs) plays a role in identifying the most suitable form for the pharmaceutical product. Polymorphism in drug formulations makes the characterization of polymorphic forms very important. This is most commonly done with X-ray powder diffraction or solid-state nuclear magnetic resonance (NMR) spectroscopy.

When improvements in the physical stability of a product are needed, choices must be based upon the nature of the problem and the desired goal. One of the first choices made is to use the most stable polymorph of the drug. This may involve an extensive polymorph screening effort to attempt to find the most stable polymorph. If the most stable polymorph is undesirable for some reason (e.g., solubility issues), then avoiding contamination of the desired polymorph with seeds of the most stable polymorph becomes very important. In a product that uses an amorphous form of a drug, it is critical to inhibit crystallization to avoid dramatic changes in stability and solubility.

4.2.2.2. Chemical Stability Chemical degradation of the drug includes reactions such as hydrolysis, dehydration, oxidation, photochemical degradation, or reaction with excipients. The constant presence of water and oxygen in our environment means that exposure to moisture or oxygen can affect the chemical stability of a compound. Chemical stability is very important, not only because a sufficient amount of the drug is needed at the time of administration for therapeutic purposes, but also because chemical degradation products may adversely affect the properties of the formulated product and may even be toxic.

Determining how a drug degrades and what factors affect degradation is very important in pharmaceutical product development. The importance of reaching (or avoiding) the activation barrier of a particular chemical process makes temperature one of the most important variables in this area.[14] A second factor in drug degradation is pH. The degradation rate depends on the pH of the formulation and/or the compartments of the body in which the drug is present. Many drug degradation pathways are catalyzed by either hydronium or hydroxide ions, reiterating the important role of water.[14] Described below (with an example or two) are several degradation reactions including hydrolysis, dehydration, oxidation, photodegradation, isomerization, racemization, decarboxylation, and elimination.

Hydrolysis is one of the most common drug degradation reactions. In hydrolysis reactions, the drug reacts with water to form two degradation products. The two most common hydrolysis reactions encountered in pharmaceutical chemistry are the hydrolysis of ester or amide functional groups. Esters hydrolyze to form carboxylic acids and alcohols, while amides form carboxylic acids and amines. For example, the ester bond in aspirin is hydrolyzed to produce salicylic acid and acetic acid, while the amide bond is hydrolyzed in acetaminophen.[15–17]

Dehydration reactions are another common degradation pathway. Ring closures are a fairly common type of dehydration, as is seen for both lactose[18,19] and glucose.[20–22] Both of these compounds dehydrate to form 5-(hydroxymethyl)-2-furfural. Batanopride is another example of a compound which can undergo a dehydration reaction.[23]

Elimination degradation pathways are also possible. Decarboxylation, in which a carboxylic acid releases a molecule of CO_2, occurs for *p*-aminosalicylic acid.[24] Oxidation is very common as well, largely due to the presence of oxygen during manufacture and/or storage. Several examples can be found in Yoshioka and Stella.[14] Isomerization and racemization reactions are other degradation pathways. Two compounds which undergo isomerization reactions are amphotericin B[25] and tirilizad.[26]

Photodegradation of pharmaceuticals has been known for decades. A complication encountered when studying photodegradation reactions is that there are many degradation pathways that each have the potential to yield different products. When oxidizers are present, photodegradation can accompany oxidation.

There are several options available to improve the stability of drugs. One is the use of cyclodextrins, in which the formation of the inclusion complex produces a more stable form of the drug. Examples of cyclodextrins inhibiting drug degradation include tauromustine,[27] mitomycin C,[28] and thymoxamine.[29] Another possibility is to generate a prodrug that has greater stability than the parent compound. Examples of prodrugs that enhance stability include [prodrug (drug)] enaloprilat (enalopril),[30–32] and dipivefrin (epinephrine).[33]

4.3. FORMULATIONS

Formulation is the stage of product manufacture in which the drug is combined with various excipients to prepare a dosage form for delivery of the drug to the patient. Excipients are defined by IPEC-America[34] as "substances other than the pharmacologically active drug or prodrug which are included in the manufacturing process or are contained in a finished pharmaceutical product dosage form." These include binders to form a tablet, aggregates to keep the tablet together, disintegrants to aid dissolution once the drug is administered, and coloring or flavoring agents. Excipients help keep the drug in the desired form until administration, aid in delivering the drug, control the release rate of the drug, or make the product more appealing in some way to the patient.

Formulation is dictated by the physicochemical properties of the drug and excipients. Each drug delivery method has specific formulation issues. As previously mentioned, the solid dosage form is the most convenient and most preferred means of administering drugs; therefore, this discussion will focus on solid dosage forms. The vast majority of solid dosage forms are tablets, which are produced by compression or molding. Powders are the most common form of both the drug and the excipients prior to processing. The process of creating tablets from bulk materials has a number of steps. Some of these are discussed below.

4.3.1. Processing Steps

First, milling is often used to ensure that the particle size distribution is adequate for mixing. Milling both reduces the particle size and produces size and shape uniformity. There are several milling options; perhaps the most common is the ball mill, in which balls are placed inside a hard cylindrical container along with the bulk drug. The cylinder is then turned horizontally along its long axis to cause the balls to repeatedly tumble over one another, thereby breaking the drug particles into smaller pieces.

Next, the drug and excipients must be blended or mixed together. It is very important at this stage that the bulk properties of the materials be conducive to mixing. This means that the materials must have good flowability characteristics. Lubricants such as magnesium stearate may be added to improve the flowability of the formulation.

Once the formulation has been blended, it must be compressed into a tablet. Flowability remains important at this stage of processing because a uniform dose of the blended ingredient mixture must be delivered to the tableting machine. Poor flowability results in poor tablet weight reproducibility. Lubricants are needed to ensure that the tablet can be removed intact from the die once it has been compressed. Finally, the tablet may require a coating. This could be as simple as a flavor coating, or it could be an enteric coating designed to avoid an upset stomach by delaying dissolution until the tablet enters the small intestine.

4.3.2. Influence of Physicochemical Properties on Drugs in Formulations

Most of the processing steps depend at least indirectly upon the physicochemical properties of the drug. Particle size, shape, and morphology often are determined by the solid form of the drug and the conditions from which the drug is crystallized. Aspirin, for example, can have multiple crystal morphologies, depending upon the conditions of recrystallization.[35] Processing can also result in changes in the form of the drug. Amorphous drug formation, changes in the polymorphic form of the drug, or the production of crystal defects can all have a negative effect upon the solubility and stability of the drug.[36] Drug–excipient interactions can affect both solubility and stability. These interactions impact the physical properties of the drug by altering the chemical nature of the drug by reactions such as desolvation, or the Maillard reaction (also known as the "browning reaction" based on the color of the products).

Physicochemical changes in the form of the drug at the formulation and processing stages are almost always undesirable. Such changes can be very costly if found only toward the end of product development. Thus, it is often desirable to perform preformulation studies to determine the optimum form for delivery.[14,37]

4.3.3. Other Issues

New excipients are needed in the industry, as not all formulation needs are satisfied by currently known excipients. This situation is likely to worsen over time as new products, each with potentially unique requirements, are brought to the development stage. Despite this need, the introduction of new excipients is becoming more difficult[38] because new excipients face regulatory requirements similar to those of new drugs themselves. Difficulty in satisfying different nations' regulatory requirements for excipients sometimes makes it more difficult for companies to make a single product that can be marketed in different countries.

4.4. DRUG DELIVERY

For many drugs, the therapeutic nature of the drug dictates the method of administration. For example, oral drug delivery may be the most logical choice for gastrointestinal diseases. If drug release is systemic, then the choice of method often relies on the physicochemical and therapeutic properties of the drug. Transdermal drug delivery, although having the advantage of being non-invasive, has several criteria that must be met by the drug in order to be delivered properly, such as high potency, ready permeability through the stratum corneum, and nonirritation.

In drug delivery, the three most important questions are: When is the drug delivered? Where is the drug delivered? and How is the drug delivered? For this reason, the rest of the section is divided into three parts which address the when, where, and how of drug delivery.

4.4.1. Duration of Release

The goal of drug delivery is to maintain the drug at the appropriate therapeutic level for a specified period of time. There are several methods to achieve this goal, some of which are demonstrated in Figure 4.3. The first is the administration of a single dose, with immediate release of drug to the site of action. This method is useful for acute therapeutic treatment requiring a short period of action. For chronic problems, the goal is to maintain the drug at the therapeutic level for a sustained period of time. Multiple-dose administration is one method for providing sustained therapeutic levels of drug. However, there are many disadvantages to multiple-dose therapy, including variations in drug levels during the treatment period and the need for patient compliance with dosage regimen requirements. To avoid this problem, nonimmediate-release devices are used to deliver the drug over an extended period of time. Nonimmediate-release devices have three types of release mechanisms: delayed release, prolonged release, and controlled release. Delayed release allows

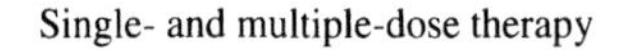

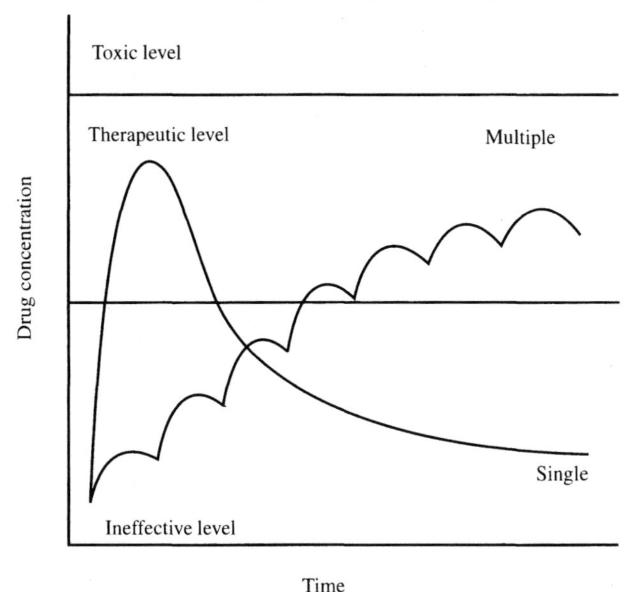

Figure 4.3. Plot of drug concentration versus time for single and multiple doses.

multiple doses to be incorporated into a single dosage form, alleviating the problems of frequent dosing and patient noncompliance. The prolonged-release device extends the release of the drug, for example, by slowing the dissolution rate of the drug compared to that of an immediate release device. The controlled-release device meters out the drug to maintain a constant release rate throughout the desired dosage period. In the prolonged and controlled-release dose, there is usually an initial release of drug to bring the drug into the therapeutic window, followed by additional drug that is released over a longer period of time. Nonimmediate-release devices maintain a more consistent level of drug than multiple doses while retaining the advantage of requiring fewer doses, which increases patient compliance. The disadvantage of nonimmediate-release delivery devices is the inability to stop delivery if adverse reactions are observed in the patient. The concentration characteristics of different nonimmediate-release systems are shown in Figure 4.4.

Nonimmediate- or sustained-release devices can be divided into two categories. The first is a reservoir device whereby the drug is loaded into the reservoir as either a solid or a liquid. Drug release occurs by diffusion through either a semipermeable membrane or a small orifice. Lasers are commonly used to generate uniform orifices through which the drug will diffuse. Osmotic pressure is commonly used to provide the driving force for drug dispersion. The second is a matrix diffusion device whereby the drug is dispersed evenly in a solid matrix. Polymers are commonly used as the matrix. Drug delivery is accomplished by either dissolution of the matrix, with corresponding release of drug, or diffusion of the drug from the insoluble matrix.

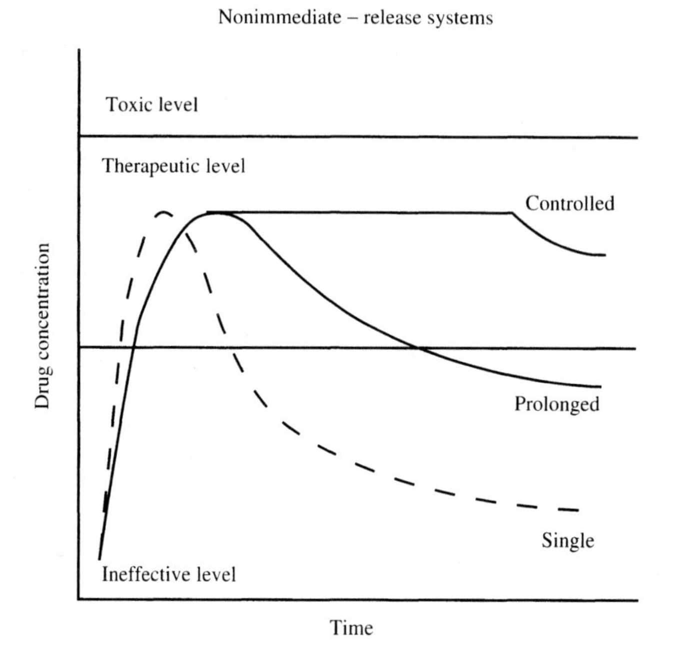

Figure 4.4. Plot of drug concentration versus time for different release systems.

The physicochemical properties of the drug are critical in the design of the dosage form. Solubility, stability, and pH can strongly affect whether a drug can be delivered effectively from a controlled delivery device. Because sustained-release devices often contain multiple doses that if released immediately would reach toxic levels, the physicochemical properties and formulation process may have to be more tightly controlled compared with immediate-release systems.

4.4.2. Site of Administration

Targeted drug delivery is often used if the desired site of action is located in a diseased organ or tissue and release of the drug systemically would produce toxic or deleterious effects. One approach to targeted drug delivery is to place the delivery device adjacent to the site of action, which is especially applicable if the device is controlled release. The other approach is to design the drug so that it has a particular receptor that is found only within the targeted tissue.

4.4.3. Methods of Administration

An increasing number of systems are available for drug delivery. The drug delivery method is chosen based upon the physicochemical properties of the drug, the desired site of action, the duration of action, and the biological barriers (including rapid drug metabolism) that must be overcome to deliver the drug. Some of the most common delivery methods are tablets (oral), parenteral, transdermal, and aerosol. The advantages and disadvantages of each of these methods are described below.

4.4.3.1. Oral Administration The oral drug delivery method is the most common and usually the most preferred drug delivery method by both the formulator and the patient for reasons discussed earlier in this chapter. If the oral delivery method is not chosen, this is primarily due to incompatibilities with the physicochemical properties, site of action, or a biological barrier. Disadvantages of oral drug administration include the low pH of the gastric juices, the first-pass effect of the liver, oral metabolism, and the fact that some patients may have difficulty swallowing the dosage form.

4.4.3.2. Parenteral Administration Parenteral dosage forms include a wide variety of delivery routes, including injections, implants, and liposomes. The advantages of parenteral delivery systems is that they avoid first-pass effects, oral metabolism, and the harsh chemical environment of the stomach's gastric juices. The disadvantage is that the delivery mechanism is invasive.

4.4.3.3. Transdermal Administration Transdermal drug delivery systems have several advantages over other drug delivery methods. These include avoiding gastrointestinal drug adsorption, first-pass effects, replacement of oral administration, and oral metabolism. These systems also provide for multiday therapy from a single dose, quick termination of drug administration, and rapid identification of the medication. The biggest disadvantage of transdermal delivery systems is that only relatively potent drugs are suitable for administration in this manner. Other disadvantages include drug irritation of the skin and adherence of the system to the skin.

4.4.3.4. Aerosol Administration Aerosols can be used for nasal, oral, and topical drug delivery. For topical delivery, aerosols have the advantages of convenient use, protection from air and moisture, and maintaining sterility of the dosage form. Metered dose inhalers (MDI) are used for oral and nasal delivery of drugs. MDIs are used most effectively for the treatment of asthma and are being developed for the delivery of insulin.[39] They have the advantages of avoiding first-pass effects and degradation within the gastrointestinal tract, and rapid onset of action. Some of the disadvantages of aerosol delivery systems for oral and nasal delivery include-lack of particle size uniformity of the drug for proper delivery.

4.4.3.5. Other Delivery Methods In addition to the methods described above, suspensions, emulsions, ointments, and suppositories are all effective drug delivery methods. New delivery methods are continually being developed, as many of the new drugs have low solubility and stability, requiring improved methods for delivery of these drugs. Improvements in the delivery methods for peptides and proteins are necessary as they continue to be developed as drugs.

4.5. CONCLUSION

Designing a successful drug delivery method for a new therapeutic agent requires a thorough understanding of the physicochemical properties of the drug. If all of the relevant physicochemical properties are not determined, the drug may not be correctly formulated, resulting in product failure at scale-up or even after the drug is on the market. In this chapter we have tried to explain some of the relevant physicochemical properties that must be considered in the proper formulation of a drug delivery method.

REFERENCES

1. Lee, T. W.-L.; Robinson, J. R. In *Remington: The Science and Practice of Pharmacy*; Limmer, D., Ed. Lippincott Williams & Wilkins: Baltimore, **2000**, p. 2077.
2. Rao, V. M.; Haslam, J. L.; Stella, V. J. *J. Pharm. Sci.* **2001**, *90*, 807–816.
3. Uekama, K.; Hieda, Y.; Hirayama, F.; Arima, H.; Sudoh, M.; Yagi, A.; Terashima, H. *Pharm. Res.* **2001**, *18*, 1578–1585.
4. Okimoto, K.; Miyake, M.; Ohnishi, N.; Rajewski, R. A.; Stella, V. J.; Irie, T.; Uekama, K. *Pharm. Res.* **1998**, *15*, 1562–1568.
5. Stella, V. J.; Martodihardjo, S.; Katsuhide, T.; Venkatramana, M. R. *J. Pharm. Sci.* **1998**, *87*, 1235–1241.
6. Stella, V. J. *Adv. Drug Deliv. Rev.* **1996**, *19*, 311–330.
7. Boucher, B. A. *Pharmacotherapy* **1996**, *16*, 777–791.
8. Acosta, E. P.; Fletcher, C. V. *Ann. Pharmacotherapy* **1997**, *31*, 185–191.
9. Perry, C. M.; Faulds, D. *Drugs* **1996**, *52*, 754–772.
10. Shinkai, I.; Ohta, Y. *Bioorg. Med. Chem.* **1996**, *4*, 1–2.
11. Budman, D. R. *Invest. New Drugs* **2000**, *18*, 355–363.
12. Bauer, J.; Spanton, S.; Henry, R.; Quick, J.; Dziki, W.; Porter, W.; Morris, J. *Pharm. Res.* **2001**, *18*, 859–866.
13. Yamanouchi Inc., 2003.
14. Yoshioka, S.; Stella, V. J. *Stability of Drugs and Dosage Forms*. Kluwer Academic/Plenum: New York, **2000**.
15. Edwards, L. J. *Trans. Faraday Soc.* **1950**, *46*, 723–735.
16. Garrett, E. R. *J. Am. Chem. Soc.* **1957**, *79*, 3401–3408.
17. Koshy, K. T.; Lach, J. L. *J. Pharm. Sci.* **1961**, *50*, 113–118.
18. Brownley, C. A., Jr.; Lachman, L. *J. Pharm. Sci.* **1964**, *53*, 452–454.
19. Koshy, K. T.; Duvall, R. N.; Troup, A. E.; Pyles, J. W. *J. Pharm. Sci.* **1965**, *54*, 549–554.
20. Wolfrom, M. L.; Schuetz, R. D.; Calvalieri, L. F. *J. Am. Chem. Soc.* **1948**, *70*, 514–517.
21. Taylor, R. B.; Jappy, B. M.; Neil, J. M. *J. Pharm. Pharmacology* **1972**, *24*, 121–129.
22. Taylor, R. B.; Sood, V. C. *J. Pharm. Pharmacol.* **1978**, *30*, 510–511.
23. Nassar, M. N.; House, C. A.; Agharkar, S. N. *J. Pharm. Sci.* **1992**, *81*, 1088–1091.
24. Jivani, S. G.; Stella, V. J. *J. Pharm. Sci.* **1985**, *74*, 1274–1282.

25. Hamilton-Miller, J. M. T. *J. Pharm. Pharmacology* **1973**, *25*, 401–407.
26. Snider, B. G.; Runge, T. A.; Fagerness, P. E.; Robins, R. H.; Kaluzny, B. D. *Int. J. Pharm.* **1990**, *66*, 63–70.
27. Loftsson, T.; Baldvinsdottir, J. *Acta Pharm. Nord.* **1992**, *4*, 129–132.
28. van der Houwen, O. A. G. J.; Teeuwsen, J.; Bekers, O.; Beijnen, J. H.; Bult, A.; Underberg, W. J. M. *Int. J. Pharm.* **1994**, *105*, 249–254.
29. Musson, D. G.; Evitts, D. P.; Bidgood, A. M.; Olejnik, O. *Int. J. Pharm.* **1993**, *99*, 85–92.
30. Allen, L. V., Jr.; Erickson, M. A., III *Am. J. Health-Sys. Pharm.* **1998**, *55*, 1915–1920.
31. Nahata, M. C.; Morosco, R. S.; Hipple, T. F. *Am. J. Health-Sys. Pharm.* **1998**, *55*, 1155–1157.
32. Stanisz, B. *J. Pharm. Biomed. Anal.* **2003**, *31*, 375–380.
33. Jarho, P.; Jarvinen, K.; Urtti, A.; Stella, V. J.; Jarvinen, T. *Int. J. Pharm.* **1997**, *153*, 225–233.
34. I. P. E. C.-Americas, http://www.ipecamericas.org. 2003.
35. Byrn, S. R.; Pfeiffer, R. R.; Stowell, J. G. *Solid-State Chemistry of Drugs*, 2nd ed. SSCI: West Lafayette, IN, 1999.
36. Zell, M. T.; Padden, B. E.; Grant, D. J. W.; Schroeder, S. A.; Wachholder, K. L.; Prakash, I.; Munson, E. J. *Tetrahedron* **2000**, *56*, 6603–6616.
37. Gibson, M., Ed. *Pharmaceutical Preformulation and Formulation: A Practical Guide from Candidate Drug Selection to Commercial Dosage Form*. Interpharm Press: Englewood, CO, 2002.
38. Moreton, R. C. In *Excipients and Delivery Systems for Pharmaceutical Formulations*; Royal Society of Chemistry: Cambridge, **1995**, pp. 12–22.
39. Owens, D. R.; Zinman, B.; Bolli, G. *Diabet. Med.* **2003**, *20*, 886–898.

5

TARGETED BIOAVAILABILITY: A FRESH LOOK AT PHARMACOKINETIC AND PHARMACODYNAMIC ISSUES IN DRUG DELIVERY

WILLIAM F. ELMQUIST

Department of Pharmaceutics, University of Minnesota, Minneapolis, MN 55455

5.1. INTRODUCTION

Pharmacokinetics and pharmacodynamics have always had an important role in drug delivery research. Investigating pharmacokinetic and pharmacodynamic issues during lead optimization is a critical part of the drug discovery and development process. The decision to move forward with a particular compound often depends on pharmacokinetic (PK) and pharmacodynamic (PD) evaluations at several stages in drug development, from early preclinical through Phase I, II, and III studies. It is

Drug Delivery: Principles and Applications Edited by Binghe Wang, Teruna Siahaan, and Richard Soltero
ISBN 0-471-47489-4

increasingly recognized that it is in the early phase of drug discovery and development that optimization of key parameters that describe the absorption, distribution, metabolism, and excretion (ADME) of a drug candidate is required to reduce the failure rate at later stages of development. With the advent of combinatorial chemistry and high-throughput screening, the bottleneck in drug development is not in the identification of new active compounds (hits), but rather in the optimization of lead compounds. Therefore, decision making at every stage of development becomes crucial. PK and PD can identify the key properties of a compound that will help in the decision to move forward with development. Because of this critical position in drug discovery and development, it is necessary to constantly review and assess the purpose of PK and PD analyses and to develop new ways to view and apply these concepts. Therefore, it is the intent of this chapter to describe an approach for considering various PD and PD issues in drug delivery and not to serve as a review of the literature in this area. In assessing PK/PD issues in drug delivery, it is necessary to carefully consider the definition of bioavailability; this chapter introduces the concept of "targeted bioavailability," a term which extends the idea that the true bioavailability of a drug is the fraction of the administered dose that reaches the site of action.

5.2. THE GOAL OF PK/PD IN DRUG DISCOVERY AND DEVELOPMENT

The overall goal of PK/PD analysis in drug development is to describe quantitatively the concentration-time course of active compounds (parent or metabolites) in the body and correlate those concentration-time profiles with the effects, either therapeutic or toxic, of those active compounds. The PK/PD modeling process can characterize crucial absorption and disposition properties of a drug *in vivo*, which then allows prediction of the intensity and duration of the response. In other words, the primary concern of drug discovery is to design molecules that have not only the desired activity but also the necessary potency and duration of action, attributes that can be somewhat predicted by the bioavailability and half-life of a compound. However, as pointed out by Levy, there can be significant variability in drug response given a particular dosage regimen, in the preclinical and especially the clinical arena, and a careful PK/PD analysis may give some insights into the source of that variability.[1–4] Therefore, it is necessary in this PK/PD analysis to include mechanistic information in PK and PD modeling and data analysis. The following discussion will outline an overall view of the variability in PK and PD and the mechanistic causes of that variability.

5.3. LOCATIONS OF VARIABILITY IN DRUG RESPONSE

Effective drug delivery to the site of action is dependent on several factors that influence the PK and PD of drugs. One way of viewing the many processes involved in the delivery of a drug from its site of administration to its site of action is to

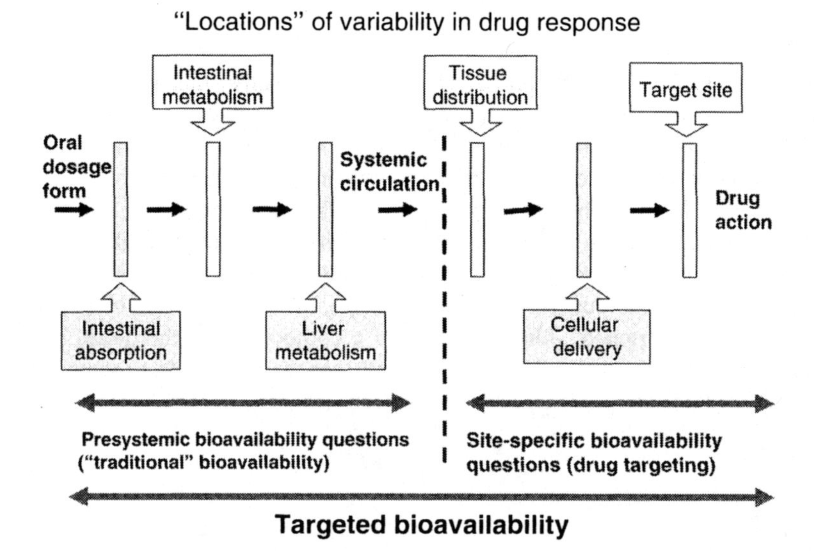

Figure 5.1. This is a schematic representation of some of the barriers that an orally administered compound must pass before reaching the site of action. The barriers that a compound must pass to reach the systemic circulation are traditionally thought to contribute to the final bioavailability of a compound, whereas the barriers that must be overcome after the drug leaves the bloodstream to reach the site of action are related to drug targeting. The overall consideration of barriers from the site of administration to the site of action, which is usually extravascular, can be thought of as related to targeted bioavailability.

consider each barrier along the delivery path as a "location" of possible variability in drug response, where one or more mechanism(s) may be affecting the rate and extent of drug that reaches the site of action (see Figure 5.1). The locations depicted in Figure 5.1 are representative examples for oral administration and are not meant to be exhaustive. The importance of various barriers will depend in large part on the site of administration, the physicochemical characteristics of the drug, and the eventual site of action.

When considering these locations of variability, it is instructive to delineate what is classically meant by the term "bioavailability." The Food and Drug Administration defines bioavailability[5] as "the rate and extent to which the active ingredient or active moiety is absorbed from a drug product and becomes available at the site of action. For drug products that are not intended to be absorbed into the bloodstream, bioavailability may be assessed by measurements intended to reflect the rate and extent to which the active ingredient or active moiety becomes available at the site of action." Typically one modifies this definition to limit the delivery path from the site of administration to the bloodstream, that is, bioavailability is defined as "the fraction of the oral dose that actually reaches the systemic circulation,"[6] and is "commonly applied to both the rate and extent of drug input into the systemic circulation."[7] Upon closer examination, one can see that these definitions inherently assume that the concentration in the systemic circulation is a reasonable surrogate for the concentration at the site of action. It is certainly possible that this assumption breaks down in several cases, which would lead to variability in PD measurements (i.e., drug response, efficacy, and toxicity) that is not reflected in the variability seen in the PK measurements (i.e., in its narrowest scope, the blood concentration-time profiles). Thus, the relevance of a concentration–effect relationship must be reconsidered when the concentration input to such a relationship is not directly related to the effect of the drug. Therefore, one must carefully take into consideration which concentration is appropriate to input into the PK and PD system analysis. As more sophisticated methods become available to measure drug concentrations at sites outside of the bloodstream, the closer we will be to determining the true or "targeted" bioavailability of a compound. Furthermore, important for drug discovery and development, we will then be more able to determine and characterize the sources of variability at each "location" in targeted bioavailability and, hence, drug efficacy and toxicity.

5.4. NOVEL MEANS TO DETERMINE SITE-SPECIFIC DRUG CONCENTRATIONS

As pointed out in the previous section, often the drug concentration of real interest for true bioavailability and drug action is not that in the blood but, rather, the drug concentration in the "biophase," or close to the site of action.[1,8] Therefore, the distribution of a compound to the biophase takes on new importance in the assessment of targeted bioavailability. In order to quantitatively assess the various mechanisms that may influence the distribution of drug from the blood to the site of action (i.e., components of targeted bioavailability; see Figure 5.1), new innovations in sampling techniques will be necessary. Molecular imaging methods may be important in the future to yield site-specific data on drug distribution.[9] Remarkable progress in the development of imaging techniques has taken place over the past 30 years, with techniques such as magnetic resonance imaging, X-ray computed tomography, and positron emission tomography becoming common in clinical use.[10] In the future, these techniques, and possibly others in development, may be modified to measure the distribution of analytes to tissues, which will allow the necessary temporal and spatial resolution to carefully characterize drug concentrations further along the chain from the site of administration to the site of action.[11] Exciting new optical methodologies like fluorescence and bioluminescence imaging techniques are of particular interest to the drug discovery and development process because of their low cost and versatility with a number of solutes.[10] Novel detection systems, such as the ultrasensitive accelerator mass spectroscopy, may be necessary to measure the concentrations at the targeted site of action.[12] Additional opportunities in preclinical studies exist for the use of novel sampling, such as *in vivo* microdialysis and quantitative whole body autoradiography.[13] Success in characterizing the mechanisms responsible for targeted bioavailability will, in large part, depend on the ability to collect the appropriate data, which will include the precise measurement of drug distribution (concentration-time profiles) outside of the blood.

5.5. MECHANISTIC VERSUS BLACK-BOX PK/PD ANALYSIS

PK and PD systems analysis (modeling) has been used as a descriptive tool to quantify drug disposition and response. The recent trend in PK/PD analysis has been to move from purely compartmental modeling to a more mechanistic approach. This is particularly useful to the current discussion of how each "location" of variability can be influenced by multiple mechanisms that affect drug disposition. This relationship is shown in Figure 5.2. Several biological and biochemical mechanisms have an effect on the eventual targeted bioavailability of a compound and, therefore, on the pharmacological and toxicological effects of the compound. To some extent, each of these mechanisms can be considered at each location of variability (barrier to drug delivery). The overall system to be analyzed then becomes more realistic, and the PK/PD analysis takes on new meaning. The system itself is composed of anatomical, physiological, and pathological components that, depending on the input (physicochemical properties of the compound and the input rate), can have several mechanisms at each barrier that affect the targeted bioavailability and the resultant pharmacological and toxicological responses (see Figure 5.2).

Many mechanisms shown in Figure 5.2 are interrelated and, therefore, would best be considered together in the final systems analysis of potential effect on drug disposition. A simple example is the effect of gene regulation on expression of a membrane transport protein, leading to the induction of the transport system that, in turn,

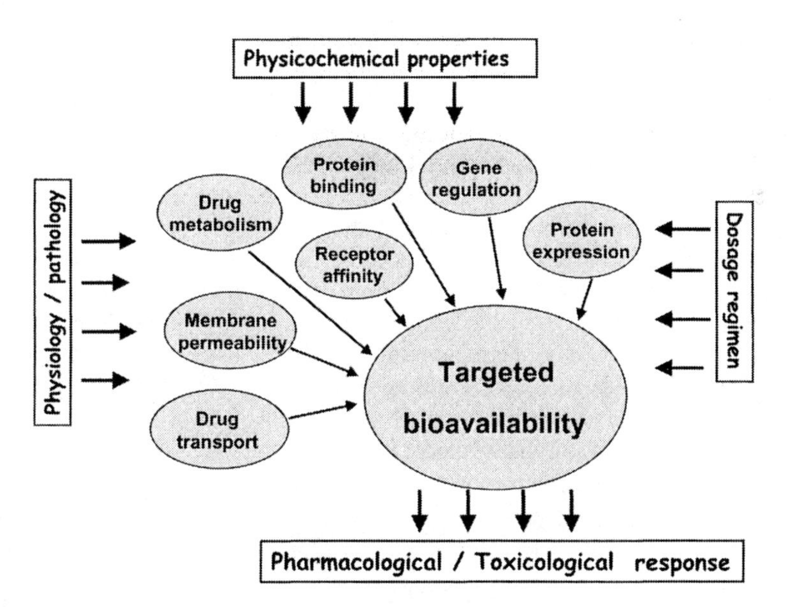

Figure 5.2. This scheme shows the interplay between the body, the compound, and the dose, considering the various mechanisms that influence the delivery of a compound to its site of action. The resultant targeted bioavailability generates the observed pharmacological response.

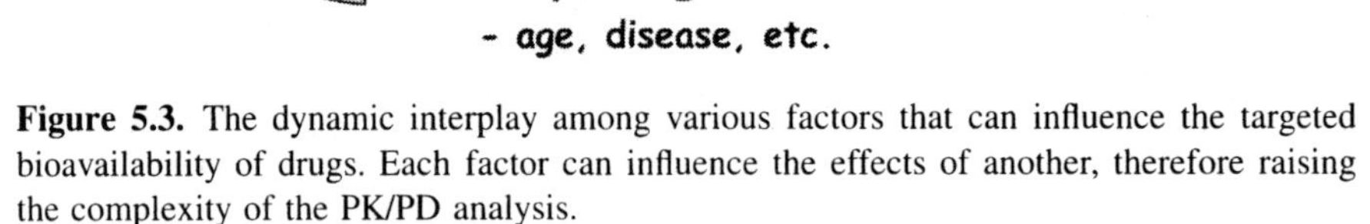

Figure 5.3. The dynamic interplay among various factors that can influence the targeted bioavailability of drugs. Each factor can influence the effects of another, therefore raising the complexity of the PK/PD analysis.

influences the metabolic activity of an enzyme that serves as a metabolic barrier (e.g., at the enterocyte or the hepatocyte) to drug delivery to the bloodstream—in many cases the first step on the journey to the site of action.

The interplay among several of the mechanisms that may affect the targeted bioavailability of drugs is depicted in Figure 5.3. The major challenge in modeling these systems will be in recognizing which delivery locations require which interdependent mechanistic processes to be included in the model to adequately predict the observations.

There have been remarkable advances in our knowledge PK/PD of the regulation and expression of many of the important factors in drug response that are depicted in Figure 5.2. Many of these parameters affecting drug delivery are discussed in other chapters in this book, including such mechanistic factors as drug metabolism, membrane permeability, receptor affinity for receptor-mediated delivery, and efflux transport. The incorporation of these parameters into predictive systems for drug discovery and development has been limited by the need for high-throughput methods to determine the parameters *in vitro* or *in silico*. Recent development of these tools should allow mechanistic PK/PD modeling to be a viable alternative for determining key parameters early in the drug discovery and development process. However, the next real challenge in this process will be incorporating these parameters into quantitative PK/PD models in a sensible and realistic fashion. The use of information developed from *in vitro* or *in silico* experiments in fitting models to *in vivo* data remains a significant hurdle. There have been some attempts to develop tools to estimate absorption, hepatic metabolism, and distribution from *in silico*/*in vitro*-based experiments. Recently, physiologically based pharmacokinetic (PB/PK) models in which *in silico* and *in vitro* prediction tools have added mechanistic ADME parameters to the model have been proposed for modeling and simulation for the prediction of PK parameters of leading drug candidates.[14–16] Poulin and Theil suggest that, through the use of generic and integrative PB/PK models of drug disposition, *a priori* simulations and mechanistic evaluations can be achieved that will aid in the improved selection and optimization of lead drug candidates.[15]

One convention to abandon when attempting to incorporate mechanistic parameters into useful PK/PD modeling is the idea that the data input into the model are generally limited to blood or plasma concentration-time data and the time course of a response. It is important to recognize that quantitative analysis for each location of variability can be applied to data collected from several types of *in vitro* experiments, as described above. It may be valuable to model in stages, using the conventional compartmental or physiologically based approach for determination of parameters that describe the time course of the plasma concentration, and then add further complexity to the system by including the mechanism-based parameters, while using the concentration-time predictions as a fixed input. As with any modeling or systems analysis exercise, the success of this approach will depend on the initial recognition of what parameters are important, how those parameters are connected to one another (model structure), the quality of the data (i.e., the level of information about the parameters that is contained within the data), and the sophistication of the statistical methods used to estimate model parameters.

5.6. SELECTED EXAMPLES OF MECHANISTIC PK/PD MODELS

Recently there have been excellent examples of hybrid mechanistic PK/PD models that address several of the questions posed by specific locations of variability in the targeted bioavailability of compounds. One of these is the integration of a quantitative structure–property relationship (QSPR) model with a PD model of the genomic effects of corticosteroids from the Jusko group.[17] In this work, the authors build on a previously developed PD model for corticosteroid action[18] by adding quantitative structure-property information.[19] In the PD model, information regarding the effects of corticosteroids on the activity of tyrosine aminotransferase is used, including the time courses of the glucocorticoid receptor (GR), GR mRNA, tyrosine aminotransferase (TAT), and TAT mRNA. Given this information about the corticosteroid receptor/gene-mediated mechanism of action, the overall modeling procedure was to use the QSPR model to predict relative receptor affinity, which is then used to determine the EC_{50} parameter in the mechanistic PD model. So, in essence, given the physicochemical and structural properties of a drug, the steroid–receptor interaction (equilibrium dissociation constant) can be predicted; this is then combined with the PD model to predict the TAT activity profiles given a particular dosing regimen (see Figure 5.4). It is instructive to compare this mechanistic analysis to the schemes outlined in Figures 5.1 and 5.2. The terminal location of variability would be the target sites within the cell (Figure 5.1). Explicit mechanisms in this model include many of those outlined in Figure 5.2, i.e., receptor affinity, gene regulation, protein binding, membrane permeability, and protein expression, with drug metabolism and drug transport implicit in the "black-box" PK

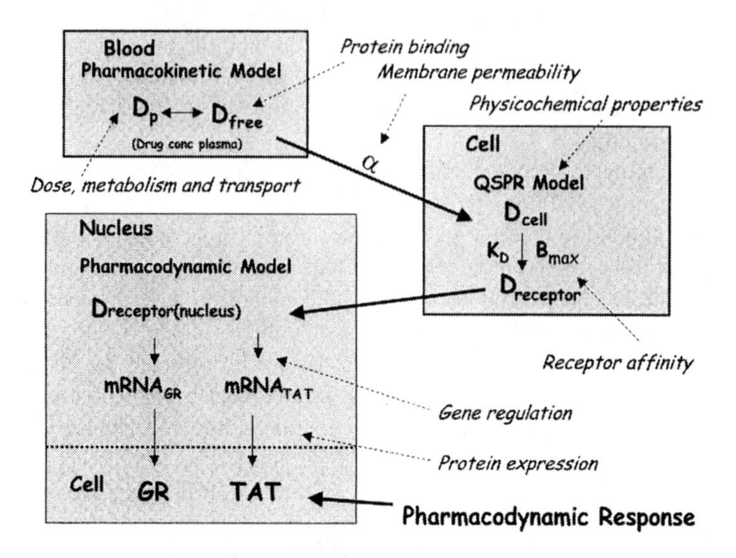

Figure 5.4. Schematic representation of the hybrid QSPR-PD model for corticosteroid action on tyrosine aminotransferase (TAT) induction in rat hepatocytes. In this model, the free intracellular concentration is a constant fraction (α) of the plasma concentration, B_{max} is the total amount of glucocorticoid receptor (GR), and K_D is an equilibrium dissociation constant.[17]

model. The use of this hybrid PK/PD model resulted in an excellent prediction of TAT dynamics following the acute dosing of methylprednisolone and provided precise estimations of the system parameters.[17] Moreover, it can be hypothesized that, given that the system parameters are well defined, the TAT activity profiles following the dosing of other corticosteroids could be predicted simply by knowing the drug pharmacokinetics and using the correct K_D value in the QSPR model. The authors concluded this study by stating that it demonstrates how pharmacokinetic/pharmacological properties and QSPR models combined with mechanistic PD models that have been developed for one drug can predict *in vivo* responses to other chemically related molecules. As these models are further refined, this example will represent a real step forward in the use of PK/PD modeling to examine critical issues in the targeted bioavailability of a compound. This should become important for successful drug discovery and development.

Another illustration of mechanistic PK/PD modeling that results in insight into the targeted bioavailability of a drug is the example recently put forth by the group of Jessie Au.[20] In this study, the interdependence of P-glycoprotein-mediated efflux and intracellular drug binding on paclitaxel PK was determined. This study, like the previous study on glucocorticoid action, was built on the previous models of cellular PK and PD.[21,22] Kuh et al. described a study to determine the mechanisms by which paclitaxel exerted its pharmacological effects and saw that the intracellular PK of paclitaxel depends on several concomitant and interdependent processes, including (1) saturable drug binding to extracellular proteins, (2) saturable and nonsaturable binding to intracellular components, (3) time- and concentration-dependent drug depletion from the culture medium, (4) cell density-dependent drug accumulation, and (5) time- and drug concentration-dependent enhancement of tubulin concentration.[21] The ensuing study by Jang et al. shows, through the use of a computational method to examine the effects of each of these factors, singly and in combination (a type of sensitivity analysis), on the cellular PK of paclitaxel, that all five factors play important roles in determining the intracellular concentration of paclitaxel.[20] Moreover, their results showed that when one or more of these mechanisms were not included in the experimental design, erroneous conclusions with regard to the importance of other factors occurred. In conclusion, this study is a good example of how a quantitative systems analysis can depict biological mechanisms (see Figure 5.2) such as drug transport, drug binding, and the affinity to binding sites that will influence the targeted bioavailability of drugs, particularly in getting an effective concentration from the blood to the site of action (see Figure 5.1). The next big challenge in this type of analysis will be the inclusion of additional barriers to drug delivery or locations of variability to quantitatively model the effects of changes in dosing and/or treatment schedule on the antitumor activity of paclitaxel.

5.7. SUMMARY

The conceptual basis of this chapter is the idea that PK and PD issues in drug delivery need to be viewed from a broad perspective, a perspective which includes an examination of the mechanisms that affect delivery of drug across biological barriers from the site of administration to the site of action. In other words, often we should view effective bioavailability not just as the fraction of a dose that reaches the systemic circulation but rather as the fraction of a dose that reaches its target, i.e., targeted bioavailability. While drug delivery research has recognized the value of mechanistic evaluations for decades, it is now time to incorporate these mechanisms into PK/PD analyses. In doing so, it is instructive to view the delivery path from administration to active site as having several locations of variability, and to understand that the causes of such variability can come from genetic, environmental, and physiological/pathological influences on each mechanistic process at each location or barrier. The determination of the extent and regulation of the specific mechanisms by which each barrier limits drug delivery will be achieved through a variety of techniques, including *in silico*, *in vitro*, biochemical, and genetic methods, and it will be necessary to incorporate the parameters that quantitatively describe these determinants into PK and PD model systems. Moreover, it is appreciated that significant advances in techniques available to measure drug distribution will be necessary to achieve the concentration-time data needed for model input. These technical advances are expected, and even now, examples of integrated mechanistic PK/PD models are can be found in the literature. The two excellent examples offered in this chapter exemplify PK/PD systems analyses that have evolved, and are currently evolving, as more information about the system becomes available. All of the advances in quantitative techniques indicate that exciting times are ahead for the pharmaceutical scientist to apply cutting-edge tools to develop quantitative models that will help predict the targeted bioavailability of novel compounds.

ACKNOWLEDGMENT

Research in the Elmquist laboratory related to determining the targeted bioavailability of drugs has been funded by NIH Grants CA71012, CA75466, and NS42549.

REFERENCES

1. Levy, G. *Clin Pharmacokinet.* **1998**, *34*, 323–333.
2. Levy, G. *Clin. Pharmacol. Ther.* **1994**, *56*, 248–252.
3. Levy, G. *Pharm. Res.* **1987**, *4*, 3–4.
4. Levy, G. *Clin. Pharmacol. Ther.* **1994**, *56*, 356–358.
5. 21CFR320.1 *Code of Federal Regulations*, Title 21, Volume 5, Parts 300 to 499, **1999**; *Guidance for Industry, CDER, FDA*, Bioavailability and Bioequivalence Studies for Orally Administered Drug Products—General Considerations.
6. Gibaldi, M.; Perrier, D. In *Pharmacokinetics*, 2nd ed., Dekker, **1982**, p. 145.
7. Rowland, M.; Tozer, T. In *Clinical Pharmacokinetics*, 3rd ed., Baltimore: Williams & Wilkins, **1995**, p. 34.
8. Eichler, H.-G.; Muller, M. *Clin. Pharmacokinet.* **1998**, *34*, 95–99.
9. Workman, P. *Curr. Pharm. Des.* **2003**, *9*, 891–901.
10. Rudin, M.; Weissleder, R. *Nature Rev. Drug Discov.* **2003**, *2*, 123–131.
11. Fischman, A. J.; Alpert, N. M.; Rubin, R. H. *Clin. Pharmacokinet.* **2002**, *41*, 581–602.
12. Lappin, G.; Garner, R. C. *Nature Rev. Drug Discov.* **2003**, *2*, 233–240.
13. Elmquist, W. F.; Sawchuk, R. J. *Pharmaceut. Res.* **1997**, *14*, 267–288.
14. Poulin, P.; Theil, F.-P. *J. Pharm. Sci.* **2002**, *91*, 129–156.
15. Poulin, P.; Theil, F.-P. *J. Pharm. Sci.* **2002**, *91*, 1358–1370.
16. Theil, F.-P.; Guentert, T. W.; Haddad, S.; Poulin, P. *Toxicol. Lett.*. **2003**, *138*, 29–49.
17. Mager, D. E.; Pyszczynski, N. A.; Jusko, W. J. *J. Pharm. Sci.* **2003**, *92*, 881–889.
18. Ramakrishnan, R.; DuBois, D. C.; Almon, R. R.; Pyszczynski, N. A.; Jusko, W. J. *J. Pharmacokinet. Pharmacodyn.* **2002**, *29*, 1–24.
19. Mager, D. E.; Jusko, W. J. *J. Pharm. Sci.* **2002**, *91*, 2441–2451.
20. Jang, S. H.; Wientjes, M. G.; Au, J. L.-S. *J. Pharmacol. Exp. Ther.* **2003**, *304*, 773–780.
21. Kuh, H.-J.; Jang, S. H.; Wientjes, M. G.; Au, J. L.-S. *J. Pharmacol. Exp. Ther.* **2000**, *294*, 761–770.
22. Jang, S. H.; Wientjes, M. G.; Au, J. L.-S. *J. Pharmacol. Exp. Ther.* **2001**, *298*, 1236–1242.

6

PRESYSTEMIC AND FIRST-PASS METABOLISM

W. GRIFFITH HUMPHREYS

Department of Metabolism and Pharmacokinetics, Bristol-Myers Squibb Pharmaceutical Research Institute, P.O. Box 5400, Princeton, NJ 08543

Correspondence should be addressed to: Dr. W. G. Humphreys Bristol-Myers Squibb PRI; P. O. Box 4000; Princeton, NJ 08543; Telephone: 609-252-3630; Fax: 609-252-6802;
E-mail: william.humphreys@bms.com

Drug Delivery: Principles and Applications Edited by Binghe Wang, Teruna Siahaan, and Richard Soltero
ISBN 0-471-47489-4

6.1. INTRODUCTION

Prior to entering the systemic circulation, drugs and xenobiotics are subject to preabsorptive and first-pass metabolism. This is often a significant limitation to the delivery of drugs via the oral route.[1,2] Although problems with drug stability and dissolution can sometimes be addressed with formulation approaches, it is difficult to use these approaches to significantly alter delivery of drugs that are subject to metabolism. Drugs that have low bioavailability (%F) due to high preabsorptive and first-pass metabolism are likely to have a high degree of inter- and intrapatient variability[3] and are more likely to suffer from drug-drug interactions. Also, drugs with low %F are likely to require suboptimal dosing regimes, i.e., b.i.d. or t.i.d., and/or relatively high doses. The former leads to poor patient compliance, suboptimal efficacy, and marketing issues, and the latter may lead to unanticipated toxicities due to a large flux of drug and drug metabolites. An additional consideration is the fact that interspecies differences in metabolic enzyme systems will make the prediction of human absorption, distribution, metabolism, and excretion (ADME) properties much more challenging than for drugs cleared via direct elimination.

These considerations make it important to understand the preabsorptive and first-pass metabolism characteristics of candidate drug molecules and to optimize these characteristics preclinically when possible. Drug metabolism plays a central role in modern drug discovery and candidate optimization, recent reviews have detailed how metabolism has impacted the discovery process and challenges that the field faces in the future.[4–8]

The special case of prodrugs requires an even greater understanding of the metabolism characteristics because of the delicate balance between preabsorptive stability and postabsorptive instability that must be achieved. Also, the potential for prodrugs to be metabolized by pathways not leading to direct bioactivation must be considered.

This chapter will review the enzyme systems responsible for preabsorptive and first-pass metabolism in the intestine and liver. Also covered are methods to study the effects of these enzyme systems *in vitro* and *in vivo*.

6.2. PHASE I ENZYME SYSTEMS

6.2.1. Esterase/Amidase Enzymes

6.2.1.1. General Esterase activity can be found in many mammalian tissues and in blood[9–11] and can be the result of catalysis by a number of distinct enzyme families, including carboxylesterases, paraoxonase, and cholinesterases. The esterase enzymes generally have wide substrate specificities and are capable of the hydrolysis of a wide range of hydrolytic biotransformations, although certain members such as cholinesterase have highly specialized functions. Substrates for biotransformations mediated by esterase enzymes include a wide variety of ester- or amide-containing compounds and also include carbamates and thioesters. Several recent papers have covered substrate preferences for different esterases and have demonstrated some structure–metabolism relationships.[12,13] The apparent differences in substrate preference found between rodents and humans should be an important consideration when designing ester-based prodrugs. The most important family of esterase enzymes is the carboxylesterase family (EC 3.1.1.1), and members of this family are responsible for the hydrolysis of a variety of drugs.[12] These enzymes are localized to the endoplasmic reticulum and are strongly inhibited by sulfonyl or phosphonyl fluorides. The enzymes employ a Ser-His-Glu triad as their catalytic domain. Improved nomenclature has recently been suggested.[12] The major forms of the enzymes are designated hCE-1, hCE-2, and hCE-3.[11,14] These enzymes have important differences in substrate recognition that can lead to dramatic interspecies differences in enzyme activity. The carboxylesterase enzymes are responsible for the hydrolysis of the majority of prodrug esters. An important substrate for hCE-1/hCE-2 is CPT-11,[15,16] which is a carbamate derivative of SN-38, a molecule much more active against the target topoisomerase enzyme. The hydrolysis to SN-38 is thought to be crucial for maximum efficacy, and one hypothesis for a poor patient response to this therapy is low conversion due to polymorphic variants of carboxylesterase enzymes.[17,18]

Paraoxonase enzymes clearly belong to a distinct enzyme family.[11,19] They are not serine or cysteine proteases but instead use divalent metal ion for catalysis. The enzymes have limited substrate specificity but do hydrolyze a variety of aromatic and aliphatic lactones, including several statin lactones.[20]

6.2.1.2. Tissue Distribution Although the highest esterase activity is normally found in liver, the intestine is also high in this activity. The carboxylase most abundant in human intestinal tissue is designated hCE-2.[11] This enzyme is found in human liver, but hCE-1 is the most abundant liver carboxylesterse enzyme.[11] In general, human blood/plasma has lower overall esterase activity relative to rodent plasma.[10]

6.2.2. Cytochrome P450 Enzymes

6.2.2.1. General The cytochrome P450 (CYP) enzymes are a large family of related enzymes expressed predominantly in the liver, but found in many tissues, that play a role in the metabolic clearance of well over 50% of drugs. Various aspects of CYP enzymology, substrate specificity, and clinical importance have been the subject of recent reviews.[21–24] The enzymes have broad and somewhat overlapping substrate specificity. Although over 50 human CYP enzymes have been characterized, only 5 seem to be are responsible for the majority of drug metabolism: CYP1A2, CYP2C9, CYP2C19, CYP2D6, and CYP3A4. Of these, by far the most important enzyme is CYP3A4. This enzyme is found in the greatest quantity (>30% of total liver CYP enzyme)[25] and has the broadest range of known substrates.[24,26,27]

The mechanism of action of the enzymes is a complex multistep process that leads to the biotransformation of substrate, most often to an oxidized product.[28–30] The process of oxidation involves high-energy intermediates and often involves the generation of reactive electrophilic intermediates at the enzyme active site that are sometimes released and can react with cellular components.[31–35] This process is thought to contribute to the acute or idiosyncratic toxicity displayed by some compounds.[32–34]

The variability in expression and activity of the CYP enzymes is a major concern in modern clinical pharmacology. This variability is due to a combination of genetic variation, environmental factors, and inhibition or induction by drugs or xenobiotics. Drugs that are cleared via metabolism by a single CYP enzyme often have significant variability in exposure that can have significant consequences for the efficacy and toxicology of the compound. This is especially problematic for drugs that are cleared primarily by CYP3A4 and CYP2D6. CYP3A4 expression is controlled by genetic factors,[36] and the enzyme is inhibited and/or induced by an extensive list of pharmaceutical agents, so new drugs that rely on this enzyme for a large percentage of their clearance will likely be the subject of multiple drug-drug interactions.[24] CYP2D6 is polymorphically expressed in the human population, and between 1% and approximately 10% of the population are functionally deficient in this enzyme activity and can be defined as poor metabolizers.[37] There are also significant populations which overexpress this enzyme.[37] Drugs that rely on CYP2D6 for a large portion of their metabolism will likely have widely variable pharmocokinetics due to these factors. Other CYP enzymes either have polymorphic distribution or are subject to induction or inhibition, but the magnitude of the interactions is not as important as those seen with CYP2D6 and CYP3A4.

6.2.2.2. CYP Enzymes in the Intestine Characterization of the CYP content of the human intestine has been challenging, and the exact complement of enzymes present remains unclear.[38–40] It is clear that the major drug-metabolizing enzymes present are CYP2C9, CYP2C19, CYP2D6, and CYP3A4. The major enzyme present in intestinal tissue is CYP3A4, and the activity of intestinal microsomes for a variety of substrates has been shown to be in good agreement with activity in liver. The importance of intestinal CYP oxidation to the first-pass metabolism of drugs has been clearly shown with experiments performed during the anhepatic phase of liver transplantation. These experiments have shown that intestinal oxidation is responsible for significant first-pass elimination of cyclosporin and midazolam.[41,42] The content of CYP3A4 in the intestine has been shown to be quite variable among subjects and contributes to interpatient variability found for CYP3A4 substrates.[42]

One aspect of intestinal metabolism that must be considered is the relatively small amount of enzyme present relative to liver and the potential for higher drug concentrations during enterocyte transit. This makes the total enzymatic capacity of the intestine lower and more susceptible to saturation than the liver, especially for high-dose drugs.[1] Also, the potential for concentration difference may lead to alterations in the enzyme responsible for metabolism in the two organs, as hypothesized for the intestinal and liver metabolism of NE-100.[43] There is growing evidence, however, that the P-gp system works in concert with CYP3A4 to effectively limit substrate concentration to the enzyme and minimize the potential for saturation.[44–47]

One area that has received a considerable amount of recent attention is the inhibition of intestinal CYP enzymes by components of grapefruit juice.[48] This effect has been shown to produce substantial increases in the plasma concentrations of a wide variety of CYP3A substrates.

6.2.2.3. CYP Enzymes in the Liver The liver contains the highest quantity of CYP enzymes of any organ and has an impressive capacity to metabolize drugs and xenobiotics. The CYP enzymes of the liver constitute the most important barrier to the entry of drug-like molecules into the systemic circulation. All major CYP2D6 enzymes are present in liver in quantities that are relatively high compared to other organs. The abundance of the major CYP enzymes in the average human liver has been reported to be CYP1A2 (13%), CYP2A6 (4.0%), CYP2B6 (0.2%), CYP2C9/CYP2C19 (18%), CYP2D6 (1.5%), CYP2E1 (6.6%), and CYP3A4/5 (29%).[25]

6.2.3. Flavin-Containing Monoxygenase Enzymes

The flavin-containing monoxygenase enzymes (FMOs) are expressed in the endoplasmic reticulum and are capable of carrying out a variety of oxidative biotransformations. Various aspects of this enzyme family have recently been reviewed.[49,50] Typical reactions are the oxidation of nitrogen or sulfur heteroatoms. The major FMO enzyme expressed in humans is FMO3 and is quite abundant in adult liver.[51] The enzyme has been shown to metabolize a large number of xenobiotics and drugs *in vitro* and *in vivo*, including drugs such as cimitidine, tamoxifen, itopride, and sulindac. There seems to be a high level of interindividual variability in the expression of FMO3, but the exact understanding of the magnitude of the variability is complicated because the enzyme can be degraded during isolation. The variability is thought to be due largely to genetic polymorphisms in the *FMO3* gene.[51,52] A rare genetic defect in the *FMO3* gene has been shown to be the cause of trimethylaminuria.[53,54] FMO1 is expressed in intestine and kidney, but the significance of this enzyme for drug metabolism has not been well established.[51] The FMO enzymes are not thought to be inducible by small molecules, and no examples of drug-drug interactions due to inhibition of the enzyme have been described.[51]

6.3. PHASE II ENZYME SYSTEMS

6.3.1. Glucuronosyl Transferases

The uridine 5′-diphosphate glucuronyltransferases (UGTs) are a family of enzymes found in the endoplasmic reticulum that catalyze the transfer of glucuronic acid to nucleophilic sites on drugs and xenobiotics.[55,56] The enzymes have broad substrate specificity and will conjugate phenols, carboxylic acids, alcohols, amines, nitrogen-containing heterocycles, and other moieties. UGT enzymes often catalyze the conjugation of metabolites of the CYP enzymes, which leads to their designation as phase II enzymes; however, it is known that the enzymes directly conjugate many substrates. Enzymes of this family play important roles in the metabolic clearance of a number of substrates, including acetaminophen, mycophenolic acid, propofol, morphine, nonsteroidal anti-inflammatories, and fibrates.

The major polymorphism in this enzyme family is located in the *UGT1A1* gene.[57] The condition characterized by a complete lack of functional UGT1A1 activity, either through lack of protein production or production of inactive protein, is called "Crigler-Najjar syndrome." The more common polymorphism results in a deficiency of UGT1A1 and is known as "Gilbert's syndrome." Polymorphisms in UGT enzymes seem to play a role in determining the efficacy and toxicity of the antitumor agent CPT-11[58,59] and may be an important determinant of the toxicity of other drugs.[60] Polymorphisms have been found for 6 of the 16 functional UGT genes characterized to date, but their functional significance is not as clear as for UGT1A1.[57,61] UGT activity that leads to acylglucuronide formation may be an important pathway for the bioactivation of some compounds to toxic agents.[62–64] Although still controversial, this is one major hypothesis for the toxicity seen with a number of nonsteriodal anti-inflammatory drugs. Acylglucuronides present an analytical challenge because of their instability in biological matrices.[63] Also, the conjugates are often hydrolyzed in the gut, which can lead to enterohepatic recirculation.

6.3.1.1. Activity in Intestinal Tissue Multiple glucuronosyl transferase enzymes have been detected in human intestinal tissue, but as with the CYP enzymes, only a subset of the liver enzymes are represented. The two enzymes that are clearly expressed are UGT1A1 and 2B7.[56] As with CYP enzymes, the most conclusive studies showing the role of intestinal metabolism involve patients in the anhepatic phase of liver transplantation. These patients were shown to produce the glucuronide conjugate of propofol,[65] showing that there was extrahepatic metabolism, although the role of the kidney could not be completely ruled out in these experiments.

6.3.1.2. Activity in Liver Tissue The liver contains the highest content of UGT enzymes, and most if not all of the 16 functional UGT genes are expressed.[56] The activities of the enzymes expressed in intestine and liver seem to be very similar.

6.3.2. Sulfotransferases

Sulfotransferase enzymes (SULTs) catalyze the sulfation of substrates through the transfer of the sulfuryl group of adenosine 3′-phospate 5′-phosphosulfate

(PAPS).[66–69] The substrates for the enzymes are similar to those of the UGTs and include alcohols, phenols, and amine-containing compounds. These enzymes are located in the cytosol and, like other xenobiotic-metabolizing enzymes are made up of a superfamily of genes.[68] In humans, approximately 11 SULTs have been identified, with the SULT1 and SULT2 families being the most important for drug metabolism.[68] The enzymes metabolize a number of drug substances but also play an important role in the metabolism of endogenous compounds. SULTs are expressed at a high level in the human fetus and may play an important role in detoxifying xenobiotics during early development.

Important drug substrates for the SULT enzymes are acetaminophen, minoxidil, and isoproterenol. The SULT enzymes play an important role in the bioactivation of some chemicals through the conjugation of alcohols to form reactive species. This reaction is most notable in the activation of hydroxylamines derived from arylamines.[70,71]

6.4. OTHER ENZYMES

In addition to the enzyme systems mentioned so far, the following enzyme families can also play a significant role in first-pass metabolism: intestinal peptidases,[72] alcohol dehydrogenases,[73] and *N*-acetyl transferases.[74,75]

6.5. TRANSPORTERS

Although this review focuses specifically on metabolism, the increasing knowledge of the intimate connections between metabolism and transport must be mentioned. There are many situations in which the interaction of transporters with drug metabolites has been explored and has led to the hypothesis that CYP, and other enzymes and transporters, work in concert to keep xenobiotics from crossing the intestinal epithelium.[44–47] Cases in which there seems to be coordinate activity to preferentially export drug metabolites have been found in Caco-2 cells and *in vivo* with esterase enzymes[76,77] and with CYP enzymes.[78]

Direct excretion of metabolites into the bile via transporter-mediated efflux can be an important determinant of clearance and first-pass extraction.[79–81] This phenomenon is probably much more common than it would seem from the literature, at least in part because there are no facile ways to study the effect other than *in vivo* using bile duct cannulated rats.

6.6. METHODS TO STUDY FIRST-PASS METABOLISM

6.6.1. General

There are several *in vivo* approaches that can be used to study first-pass metabolism in preclinical species, and these, along with *in vitro* results, can often shed mechanistic insight on the problem of incomplete oral bioavailability. Modern liquid chromatography tandem mass spectrometry (LC-MS/MS) measurement of plasma drug concentrations is a tool to rapidly assess oral bioavailability of new candidate compounds and allows for early definition of bioavailability problems. When bioavailability concerns do arise, there are several procedures, discussed below, that can be followed to isolate the factor(s) limiting oral delivery of the compound.

6.6.2. *In Vitro* Methods to Study First-Pass Metabolism

In vitro experiments can sometimes provide valuable insight into what is happening *in vivo* that is limiting oral bioavailability. The typical experiments, often employed in tandem, to understand bioavailablilty are determinations of compound solubility, membrane permeability, and stability in subcellular fractions. The membrane permeability assays that are most often employed are either a measurement of permeability through an artificial membrane (Parallel artificial membrane permeability assay, PAMPA, is the most common technique) or a cell monolayer (Caco-2, a human colon carcinoma–derived cell line, is the most common cell monolayer). The subcellular fractions most often employed are plasma (for ester-containing compounds) and liver microsomes with the addition of either reduced nicotinamide adenine dinucleotide phosphate (NADPH) or uridine diphosphoglucuronic acid (UDPGA) as cofactor.

Hepatocytes are also a very useful *in vitro* tool that provide a more complete system for studying metabolism. The low availability of high-quality human hepatocytes for study remains a drawback of this approach, although cryopreserved human cells provide an attractive alternative to freshly isolated cells for performing biotransformation studies.

All of the assays mentioned can be set up in an automated, medium-throughput system; however, all require a specific assay to measure the compound concentration at the end of the assay, which places significant limitations on throughput. There have been recent attempts to solve the throughput problems inherent in this type of assay by developing generic endpoint assays for metabolic stability, but there has been no clear solution to date.

6.6.3. Prediction of Human Clearance

6.6.3.1. *In Vitro–In Vivo Correlations*

6.6.3.1.1. Oxidative Biotransformation in Microsomes The rapid determination of pharmacokinetic parameters, solubility, permeability, and *in vitro* stability in plasma or liver tissue can often provide a reasonable explanation of the mechanisms limiting oral bioavailability. An approach that is often used is to extrapolate the *in vitro* rate of metabolism to estimate the hepatic clearance using *in vitro–in vivo* correlation methods.[82–86] These methods use *in vitro* kinetic parameters, usually V_{max}/K_m or *in vitro* $t_{1/2}$, to determine the intrinsic clearance, which is then scaled to hepatic clearance using the amount of tissue in the *in vitro* incubation, the weight of the liver, and the well-stirred model for hepatic clearance.

Care must be used when determining *in vitro* parameters.[87–92] The methods used to extrapolate to the *in vivo* situation all assume that the drug concentration at the active site of the CYP enzymes will be much less than the K_m value. This is a reasonable assumption for most drugs used clinically under typical *in vitro* experimental conditions; however, experimental design, especially *in vitro* $t_{1/2}$ determinations, should take this into account. The incorporation of protein-binding corrections in these calculations has been somewhat controversial. Clearance of "free drug" was included in the first theoretical models for predicting hepatic extraction from *in vitro* data[93] and was the favored method in early *in vitro–in vivo* correlation attempts. However, the inclusion of protein binding into the scaling equations tends to underpredict actual *in vivo* values, and generally, better results are achieved when protein binding is left out of the equations.[87–89] The reasons for this could be that (1) the intrahepatic concentration of most drugs is closer to the total plasma concentration than the free plasma concentration due to efficient uptake by the liver or (2) protein binding to microsomal proteins may serve to "cancel out" protein binding in plasma, i.e., the free concentration in microsomal incubations is closer to the free concentration in plasma, so the effect of protein binding is already accounted for in the *in vitro* parameters. Methods used to account for protein binding or to lessen its effect are (1) to determine the free concentration in microsomal incubations and then calculate *in vitro* parameters based on free drug (this would then be compared to free plasma clearance values) or (2) to use as little protein in microsomal incubations as possible to lessen the effect of binding. The scale-up from *in vitro* parameters to predict *in vivo* human clearance of new chemical entities is still difficult prospectively, especially in cases where the CYP kinetic data do not follow classical Michaelis-Menten behavior.[94] The uncertainty of this calculation can be somewhat lessened if the same scale-up methods provide accurate results when applied to preclinical animal data.

6.6.3.1.2. In Vitro–in Vivo Correlations: Other Systems This method has most often been applied to scale rates of CYP-mediated oxidative metabolism in microsomal systems but can also be applied to conjugation[95,96] or FMO-catalyzed reactions[97] or data derived from hepatocytes[98] or other *in vitro* systems. The scaled clearance values can be compared to the values determined *in vivo* for a single compound or a set of analogs to give some idea of the predictive capacity of the *in vitro* systems.

The use of microsomes along with UDPGA as a cofactor assay to measure UGT enzyme activity has been hampered historically by the fact that this enzymatic activity in microsomes is often in a "latent" form and requires activation by physical or detergent-induced disruption of the membrane matrices. Recently, a generic method involving the addition of the pore-forming peptide alamethicin to overcome the latency exhibited by this enzyme system has been described.[99] The inclusion of alamethicin seems to provide a more consistent method of assessing UGT enzyme activity.

6.6.3.2. Allometric Scaling A second method for predicting clearance is through allometric scaling.[100–103] The method has been employed for many years to predict human pharmacokinetic parameters and has been shown to provide excellent results in cases where the compound is cleared predominantly through renal elimination. The basic method attempts to derive a relationship between animal pharmacokinetic parameters and body weight and then scale that to derive human parameters. As mentioned above, the method is usually employed with renally eliminated drugs because the allometric relationship often breaks down with drugs cleared through metabolism. This is because there is no accounting for interspecies differences in metabolic rates for the compound of interest. Factors such as maximum lifetime potential have been added to the allometric relationship to try to improve the predictions for highly metabolized drugs, but they have met with only limited success. A hybrid method to calculate human clearance using *in vitro* data as well as animal pharmacokinetic parameters has been described by Lave et al.[104] In this method, the authors correct each animal clearance parameter by multiplying the ratio of the *in vitro* rate of metabolism in that species by the human *in vitro* rate. The allometric relation is then derived with the corrected animal pharmacokinetic parameters.

6.6.4. *In Vivo* Methods to Study First-Pass Metabolism

The methods outlined above may not lead to a satisfactory determination of the mechanism of incomplete bioavailability, and additional methods may be necessary in some cases to fully characterize the factors responsible. Bile duct cannulated animals provide a powerful model to examine incomplete bioavailability issues, and the rat provides the most flexibility because terminal studies can routinely be done. For compounds that are thought to have dissolution limitations or instability in the gastrointestinal (GI) tract, the GI tract can be removed at the end of the experiment and the contents assayed for drug and metabolites. The amount of parent in bile and urine can be quantitated by LC-MS/MS, and this method can also be used for any metabolites when authentic standards have been prepared. Alternatively, bile and urine metabolites can be estimated with ultraviolet high-performance liquid chromatography (HPLC-UV) using the extinction coefficient of the parent. These measurements will begin to define the total absorption of the compound and how it relates to the systemic bioavailabilty. The availability of radiolabeled compound at this point makes the bile duct cannulated rat experiment especially powerful. An interesting new methodology for getting early estimates on total drug-related material is through the use of quantitative ^{19}F-nuclear magnetic resonance (^{19}F-NMR).[105] In this technique, the ^{19}F-NMR signals in the chemical shift region of the parent signals are recorded and integrated for intact samples of urine and bile. The integrated signals are compared with standards and allow for quantitation of drug-related material. The technique has been shown to produce results that compare favorably to those achieved with standard scintillation counting techniques and, with the prevalence of fluorine-containing compounds in modern drug discovery, could become a useful technique for discovery work.

A second *in vivo* model system that is very useful in dealing with the problem of low oral bioavailability is portal vein cannulated animals. There are two ways this experiment can be conducted to determine hepatic extraction: (1) measure the systemic plasma concentration after oral, portal vein, and systemic administration or (2) measure portal vein and hepatic vein concentrations after an oral dose. Both methods yield information on hepatic extraction and the percentage of dose reaching the portal circulation (the product of the fraction absorbed and the fraction metabolized by the gut wall).

Both the bile duct cannulated model and the portal vein cannulated model can be combined with a number of methods for modulating absorption or metabolism to ask specific mechanistic questions regarding stability, permeability, and metabolism. The following are several methods that can be used to modulate metabolism *in vivo*. The most commonly used method to modulate oxidative metabolism *in vivo* is to coadminister either ketoconazole or 1-aminobenzotriazole[106] to inhibit CYP enzymes. Both of these compounds will inhibit intestinal and liver CYP enzymes after oral administration. Care must be exercised when using ketoconazole for this purpose because the compound also has effects on transporters, and these effects may make interpretation of results ambiguous. Alternatively, a CYP inducer can be coadministered to determine the effect on the clearance of the compound of interest. This can be accomplished with one of many known inducers of CYP enzymes. For inhibition of esterase enzyme activity, *bis*-[*p*-nitrophenyl] phosphate (BNPP) can be coadministered because this compound does not inhibit cholinesterase activity.[107] Intestinal peptidases can be inhibited by a variety of specific inhibitors[108] such as bestatin for aminopeptidase[109] or aprotinin for chymotrypsin[110] in *in vitro* or *in vivo* experiments, but the studies must be done by examining one peptidase at a time or by using inhibitor "cocktails." Methods for inhibiting phase II enzymes *in vivo* are not as well defined, since in most cases good inhibitors of the enzymes have not been identified. It is possible to deplete co-factor stores by co-administering large amounts of substrate to ask mechanistic questions, although this method is most easily applied to SULT enzymes.[111] Many phase II enzymes are subject to induction, which may be an avenue available for modulating enzyme activities *in vivo*.

6.6.5. Screening Strategies

The approach often taken in candidate optimization is to try to isolate bioavailability problems for a member of a chemotype of interest with a combination of *in vitro* and *in vivo* experiments and then to try to devise rapid techniques to screen for the liability (Figure 6.1). This approach relies on a good deal of up-front work to fully understand the bioavailability limitations and periodic checking of the property to ensure that the screen is providing reliable results (Figure 6.1). Systematic decision trees can be employed to allow clear pathways for evaluation of new compounds (Figure 6.2). Alternatively, screens can be run in parallel so that all information is generated for all compounds (Figure 6.2). This strategy makes workflows simpler and increases efficiency but can lead to information overload and complex decision making.

For metabolic stability screens to be most effective, the screens must be tightly linked to some means of gathering information on metabolite structure. Methods

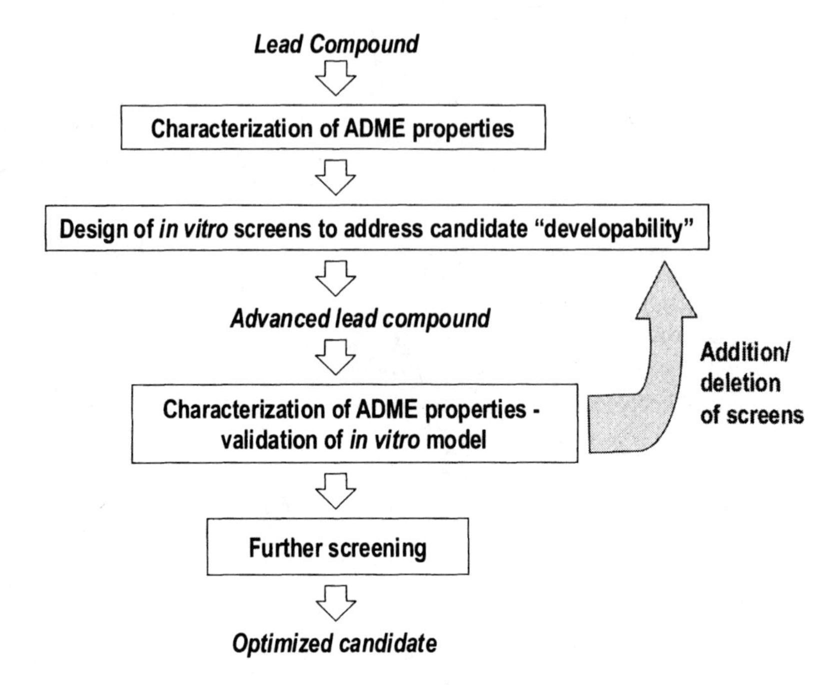

Figure 6.1. Scheme for incorporation of ADME-based "developability" screens during the candidate optimization phase of the drug discovery process.

for rapidly determining metabolite molecular weight and limited structural information have improved dramatically and allow this approach to be routinely employed.[112] The purpose of this approach is to allow the identification of metabolic "soft spots," which can then be altered to produce compounds with improved metabolic stability. The literature on successful structural modification to increase stability has recently been reviewed.[7]

6.6.6. *In Silico* Methods to Study First-Pass Metabolism

Metabolism prediction through *in silico* methods may be possible in the future, although a great deal of technical development is needed before it is a viable alternative.[113] Predictive metabolism programs may allow metabolism scientists and medicinal chemists to obtain metabolic rate or site of metabolism information prior to the first synthesis of compounds. The majority of the methods in this area that have been published to date fall into four categories: (1) expert systems designed to identify probable sites of metabolism based on comparison to database information,[114,115] (2) programs to predict the site of metabolism based on molecular orbital theory,[116] (3) programs to predict the rate of metabolism based on physical chemical attributes of the molecule and quantitative structure–activity relationship (QSAR) development,[117–123] and (4) prediction of the site of metabolism based on knowledge of the CYP active site.[124,125]

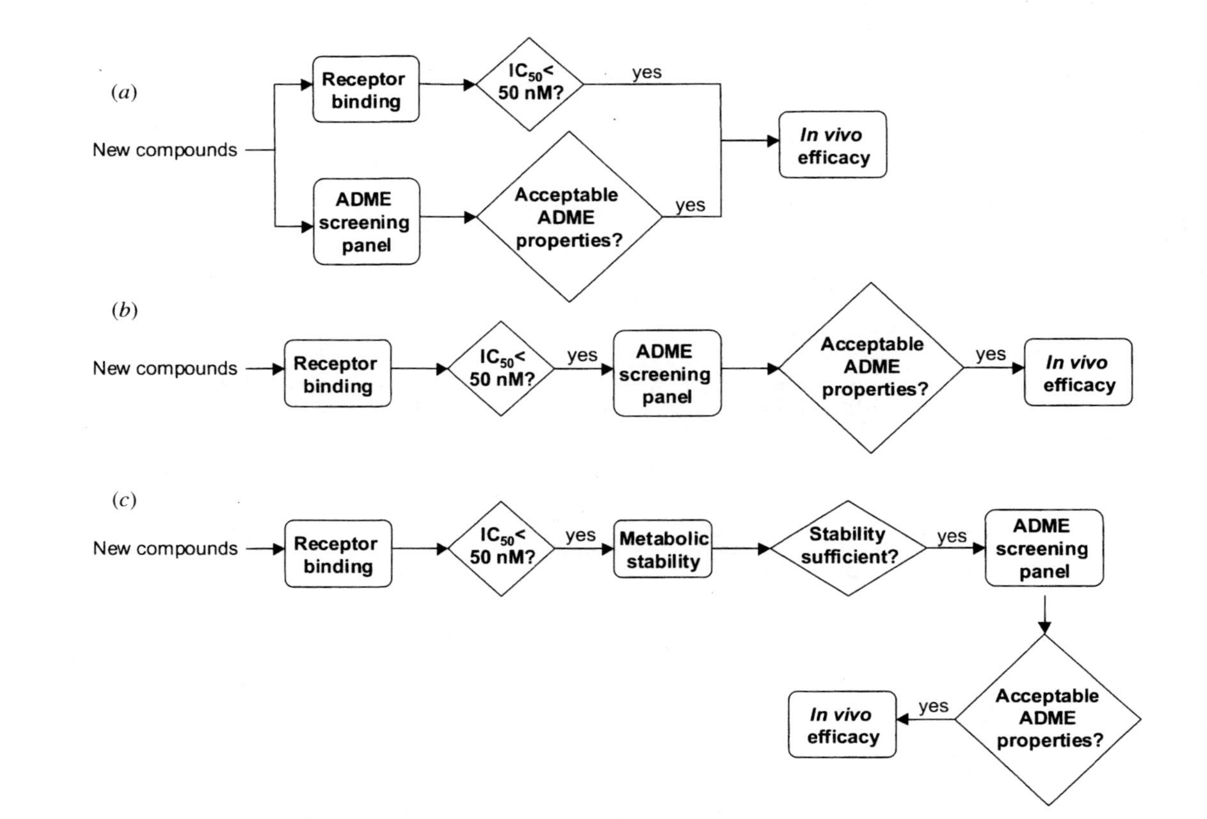

Figure 6.2. Strategies for the design of decision trees for new compound evaluation. ADME-based screens can be incorporated in many places in a decision tree, three examples are illustrated: A, ADME screens are run as a panel in parallel with early activity screens, compounds proceed to further biological screens after meeting set criteria; B, ADME screens are run after a compound meets set potency criteria and compounds proceed to further biological screens after meeting set ADME criteria; C, critical ADME screen (metabolic stability in this case) is run on compounds that meet potency criteria and then full ADME screen panel is run before compounds proceed.

Attempts to calculate the site of metabolism based on molecular orbital theory have met with some success. Currently the state-of-the-art method is to calculate an electron density map of molecules using quantum mechanics and then calculate a steric factor using knowledge of the CYP active site.[116] Taken together these two parameters have successfully predicted the sites of oxidation for limited sets of molecules.[116] This approach may be dramatically enhanced as the crystal structures of more mammalian CYP enzymes become available.[126–128]

The QSAR approach revolves around the calculation of molecular descriptors as a means of calculating the rate at which a molecule will be metabolized.[117–123] This QSAR approach to help define the drug-like properties of a molecule could have a dramatic impact on early drug discovery and allow refinement of early chemical synthesis and library design.

6.7. RATIONALE FOR OPTIMIZATION OF CLEARANCE PROPERTIES OF NEW CHEMICAL ENTITIES

The major reason to optimize the metabolism of a new chemical entity is that the clearance properties can be matched to the indication. For most indications involving chronic oral therapy, that means minimizing clearance through minimization of metabolism (this is not true in all cases; some indications may require a short duration of action and thus rapid metabolism). Other benefits of understanding and reducing metabolism are:

1. decreased clearance, which may translate into a lower overall dose;
2. lower rates of formation and overall amounts of reactive intermediates, which may mediate acute or idiosyncratic toxicities;
3. an increased pharmacokinetic half-life, which will hopefully translate into a longer duration of action, less frequent dosing, and better patient compliance;
4. better understanding of the extrapolation of animal data to humans, making human dose projections more reliable and reducing the risk upon entry into clinical development (this is especially important in cases where prodrugs are being developed);
5. lower risk of drug-drug interactions; even if the compound is still dependent on metabolism for the bulk of its clearance, low-clearance drugs are less susceptible in drug-drug interaction caused by the coadministration of inhibitor;
6. lower risk of drug-food interactions due to the reduced dose;
7. decreased formation of metabolites that may have pharmacological activity against the target or may have significant off-target activity.

6.8. SUMMARY

The past several decades have witnessed an explosion in our knowledge of drug-metabolizing enzymes. It is now possible to fully characterize and predict the metabolic fate of new chemical entities in humans with reasonable certainty. The *in vitro* methods used to do this, i.e., human liver microsomes, expressed enzymes, and cryopreserved or freshly isolated hepatocytes, can be adapted to be run in medium-throughput fashion and allow the metabolic properties of new compounds to be optimized during the discovery phase. Also, medium-throughput screens for bioavailability are now possible with rapid quantitation by LC-MS/MS techniques. These techniques must continue to evolve to allow metabolism scientists to keep pace with the increasing speed of drug discovery. The increasing complexity of the process will place ever greater dependence on information management, and the ability to turn that information into knowledge will be critical. This is especially true when one considers the advances in other fields of ADME-related research, such as transporters, interactions with nuclear hormone receptors, and so on, that need to be characterized during the discovery of a drug. This complexity ensures that the field of drug metabolism will continue to play an increasing role in the endeavor to provide high-quality drug candidates that will become safe, highly efficacious medicines.

REFERENCES

1. Shen, D. D.; Kunze, K. L.; Thummel, K. E. *Adv. Drug Deliv. Rev.* **1997**, *27*, 99–127.
2. Hall, S. D.; Thummel, K. E.; Watkins, P. B.; Lown, K. S.; Benet, L. Z.; Paine, M. F.; Mayo, R. R.; Turgeon, D. K.; Bailey, D. G.; Fontana, R. J.; Wrighton, S. A. *Drug Metab. Dispos.* **1999**, *27*, 161–166.
3. Hellriegel, E. T.; Bjornsson, T. D.; Hauck, W. W. *Clin. Pharmacol. Ther.* **1996**, *60*, 601–607.
4. Smith, D. A.; Jones, B. C.; Walker, D. K. *Med. Res. Rev.* **1996**, *16*, 243–266.
5. Lin, J. H.; Lu, A. Y. *Pharmacol. Rev.* **1997**, *49*, 403–449.
6. White, R. E. *Annu. Rev. Pharmacol. Toxicol.* **2000**, *40*, 133–157.
7. Thompson, T. N. *Med. Res. Rev.* **2001**, *21*, 412–449.
8. Smith, D.; Schmid, E.; Jones, B. *Clin. Pharmacokinet.* **2002**, *41*, 1005–1019.
9. McCracken, N. W.; Blain, P. G.; Williams, F. M. *Biochem. Pharmacol.* **1993**, *45*, 31–36.
10. McCracken, N. W.; Blain, P. G.; Williams, F. M. *Biochem. Pharmacol.* **1993**, *46*, 1125–1129.
11. Satoh, T.; Taylor, P.; Bosron, W. F.; Sanghani, S. P.; Hosokawa, M.; La Du, B. N. *Drug Metab. Dispos.* **2002**, *30*, 488–493.
12. Satoh, T.; Hosokawa, M. *Annu. Rev. Pharmacol. Toxicol.*, **1998**, *38*, 257–288.
13. Buchwald, P.; Bodor, N. *Pharmazie* **2002**, 57, 87–93.
14. Xie, M.; Yang, D.; Liu, L.; Xue, B.; Yan, B. *Drug Metab. Dispos.* **2002**, *30*, 541–547.
15. Satoh, T.; Hosokawa, M.; Atsumi, R.; Suzuki, W.; Hakusui, H.; Nagai, E. *Biol. Pharm. Bull.* **1994**, *17*, 662–664.
16. Rivory, L. P.; Bowles, M. R.; Robert, J.; Pond, S. M. *Biochem. Pharmacol.* **1996**, *52*, 1103–1111.
17. Ma, M. K.; McLeod, H. L. *Curr. Med. Chem.* **2003**, *10*, 41–49.

18. Toffoli, G.; Cecchin, E.; Corona, G.; Boiocchi, M. *Curr. Med. Chem. Anti-Cancer Agents* **2003**, *3*, 225–237.
19. La Du, B. N.; Billecke, S.; Hsu, C.; Haley, R. W.; Broomfield, C. A. *Drug Metab. Dispos.* **2001**, *29*, 566–569.
20. Billecke, S.; Draganov, D.; Counsell, R.; Stetson, P.; Watson, C.; Hsu, C.; La Du, B. N. *Drug Metab. Dispos.* **2000**, *28*, 1335–1342.
21. Danielson, P. B. *Curr. Drug Metab.* **2002**, *3*, 561–597.
22. Nebert, D. W.; Russell, D. W. *Lancet* **2002,** *360*, 1155–1162.
23. Pelkonen, O. *Drug Metab. Rev.* **2002**, *34*, 37–46.
24. Rendic, S. *Drug Metab. Rev.* **2002**, *34*, 83–448.
25. Shimada, T.; Yamazaki, H.; Mimura, M.; Inui, Y.; Guengerich, F. P. *J. Pharmacol. Exp. Ther.* **1994**, *270*, 414–423.
26. Gibbs, M.; Hosea, N. *Clin. Pharmacokinet.* **2003**, *42*, 969–984.
27. Lewis, D. F. *Curr. Med. Chem.* **2003**, *10*, 1955–1972.
28. Guengerich, F. P. *Chem. Res. Toxicol.* **2001**, *14*, 611–650.
29. Newcomb, M.; Hollenberg, P. F.; Coon, M. J. *Arch. Biochem. Biophys.* **2003**, *409*, 72–79.
30. Makris, T. M.; Davydov, R.; Denisov, I. G.; Hoffman, B. M.; Sligar, S. G. *Drug Metab. Rev.* **2002**, *34*, 691–708.
31. Guengerich, F. P.; Cai, H.; Johnson, W. W.; Parikh, A. *Adv. Exp. Med. Biol.* **2001**, *500*, 639–650.
32. Ju, C.; Uetrecht, J. P. *Curr. Drug Metab.* **2002**, *3*, 367–377.
33. Uetrecht, J. *Drug Metab. Rev.* **2002**, *34*, 651–665.
34. Uetrecht, J. *Drug Discov. Today* **2003**, *8*, 832–837.
35. Guengerich, F. P. *Arch. Biochem. Biophys.* **2003**, *409*, 59–71.
36. Lamba, J. K.; Lin, Y. S.; Schuetz, E. G.; Thummel, K. E. *Adv. Drug Deliv. Rev.* **2002**, *54*, 1271–1294.
37. Dorne, J. L.; Walton, K.; Slob, W.; Renwick, A. G. *Food Chem. Toxicol.* **2002**, *40*, 1633–1656.
38. Paine, M. F.; Khalighi, M.; Fisher, J. M.; Shen, D. D.; Kunze, K. L.; Marsh, C. L.; Perkins, J. D.; Thummel, K. E. *J. Pharmacol. Exp. Ther.* **1997**, *283*, 1552–1562.
39. Obach, R. S.; Zhang, Q. Y.; Dunbar, D.; Kaminsky, L. S. *Drug Metab. Dispos.* **2001**, *29*, 347–352.
40. Ding, X.; Kaminsky, L. S. *Annu. Rev. Pharmacol. Toxicol.* **2003**, *43*, 149–173.
41. Watkins, P. B.; Turgeon, D. K.; Saenger, P.; Lown, K. S.; Kolars, J. C.; Hamilton, T.; Fishman, K.; Guzelian, P. S.; Voorhees, J. J. *Clin. Pharmacol. Ther.* **1992**, *52*, 265–273.
42. Paine, M. F.; Shen, D. D.; Kunze, K. L.; Perkins, J. D.; Marsh, C. L.; McVicar, J. P.; Barr, D. M.; Gillies, B. S.; Thummel, K. E. *Clin. Pharmacol. Ther.* **1996**, *60*, 14–24.
43. Yamamoto, T.; Hagima, N.; Nakamura, M.; Kohno, Y.; Nagata, K.; Yamazoe, Y. *Drug Metab. Dispos.* **2003**, *31*, 60–66.
44. Suzuki, H.; Sugiyama, Y. *Eur. J. Pharm. Sci.* **2000**, *12*, 3–12.
45. Wacher, V. J.; Salphati, L.; Benet, L. Z. *Adv. Drug Deliv. Rev.* **2001**, *46*, 89–102.
46. Patel, J.; Mitra, A. K. *Pharmacogenomics* **2001**, *2*, 401–415.

47. Cummins, C. L.; Salphati, L.; Reid, M. J.; Benet, L. Z. *J. Pharmacol. Exp. Ther.* **2003**, *305*, 306–314.
48. Bailey, D. G.; Dresser, G. K.; Bend, J. R. *Clin. Pharmacol. Ther.* **2003**, *73*, 529–537.
49. Cashman, J. R. *Drug Metab. Rev.* **2002**, *34*, 513–521.
50. Ziegler, D. M. *Drug Metab. Rev.* **2002**, *34*, 503–511.
51. Cashman, J. R.; Zhang, J. *Drug Metab. Dispos.* **2002**, *30*, 1043–1052.
52. Krueger, S. K.; Williams, D. E.; Yueh, M. F.; Martin, S. R.; Hines, R. N.; Raucy, J. L.; Dolphin, C. T.; Shephard, E. A.; Phillips, I. R. *Drug Metab. Rev.* **2002**, *34*, 523–532.
53. Cashman, J. R.; Camp, K.; Fakharzadeh, S. S.; Fennessey, P. V.; Hines, R. N.; Mamer, O. A.; Mitchell, S. C.; Nguyen, G. P.; Schlenk, D.; Smith, R. L.; Tjoa, S. S.; Williams, D. E.; Yannicelli, S. *Curr. Drug. Metab.* **2003,** *4*, 151–170.
54. Hernandez, D.; Addou, S.; Lee, D.; Orengo, C.; Shephard, E. A.; Phillips I. R. *Hum. Mutat.* **2003**, *22*, 209–213.
55. King, C. D.; Rios, G. R.; Green, M. D.; Tephly, T. R. *Curr. Drug Metab.* **2000**, *1*, 143–161.
56. Fisher, M. B.; Paine, M. F.; Strelevitz, T. J.; Wrighton, S. A. *Drug Metab. Rev.* **2001**, *33*, 273–297.
57. Burchell, B. *Am. J. Pharmacogenomics.* **2003**, *3*, 37–52.
58. Tukey, R. H.; Strassburg, C. P.; Mackenzie, P. I. *Mol. Pharmacol.* **2002**, *62*, 446–450.
59. Gagne, J. F.; Montminy, V.; Belanger, P.; Journault, K.; Gaucher, G.; Guillemette, C. *Mol. Pharmacol.* **2002**, *62*, 608–617.
60. Court, M. H.; Duan, S. X.; von Moltke, L. L.; Greenblatt, D. J.; Patten, C. J.; Miners, J. O.; Mackenzie, P. I. *J. Pharmacol. Exp. Ther.* **2001**, *299*, 998–1006.
61. Miners, J. O.; McKinnon, R. A.; Mackenzie, P. I. *Toxicology* **2002**, *181–182*, 453–456.
62. Boelsterli, U. A. *Curr. Drug Metab.* **2002**, *3*, 439–450.
63. Shipkova, M.; Armstrong, V. W.; Oellerich, M.; Wieland, E. *Ther. Drug Monit.* **2003**, *25*, 1–16.
64. Bailey, M. J.; Dickinson, R. G. *Chem. Biol. Interact.* **2003**, *145*, 117–137.
65. Veroli, P.; O'Kelly, B.; Bertrand, F.; Trouvin, J. H.; Farinotti, R.; Ecoffey, C. *Br. J. Anaesth.* **1992**, *68*, 183–186.
66. Negishi, M.; Pedersen, L. G.; Petrotchenko, E.; Shevtsov, S.; Gorokhov, A.; Kakuta, Y.; Pedersen, L. C. *Arch. Biochem. Biophys.* **2001**, *390*, 149–157.
67. Duffel, M. W.; Marshal, A. D.; McPhie, P.; Sharma, V.; Jakoby, W. B. Enzymatic aspects of the phenol (aryl) sulfotransferases. *Drug Metab. Rev.* **2001**, *33*, 369–395.
68. Coughtrie, M. W. *Pharmacogenom. J.* **2002**, *2*, 297–308.
69. Chen, G.; Zhang, D.; Jing, N.; Yin, S.; Falany, C. N.; Radominska-Pandya, A. *Toxicol. Appl. Pharmacol.* **2003**, *187*, 186–197.
70. Nagata, K.; Yoshinari, K.; Ozawa, S.; Yamazoe, Y. *Mutat. Res.* **1997**, *376*, 267–272.
71. King, R. S.; Teitel, C. H.; Kadlubar, F. F. *Carcinogenesis* **2000**, *21*, 1347–1354.
72. Bai, J. P.; Amidon, G. L. *Pharm. Res.* **1992**, *9*, 969–978.
73. Jornvall, H.; Hoog, J. O.; Persson, B.; Pares, X. *Pharmacology* **2000**, *61*, 184–191.
74. Hein, D. W. *Mutat Res.* **2002**, *506–507*, 65–77.
75. Meisel, P. *Pharmacogenomics* **2002**, *3*, 349–366.

76. Okudaira, N.; Tatebayashi, T.; Speirs, G. C.; Komiya, I.; Sugiyama, Y. *J. Pharmacol. Exp. Ther.* **2000**, *294*, 580–587.
77. Okudaira, N.; Komiya, I.; Sugiyama, Y. *J. Pharmacol. Exp. Ther.* **2000**, *295*, 717–723.
78. Humphreys, W. G.; Obermeier, M. T.; Chong, S.; Kimball, S. D.; Das, J.; Chen, P.; Moquin, R.; Han, W. C.; Gedamke, R.; White, R. E.; Morrison, R. A. *Xenobiotica* **2003**, *33*, 93–106.
79. Kato, Y.; Suzuki, H.; Sugiyama, Y. *Toxicology* **2002**, *181–182*, 287–290.
80. Masuda, M.; I'izuka, Y.; Yamazaki, M.; Nishigaki, R.; Kato, Y.; Ni'inuma, K.; Suzuki, H.; Sugiyama, Y. *Cancer Res.* **1997**, *57*, 3506–3510.
81. Humphreys, W. G.; Obermeier, M. T.; Barrish, J. C.; Chong, S.; Marino, A. M.; Murugesan, N.; Wang-Iverson, D.; Morrison, R. A. *Xenobiotica*, in press.
82. Houston, J. B.; Carlile, D. J. *Drug Metab. Rev.* **1997**, *29*, 891–922.
83. Ito, K.; Iwatsubo, T.; Kanamitsu, S.; Nakajima, Y.; Sugiyama, Y. *Annu. Rev. Pharmacol. Toxicol.* **1998**, *38*, 461–499.
84. Lave, T.; Coassolo, P.; Reigner, B. *Clin. Pharmacokinet.* **1999**, *36*, 211–231.
85. Lin, J. H. *Drug Metab. Dispos.* **1998**, *26*, 1202–1212.
86. Obach, R. S. *Curr. Opin. Drug Discov. Dev.* **2001**, *4*, 36–44.
87. Obach, R. S.; Baxter, J. G.; Liston, T. E.; Silber, B. M.; Jones, B. C.; MacIntyre, F.; Rance, D. J.; Wastall, P. *J. Pharmacol. Exp. Ther.* **1997**, *283*, 46–58.
88. Obach R. S. *Drug Metab. Dispos.* **1997**, *25*, 1359–1369.
89. Obach, R. S. *Drug Metab. Dispos.* **1999**, *27*, 1350–1359.
90. Austin, R. P.; Barton, P.; Cockroft, S. L.; Wenlock, M. C.; Riley, R. J. *Drug Metab. Dispos.* **2002**, *30*, 1497–1503.
91. Margolis, J. M.; Obach, R. S. *Drug Metab. Dispos.* **2003**, *31*, 606–611.
92. Tran, T. H.; Von Moltke, L. L.; Venkatakrishnan, K.; Granda, B. W.; Gibbs, M. A.; Obach, R. S.; Harmatz, J. S.; Greenblatt, D. J. *Drug Metab. Dispos.* **2002**, *30*, 1441–1445.
93. Rane, A.; Wilkinson, G. R.; Shand, D. G. *J. Pharmacol. Exp. Ther.* **1977**, *200*, 420–424.
94. Houston, J. B.; Kenworthy, K. E. *Drug Metab. Dispos.* **2000**, *28*, 246–254.
95. Soars, M. G.; Burchell, B.; Riley, R. J. *J. Pharmacol. Exp. Ther.* **2002**, *301*, 382–390.
96. Lin, J. H.; Wong, B. K. *Curr. Drug Metab.* **2002**, *3*, 623–646.
97. Fisher, M. B.; Yoon, K.; Vaughn, M. L.; Strelevitz, T. J.; Foti, R. S. *Drug Metab. Dispos.* **2002**, *30*, 1087–1093.
98. Lau, Y. Y.; Sapidou, E.; Cui, X.; White, R. E.; Cheng, K. C. *Drug Metab. Dispos.* **2002**, *30*, 1446–1454.
99. Fisher, M. B.; Campanale, K.; Ackermann, B. L.; VandenBranden, M.; Wrighton, S. A. *Drug Metab. Dispos.* **2000**, *28*, 560–566.
100. Zuegge, J.; Schneider, G.; Coassolo, P.; Lave, T. *Clin. Pharmacokinet.* **2001**, *40*, 553–563.
101. Mahmood, I. *Am. J. Ther.* **2002**, *9*, 35–42.
102. Mahmood, I. *Drug Metab. Drug. Interact.* **2002**, *19*, 49–64.
103. Hu, T. M.; Hayton, W. L. *AAPS PharmSci.* **2001**, *3*, E29.
104. Lave, T.; Dupin, S.; Schmitt, C.; Chou, R. C.; Jaeck, D.; Coassolo, P. *J. Pharm. Sci.* **1997**, *86*, 584–590.

105. Lenz, E. M.; Wilson, I. D.; Wright, B.; Partridge, E. A.; Rodgers, C. T.; Haycock, P. R.; Lindon, J. C.; Nicholson, J. K. *J. Pharm. Biomed. Anal.* **2002**, *28*, 31–43.
106. Balani, S. K.; Zhu, T.; Yang, T. J.; Liu, Z.; He, B.; Lee, F. W. *Drug Metab. Dispos.* **2002**, *30*, 1059–1062.
107. Buch, H.; Buzello, W.; Heymann, E.; Krisch K. *Biochem. Pharmacol.* **1969**, *18*, 801–811.
108. Aungst, B. J. *J. Pharm. Sci.* **1993**, *82*, 979–987.
109. Agu, R. U.; Vu Dang, H.; Jorissen, M.; Willems, T.; Kinget, R.; Verbeke, N. *Int. J. Pharm.* **2002**, *237*, 179–191.
110. Tozaki, H.; Emi, Y.; Horisaka, E.; Fujita, T.; Yamamoto, A.; Muranishi, S. *J. Pharm. Pharmacol.* **1997**, *49*, 164–168.
111. Kim, H. J.; Cho, J. H.; Klaassen, C. D. *J. Pharmacol. Exp. Ther.* **1995**, *275*, 654–658.
112. Watt, A. P.; Mortishire-Smith, R. J.; Gerhard, U.; Thomas, S. R. *Curr. Opin. Drug Discov. Dev.* **2003**, *6*, 57–65.
113. van de Waterbeemd, H.; Gifford, E. *Nat. Rev. Drug Discov.* **2003**, *2*, 192–1204.
114. Erhardt, P. W. *Drug Metabolism: Databases and High-Throughput Testing During Drug Design and Development.* IUPAC and Blackwell Science: Boston, **1999**.
115. Hawkins, D. R. *Drug Discov. Today* **1999**, *4*, 466–471.
116. Jones, J. P.; Mysinger, M.; Korzekwa, K. R. *Drug Metab. Dispos.* **2002**, *30*, 7–12.
117. Andrews, C. W.; Bennett, L.; Yu, L. X. *Pharm. Res.* **2000**, *17*, 639–644.
118. Ekins, S.; Obach, R. S. *J. Pharmacol. Exp. Ther.* **2000**, *295*, 463–473.
119. Yoshida, F.; Topliss, J. G. *J. Med. Chem.* **2000**, *43*, 2575–2585.
120. Ekins, S.; de Groot, M. J.; Jones, J. P. *Drug Metab. Dispos.* **2001**, *29*, 936–944.
121. Lewis, D. F.; Modi, S.; Dickins, M. *Drug Metab. Rev.* **2002**, *34*, 69–82.
122. Long, A.; Walker, J. D. *Environ. Toxicol. Chem.* **2003**, *22*, 1894–1899.
123. Taskinen, J.; Ethell, B. T.; Pihlavisto, P.; Hood, A. M.; Burchell, B.; Coughtrie, M. W. *Drug Metab. Dispos.* **2003**, *31*, 1187–1197.
124. Lewis, D. F. *Drug Metab. Rev.* **2002**, *34*, 55–67.
125. Lewis, D. F. *Pharmacogenomics* **2003**, *4*, 387–395.
126. Williams, P. A.; Cosme, J.; Sridhar, V.; Johnson, E. F., McRee, D. E. *J. Inorg. Biochem.* **2000**, *81*, 183–190.
127. Williams, P. A.; Cosme, J.; Ward, A.; Angove, H. C.; Vinkovic, D. M.; Jhoti, H. *Nature* **2003**, *424*, 464–468.
128. Tennant, M.; McRee, D. E. *Curr. Opin. Drug Discov. Dev.* **2001**, *4*, 671–677.

7

CELL CULTURE MODELS FOR DRUG TRANSPORT STUDIES

D. Nedra Karunaratne, Peter S. Silverstein, Veena Vasandani, Amber M. Young, Erik Rytting, Bradley Yops, and Kenneth L. Audus

Department of Pharmaceutical Chemistry, School of Pharmacy
The University of Kansas, 2095 Constant Avenue, Lawrence, KS 66047

Drug Delivery: Principles and Applications Edited by Binghe Wang, Teruna Siahaan, and Richard Soltero
ISBN 0-471-47489-4

7.1. INTRODUCTION

In pharmaceutical research there is a need for cell-based assays that have some predictive capability with regard to tissue permeability properties of lead compounds. These assays are generally used after high-throughput screening procedures have identified compounds with the desired pharmacological properties. Cell-based assays can be used to identify those leads that have the desired properties with regard to tissue permeability, and then animal models may be used as a final screen. This approach is not only more cost-effective but also minimizes the use of animal resources.

The purpose of this chapter is to present overviews of a selection of the major endothelial and epithelial barriers to drug delivery for which there are either primary culture or cell line systems that recapitulate the characteristics of the *in vivo* barrier. Our objective is to define some general characteristics of cell culture models and highlight the more commonly applied primary cell cultures and cell lines in use today. Specifically, we focus on cell culture models for the intestinal epithelium, blood-brain barrier, pulmonary and nasal epithelium, ocular epithelium, placental barrier, and renal epithelium. Renal epithelium was included here primarily because some cell lines derived from this tissue [e.g., Madin-Darby canine kidney cells (MDCK)] are often used as surrogates for other barriers by pharmaceutical scientists. We have arbitrarily chosen to exclude the skin and liver from the scope of this overview. However, it should be noted that hepatocyte cell culture models, for example, are becoming more widely available and have been the subject of recent reviews.[1,2]

7.2. GENERAL CONSIDERATIONS

Technologies for isolating and cultivating endothelial and epithelial cells from many of the tissue interfaces influencing drug distribution have rapidly evolved in the past 20 years. As a consequence, the pharmaceutical scientist now has an assortment of cell culture systems to characterize fundamental drug delivery processes at the biochemical and molecular levels in major epithelial and endothelial barriers.

Cell culture models are described as either primary cultures or cell lines. Primary cultures are generated by the growth of cells that migrate out from a fragment of tissue or by growth of cells isolated by either enzymatic or mechanical dispersal of tissue.[3] A primary culture may be subcultured or passaged by harvesting the cells

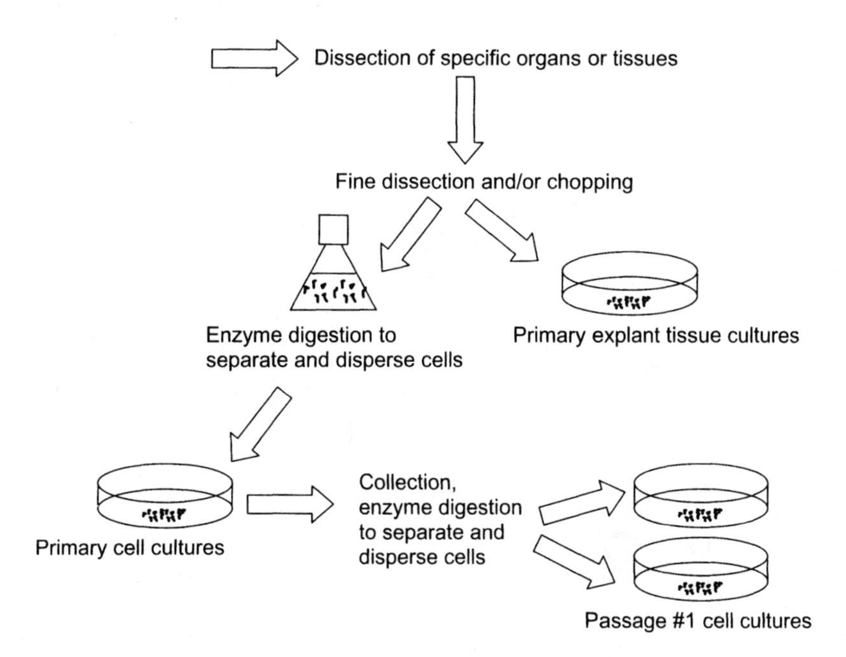

Figure 7.1. General methods for establishing primary explant tissue, primary cell, and passaged cell culture models from animal tissues.

and redistributing (i.e., splitting) them on new growth surfaces. Once a primary culture is passaged, it is known as a "cell line" (Figure 7.1). Generally, subsequent passaging of the cells selects for cells with rapid growth, and by the third passage, the cultures are more stable. Most cell lines may be stable for a limited number of passages. A continuous cell line that may be grown for essentially unlimited passages may occur through *in vitro* transformation and may arise spontaneously or through chemical or viral induction. Cell cultures derived from most normal tissues rarely give rise to continuous cell lines spontaneously.[3] Researchers prefer to work with continuous cell lines because of the convenience and probability of working with a homogeneous cell population versus primary cultures that have to be continuously generated and can vary significantly due to inconsistencies among different preparations.

The appropriateness of any cell culture model used to study drug delivery processes, whether a primary or a cell line, should be based on certain basic criteria. These include but are not limited to the presence of a restrictive paracellular pathway that allows effective characterization of transcellular permeability, the presence of physiologically realistic cell architecture reflective of the tissue barrier of interest, the expression of functional transporter mechanisms representative of the tissue barrier of interest, and the ease, convenience, and reproducibility of the culture methods.[4] However, pharmaceutical scientists must realize that cell culture models in use today are not ideal with respect to these criteria, and there will be advantages and limitations with any choice.

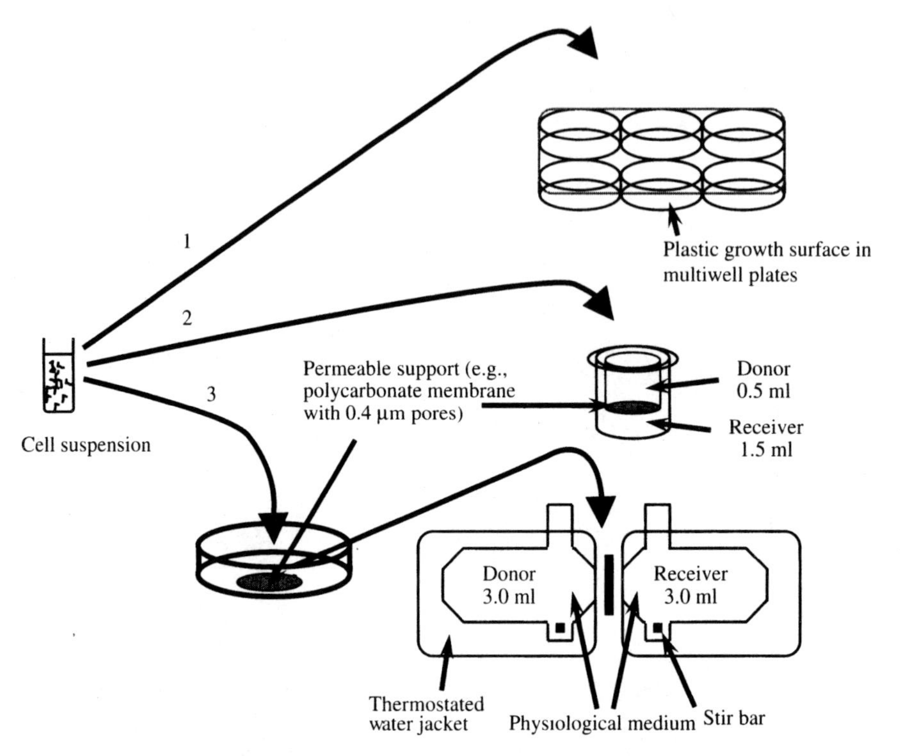

Figure 7.2. General uptake and transport configurations for cell culture models. Depending on the cell type, cell suspensions may be (1) seeded into and grown on plastic surfaces for uptake studies, (2) seeded and grown on permeable supports in Transwell™-type systems, or (3) seeded and grown on permeable supports that can be placed in a side-by-side™ diffusion apparatus.

Part of the selection of any cell culture model will also include consideration of an appropriate transport configuration (Figure 7.2) and the quantitative processes necessary to appropriately analyze transport data.[5] Options generally available to the pharmaceutical scientist include growth of cells on multiwell plastic dishes for assessment of cellular uptake; this provides for analysis of only one-half of the transcellular transport process. This format is favored for high-throughput types of assays where quick assessment of cellular uptake, retention, or efflux of materials can be quickly assayed with the appropriate markers. More likely, one will consider either the Transwell or side-by-side types of configurations, which provide the advantage of allowing evaluation of transcellular transport and access to both apical and basolateral surfaces of cells grown on permeable supports. Criteria for selection of an appropriate transport model include the solute's properties, sufficient stirring conditions, filter (permeable support) properties, assay conditions, matrix (growth surface treatments), assay medium properties, and the cell type chosen. Further elaboration on each of these points and the quantitative evaluation of transport processes are beyond the scope of this short review. Readers are directed to the extensive review by Ho et al.[5] for further details.

7.3. INTESTINAL EPITHELIUM

7.3.1. The Intestinal Epithelial Barrier

The general desire for orally administered drugs that are convenient, affordable, and highly bioavailable dictates that the basic characteristics of the intestinal barrier are fully elucidated. Many promising drug candidates are developed at great expense, only to be removed from consideration due to poor oral bioavailability; this is due, in large part, to poor intestinal absorption. Implementation of an accurate model of this barrier at the earliest possible stage of drug design and development will provide valuable information and result in significant economic advantage.

A monolayer of multiple cell types comprises the intestinal epithelium. Enterocytes, the absorptive cells, are the main focus in drug delivery issues. In addition to enterocytes, there are undifferentiated crypt cells, mucus-secreting goblet cells, and M cells, which sample the gut lumen for the lymphatic system.[6] However, it is the columnar villar enterocyte that forms the primary barrier to absorption. An extensive network of microvilli on the apical surface of villar enterocytes provides a vast amount of surface area for absorption of nutrients, pharmaceuticals, and xenobiotics.[6,7]

Drugs can cross the intestinal epithelial barrier in a number of ways. They may permeate either through the cell (transcellular) or between adjacent cells (paracellular). Enterocytes have tight intercellular junctions that restrict paracellular transport to small hydrophilic molecules.[7] These cells possess active and facilitative transporters for nutrients, as well as an array of efflux transporters [e.g., P-glycoprotein (P-gp) and related transporters] and enzymes (e.g., cytochrome P450 type 3A4) that restrict transcellular absorption. Transcytotic transport of macromolecules is possible, but compounds are often destroyed in lysosomes. With the exception of M-cells, transcytosis is not considered a major mechanism of the transcellular pathway for absorption of macromolecules across gastrointestinal epithelium.[6]

7.3.2. Intestinal Epithelial Cell Culture Models

Many *in vivo* and *in vitro* techniques exist to model the intestinal barrier, but recent trends are toward an inexpensive *in vitro* model that accurately represents human oral absorption of pharmaceuticals.[6,8] Multiple cell culture models have been developed, each with its own benefits and drawbacks.

In order to effectively assess drug transport, a cell line must form polarized monolayers of differentiated enterocytes. In addition, the cells must retain the morphological and biochemical characteristics of the human intestine.[8,9] Primary cultures of human enterocytes have low viability and do not differentiate or form monolayers.[10] Immortalized cell lines have been developed with mixed results.[11] The Caco-2 cell line, derived from a human colon carcinoma, is the cell culture model most extensively utilized for studies of oral bioavailability. Isolated in 1974, it was not employed as a model of oral drug delivery until the late 1980s.[12] Caco-2 cells are easily grown in tissue culture, and they form differentiated, polarized monolayers on polycarbonate membranes. The monolayer formed differs from the human intestinal barrier in several ways. Since goblet cells are absent from culture, there is no mucus layer on the apical surface. Intestinal mucosa is considered an additional barrier to drug delivery, so Caco-2 permeability studies may yield artificially high values for certain compounds.[9] Biochemical and morphological characteristics are similar to those of enterocytes but are not identical.[11–17]

Caco-2 cells must be grown for 20 days to form a fully polarized monolayer suitable for drug transport studies. Compared to other cells lines such as MDCK,[18,19] which can be established and used for studies in less than a week, Caco-2 cells are labor-intensive and expensive to maintain. For that reason, MDCK cells have begun to be used as a surrogate for Caco-2 cells in screening compounds that have enhanced permeation by simple passive diffusion. There have been attempts to reduce the time necessary to form competent Caco-2 monolayers to 3–5 days, but these efforts have met with limited success.[20] At this time, Caco-2 cells are still the most convenient model that provides reasonable *in vivo–in vitro* correlations. Caco-2 is particularly helpful in predicting the apparent permeabilities of passively absorbed drugs. Carrier-mediated processes will differ among cell lines due to differences in expression levels of transporters. Even if permeability values differ from the actual permeability in the intestine, Caco-2 is a valuable tool for predicting the rank order of permeability for a class of compounds. The differences in tight junctions between Caco-2 and enterocytes *in vivo* make prediction of paracellular permeability difficult.[1]

Caco-2 also expresses efflux transporters, such as P-glycoprotein, and CYP450 enzymes that limit oral bioavailability and cause potential drug-drug interactions. Such interactions are not predictable based on current models of transporter-substrate interactions, so Caco-2 cells represent a valuable tool with which to obtain this information.[20]

Caco-2 cells are currently used at all levels of pharmaceutical research and development. Automation technologies allow tissue culture labs to easily maintain a large number of Caco-2 cells as well as to perform numerous permeability studies without the introduction of many common human errors. Combinatorial chemistry provides vast arrays of compounds, and Caco-2 assays can be used to assess potential permeability and metabolic issues before much money is invested in the candidate.[6]

These cells are also helpful at the formulation stages. Very little is known about the effect of excipients on intestinal permeability; therefore, it is advantageous to measure the permeation of the active ingredients in the presence of excipients,[20] as revealed in recent studies.[21]

It is apparent that no single method is sufficient for studying or predicting drug absorption across the intestinal epithelium.[6] Consequently, simple algorithms based on physicochemical values are only marginally successful in predicting permeability, but the use of Caco-2 data increases predictive capabilities.[20] Thus, Caco-2 permeability values are also helpful in developing *in silico* models and, in combination, are more useful in predicting drug absorption.[6]

7.4. THE BLOOD-BRAIN BARRIER (BBB)

7.4.1. The Blood-Brain Endothelial Barrier

The BBB controls the exchange of nutrients, xenobiotics, and drugs between the brain and the systemic circulation. The BBB is a significant obstacle to achieving optimal concentrations of desirable therapeutics in the central nervous system, including agents for the treatment of stroke, cancer, Alzheimer's disease, and human immunodeficiency virus-1 (HIV-1) infection.[22]

The BBB is localized to the cerebral microvascular endothelial cells and is composed of at least four major components. The first component is the endothelial cells, which differ from their counterparts in peripheral organs in that those in the brain have few pinocytic vesicles and exhibit limited transcytotic activity. Tight junctions effectively seal the spaces between the endothelial cells and restrict paracellular movement of molecules. Thus, the tight junctions form another component of the BBB.[23] These intercellular junctions are considered to be dynamic, and permeability can be regulated by several mechanisms.[24] Another component of the BBB, and probably the least understood, is the metabolic activity present in the endothelial cell. The presence in endothelial cells of some of the members of the cytochrome P450 family of drug-metabolizing enzymes could restrict access of pharmaceutical agents to the central nervous system (CNS).[25]

Numerous transmembrane transporters in the endothelial cells are part of the fourth component of the BBB, and are probably the component of greatest interest to researchers in the pharmaceutical sciences. Generally facilitative in nature, transporters at the BBB are involved in the uptake of substances (such as glucose, biotin, amino acids, and various other nutrients and xenobiotics) from the extracellular environment.[26] Their importance is emphasized by studies targeting a select group of these transporters to deliver agents that would not normally penetrate the BBB to the CNS; this involves a strategy of conjugation of the agents to natural substrates.[22] Other transporters are involved in the active efflux of substances from the endothelial cells. The best-characterized efflux transporter present in the BBB is P-gp, an efflux transporter that was first identified because of its association with multidrug-resistant cancer.[27] P-gp is located on the circulation (blood) side of the endothelial cell. It has a very broad substrate specificity and has been shown to be responsible for the exclusion of many compounds from the brain, including HIV-1 protease inhibitors, vincristine, vinblastine, and taxol.[27,28] Members of the multidrug resistance–associated protein (MRP) family of transporters are also present in brain microvascular endothelial cells. The mRNAs from MRP 1, 4, 5, and 6 have been identified in bovine brain microvascular endothelial cells (BBMECs), but their functional activities in this tissue remain to be confirmed.[29]

7.4.2. BBB Cell Culture Models

Currently, there are two primary cell cultures that are used in both academic and industrial settings: BBMECs and porcine brain microvascular endothelial cells (PBMECs).[4] Over the past two decades, BBMECs have become a very widely used *in vitro* cell culture model of the BBB. These cells form monolayers with tight junctions and retain expression of many transporters and enzymes typical of the BBB.[4] Large quantities of the cells can be isolated from beef brains using material obtained from freshly sacrificed animals. Brain microvessel-enriched gray matter serves as a source of cells that are isolated by enzymatic digestion and purified by centrifugation. The cells can be stored at −80°C for several weeks; for longer periods they should be stored in liquid nitrogen. For uptake studies the cells can be seeded on tissue culture plates, whereas for transport studies the cells can be seeded on polycarbonate membranes. The cells isolated using this procedure form monolayers with relatively tight junctions.[4,30] Detailed protocols for these procedures have been provided by Audus et al.[30]

In primary culture, BBMECs are not exposed to the influence of other cells that may impact their development. In this regard, glial cells and astrocytes are thought to be of particular importance.[24,31] Two basic strategies have evolved to try to improve the faithfulness of BBMECs in primary culture to the BBB: growth of BBMECs in astrocyte-conditioned media and growth of BBMECs in proximity to astrocytes in such a way that soluble factors secreted by the astrocytes can influence the BBMECs. One way in which this can be accomplished is by growth of the BBMECs in a Transwell, and growth of the astrocytes in the bottom of the well into which the Transwell is placed. Most of the changes reported by those using astrocyte coculture, or astrocyte-conditioned media, relate to a decrease in paracellular permeability, as demonstrated by increases in transepithelial electrical resistance (TEER) or decreases in sucrose permeability.[32]

The presence of bovine spongiform encephalopathy has limited the use of cow brains in Europe. However, Galla and coworkers have pioneered the use of PBMECs.[33] Although these cells are not as well characterized as BBMECs, they do appear to have characteristics similar to those seen in BBMECs.[4] Relevant transporters, such as P-gp and at least one MRP, are expressed in PBMECs.[34,35] The procedures for isolating PBMECs are similar to those for isolating BBMECs.[36,37] However, with PBMECs, cells are passaged once prior to use, and there is a report that addition of hydrocortisone to the medium results in a significant decrease in paracellular permeability.[38] Although the literature on PBMECs is not as extensive as that on BBMECs, it is clear that PBMECs are suitable for pharmaceutical research.

Although both BBMECs and PBMECs are widely accepted as suitable *in vitro* models of the BBB, there are some limitations to the utility of these models. Over the years, the *in vitro* cell culture models have proven to be fairly reliable predictors of BBB permeability. However, as with any *in vitro* model, confirmation of results using an animal model is recommended. One should also be aware that substrate-transporter interactions have been shown to be altered in allelic variants of the transporter P-gp.[39] Thus, results need to be interpreted with the caveat that the transporters whose activity was assayed in the experiment may be qualitatively or quantitatively different from those in the human brain. Regulation of the transporters, along with any receptors that may be affected, may also be different in other species. A prime example of this is the pregnane-X receptor, which regulates the expression of P-gp and which has a very species-specific activation profile.[40]

BBMECs and PBMECs have been the models of choice for *in vitro* cell culture models of the BBB. However, it can be assumed that these models will be refined in the near future. One advance that may not be far off is the development of immortalized cell lines that are suitable for uptake and transport studies. Such cell lines would promote standardization of experimental conditions and would eliminate the need for the constant preparation of primary cells. Thus far, the immortalized cell lines that have been generated are not a suitable replacement for either BBMECs or PBMECs. One of the recurrent themes with immortalized cells is that their characteristics, especially with regard to transport, appear to change with increasing passage number.[4] The generation of immortalized cell lines would allow new technologies to be employed in studies of transporters. For example, the overlap of transporter specificity with regard to substrates and inhibitors has been a major problem for researchers. Interfering RNA technology, which has the capability of knocking out either single or multiple genes with high specificity,[41] could be used to determine transporter specificity. If immortalized cell lines could be produced from mice, it might be possible to take advantage of genetic knockout (KO) mice to generate cell lines that are genetically deficient in transporters. Clearly, the next few years will be an interesting time for those working with *in vitro* models of the BBB.

7.5. NASAL AND PULMONARY EPITHELIUM

7.5.1. The Respiratory Airway Epithelial Barrier

Drug delivery by inhalation is attractive because of improved patient compliance and ease of administration compared to injection.[42] In addition, macromolecules have shown surprisingly high absorption rates, particularly from the lower respiratory tract compared to other routes.[43,44] As air passes through the nasopharyngeal region it is filtered, warmed, and humidified before entering the lungs. The trachea bifurcates into main bronchi, and branching continues through the small bronchi, bronchioles, terminal bronchioles, respiratory bronchioles, alveolar ducts, and the alveoli.[45] The epithelial lining of the airways presents the primary barrier to drug transport throughout the airways; it is composed of a mix of ciliated and nonciliated pseudostratified columnar epithelial cells, secretory (mucous, goblet, serous, or Clara), and basal cells in the nasal, tracheal, and bronchial regions. Drugs and drug particles in the lowest portion of the respiratory tract, the alveolar region, encounter the thinnest cellular barrier, a large surface area for possible absorption and high blood flow.[43,44]

Discussion of specific cell types characteristic of the epithelial cells along the respiratory tract and corresponding cell culture models for studying nasal and pulmonary drug transport are separated below, with generally more focus on the airway (tracheal and bronchial) epithelium and the alveolar epithelium.

7.5.2. The Nasal Epithelial Barrier and Cell Culture Models

Upon intranasal administration, a drug is not as susceptible to dilution and first-pass effects as in oral delivery.[46,47] The nasal route may also be an effective means of delivering drugs to the brain.[46] Barriers to nasal delivery include the enzymes of the nasal mucosa, the epithelial barrier, the mucus layer, and limited absorption time resulting from mucociliary clearance.[48]

In primary culture of human nasal cells, the cells most relevant for transport studies are pseudostratified ciliated columnar epithelial cells from the medium and inferior turbinates.[48] These cells show a high degree of differentiation, grow to a confluent monolayer in 6 to 8 days, and form tight junctions; however, they are expensive, their availability and life span are limited, and the model does not reflect the heterogeneity of cells *in vivo*.[47–49]

In a comparison of human nasal epithelial cells in primary culture with the RPMI 2650 and bovine turbinate cell lines, Werner and Kissel concluded that only the primary cultured cells appear suitable for drug transport studies because the cell lines lack differentiation and do not form a confluent monolayer with tight junctions.[49] Another option for studying nasal transport is the use of excised mucosa of human or animal origin. Wadell and coworkers observed a good correlation for permeability data for seven of eight drugs tested in porcine nasal mucosa and the fraction absorbed upon nasal administration in humans.[50] Excised mucosa have a lower TEER than primary cultured cells, which can make it difficult to distinguish between intact and damaged tissue; one must monitor tissue integrity and viability with alternative markers.[47,48]

7.5.3. The Airway Epithelial Barrier and Cell Culture Models

Drug transport in the airway epithelium is by passive paracellular diffusion, transcytosis, and specific receptor binding.[51] Only 4% of inhaled air is in the airway, which has less surface area available than the alveolar region, and airway cells have tighter junctions than alveolar cells; nevertheless, solute permeation in both types of cells is similar.[43,51] The airway epithelium consists of ciliated, secretory (mucous, goblet, serous, or Clara), and basal cells. The "mucociliary escalator" serves to remove foreign material from the lungs. The epithelium thickness decreases from about 60 μm in the trachea to about 10 μm in the bronchioles, and the ciliated, pseudostratified cells more populous in the trachea and bronchi give way to an increasing number of secretory cells in the distal airways.[43,51,52]

Primary cell cultures of the airway epithelium are not as easy to maintain as cell lines, but they better represent the biological barrier. Primary cultured cells resemble *in vivo* cell layers in terms of tight junctions, apical cilia, and mucus production, but they are prone to infection and the quantity is often limited.[43,51]

The Calu-3 human submucosal gland cell line forms polarized cell monolayers with tight junctions, produces mucus, and develops apical cilia when grown at an air interface. Transport studies can be performed after 10–14 days in culture, and it has been shown that Calu-3 cells express P-gp and actively transport amino acids, nucleosides, and dipeptide analogs; organic anions, organic cations, polyamines, and efflux pump substrates are not actively transported.[43,51,53–56] Because Calu-3 cells are not subject to the influence of multiple *in vivo* cell types, the expression of carrier proteins and enzymes may not reflect *in vivo* levels. Nevertheless, values obtained in Calu-3 permeability studies correlate well with those obtained from primary cultured rabbit tracheal epithelial cells and *in vivo* rat lung absorption studies.[54] Mannitol permeation in Calu-3 cells is about 10 times less than that *in vivo*, but this is the same ratio difference between Caco-2 cells and *in vivo* intestinal epithelium.[51]

The 16HBE14o- and BEAS-2B cells are viral transformed bronchial cell lines.[52,57] P-gp is expressed in 16HBE14o- cells, tight junctions form, and transport experiments can be carried out in 8 to 10 days, but the cells lack a representative mucus layer. Apical microvilli and cilia can be produced when 16HBE14o- cells are grown at an air interface, but this may not be favorable due to the basal origin of the cells.[51,57] The use of BEAS-2B cells for transport studies is limited by their low TEER.[57]

7.5.4. The Alveolar Epithelial Barrier and Cell Culture Models

In the separation of alveolar air space and blood circulation, the alveolar epithelium is a more restrictive paracellular barrier than the capillary endothelium. About one-third of the alveolar epithelial cells are type I cells, but these cells make up approximately 95% of the cellular surface area. The remaining two-thirds of the alveolar epithelial cells that comprise the remaining 5% of the cellular surface area are the surfactant-producing cuboidal type II cells.[43,45,52] Type I cells have thin cytoplasmic extensions and exhibit a large number of plasmalemmal invaginations called "caveolae," which may play a role in macromolecular and protein transport across the "blood-air barrier" of the lung.[44,45,58,59]

Although interspecies differences can be circumvented through the use of primary cultures of human alveolar epithelial cells, routine use of these cells is limited by availability.[58,60] Primary culture of rat type II alveolar epithelial cells is economical and demonstrates high reproducibility, and its phospholipid secretion is similar to that of human type II cells.[43] P-gp is expressed in both human and rat type I cells but not in freshly isolated rat type II cells.[61] It is assumed that alveolar type II cells are progenitors for regenerating type I cells *in vivo*, and type II cells in culture lose their cuboidal appearance, lamellar bodies, and microvilli, and the number of surfactant proteins decrease. Monolayers are formed in 5 to 8 days, and the trans-differentiation to type I-like cells is complete within 7–8 days, characterized by the development of attenuations, tighter junctions, and increasing expression of

P-gp.[43,45,61] If necessary, cell culture conditions can be controlled to maintain the type II cell phenotype.[60]

The A549 alveolar epithelial cell line exhibits metabolic and transport properties consistent with those of type II cells *in vivo.* A549 cells have microvilli and lamellar bodies and synthesize phospholipids. However, A549 cells do not form domes like type II cells, they do not differentiate into type I-like cells, and they do not form tight junctions.[42,43,45,62] Because no appreciable TEER is observed in the A549 cells, performing transport studies with low molecular weight drugs and xenobiotics is not feasible.[62]

7.6. THE OCULAR EPITHELIAL AND ENDOTHELIAL BARRIERS

7.6.1. The Corneal Epithelial Barrier and Cell Culture Models

The anterior ocular barrier consists of two distinct components—the cornea and conjunctiva. The anterior surface of the eyeball is formed by the optically clear cornea and is composed of multiple cellular and acellular layers. However, it is the corneal epithelium on the surface of the eye that generally forms the primary barrier to drug delivery to pharmacological targets within the eye. The stratified corneal epithelium has tight intercellular junctions between the cells of the outermost layer and restricts drug diffusion to the transcorneal epithelial route. Transcorneal epithelial passage of drugs is largely dependent on their lipophilic nature. Therefore, the mechanism for drug distribution across the cornea is considered to be purely passive diffusion.[63] The current method for culturing corneal epithelial cells is based on that developed by McPherson et al. in the 1950s while studying the cellular nutritional requirements for the preservation of corneal tissues.[64] Primary corneal epithelial cells from both humans and rabbits have been successfully cultured.[65–69] The advantage of the primary culture model is that the tissue has recently been removed from an *in vivo* condition and thus more closely resembles the structure and function of the epithelium *in vivo.* The disadvantage is that these cells degrade quickly *in vitro* as a result of the physical removal of the cells, as well as the constantly changing environment. Corneal epithelial cells have been immortalized in a variety of ways.[70–76] Results obtained with immortalized cells are highly variable and cell-type dependent; the expression of different phenotypes is altered or absent altogether. Thus, primary lines propagated by extended subculture have been more successful.[77–82] The latter lines are advantageous in that they very closely resemble normal, freshly excised cells. A newer, more useful culture model incorporates an air-interface condition.[81]

The conjunctiva provides a mucosal covering for the eye and is composed of a pseudostratified columnar epithelium. The conjunctiva generally has a higher permeability to hydrophilic compounds than does the cornea, and is also more permeable to macromolecules due to the larger and more numerous paracellular spaces.[83,84] Passage of drugs across the conjunctiva into the systemic circulation is generally considered nonproductive absorption because delivery to pharmacological sites in the eye (i.e., within the aqueous humor) does not occur.[63] Primary conjunctival cells from rabbits have been cultured. However, these cells lose viability after only three passages.[63] Despite these disadvantages, primary conjunctival cells have proven useful for studying drug transport in liquid-covered cultures (LCC)[85] as well as in air-interface cultures (AIC).[86,87]

Corneal and conjunctival cell cultures both exhibit some potential for application in drug transport studies. However, neither cell culture system has been extensively exploited for this purpose.[63]

7.6.2. The Blood-Retinal Epithelial Barrier and Cell Culture Models

Accessibility of drugs to the posterior portion of the eye is necessary to treat a number of autoimmune and inflammatory diseases associated especially with aging. As a consequence, drug delivery across the blood-retinal barrier is important to the pharmaceutical chemist. The blood-retinal barrier (BRB) limits the passage of materials between the systemic circulation and the retina[88] and is formed by the tight junctions of a monolayer of retinal pigment epithelium (RPE). The epithelium comprises the outer portion of the BRB, while the retinal vascular endothelium (RVE) makes up the inner portion.[89] The tight junctions of the RPE retard diffusion between the RPE cells, delineate the apical and basolateral membrane domains, and maintain the surface specific distribution of the RPE membrane proteins. The asymmetric nature of the RPE is essential for the transport of nutrients and the regulation of fluid flow between the neural retina and the fenestrated capillaries of the choroids.[88] This outer barrier also prevents the paracellular passage of ions and metabolites that escape the choriocapillaris.[80–93] Thus, the RPE is strategically located within the retina and serves as a model with high interest. The RVE is very comparable to the BBB.[94,95]

There are several ways to purify a primary RPE culture.[88,96–100] The process of isolation, culture technique, and cryopreservation is well documented.[98,99] Grown on porous supports, RPE cells have been applied in a limited number of permeability and toxicity studies.[88] The isolation and culture technique of primary RVE cells from *Macaca* monkeys is also documented;[101] however, their application in drug permeability studies has been minimal.[88]

7.7. THE PLACENTAL BARRIER

7.7.1. The Syncytiotrophoblast Barrier

The placenta separates maternal and fetal circulations with the primary function of allowing and promoting appropriate metabolic exchange.[102] Appropriate placental function is very important for two reasons. First, respiratory gases and nutrients necessary for fetal growth are transported from the maternal circulation across the placenta to the fetus. Second, the metabolic waste products from the fetus enter the maternal circulation through the placenta. This exchange can occur by both passive diffusion and active transport involving transport proteins. Nutrients, drugs, and xenobiotics require the presence of specific transporters for their transfer.[103]

The human placental barrier is unique among mammalian species and is composed simply of a single layer of multinucleated cells, the syncytiotrophoblasts. The syncytial trophoblast forms the rate-limiting barrier for drug permeation across the human placenta and between maternal and fetal compartments.[103] Asymmetry in the distribution in the expression of transporters in the syncytiotrophoblast results in polarized transport of substrates. As such, transporters specific for transport of substrates from the maternal circulation to the fetal circulation are expressed in the maternal facing brush border membrane (e.g., serotonin, norepinephrine, carnitine, monocarboxylate, folate, and P-gp transporters). Likewise, the fetal facing surface has a distinct array of transporters (e.g., extraneuronal monoamine transporter).[104,105]

7.7.2. The Syncytiotrophoblast Barrier and Cell Culture Models

By comparison with other tissues discussed in this review, the identification of representative cell culture models of the placental barrier for drug delivery applications is more recent. Prior to the development of cell culture techniques, isolated tissues and cells of human placenta were used; these methods have been described elsewhere.[102]

Recent advances in cell and tissue culture techniques provide the potential for evaluation of drug transport or metabolism processes at the placenta. Techniques are available for culturing trophoblasts of both animal and human origin.[106] However, our focus here is primarily on human systems. Primary explant and isolated cell cultures of human cytotrophoblasts have been well described;[106–109] however, these systems do not form confluent monolayer systems adequate for transcellular transport studies.[105]

Three cell lines (BeWo, JAr, and JEG) have been generated from malignant trophoblastic cells and are available from the American Tissue Culture Collection (ATCC). The BeWo cell line, which is heterogeneous by several criteria,[110] is comprised of cytotrophoblasts with minimal differentiation to syncytium[111] under nonactivated conditions.[112] This is in contrast to primary cultures of term human cytotrophoblasts, which spontaneously aggregate and form syncytia.[107] The JEG3 cell line was originally derived from BeWo.[113] It expresses human chorionic gonadotrophin and placental lactogen, a characteristic of normal trophoblasts,[114] and forms large multinucleated syncytia in culture similar to syncytiotrophoblasts *in vivo*.[115] The JAr cells synthesize human chorionic gonadotrophin and steroids, characteristic of early placental trophoblasts,[116] and also have the ability to differentiate into syncytiotrophoblast-like cells *in vitro*.[117] While the JEG3 and JAr cells are adequate for drug uptake studies, only a clone of the BeWo cell line forms a confluent monolayer suitable for transcellular drug delivery studies and has been applied to study a variety of nutrient and metabolic pathways in the trophoblast. The BeWo cell line has also proven useful in characterizing drug transport and drug efflux mechanisms present in the normal trophoblast.[106,118,119]

7.8. THE RENAL EPITHELIUM

7.8.1. The Renal Epithelial Barrier

Nephrons are the functional units of the kidney. Each nephron consists of a filtering component, the glomerulus, and a tubular portion made up of three distinct segments—proximal tubule, Henle's loop, and distal convoluted tubule. The net renal elimination is a sum of three distinct mechanisms: glomerular filtration, tubular reabsorption, and tubular secretion. Glomerular filtration consists of passive removal of substances of molecular weight less than 5 kDa from the blood. This forms the ultrafiltrate, which contains glucose, amino acids, vitamins, and proteins. The ultrafiltrate then passes into the lumen of the tubular portion.[120] The tubule is made up of a single layer of epithelial cells resting on a basement membrane. The apical side of the cell layer faces the lumen of the tubule, while the basolateral surface lies adjacent to the blood vessel. In the tubular portion, nutrients like glucose, proteins, vitamins, and amino acids, ions, and water are salvaged from the ultrafiltrate and are either transported across the tubular epithelium into blood or metabolized in the cells. Waste products on the blood side are taken up the tubule and secreted into the lumen to be eliminated in the urine. The transport across the epithelial cells occurs either by passive diffusion or by active carrier-mediated transport and can include a number of drugs.[120–124]

Various *in vivo* and *in vitro* models have been described to study the elimination of drugs from the kidneys. Glomerular filtration and clearance rates require *in vivo* studies or are performed on perfused whole kidneys.[121,122] Tubular secretion and reabsorption of substances have been studied in kidney slices, tubular segments, tubular suspensions, primary cultures, and established cell lines.

7.8.2. Renal Epithelial Cell Culture Models

Isolation and culture of primary cells from tubular segments of the nephron from different species such as rabbit, rat, flounder, and human have been described.[125–129] Although the methods differ, they basically involve homogenization of the organ followed by protease treatment to dissociate cells. The major drawback of primary cultures is that cell types other than the intended one are also copurified. In the case of renal epithelium, fibroblasts are major contaminants; as these grow at a faster rate than the epithelial cells, they form the majority population in the culture. Taub et al. have described a defined medium containing hormones and epithelial cell growth additives which, when used in place of serum, permit selective growth of only the epithelial cells.[125]

Primary cultures retain the differentiated properties of the tissues of origin. The continuous cell lines produce a homogeneous population of cells and are easier to work with in general. Therefore, a number of renal epithelial cell lines have been developed. Two of the more widely used and well-characterized cell lines are MDCK and LLC-PK1. MDCK cells were derived from the distal tubule of dog kidney, while LLC-PK1 cells originated from the proximal tubule of the pig kidney.[18,19] One of the critical requirements for a cell line to be useful for drug transport studies is that the cells form a polarized monolayer with respect to expression and function of transporters and enzymes when grown onto a microporous membrane. Both MDCK and LLC-PK1 cells form well-defined junction structures at confluency, as well as exhibit a significant TEER and a measurable potential difference.[18,130] In comparison to MDCK cells, LLC-PK1 cells express higher activities of proximal tubular brush-border membrane enzymes—gamma-glutamyltranspeptidase, alkaline phosphatase, and leucine aminopeptidase as well as lysosomal enzymes—acid phosphatase and *N*-acetyl glucosaminidase. Conversely, the distal tubular membrane marker Na^+-K^+ ATPase shows higher activity in MDCK cells.[131]

Many transporters display overlapping substrate specificity and, as most cells express a number of different transporters, it is difficult to evaluate the specific interactions between a given drug and transporter. Overexpression of these proteins in cell lines is a way to study the substrate specificity and transport characteristics. MDCK and LLC-PK1 cells, due to their polarized nature, are capable of sorting membrane proteins to either the apical (seen with influenza viral proteins) or basolateral surface (vesicular stomatitis viral proteins).[132] Human and rat multidrug resistance–associated protein (MRP-2) are located on the apical surface and, when expressed in MDCK cells, they are predominantly routed to the luminal membrane.[133] When expressed in HEK cells these proteins were localized in intracellular membranes, but they do retain full functional activity. The apical transporter P-gp, when expressed in MDCK cells, is trafficked to the luminal membrane.[134] Overexpression of canalicular multispecific organic anion transporter (cMOAT, also known as MRP2) in MDCK cells shows that it can transport organic anions S-(2,4-dinitrophenyl)-glutathione and S- (PGA1)-glutathione, compounds which were thought to be specific for MRP1. The transport of these compounds is not completely abolished in the presence of inhibitors specific to MRP1.[135] MDCK cells overexpressing P-gp and canalicular multispecific organic anion transporter have been shown to transport the anticancer drug etopside.[136] Although MDCK cells are easily transfected with specific transporters, their widespread use for high-throughput screening may be subject to some limitations due to cell type–dependent expression and functional differences.[137–139]

7.9. CONCLUSIONS

Current technologies provide for availability of cell culture models from a selection of the significant tissue barriers as either primary cell culture systems or continuous cell line cultures. A number of endothelial and epithelial cell systems can now be grown onto permeable supports as monolayer systems to facilitate transcellular transport. These cell culture models provide powerful tools for the pharmaceutical chemist to characterize fundamental cellular transport mechanisms at the biochemical and molecular levels and effective screening systems to facilitate appropriate drug design and development. Moreover, they all have applications in pharmacological and toxicological investigations. Expectations are that cell culture models or cell-based assays will continue to evolve into better representatives of *in vivo* tissues and, therefore, will be even more valuable to pharmaceutical scientists in more rapidly evaluating new chemical entities as potential drugs.

ACKNOWLEDGMENTS

Work in the Audus laboratory described in this review has been supported in part by the National Institute on Child Health and Development (P01 HD39878), the National Cancer Institute (RO1 CA82801), and Boehringer Ingelheim Pharmaceutics.

REFERENCES

1. Battle, T.; Stacey, G. *Cell Biol. Toxicol.* **2001**, *17*, 287–299.
2. LeCluyse, E. L. *Eur. J. Pharm.* **2001**, *13*, 343–368.
3. Freshney, R. I. In *Culture of Animal Cells*. Wiley-Liss; New York, 1987, pp. 7–13.
4. Gumbleton, M.; Audus, K. L. *J. Pharm. Sci.* **2001**, *90*, 1681–1698.
5. Ho, N. F. H.; Raub, T. J.; Burton, P. S.; Barsuhn, C. L.; Adson, A.; Audus, K. L.; Borchardt, R. T. In *Transport Processes in Pharmaceutical Systems*, Amidon, G. L., Lee, P. I.; Topp, E. M., Eds. Marcel Dekker: New York, **2000**, pp. 219–316.
6. Hidalgo, I. J. *Curr. Top. Med. Chem.* **2001**, *1*, 385–401.
7. Hillgren, K. M.; Kato, A.; Borchardt, R. T. *Med. Res. Rev.* **995**, *15*, 83–109.
8. Artursson, P.; Borchardt, R. T. *Pharm. Res.* **1997**, *14*, 1655–1658.
9. Wilson, G. *Eur. J. Drug Metab. Pharmacokinet.* **1990**, *15*, 159–163.
10. Barthe, L.; Woodley, J.; Houin, G. *Fundam. Clin. Pharmacol.* **1999**, *13*, 154–168.
11. Artursson, P.; Palm, K.; Luthman, K. *Adv. Drug Deliv. Rev.* **1996**, *22*, 67–84.
12. Hidalgo, I. J.; Raub, T. J.; Borchardt, R. T. *Gastroenterology* **1989**, *96*, 736–749.
13. Audus, K. L.; Bartel, R. L.; Hidalgo, I. J.; Borchardt, R. T. *Pharm. Res.* **1990**, *7*, 435–451.
14. Lennernas, H. *J. Pharm. Sci.* **1998**, *87*, 403–410.
15. Stewart, B. H.; Chan, O. H.; Lu R. H.; Reyner, E. L.; Schmid, H. L.; Hamilton, H. W.; Steinbaugh, B. A.; Taylor, M. D. *Pharm. Res.* **1995**, *12*, 693–699.
16. Artursson, P.; Karlsson, J. *Biochem. Biophys. Res. Commun.* **1991**, *175*, 880–885.
17. Bailey, C. A.; Bryla, P.; Malick, A. W. *Adv. Drug Deliv. Rev.* **1996**, *22*, 85–103.
18. Handler, J. S.; Perkins, F. M.; Johnson, J. P. *Am. J. Physiol.* **1980**, *238*, F1–F9.
19. McRoberts, J. A.; Taub, M.; Saier, M. H. In *Functionally Differentiated Cell Lines*; Sato, G., Ed. Alan R. Liss New York, **1981**, pp.117–139.
20. Bohets, H.; Annaert, P.; Mannens, G.; van Beijsterveldt, L.; Anciaux, K.; Verboven, P.; Meuldermans, W.; Lavrijsen, K. *Curr. Top. Med. Chem.* **2001**, *1*, 367–383.
21. Hugger, E. D.; Novak, B. L.; Burton, P. S.; Audus, K. L.; Borchardt, R. T. *J. Pharm Sci.* **2002**, *91*, 1991–2002.
22. Pardridge, W. M. *Nat. Rev. Drug Discov.* **2002**, *1*, 131–139.
23. Audus, K. L.; Borchardt, R. T. *Handbook Exp. Pharmacol.* **1991**, *100*, 43–70.

24. Rubin, L. L.; Staddon, J. M. *Annu. Rev. Neurosci.* **1999**, *22*, 11–28.
25. El Bacha, R. S.; Minn, A. *Cell Mol. Biol. (Noisy.-le-grand)* **1999,** *45*, 15–23.
26. Rochat, B.; Audus, K. L. In *Membrane Transporters as Drug Targets;* Amidon, G., Sadée, W., Eds. Kluwer Academic/Plenum Publishers: New York, **1999**, pp. 181–200.
27. Gottesman, M. M.; Pastan, I.; Ambudkar, S. V. *Curr. Opin. Genet. Dev.* **1996**, *6*, 610–617.
28. Borst, P.; Elferink, R. O. *Annu. Rev. Biochem.* **2002**, *71*, 537–592.
29. Zhang, Y.; Han, H.; Elmquist, W. F.; Miller, D. W. *Brain Res.* **2000**, *876*, 148–153.
30. Audus, K. L.; Rose, J. M.; Wang, W.; Borchardt, R. T. In *An Introduction to the Blood-Brain Barrier: Methodology and Biology*; Pardridge, W., Ed. Cambridge University Press: Cambridge, **1998**, pp. 86–93.
31. Bauer, H. C.; Bauer, H. *Cell Mol. Neurobiol.* **2000**, *20*, 13–28.
32. Raub, T. J. *Am. J. Physiol.* **1996**, *271*, C495–C503.
33. Franke, H.; Galla, H.-J.; Beuckmann, C. T. *Brain Res.* **1999**, *818*, 65–71.
34. Gutmann, H.; Fricker, G.; Drewe, J.; Toeroek, M.; Miller, D. S. *Mol. Pharmacol.* **1999**, *56*, 383–389.
35. Miller, D. S.; Nobmann, S. N.; Gutmann, H.; Toeroek, M.; Drewe, J.; Fricker, G. *Mol. Pharmacol.* **2000**, *58*, 1357–1367.
36. Tewes, B.; Franke, H.; Hellwig, S.; Hoheisel, D.; Decker, S.; Griesche, D.; Tilling, T.; Wegener, J.; Galla, H.-J. In *Transport Across the Blood-Brain Barrier: In Vitro and in Vivo Techniques*, Harwood Academic Publishers: Amsterdam, **1997**, pp. 91–97.
37. Franke, H.; Galla, H.-J.; Beuckmann, C. T. *Brain Res. Protoc.* **2000**, *5*, 248–256.
38. Hoheisel, D.; Nitz, T.; Franke, H.; Wegener, J.; Hakvoort, A.; Tilling, T.; Galla, H.-J. *Biochem. Biophys. Res. Commun.* **1998**, *247*, 312–315.
39. Hoffmeyer, S.; Burk, O.; von Richter, O.; Arnold, H. P.; Brockmoller, J.; Johne, A.; Cascorbi, I.; Gerloff, T.; Roots, I.; Eichelbaum, M.; Brinkmann, U. *Proc. Natl. Acad. Sci. USA* **2000**, *97*, 3473–3478.
40. Willson, T. M.; Kliewer, S. A. *Nat. Rev. Drug Discov.* **2002**, *4*, 259–266.
41. McManus, M. T.; Sharp, P. A. *Nat. Rev. Genet.* **2002**, *10*, 737–747.
42. Ehrhardt, C.; Fiegel, J.; Fuchs, S.; Abu-Dahab, R.; Schäfer, U. F.; Hanes, J.; Lehr, C.-M. *J. Aerosol. Med.* **2002**, *15*, 131–139.
43. Mathias, N. R.; Yamashita, F.; Lee, V. H. L. *Adv. Drug Deliv. Rev.* **1996**, *22*, 215–249.
44. Gumbleton, M. *Adv. Drug Deliv. Rev.* **2001**, *49*, 281–300.
45. Fuchs, S.; Gumbleton, M.; Schäfer, U. F.; Lehr, C.-M. In *Cell Culture Models of Biological Barriers*, Lehr, C.-M., Ed. Taylor & Francis: New York, **2002**, pp. 189–210.
46. Hussain, A. A. *Adv. Drug Deliv. Rev.* **1998**, *29*, 39–49.
47. Koch, A. M.; Schmidt, M. C.; Merkle, H. P. In *Cell Culture Models of Biological Barriers*; Lehr, C.-M., Ed. Taylor & Francis: New York, **2002**, pp. 228–252.
48. Schmidt, M. C.; Peter, H.; Lang, S. R.; Ditzinger, G.; Merkle, H. P. *Adv. Drug Deliv. Rev.* **1998**, *29*, 51–79.
49. Werner, U.; Kissel, T. *Pharm. Res.* **1996**, *13*, 978–988.
50. Wadell, C.; Bjork, E.; Camber, O. *Eur. J. Pharm. Sci.* **2003**, *18*, 47–53.
51. Meaney, C.; Florea, B. I.; Ehrhardt, C.; Lehr, C.-M.; Lehr, C.-M.; Junginger, H. E.; Borchard, G. In *Cell Culture Models of Biological Barriers*; Lehr, C.-M., Ed. Taylor & Francis: New York, **2002**, pp. 211–227.

52. Hamilton, K. O.; Yazdanian, M. A.; Audus, K. L. *Curr. Drug Metab.* **2002**, *3*, 1–12.
53. Foster, K. A.; Avery, M. L.; Yazdanian, M.; Audus, K. L. *Int. J. Pharm.* **2000**, *208*, 1–11.
54. Mathias, N. R.; Timoszyk, J.; Stetsko, P. I.; Megill, J. R.; Smith, R. L.; Wall, D. A. *J. rug Target.* **2002**, *10*, 31–40.
55. Hamilton, K. O.; Backstrom, G.; Yazdanian, M. A.; Audus, K. L. *J. Pharm. Sci.* **2001**, *90*, 647–658.
56. Florea, B. I.; van der Sandt, I. C. J.; Schrier, S. M.; Kooiman, K.; Deryckere, K.; de Boer, A. G.; Junginger, H. E.; Borchard, G. *Br. J. Pharmacol.* **2001**, *134*, 1555–1563.
57. Forbes, B. *Pharm. Sci. Tech. Today* **2000**, *3*, 18–27.
58. Fuchs, S.; Hollins, A. J.; Laue, M.; Schäfer, U. F.; Roemer, K.; Gumbleton, M.; Lehr, C.-M. *Cell Tissue Res.* **2003**, *311*, 31–45.
59. Kim, K.-J.; Malik, A. B. *Am. J. Physiol.* **2003**, *284*, L247–L259.
60. Forbes, B. *Methods Mol. Biol.* **2002**, *188*, 65–75.
61. Campbell, L.; Abulrob, A. N.; Kandalaft, L. E.; Plummer, S.; Hollins, A. J.; Gibbs, A.; Gumbleton, M. *J. Pharmacol. Exp. Ther.* **2003**, *304*, 441–452.
62. Foster, K. A.; Oster, C. G.; Mayer, M. M.; Avery, M. L.; Audus, K. L. *Exp. Cell. Res.* **1998**, *243*, 359–366.
63. Suhonen, P.; Sporty, J.; Lee, V. H. L.; Urtti, A. In *Cell Culture Models of Biological Barriers*; Lehr, C.-M., Ed. Taylor & Francis: New York, **2002**, pp. 253–270.
64. McPherson, S. D.; Draheim, J. W. J.; Evans, V. J.; Earle, W. R. *Am. J. Ophthalmol.* **1956**, *41*, 513–521.
65. Chan, K. Y.; Haschke, R. H. *Exp. Eye Res.* **1983**, *36*, 231–246.
66. Chang, J. E.; Basu, S. K.; Lee, V. H. L. *Pharm. Res.* **2000**, *17*, 670–676.
67. Doran, T. I.; Vidrich, A.; Sun, T. T. *Cell* **1980**, *22*, 17–25.
68. Jumblatt, M. M.; Neufeld, A. H. *Invest. Ophthalmol. Vis. Sci.* **1983**, *24*, 1139–1143.
69. Sun, T. T.; Green, H. *Nature*, **1963**, *269*, 489–493.
70. Houweling, A.; van den Elsen, P. J.; van der Eb, A. J. *Virology* **1980**, *105*, 537–550.
71. Cone, R. D.; Grodzicker, T.; Jaramillo, M. *Mol. Cell. Biol.* **1988**, *8*, 1036–1044.
72. Hronis, T. S.; Steinberg, M. L.; Defendi, V.; Sun, T. T. *Cancer Res.* **1984**, *44*, 5797–5804.
73. Steinberg, M. L.; Defendi, V. *J. Invest. Dermatol.* **1983**, *81*, 131S–136S.
74. Agarwal, C.; Eckert, R. L. *Cancer Res.* **1990**, *50*, 5947–5953.
75. Halbert, C. L.; Demers, G. W.; Galloway, D. A. *J. Virol.* **1991**, *65*, 473–478.
76. Kuppuswamy, M.; Chinnadurai, G. *Oncogene* **1988**, *2*, 567–572.
77. Araki, K.; Ohashi, Y.; Sasabe, T.; Kinoshita, S.; Hayashi, K.; Yang, X. Z., Hosaka, Y.; Aizawa, S.; Handa, H. *Invest. Ophthalmol. Vis. Sci.* **1993**, *34*, 2665–2671.
78. Aizawa, S.; Yaguchi, M.; Nakano, M.; Inokuchi, S.; Handa, H.; Toyama, K. *J. Cell. Physiol.* **1991**, *148*, 245–251.
79. Van Doren, K.; Gluzman, Y. *Mol. Cell. Biol.* **1984**, *4*, 1653–1656.
80. Kahn, C. R.; Young, E.; Lee, I. H.; Rhim, J. S. *Invest. Ophthalmol. Vis. Sci.* **1993**, *34*, 3429–3441.
81. Araki-Sasaki, K.; Ohashi, Y.; Sasabe, T.; Hayashi, K.; Watanabe, H.; Tano, Y.; Handa, H. *Invest. Ophthalmol. Vis. Sci.* **1995**, *36*, 614–621.

82. Griffith, M.; Osborne, R.; Munger, R.; Xiong, X.; Doillon, C. J.; Laycock, N. L.; Hakim, M.; Song, Y.; Watsky, M. A. *Science* **1999**, *286*, 2169–2172.
83. Wang, W.; Sasaki, H.; Chien, D. S.; Lee, V. H. *Curr. Eye Res.* **1991**, *10*, 571–579.
84. Hamalainen, K. M.; Kontturi, K.; Murtomaki, L.; Auriola, S.; Urtti, A. *J. Control Release* **1997**, *49*, 97–104.
85. Saha, P.; Kim, K. J.; Lee, V. H. *Curr. Eye Res.* **1996**, *15*, 1163–1174.
86. Mathias, N. R.; Kim, K. J.; Robison, T. W.; Lee, V. H. *Pharm. Res.* **1995**, *12*, 1499–1505.
87. Yang, J. J.; Ueda, H.; Kim, K.; Lee, V. H. *J. Control Release* **2000**, *65*, 1–11.
88. Eldem, T.; Durlu, Y.; Eldem, B.; Ozguc, M. In *Cell Culture Models of Biological Barriers*; Lehr, C.-M., Ed. Taylor & Francis: New York, **2002**, pp. 271–288.
89. Pederson, J. E. In *Fluid Physiology of the Subretinal Space in Retina*, 2nd ed.; Ryan, S. J., Ed. Mosby–Year Book: St. Louis, **1994**, pp. 1955–1967.
90. Hewitt, A. T.; Adler, R. In *The Retinal Pigment Eptithelium and Interphotoreceptor Matrix: Structure and Specialized Functions in Retina*, 2nd ed.; Ryan, S. J., Ed. Mosby–Year Book: St. Louis, **1994**, pp. 58–71.
91. Cunha-Vaz, J. G.; Faria de Abreu, J. R.; Campos, A. J.; Figo, G. M. *Br. J. Ophthalmol.* **1975**, *59*, 649–656.
92. Do Cormo, A. D.; Ramos, P.; Reis, A.; Proenca, R.; Cunha-Vaz, J. G. *Exp. Eye Res.* **1998**, *67*, 569–575.
93. Jampol, L. E.; Po, S. M. In *Macular Edema in Retina*, 2nd ed.; Ryan, S. J., Ed. Mosby–Year Book: St. Louis, **1994**, pp. 999–1008.
94. Lee, V. H. L.; Pince, K. J.; Frambach, D. A.; Martenhed, B. In *Drug Delivery to the Posterior Segment in Retina*, 2nd ed.; Ryan, S. J., Ed. Mosby–Year Book: St. Louis, **1994**, pp. 533–551.
95. Greenwood, J.; Howes, R.; Lightman, S. *Lab. Invest.* **1994**, *70*, 39–52.
96. Durlu, K.; Tamai, M. In *In Vitro Expression of Epidermal Growth Factor Receptor by Human Retinal Pigment Epithelial Cells in Degenerative Diseases of the Retina*; Anderson, R. E., Ed. Plenum Press: New York, **1995**, pp. 69–76.
97. McKay, B. S.; Burke, J. M. *Exp. Cell Res.* **1994**, *213*, 85–92.
98. Song, M.-K.; Lui, G. M. *J. Cell Physiol.* **1990**, *143*, 196–203.
99. Durlu, K.; Tamai, M. *Cell Transplant.* **1997**, *6*, 149–162.
100. Durlu, Y. K.; Ishiguro, S.-I.; Akaishi, K.; Abe, T.; Chida, Y.; Shibahara, S. In *Retinal Degenerative Diseases and Experimental Therapy*; Hollyfield, J., Anderson, R. E., LaVail, M. M., Eds. Plenum Press: New York, **1999**, pp. 559–567.
101. Yan, Q.; Vernon, R. B.; Hendrickson, A. E.; Sage, E. H. *Invest. Ophthalmol. Vis. Sci.* **1996**, *37*, 2175–2184.
102. Enders, A. C.; Blankenship, T. N. *Adv. Drug Deliv. Rev.* **1999**, *38*, 1–15.
103. Sastry, B. V. R. *Adv. Drug Deliv. Rev.* **1999**, *38*, 17–39.
104. Ganapathy, V.; Prasad, P. D.; Ganapathy, M. E.; Leibach, F. H. *J. Pharmacol. Exp. Ther.* **2000**, *294*, 413–420.
105. St.Pierre, M. V.; Serrano, M. A.; Macais, R. I. R.; Dubs, U.; Hoechli, M.; Lauper, U.; Meier, P. J.; Marin, J. J. G. *Am. J. Physiol.* **2000**, *279*, R1495–R1503.
106. Liu, F.; Soares, M. J.; Audus, K. L. *Am. J. Physiol.* **1997**, *273*, C1596–C1604.

107. Kliman, H. J.; Nestler, J. E.; Sermasi, E.; Sanger, J. M.; Straus, J. F., 3rd. *Endocrinology* **1986**, *118*, 1567–1582.

108. Cimerikic, B.; Zamah, R.; Ahmed, M. S. *Trophoblast Res.* **1994**, *8*, 405–419.

109. Hall, C. S.; James, T. E.; Goodyear, C.; Branchard, C.; Guyda, H.; Giroud, C. J. P. *Steroids* **1977**, *30*, 569–580.

110. Aplin, J. D.; Sattar, A.; Mould, A. P. *J. Cell Sci.* **1992**, *103*, 435–444.

111. Pattillo, R. A.; Gey, G. O.; Delfs, E.; Mattingly, R. F. *Science* **1968**, *159*, 1467–1469.

112. Speeg, K. V., Jr.; Azizkhan, J. C.; Stromberg, K. *Cancer Res.* **1976**, *36*, 4570–4576.

113. Tuan, R. S.; Moore, C. J.; Brittingham, J. W.; Kirwin, J. J.; Atkins, R. E.; Wong, M. *J. Cell Sci.* **1991**, *98*, 333–342.

114. Kohler, P. O.; Bridson, W. E. *J. Clin. Endocrinol. Metab.* **1971**, *32*, 683–687.

115. Babalola, G. O.; Coutifaris, C.; Soto, E. A.; Kliman, H. J.; Shuman, H.; Straus, J. F., 3rd, *Dev. Biol.* **1990**, *137*, 100–108.

116. White, T. E.; Saltzman, R. A.; Di Sant'Agnese, P. A.; Keng, P. C.; Sutherland, R. M.; Miller, R. K. *Placenta* **1988**, *9*, 583–598.

117. Grummer, R.; Hohn, H. P.; Mareel, M. M.; Deker, H. W. *Placenta* **1994**, *15*, 411–429.

118. Young, A. M.; Fukuhara, A.; Audus, K. L. In *Cell Culture Models of Biological Barriers*; Lehr, C.-M., Ed. Taylor & Francis: New York, **2002**, pp. 337–349.

119. Young, A. M.; Allen, C. E.; Audus, K. L. *Adv. Drug Deliv. Rev.* **2003**, *55*, 125–132.

120. Vander, A. J. In *Renal Physiology*; McGraw-Hill: New York, 1995,

121. Bonate, P. L.; Reith, K.; Weir, S. *Clin. Pharmacokinet.* **1998**, *34*, 375–404.

122. Zhang, L.; Brett, C. M.; Giacomini, K. M. *Annu. Rev. Pharmacol. Toxicol.* **1998**, *38*, 431–460.

123. Inui, K.; Masuda, S.; Saito, H. *Kidney Int.* **2000**, *58*, 944–958.

124. Van Aubel, R. A.; Masereeuw, R.; Russel, G. *Am. J. Physiol. Renal Physiol.* **2000**, *279*, F216–F232.

125. Chung, S. D.; Alavi, N.; Livingston, D.; Hiller, S.; Taub, M. *J. Cell Biol.* **1982**, *95*, 118–126.

126. Sakhrani, L. M.; Fine, L. G. *Miner Electrolyte Metabol.* **1983**, *9*, 276–281

127. Wilson, P. D.; Dillingham, M. A.; Breckon, R.; Anderson, R. J. *Am. J. Physiol.* **1985**,*248*, F436–F443.

128. Dickman, K. G.; Renfro, J. L. *Am. J. Physiol.* **1986**, *251*, F424–F432.

129. Elliget, K. A.; Trump, B. F. *In Vitro Cell Dev. Biol.* **1991**, *27A*, 739–748.

130. Misfeldt, D. S.; Hamamoto, S. T.; Pitelka, D. R. *Proc. Natl. Acad. Sci. USA* **1976**, *73*, 1212–1216.

131. Gstraunthaler, G.; Pfaller, W.; Kotanko, P. *Am. J. Physiol.* **1985**, *248*, F536–F544.

132. Boulan, E. R.; Sabatini, D. *Proc. Natl. Acad. Sci. USA* **1978**, *75*, 5071–5075.

133. Cui, Y.; Konig, J.; Buchholz, J. K.; Spring, H.; Leier, I.; Keppler, D. *Mol. Pharmacol.* **1999**, *55*, 929–937.

134. Pastan, I.; Gottesman, M. M.; Ueda, K.; Lovelace, E.; Rutherford, A. V.; Willingham, M. C. *Proc. Natl. Acad. Sci. USA* **1988**, *85*, 4486–4490.

135. Evers, R.; Kool, M.; van Deemter, L.; Janssen, H.; Calafat, J.; Oomen, L. C.; Paulusma, C. C.; Oude Elferink, R. P.; Baas, F.; Schinkel, A. H.; Borst, P. *J. Clin. Invest.* **1998**, *101*, 1310–1319.

136. Guo, A.; Marinaro, W.; Hu, P.; Sinko, P. J. *Drug Metab. Dispos.* **2002**, *30*, 457–463.
137. Putnam, W. S.; Ramanathan, S.; Pan, L.; Takahashi, L. H.; Benet, L. Z. *J. Pharm. Sci.* **2002**, *91*, 2622–2635.
138. Tang, F.; Horie, K.; Borchardt, R. T. *Pharm. Res.* **2002**, *19*, 765–772.
139. Tang, F.; Horie, K.; Borchardt, R. T. *Pharm. Res.* **2002**, *19*, 773–779.

8

PRODRUG APPROACHES TO DRUG DELIVERY

LONGQIN HU

Department of Pharmaceutical Chemistry, Ernest Mario School of Pharmacy, Rutgers, The State University of New Jersey, Piscataway, NJ 08854

Drug Delivery: Principles and Applications Edited by Binghe Wang, Teruna Siahaan, and Richard Soltero
ISBN 0-471-47489-4

8.1. INTRODUCTION

Most drugs, in order to produce their desired pharmacological action, have to overcome many hurdles before reaching the desired site of action. These hurdles include the intestinal barrier, the blood-brain barrier (BBB), and metabolic reactions that could render them inactive. These three subjects are covered in Chapters 2, 3, and 6, respectively, and therefore will not be discussed in detail here. Most drugs distribute randomly throughout the body, and the amount of drugs reaching the site of action is relatively small. For an effective amount to reach the site of action and not cause severe systemic side effects, a drug must possess certain physicochemical properties that make it conducive to penetration through various biological membranes (i.e., sufficiently bioavailable), to avoid metabolic inactivation by various enzymes, and to avoid retention in body depot tissues that could lead to undesirable long-lasting effects. These desired physicochemical properties are not always present in pharmacologically active compounds.

With the advance of new technologies such as combinatorial and computational chemistry, more and more compounds are being identified with extremely potent *in vitro* activity but are found to be inactive *in vivo*. They may have the optimal configuration and conformation needed to interact with their target receptor or enzyme, but they do not necessarily possess the best molecular form and physicochemical properties needed for their delivery to the site of action. Some of the problems often encountered include (1) limited solubility and poor chemical stability preventing the drug from being adequately formulated, (2) low or variable bioavailability due to incomplete absorption across biological membranes or extensive first-pass metabolism, and (3) lack of site specificity. Further structural modifications are often performed but do not always solve all the problems. Another approach that is often effective in solving some of these delivery problems is the design of prodrugs by attaching a promoiety to the active drug.[1–3] This chapter will focus on the various prodrug approaches that have been used to overcome many of the pharmaceutical and pharmacokinetic barriers that hinder optimal delivery of the active drug.

8.2. BASIC CONCEPTS: DEFINITION AND APPLICATIONS

A prodrug by definition is inactive or much less active and has to be converted to the active drug within the biological system. There are a variety of mechanisms by which a prodrug can be activated. These include metabolic activation mediated by enzymes present in the biological system as well as the less common, simple chemical means of activation such as hydrolysis.

Prodrugs occur in nature. One example is proinsulin, which is synthesized in the pancreas and releases its active moiety, insulin, and an inactive peptide. Codeine is another example; it can be regarded as a prodrug of morphine, which is responsible for its analgesic effect.

Most synthetic prodrugs are prepared by attachment of the active drug through a metabolically labile linkage to another molecule, the "promoiety". The promoiety is not necessary for activity but may impart some desirable properties to the drug, such as increased lipid or water solubility or site specificity. Advantages that can be gained with such a prodrug include increased bioavailability, alleviation of pain at the site of injection, elimination of an unpleasant taste, decreased toxicity, decreased metabolic inactivation, increased chemical stability, and prolonged or shortened duration of action.

8.2.1. Increasing Lipophilicity to Increase Systemic Bioavailability

This is the most successful application of prodrugs. Because of the lipid bilayer nature of biological membranes, the rate of passive drug transport is affected by both lipophilicity and aqueous solubility (also called "hydrophilicity"). The rate of passive diffusion across the biological membrane will increase exponentially with increasing lipophilicity and then level off at higher lipophilicity. This is due to the fact that an increase in lipophilicity is usually accompanied by a decrease in water solubility and will eventually decrease the flux over the membrane due to poor water solubility. The design of prodrugs aims to achieve a balance between lipophilicity and aqueous solubility in order to improve passive drug transport across various biological membranes. Using data drawn from U.S. Adopted Names, the World Drug Lists, and Pfizer internal compound collections, it was concluded that, to have good membrane permeability, drugs should have a relatively low molecular weight ($\leq$500), be relatively nonpolar, and partition between an aqueous and a lipid phase in favor of the lipid phase but, at the same time, possess certain water solubility ($-1 \leq \mathrm{Log}P \leq +5$).[4] The majority of effective oral drugs obey this so-called Lipinski's rule of 5.

Since most drugs are either weak acids or weak bases, they are often given and present in the salt form under relevant physiological conditions. Therefore, dissociation constants also affect membrane permeability, and thus bioavailability. It is generally accepted that the neutral, unionized and thus most lipophilic form of an acidic or basic drug is absorbed far more efficiently than the ionized species. In these cases, the distribution between the ionized and neutral form depends strongly on pH. The effective partition coefficient for a dissociative system (Log*D*) gives the correct description of such complex partitioning equilibria.

$$\mathrm{Log}D = \mathrm{Log}P_{HA} - \log(1 + 10^{(\mathrm{pH}-\mathrm{p}Ka)}) \quad \text{for an acid}$$

$$\mathrm{Log}D = \mathrm{Log}P_{B} - \log(1 + 10^{(\mathrm{p}Ka-\mathrm{pH})}) \quad \text{for a base}$$

where P_{HA} and P_B are the intrinsic partition coefficients of the weak acid and weak base, respectively. Programs such as ACD/Log*P* and cLog*P* are available to calculate with reasonable accuracy the Log*P* and Log*D* values using a structure-fragment approach as well as internal structure databases. To illustrate the principles discussed in this chapter, examples will be given with their Log*P* and/or Log*D* values calculated using the Advanced Chemistry Development (ACD) Software.

Many prodrugs feature the addition of a hydrophobic group in order to increase their lipid solubility to improve their gastrointestinal absorption. Bacampicillin (**2**), pivampicillin (**3**), and talampicillin (**4**) are more lipophilic esters of ampicillin (**1**), and pivmecillinam (**6**) is a more lipophilic ester prodrug of mecillinam (**5**), all with improved oral bioavailability. For example, absolute oral bioavailability in horses was 39%, 31%, and 23% for bacampicillin, pivampicillin, and talampicillin, respectively, compared to only 2% for ampicillin sodium.[5] Esterification of carboxylic acid in ampicillin (**1**) resulted in an increase of 0.8-1.4 unit in Log*P*. More significant are the increases in Log*D* values for the prodrugs when ionization of the amino group is taken into consideration; as much as a 4-unit difference in Log*D* is estimated at pH values in the intestines where the prodrugs are believed to be absorbed. Other prodrugs of antibiotics include esters of carbenicillin (for urinary tract infection), cefotiam, and erythromycin.

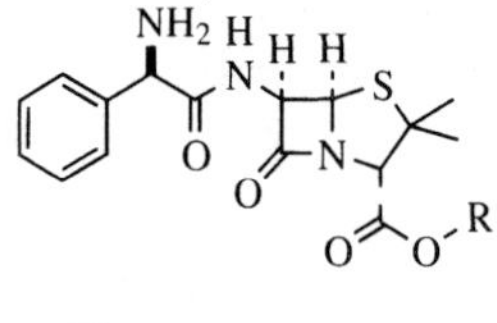

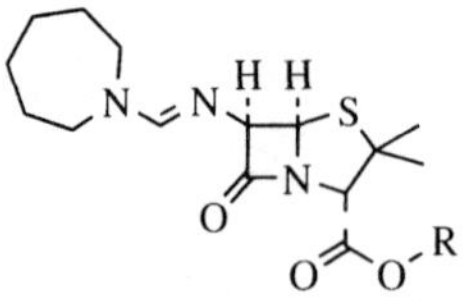

		ACD/Log*P*	ACD/Log*D* pH 7	ACD/Log*D* pH 1
(**1**) Ampicillin	R = H	1.350±0.320	−1.54	−1.72
(**2**) Bacampicillin	R = -CH(CH$_3$)OCOOC$_2$H$_5$	2.172±0.894	1.99	−0.93
(**3**) Pivampicillin	R = -CH(CH$_3$)OCOC(CH$_3$)$_3$	2.552±0.884	2.37	−0.55
(**4**) Talampicillin	R 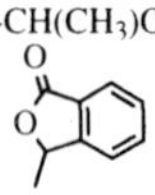 = (phthalidyl)	2.789±0.882	2.61	−0.31
(**5**) Mecillinam	R = H	1.493±0.866	−1.01	−1.58
(**6**) Pivmecillinam	R = -CH(CH$_3$)OCOC(CH$_3$)$_3$	3.453±0.921	1.77	0.35

Enalapril (**8**) is an ester prodrug of enalaprilat (**7**). The latter binds tightly to the angiotensin-converting enzyme (ACE) but is transported with low efficacy by the peptide carrier in the gastrointestinal tract. The prodrug enalapril has a higher affinity for the peptide carrier[6] and is much better absorbed, with about 60% oral bioavailability.[7,8] As a matter of fact, all ACE inhibitors except captopril and lisinopril are administered as prodrugs; other commercialized ACE inhibitor prodrugs include perindopril, quinapril, ramipril, cilazapril, benazepril, spirapril, and trandolapril, all based on esterification of the same carboxylic acid group.[9] The esters are hydrolyzed *in vivo*, after absorption, to the corresponding active but poorly absorbed dicarboxylate forms.

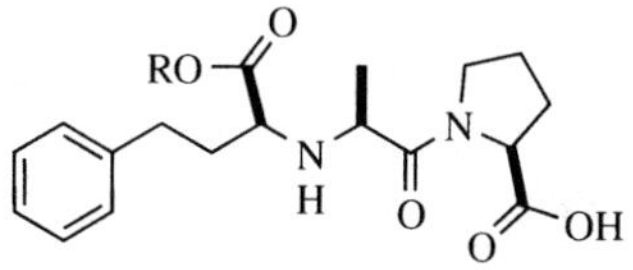

			ACD/Log*D*	
		ACD/Log*P*	pH 7	pH 1
(**7**) Enalaprilat	R = H	2.102±0.574	−1.45	−0.92
(**8**) Enalapril	R = Et	2.983±0.580	−0.12	−0.10

Valacyclovir (**10**)[10,11] and famciclovir (**12**) are ester prodrugs of acyclovir (**9**) and penciclovir (**11**), respectively, for the treatment of viral infections. Both acyclovir and penciclovir exhibit site-specific conversion to the active triphosphate species by viral thymidine kinase. They show remarkable antiviral selectivity and specificity. However, their oral bioavailability is quite low, 15–20% of an oral dose being absorbed in humans for acyclovir and 5% for penciclovir.[12] Both valacyclovir and famciclovir have no intrinsic antiviral activity, and both are rapidly hydrolyzed to acyclovir and penciclovir by esterases present in the liver and gut wall. Valacyclovir displays a mean absolute bioavailability of 54%, a threefold increase in oral bioavailability over acyclovir, while famciclovir has an absolute bioavailability of 77% in humans.[13–15] Famciclovir's better bioavailability could be explained by the increase in lipophilicity; the high oral bioavailability of valacyclovir was also partly attributed to the involvement of an active transport mechanism through PEPT1.[16] Therefore, in addition to increasing lipophilicity, prodrug design can utilize active transport mechanisms as a means of enhancing bioavailability.

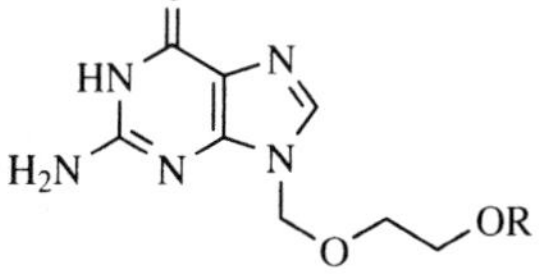

			ACD/Log*D*	
		ACD/Log*P*	pH 7	pH 1
(**9**) Acyclovir	R = H	−1.760±0.489	−1.76	−3.76
(**10**) Valacyclovir	R = L-Valyl-	0.040±0.577	−0.78	−4.76
(**11**) Penciclovir	R = H	−2.031±0.584	−2.03	−4.44
(**12**) Famciclovir	R = CH_3CO-	−0.088±0.265	−0.09	−3.27

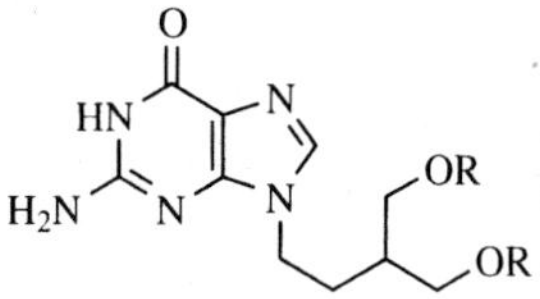

8.2.2. Sustained-Release Prodrug Systems

Antipsychotic drugs are the mainstay treatment for schizophrenia and similar psychotic disorders. Long-acting depot injections of antipsychotic drugs are extensively used as a means of long-term maintenance treatment. The duration of action for many antipsychotic drugs with a free hydroxyl group can be considerably prolonged by the preparation of long-chain fatty acid esters with very high Log*P* values (usually 7 or above). Fluphenazine enanthate (**14**) and fluphenazine decanoate (**15**) were the first of these esters to appear in clinical use and are longer-acting, with fewer side effects than the parent drug. The ability to treat patients with a single intramuscular injection every 1–2 weeks with the enanthate or every 2–3 weeks with the decanoate esters means that problems associated with patient compliance with the drug regimens and with drug malabsorption can be reduced.[17] Esterification of antipsychotic drugs with decanoic acid yields very lipophilic prodrugs which are dissolved in a light vegetable oil such as Viscoleo or sesame oil. Intramuscular injection creates an oily depot from which the prodrug molecules slowly diffuse into the systemic circulation, where they are hydrolyzed quickly by esterases to the active moieties. These depot forms allow these drugs to be given only once or twice a month, permitting the long-term treatment of schizophrenia. Antipsychotic drugs available in depot formulation include fluphenazine (**13**), flupenthixol, haloperidol, and zuclopenthixol in their enanthate or decanoate esters.

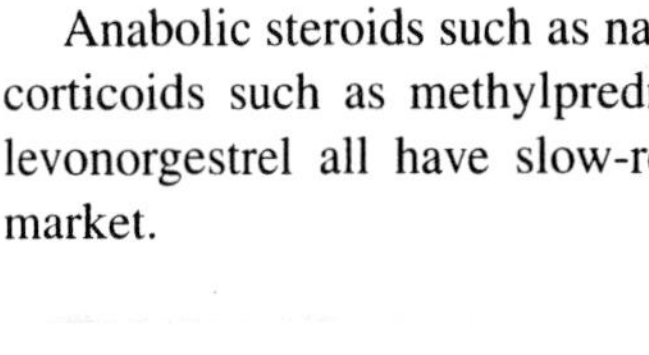

		ACD/Log*P*
(**13**) Fluphenazine	R= H	4.841±0.460
(**14**) Fluphenazine enanthate	R = -CO$(CH_2)_5CH_3$	8.034±0.453
(**15**) Fluphenazine decanoate	R = -CO$(CH_2)_8CH_3$	9.628±0.453

Anabolic steroids such as nandrolone and testosterone, anti-inflammatory glucocorticoids such as methylprednisolone, and contraceptives such as estradiol and levonorgestrel all have slow-release formulations of their ester prodrugs in the market.

8.2.3. Improving Gastrointestinal Tolerance

Temporary masking of carboxylic acid groups in nonsteroidal anti-inflammatory drugs was proposed as a promising means of reducing gastrointestinal toxicity resulting from direct mucosal contact mechanisms. Morpholinoalkyl esters (**17** and **19**, HC1 salts) of naproxen (**16**) and indomethacin (**18**) were evaluated *in vitro* and *in vivo* for their potential use as prodrugs for oral delivery.[18] The prodrugs were freely soluble in simulated gastric fluid and pH 7.4 phosphate buffer and showed a minimum of a 2000-fold increase in solubility over the parent drugs. The prodrugs were more lipophilic than the parent drugs and were quantitatively hydrolyzed to their respective parent drugs *in vivo*. The prodrugs were 30–36% more bioavailable orally than the parent drugs following a single dose in rats. They were significantly less irritating to gastric mucosa than the parent drugs following a single dose as well as chronic oral administration in rats.

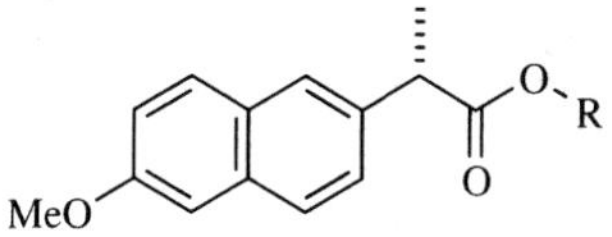

(**16**) Naproxen (R = H)

(**17**) R = $-(CH_2)_n-$N(morpholino)

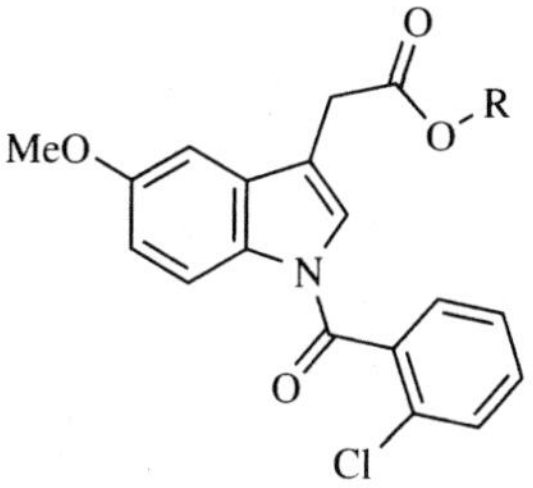

(**18**) Indomethacin (R = H)

(**19**) R = $-(CH_2)_n-$N(morpholino)

8.2.4. Improving Taste

Oral drugs with a markedly bitter taste may lead to poor patient compliance if administered as a solution or syrup. The prodrug approach has been used to improve the taste of chloramphenicol (**20**), clindamycin, erythromycin, and metronidazole.[19] A prodrug such as chloramphenicol palmitate (**21**), with Log*P* of around 10, does not dissolve in an appreciable amount in the mouth and, therefore, does not interact with the taste receptors.

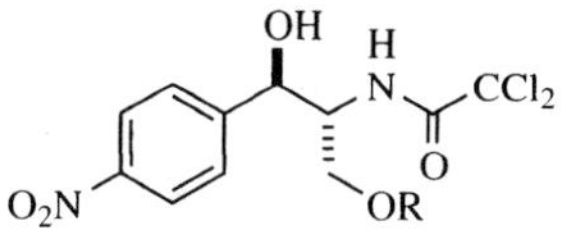

		ACD/Log*P*	ACD/Log*D* pH 7	ACD/Log*D* pH 1
(**20**) Chloramphenicol	R = H	1.018±0.321	1.02	1.02
(**21**) Chloramphenicol palmitate	R = $-CO(CH_2)_{14}CH_3$	9.920±0.756	9.92	9.92
(**22**) Chloramphenicol sodium succinate	R = $-COCH_2CH_2COO^-Na^+$	2.287±0.849	−0.34	2.29

8.2.5. Diminishing Gastrointestinal Absorption

Many prodrugs have been evaluated in this context for colon-specific drug delivery. Colon targeting is of value for the topical treatment of diseases of the colon such as Crohn's disease, ulcerative colitis, and colorectal cancer. Sustained colonic release of drugs can be useful in the treatment of nocturnal asthma, angina, and arthritis. Prodrugs have been designed to pass intact and unabsorbed from the upper gastrointestinal tract and undergo biotransformation in the colon, releasing the active drug molecule. Prodrug activation can be carried out by microflora and distinct enzymes present in the colon (such as azoreductase, glucuronidase, glycosidase, dextranase, esterase, nitroreductase, and cyclodextranase).[20,21] Balsalazide (**23**), ipsalazide (**24**), olsalazine (**25**), and sulfasalazine (**26**) are azo-containing prodrugs developed for colon-specific delivery of an anti-inflammatory agent in the treatment of inflammatory bowel disease. As shown in Scheme 1, they can undergo azoreduction in the colon to release the active 5-aminosalicylic acid (5-ASA or mesalazine, **27**). Other prodrugs evaluated for colon-specific delivery include conjugates of amino acids, glucuronide, glycoside, dextran, and cyclodextrin.[22]

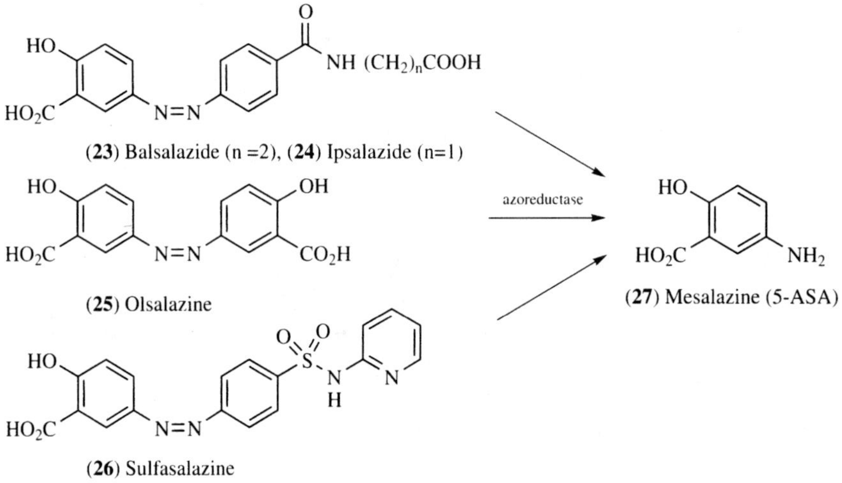

(**23**) Balsalazide (n =2), (**24**) Ipsalazide (n=1)

(**25**) Olsalazine

(**26**) Sulfasalazine

azoreductase

(**27**) Mesalazine (5-ASA)

Scheme 1

8.2.6. Increasing Water Solubility

Poorly water-soluble, lipophilic drugs also have difficulty getting absorbed, as discussed earlier. The prodrug approach has been applied to circumvent solubility problems by introduction of an ionizable functional group such as phosphate esters, amino acid esters, and hemiesters of dicarboxylic acids, allowing various salts of such prodrugs to be formed. Prodrugs can also be used to increase water solubility in order to increase the amount of drug that will reach the systemic circulation through parenteral administration. Examples include chloramphenicol sodium succinate (**22**), hydrocortisone sodium succinate, methylprednisolone sodium succinate, betamethasone sodium phosphate, clindamycin phosphate, and prednisolone phosphate.

In addition to the use of ionizable groups, disruption of the crystal lattice can also result in a significant increase in aqueous solubility, as illustrated by the antiviral agent vidarabine (**28**). The 5′-formate ester derivative (**29**) of vidarabine is 67-fold more soluble in water than vidarabine itself and has been attributed to disruption of the strong intermolecular interactions in the crystal, as indicated by the 85°C drop in the melting point.[23]

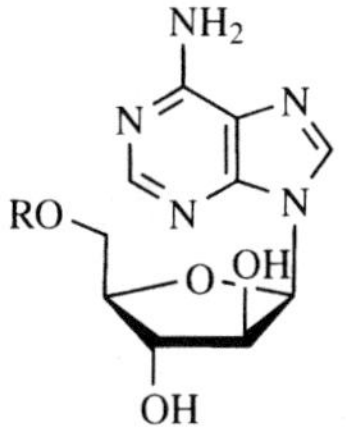

	R	ACD/Log*P*	m.p.	solubility$_{H_2O}$ (25 °C)
(**28**) Vidarabine	R = H	−1.458±0.470	260 °C	0.0018 M
(**29**) 5′-formate	R = C(=O)-H	−0.364±0.601	175 °C	0.12 M

8.2.7. Tissue Targeting and Activation at the Site of Action

Prodrugs can be designed to target specific tissues. This is especially useful in improving the therapeutic effectiveness and decreasing the systemic toxicity of anticancer agents in the treatment of cancer. Anticancer agents are usually highly toxic, with a very small therapeutic index, and their therapeutic effectiveness is often limited by their dose-limiting side effects. Here, several strategies for targeting chemotherapeutic agents to cancers will be briefly discussed to illustrate the applications of prodrugs. For details, refer to Chapter 11 on metabolic activation and drug targeting.

8.2.7.1. Tumor Hypoxia and Bioreductive Activation of Anticancer Prodrugs
Solid tumors often contain regions which are subject to chronic or transient deficiencies of blood flow and, therefore, to the development of chronic or acute hypoxia owing to the primitive state of tumor vasculature.[24] Hypoxic cells in a solid tumor frequently constitute 10–20% and occasionally over half of the total viable tumor cell population. Agents that are active against proliferating cells are relatively ineffective against these hypoxic tumor cells, which are not actively replicating at the time of treatment but are capable of commencing proliferation at a later time and causing the tumor to regrow. Hypoxic cells also may be resistant to conventional chemotherapy due to pharmacodynamic considerations.[24] To produce a therapeutic response, appropriate drug concentrations must be reached. Drugs that have physicochemical properties not conducive to diffusion into tumor tissue, or that are unstable or metabolized rapidly, may not reach chronically hypoxic tumor cells located in regions of severe vascular insufficiency. Therefore, the presence of hypoxic cells in solid tumors is an obstacle to effecting a cure.

Since hypoxic cells located remotely from the vascular supply of a tumor mass may have a greater capacity for reductive reactions than their normal, well-oxygenated counterparts, hypoxia could provide an opportunity for the design of selective cancer chemotherapeutic agents that could be reductively activated in these hypoxic cells.[24] Several classes of agents are presently known which exhibit preferential cytotoxicity toward hypoxic cells through reductive activation. They include nitro compounds, quinones, and aromatic *N*-oxides.[25,26] Examples include mitomycin C (**30**), CB1954 (**31**), EO9 (**32**), and AQ4N (**33**).

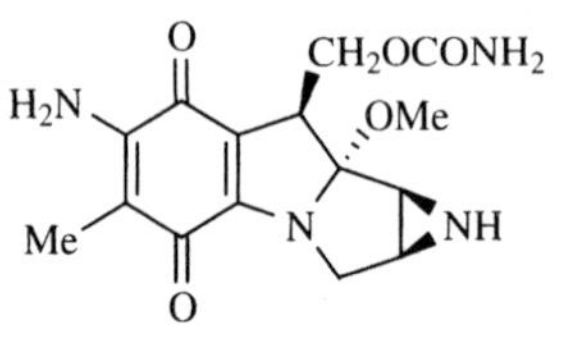

(**30**) Mitomycin C 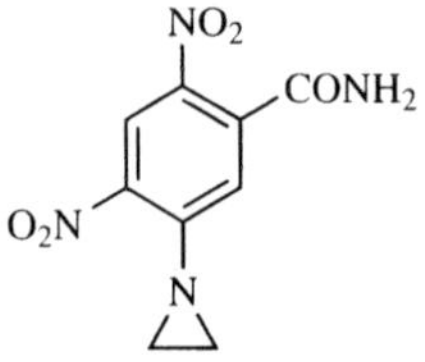 (**31**) CB1954

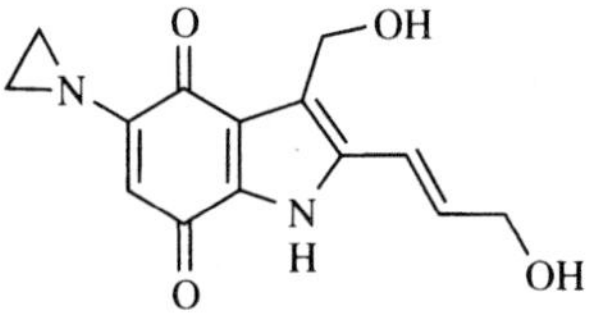

(**32**) EO9 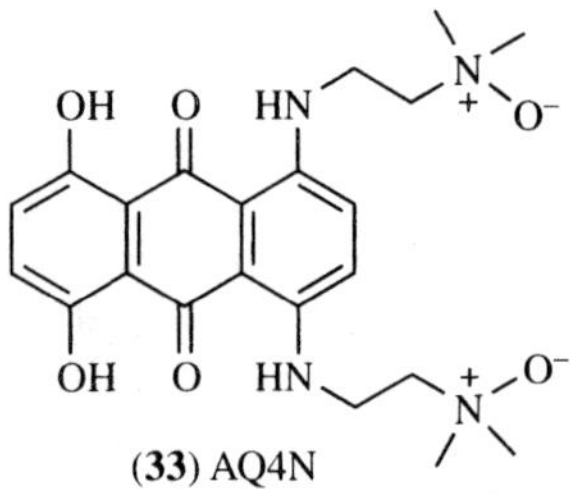 (**33**) AQ4N

8.2.7.2. Activation of Prodrugs by Tissue- or Tumor-Specific Enzymes Investigations of the biochemistry and molecular biology of cancer have also identified several reductive or proteolytic enzymes that are unique to tumors or tissues and could be used as potential therapeutic targets or prodrug-converting enzymes for novel cancer therapy. These include DT-diaphorase,[27] prostate specific antigen (PSA),[28] plasminogen activator,[29] and members of matrix metalloproteinases.[30]

One such example is the peptide doxorubicin conjugate, glutaryl-Hyp-Ala-Ser-Chg-Gln-Ser-Leu-Dox, L-377202 (**34**), which was reported to have the profile of physical and biological properties needed for further clinical development.[31] Conjugate **34** was found to have a greater than 20-fold selectivity against PSA-secreting LNCaP cells relative to non-PSA-secreting DuPRO cells. In nude mouse xenograft studies, it reduced PSA levels by 95% and tumor weight by 87% at 21 μmol/kg, a dose below its maximal tolerated dose (MTD). On the basis of these results, this conjugate was selected for further studies in clinical trials to assess its ability to inhibit human prostate cancer cell growth and tumorigenesis. It was believed that PSA cleavage in and around prostate cancer cells would release, as shown in Scheme 2, dipeptide-doxorubicin conjugate (**35**), which would be further cleaved by aminopeptidases to the cytotoxic Leu-doxorubicin (**36**) and doxorubicin (**37**).

8.2.7.3. Antibody- or Gene-Directed Enzyme Prodrug Therapy Besides targeting hypoxic tumor cells and using tumor- or tissue-specific enzymes like PSA to activate prodrugs, other specific enzymes can be delivered to tumor tissues using antibodies or expressed by tumor cells through gene therapy and can be used as prodrug converting enzymes. These strategies are called "antibody-directed

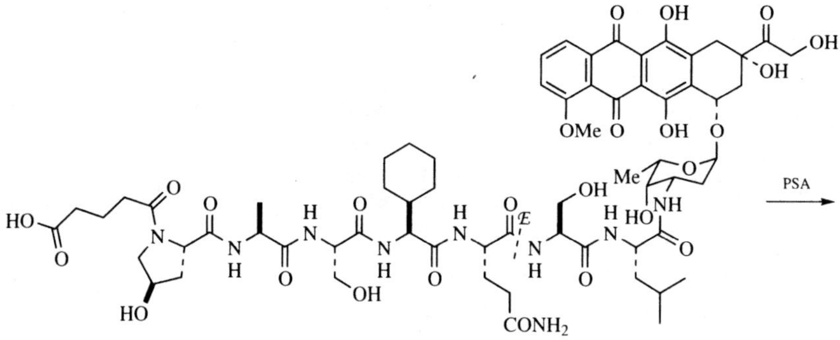

(**34**) L-377202

H-Ser-Leu-doxorubicin (**35**) → H-Leu-doxorubicin (**36**) → Doxorubicin (**37**)

E indicates enzyme cleavage site

Scheme 2

enzyme prodrug therapy" (ADEPT) or "gene-directed enzyme prodrug therapy" (GDEPT). In these approaches, an enzyme is delivered site specifically by chemical conjugation or genetic fusion to a tumor-specific antibody or by enzyme gene delivery systems into tumor cells. This is followed by the administration of a prodrug, which is selectively activated by the delivered enzyme at the tumor cells. A number of these systems are in development and have been reviewed.[32] Among the enzymes under evaluation is a bacterial nitroreductase from *Escherichia coli*. This is a flavin mononucleotide (FMN)–containing flavoprotein capable of reducing certain aromatic nitro groups to the corresponding amines or hydroxylamines in the presence of a cofactor NADH or NADPH. The nitroaromatics that were found to be good substrates of *E. coli* nitroreductase include dinitroaziridinylbenzamide CB1954 (**31**), dinitrobenzamide mustards SN 23862 (**38**), 4-nitrobenzylcarbamates (**39**), and nitrophenyl phosphoramides (**40** and **41**).[33]

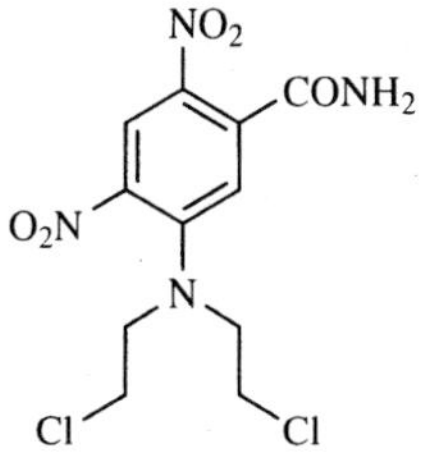

(**38**) SN 23862

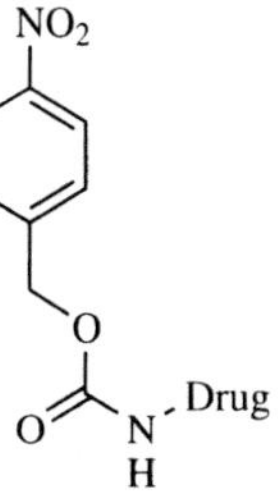

(**39**) 4-nitrobenzylcarbamates

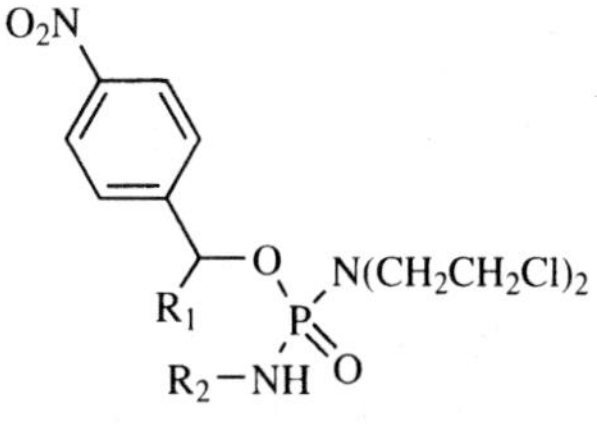

nitrophenyl phosphoramides
(**40**) $R_1 = R_2 = H$ (LH7)
(**41**) $R_1, R_2 = CH_2CH_2$

8.2.7.4. Tumor-Specific Transporters Antibody-drug conjugates would have to overcome problems inherent in proteins such as susceptibility to proteolytic cleavage and high immunogenicity; the latter could lead to an antibody response against the conjugate, thereby precluding further use. To increase the selectivity of chemotherapeutic agents, considerable efforts have also been made to identify biochemical characteristics unique to malignant tumor cells that could be exploited in a therapeutic intervention. The small and nonimmunogenic, tumor-specific molecules like folic acid are among the promising alternatives to antibody molecules as targeting agents for drug delivery. Folate conjugates of radiopharmaceuticals, magnetic resonance imaging (MRI) contrast agents, antisense oligonucleotides and ribozymes, proteins and protein toxins, immunotherapeutic agents, liposomes with entrapped drugs, and plasmids have all been successfully delivered to folate receptor–expressing cells.[34] More details can be found in Chapter 9 on receptor-mediated endocytosis.

8.3. PRODRUG DESIGN CONSIDERATIONS

Often medicinal chemists encounter a situation where a structure has adequate pharmacological activity but an inadequate pharmacokinetic profile (i.e., absorption, distribution, metabolism, and excretion). Prodrugs can be designed to improve physicochemical properties, resulting in improvement in pharmacokinetic as well as pharmaceutical properties. The pharmaceutical properties that could be improved, as discussed earlier, include drug product stability, taste and odor, pain on injection, and gastrointestinal irritation. These are great benefits that can be achieved through the design of prodrugs. However, regulatory issues should also be considered in the design process. In general, regulatory agencies are reluctant to register this type of product. Of particular concern is the fact that toxicological studies might not be relevant for human use of the drug because of differences in the rate and/or extent of formation of the active moiety—metabolic aspects. Experiments should thus be designed early to address these concerns. As examples of interspecies differences, the pivaloyloxyethyl ester of methyldopa was essentially hydrolyzed presystematically to pivalic acid and methyldopa at the same rate in human, dog, and rat, while the succinimidoethyl derivative was hydrolyzed faster in rat than in man and dog.[35] This suggests that the succinimidoethyl ester of methyldopa was more resistant to extrahepatic esterase action in man and dog but not in rat. For different ester prodrugs of dyphylline, the relative rates of release were 1.3 to 13 times faster in rabbit plasma than in human plasma.[36]

The bond between the active moiety (parent drug) and the promoiety plays a major role in determining the pharmacokinetic properties of a prodrug. Knowledge about the nature of the bond and the promoiety may help explain the nature of the biotransformation process and its location in specific tissues or cells. The study of the fate in the body of the promoiety is particularly important from the safety point of view and should be investigated just as thoroughly as that of the active moiety. In some cases, the fate of the released carrier moiety is well known, such as the esters of methanol or ethanol; no extra study is needed during drug development. In other cases, additional pharmacokinetic investigations may be necessary.

Rational design of a prodrug should begin with identification of the problem(s) encountered with the delivery of the parent compound/drug and the physicochemical properties needed to overcome the delivery problem(s). Only then can the appropriate promoiety be selected to construct a prodrug with the proper physicochemical properties that can be effectively transformed to the active drug in the desired biological compartment.

The most important requirement in prodrug design is naturally the adequate reconversion of the prodrug to the active drug *in vivo* at the intended compartment. This prodrug-drug conversion may take place before absorption (e.g., in the gastrointestinal system), during absorption (e.g., in the gastrointestinal wall or in the skin), after absorption, or at the specific site of drug action. It is important that the conversion be essentially complete because the intact prodrug, being usually inactive, represents unavailable drug. However, the rate of conversion would depend on the specific goal of the prodrug design. A prodrug designed to overcome poor solubility for an intravenous drug formulation should be converted very quickly to the active moiety after injection. If the objective of the prodrug is to produce sustained drug action through rate-limiting conversion, the rate of conversion should not be too fast.

Prodrugs can be designed to use a variety of chemical and enzymatic reactions to achieve cleavage to generate their active drug at the desired rate and place. The design is often limited by the availability of a suitable functional group in the active drug for the attachment of a promoiety. Table 8.1 lists some of the common reversible prodrug forms for various functional groups that are often present in biologically active substances.

The most common prodrugs are those that require hydrolytic cleavage, but reductive and oxidative reactions have also been used for the *in vivo* regeneration of the active drug. Besides using the various enzyme systems for the necessary activation of prodrugs, the buffered and relatively constant physiological pH may be used to trigger their release.

Enzymes considered important to orally administered prodrugs are found in gastrointestinal walls, liver, and blood. In addition, enzyme systems present in the gut microflora may be important in metabolizing prodrugs before they reach the intestinal cells. In addition, site-specific delivery can be accomplished by exploiting enzymes that are present specifically or at high concentrations in the targeted tissues relative to nontarget tissues. A number of enzymes can also be delivered to targeted tissues through antibodies or gene-delivery approaches for the activation of subsequently administered prodrugs, as discussed earlier in this chapter.

8.4. PRODRUGS OF VARIOUS FUNCTIONAL GROUPS

8.4.1. Ester Prodrugs of Compounds Containing –COOH or –OH

Due to the presence of a wide variety of esterases in various body tissues, it is not surprising that esters are the most common prodrugs used to improve gastrointestinal absorption. By appropriate esterification of molecules containing a carboxylic acid or hydroxyl group, it is possible to obtain derivatives with almost any desirable

TABLE 8.1 Reversible Prodrug Forms for Various Functional Groups Present in Biologically Active Substances

Funtional Group	Reversible Prodrug Forms
—COOH	—C(=O)O—R; —C(=O)O—CH(R)—O—C(=O)—R′; —C(=O)O—CH(R)—O—C(=O)—O—R′; —C(=O)NH—R; —CH₂C(=O)—R; —CH₂OH
—OH	—O—C(=O)—R; —O—C(=O)—O—R; —O—P(=O)(OH)—OH; —O—CH(R)—O—C(=O)—R′; —O—R
—SH	—S—C(=O)—R; —S—CH(R)—O—C(=O)—R′; —S—S—R
$-NH_2$	—NH—C(=O)—R; —NH—C(=O)—O—R; —NH—C(=O)—O—CH(R)—O—C(=O)—R′; —N=C(R)—R′; —NH—C(R)=C(R′)—R″; Ar—N=N—Ar; $Ar-NO_2$
—N<	>N⁺(—)—CH(R)—O—C(=O)—R′
>C=O	>C=N—R; >C=N—OH; >C=N—O-R; >C=C—O—C(=O)—R; >C(OR)(OR′); oxazolidine (O, NH); thiazolidine (S, NH)
>C=N—OH	>C=N—O-R; >C=N—O—C(=O)—R
>P(=O)OH	>P(=O)—O—R; >P(=O)—O—CH(R)—O—C(=O)—R′; >P(=O)—O—CH(R)—O—C(=O)—O—R′
$X(=O)_n$—NH— (acidic)	$X(=O)_n$—N(—)—CH₂—N(R)(R′); $X(=O)_n$—N(—)—CH₂—OH; X = C, n = 1; X = S, n = 2
XH / NH_2 (1,2-)	oxazolidine/thiazolidine (X, NH); X = O or S
catechol (OH, OH)	cyclohexenone (=O); =N—O—R

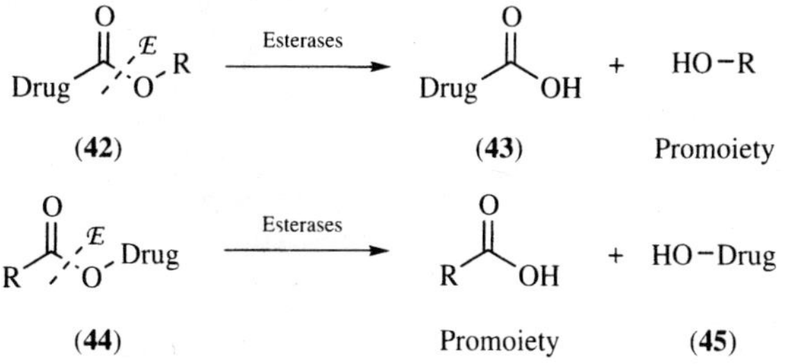

Drug–C(=O)–O–R (**42**) —Esterases→ Drug–C(=O)–OH (**43**) + HO−R (Promoiety)

R–C(=O)–O–Drug (**44**) —Esterases→ R–C(=O)–OH (Promoiety) + HO−Drug (**45**)

Scheme 3

As shown in Scheme 3, esters in the form of **42** can be used as prodrugs for acid drugs (**43**), and the alcohol would serve as a promoiety. Esters in the form of **44** can be used as prodrugs for alcohol drugs (**45**), and here the acid would serve as a promoiety.

Both the acyl and the alcohol portion surrounding the cleavable ester bond affect the enzyme-catalyzed ester hydrolysis. Sometimes because of steric hindrance in the active drug, direct ester formation with the existing functional group might not produce a prodrug that is sufficiently labile *in vivo*. This problem can be solved by designing the so-called cascade prodrugs containing double esters using α-acyloxyalkyl, carbonate, or alkoxycarbonyloxyalkyl esters (**46**, **47**, or **48**), where the terminal ester group is accessible for enzymatic cleavage (Scheme 4). Cascade prodrugs are those prodrugs that require a sequence of two or more reactions for drug release and activation, usually triggered by a first enzymatic-catalyzed reaction followed by a spontaneous chemical release/activation step(s). A number of such examples are known, including several prodrugs of β-lactam antibiotics, corticosteroids, and angiotensin II receptor antagonists. The 2-carboxylic acid on the thiazolidine ring of β-lactam antibiotics is required for antibacterial activity,

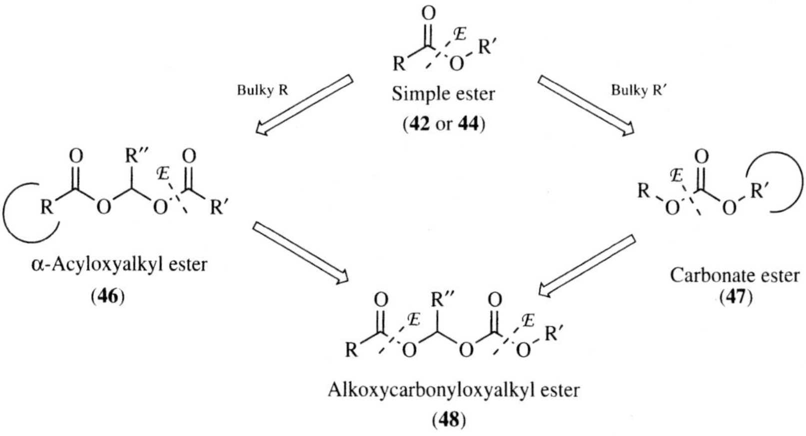

Simple ester (**42** or **44**)

Bulky R → α-Acyloxyalkyl ester (**46**)

Bulky R′ → Carbonate ester (**47**)

Alkoxycarbonyloxyalkyl ester (**48**)

Scheme 4

providing an ideal site for attaching a promoiety in the design of ester prodrugs. But, because of steric hindrance, simple esters of this carboxylic acid group would resist enzymatic hydrolysis. Thus, a number of cascade prodrugs were made to extend the chain and render the terminal ester group easily accessible to hydrolytic enzymes. Examples of α-acyloxyalkyl ester prodrugs include bacampicillin (**2**), pivampicillin (**3**), pivmecillinam (**6**), and cefuroxime axetil (**49**).

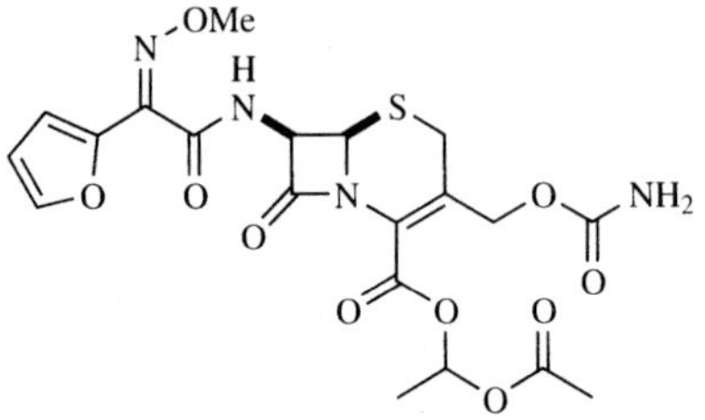

(**49**) Cefuroxime axetil

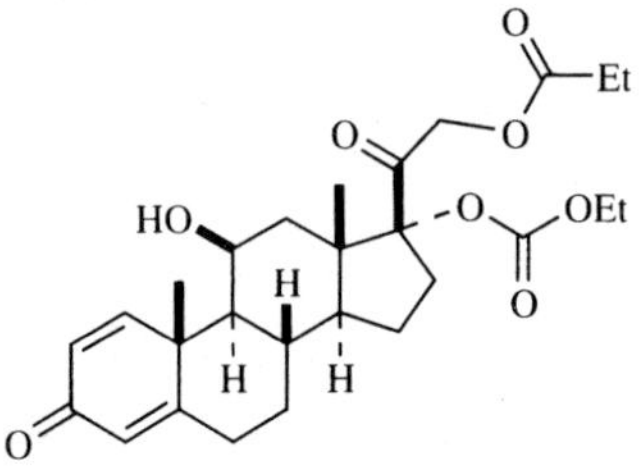

(**50**) Prednicarbate

Prednicarbate (**50**) is an example of a carbonate prodrug of corticosteroids, while candesartan cilexetil (**51**) is a racemic mixture of an alkoxycarbonyloxyalkyl ester of candesartan (**53**) with a chiral center at the carbonate ester group. Following oral administration, candesartan cilexetil (**51**) undergoes rapid and complete hydrolysis during absorption from the gastrointestinal tract to form, as shown in Scheme 5, the

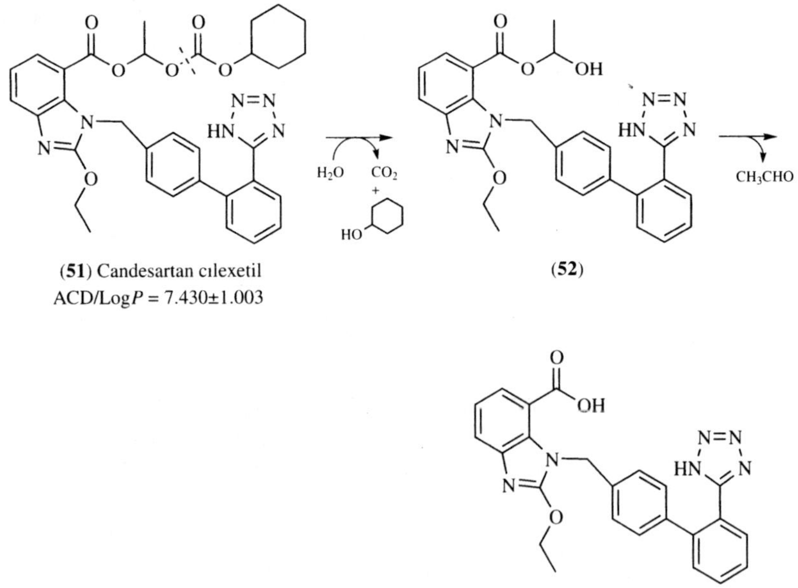

(**51**) Candesartan cilexetil
ACD/Log$P$ = 7.430±1.003

H_2O → CO_2 + cyclohexanol

(**52**)

→ CH_3CHO

(**53**) Candesartan
ACD/Log$P$ = 4.651±0.930
ACD/Log$D$ = 0.54 at pH 7

Scheme 5

active drug candesartan (**53**), which is an achiral selective AT1 subtype angiotensin II receptor antagonist.[37]

For a more recent effort to find orally active aminomethyl-THF 1β-methylcarbapenems (Scheme 6), a number of mono and bis double esters were investigated as

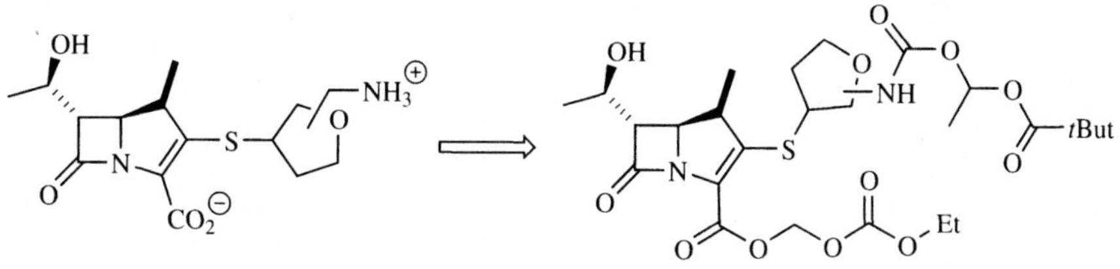

Aminomethyl-THF1β-methylcarbapenems (**54**) A bis double ester prodrug (**55**)

Scheme 6

potential prodrugs. The bis double ester derivatives such as (**55**) demonstrated enhanced oral activity, while the mono double ester derivatives did not demonstrate significantly improved oral activity due to the presence of a charged group.[38]

The α-acyloxylalkyl esters have also been extended to include the phosphate group, phosphonic acids, and phosphinic acids. One such example is fosinopril (**57**), an ACE inhibitor, where the phosphinic acid is *O*-α-acyloxyalkylated to increase lipophilicity to provide better absorption.

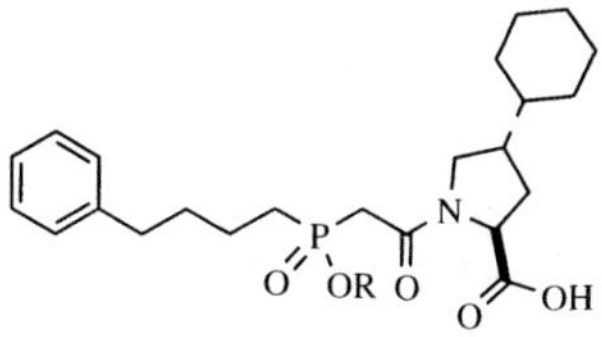

		ACD/Log*P*	ACD/Log*D* pH 7	ACD/Log*D* pH 1
(**56**) Fosinoprilat	R = H	3.069±0.660	−1.93	3.06
(**57**) Fosinopril	R = 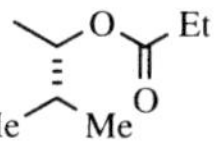	5.810±0.679	2.65	5.81

O-α-acyloxyalkyl ethers are also a useful prodrug type for compounds containing a phenol group. Such derivatives (**58**) are hydrolyzed by a sequential reaction involving formation of an unstable hemiacetal intermediate (**59**), as shown in Scheme 7. These kinds of ethers might be better prodrugs than normal phenol esters

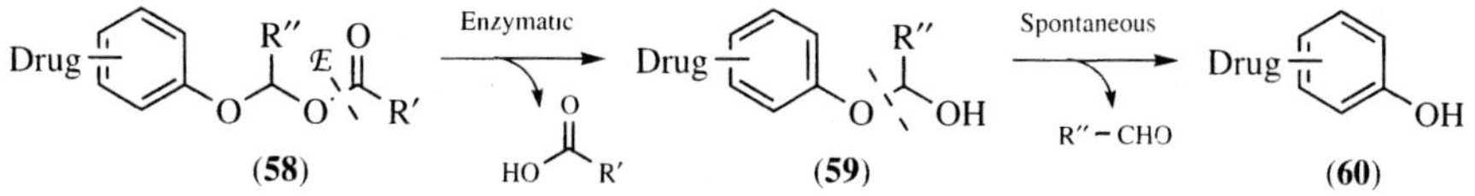

(**58**) Enzymatic (**59**) Spontaneous $R''-CHO$ (**60**)

Scheme 7

because they are more stable against chemical hydrolysis, but they are still susceptible to enzymatic hydrolysis by human plasma esterases.

Carboxylic acids have also been masked as ketones and alcohols, which would require oxidation to convert to the active acid drugs. Nabumetone (**61**) is a nonacidic nonsteroidal anti-inflammatory prodrug (NSAID).[39] After absorption, nabumetone undergoes extensive metabolism, the main circulating active form is 6-methoxy-2-naphthylacetic acid (**62**), a potent COX-2 inhibitor (Scheme 8). Since

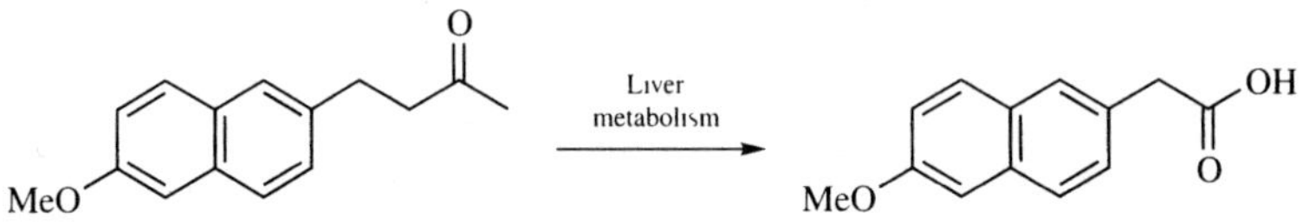

Liver metabolism

(**61**) (**62**)

Scheme 8

nabumetone is not acidic and the active acid metabolite does not undergo enterohepatic circulation, nabumetone does not cause gastric irritation and is the most widely prescribed NSAID in the United States.

8.4.2. Prodrugs of Compounds Containing Amides, Imides, and Other Acidic NH

8.4.2.1. Mannich Bases Mannich base prodrugs could enhance the delivery of their parent drugs through the skin because of their enhanced water solubility as well as enhanced lipid solubility. *N*-Mannich bases, or *N*-acyl *gem*-diamines (**67**), are generally formed, as shown in Scheme 9, by reaction of an acidic NH com-

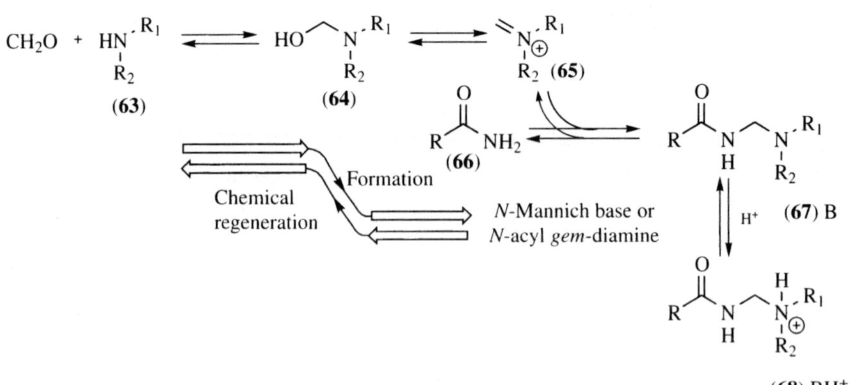

CH_2O + (**63**) ⇌ (**64**) ⇌ (**65**)

(**66**) Formation / Chemical regeneration

N-Mannich base or *N*-acyl *gem*-diamine

(**67**) B ⇌ H^+ (**68**) BH^+

Scheme 9

pound (**66**) with an aldehyde, usually formaldehyde, and a primary or secondary aliphatic amine (**63**). Aromatic amines do not usually undergo this reaction. Mannich base prodrugs are regenerated by chemical hydrolysis without enzymatic catalysis in the reverse direction of their formation.[40]

Transformation of an amide to an *N*-Mannich base introduces a readily ionizable amino functional group (**67** $\rightleftharpoons$ **68**) that would allow the preparation of sufficiently stable derivatives with greatly enhanced water solubility at slightly acidic pH. Clinically useful *N*-Mannich base prodrugs include rolitetracycline and hetacillin. The highly water-soluble rolitetracycline (**70**) is an *N*-Mannich base of tetracycline (**69**) with pyrrolidine and is decomposed to tetracycline quantitatively with a half-life of 40 minutes at pH 7.4 and 35°C.[41] Since the decomposition of *N*-Mannich bases does not rely on enzymatic catalysis, the rate of hydrolysis is the same in plasma and in buffer. Hetacillin (**71**) is an example of a cyclic *N*-Mannich base-type prodrug, which is formed by condensation of ampicillin with acetone. The prodrug is readily converted back to the active ampicillin and acetone, with a half-life of 15–20 minutes at pH 4–8 and 35°C.[42,43] The advantage of hetacillin is its higher stability in concentrated aqueous solutions as compared to ampicillin, which has a more nucleophilic amine that would react with the strained β-lactam ring.

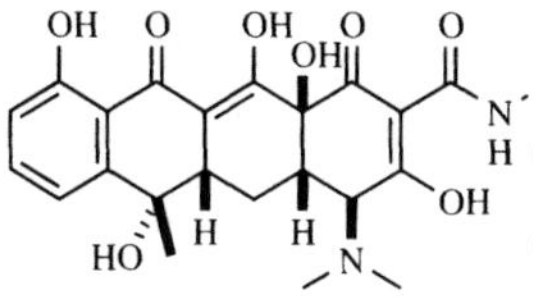

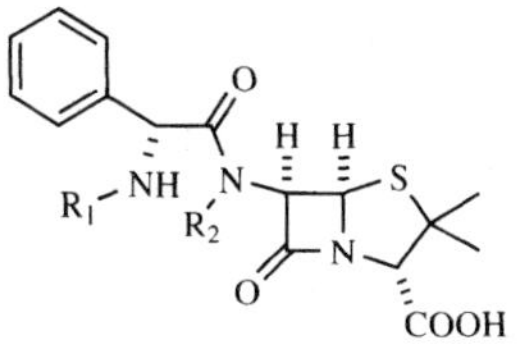

Compound	R	ACD/Log*P*	ACD/Log*D* pH 7	ACD/Log*D* pH 1
(**69**) Tetracycline	R = H	−1.187±0.750	−4.78	−4.29
(**70**) Rolitetracycline	R = $-CH_2$-N(pyrrolidine)	0.150±0.750	−3.02	−4.84
(**1**) Ampicillin	$R_1 = R_2 = H$	1.350±0.320	−1.54	−1.72
(**71**) Hetacillin	$R_1,R_2 = >C(CH_3)_2$	2.301±0.901	−1.61	−0.29

8.4.2.2. N-α-Acyloxyalkyl Derivatives *N*-α-Acyloxyalkylation has become a commonly used approach to obtain prodrug forms of various NH-acidic drug substances such as carboxamides, carbamates, ureas, and imides. This is because *N*-α-acyloxyalkyl derivatives (**72**) combine high *in vitro* stability with enzymatic lability. The derivatives are cascade or double prodrugs. The regeneration of the parent drug occurs via a two-step mechanism, the enzymatic cleavage of the ester group followed by spontaneous decomposition of the *N*-α-hydroxyalkyl intermediate (**73**) (Scheme 10). The usefulness of this approach depends on the stability of the *N*-α-acyloxyalkyl derivative **72**, its susceptibility to esterase-catalyzed hydrolysis, and the rate of decomposition of the intermediate **73**. *N*-α-acyloxyalkyl derivatives of imides and secondary amides, as well as ring structures containing such moieties, showed normal ester stability.[44] To make such *N*-α-acyloxyalkyl derivatives useful as prodrugs, the α-hydroxyalkyl intermediate (**73**) formed after the enzyme-catalyzed hydrolysis must decompose quickly to release the original

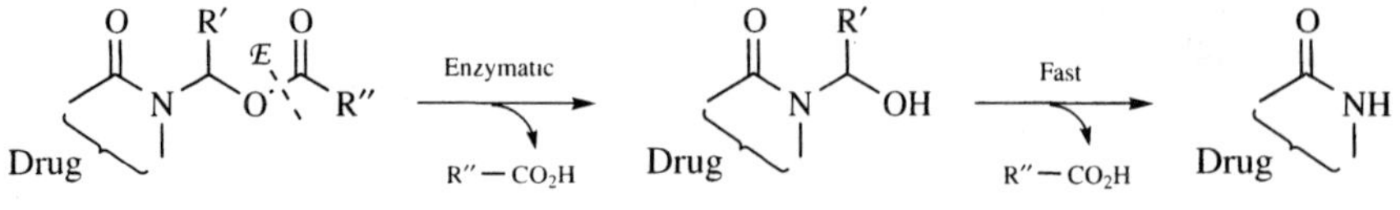

Scheme 10

drug molecule (**74**). The rate of the chemical decomposition step was found to correlate with the pK_a of the acidic NH group; a pK_a of less than 10.5 is required for instantaneous decomposition of *N*-hydroxymethyl derivatives.

However, *N*-α-acyloxyalkyl derivatives (**75**) of primary amides, and other primary amide-type structures such as carbamates and sulfonamides, are extremely unstable in aqueous solution and quickly undergo decomposition to the corresponding *N*-hydroxymethyl derivatives, which are stable. Such derivatives of simple primary amides decompose by an elimination-addition mechanism involving a reactive *N*-acylimine intermediate (**76**) (Scheme 11). For imides and secondary

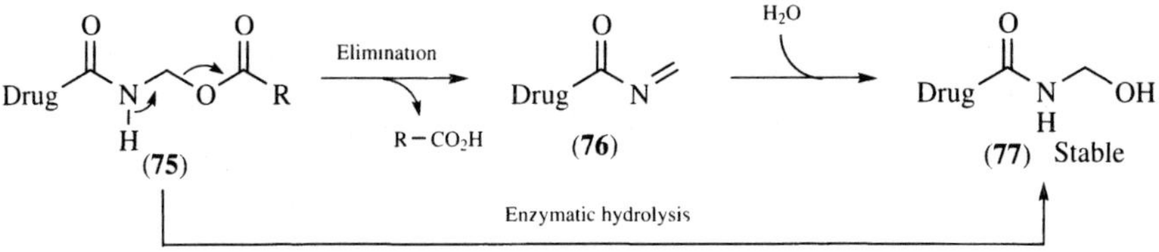

Scheme 11

amides, their inability to form an *N*-acylimine is believed to contribute to the stability of their *N*-α-acyloxyalkyl derivatives (**72**). At pH 4 and 37°C, half-lives of hydrolysis of *N*-α-acyloxyalkyl derivatives (**75**) range from 1 to 90 minutes, whereas at pH 7.4 the half-lives of hydrolysis are < 1 minute.[45] The resulting *N*-hydroxymethyl derivatives (**77**) are rather stable; the half-life for the decomposition of *N*-(hydroxymethyl) benzamide is 183 hours at pH 7.4 and 37°C.[45] However, aldehydes other than formaldehyde can be used to from *N*-α-hydroxyalkyl derivatives that are more unstable than *N*-hydroxymethyl analogs. For example, the half-life for *N*-(α-hydroxybenzyl)benzamide is only 6.5 minutes at pH 7.4 and 37°C. The use of aldehydes other than formadehyde may further expand the applicability of this approach to simple amides with pK_a above 11.[45]

8.4.3. Prodrugs of Amines

The presence of an amino group can affect a drug's physicochemical and biological properties in several ways. These include (1) intermolecular or intramolecular aminolysis leading to reactive and/or potentially toxic substances, (2) solubility problems when the drug is present with another ionizable functionality such as COOH (zwitterionic nature under physiological pH, potentially limiting its dissolution rate and/or its passive permeability), and (3) terminal free amino acid groups providing recognition sites for proteolytic enzymes, such as aminopeptidase and trypsin, present in the gastrointestinal tract lumen, the brush border region, and the cytosol of the intestinal mucosa cells. For all these reasons, prodrug approaches have been advocated for improving *in vivo* behavior of active compounds containing amino groups.

8.4.3.1. Amides Because of the relatively high stability of amides *in vivo*, *N*-acylation of amines was formerly of limited use in prodrug design. Only a few examples of simple amide prodrugs are known that are sufficiently labile *in vivo*; these include the *N*-L-isoleucyl derivative of dopamine[46] and the *N*-glycyl derivative midodrine.[47,48] With the use of proteases as prodrug-converting enzymes, amines can be coupled to peptide carboxylates, resulting in amide bonds cleavable by proteases (e.g., **34**).

Midodrine (**79**) is a glycinamide prodrug, and the therapeutic effect of orally administered midodrine is due to the major metabolite desglymidodrine (**78**), an α-agonist formed by deglycination of midodrine. Midodrine is rapidly absorbed after oral administration. The plasma level of the prodrug peaks after about half an hour, and declines with a half-life of approximately 25 minutes, while the metabolite reaches peak blood concentrations about 1 to 2 hours after a dose of midodrine and has a half-life of about 3 to 4 hours. The absolute bioavailability of midodrine (measured as desglymidodrine) is 93% and is not affected by food. Approximately the same amount of desglymidodrine is formed after intravenous and oral administration of midodrine. Midodrine has been used successfully in the treatment of neurogenic orthostatic hypotension and, more recently, in the treatment of dialysis hypotension. It acts through vasoconstriction of the arterioles and the venous capacitance vessels, thereby increasing peripheral vascular resistance and augmenting venous return, respectively. The prodrug is a unique agent in the armamentarium against orthostatic hypotension since it has minimal cardiac and central nervous system (CNS) effects.[47,48]

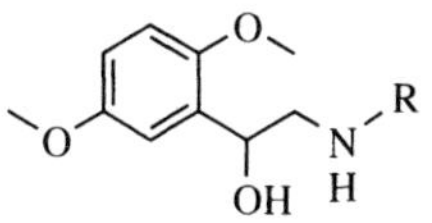

		ACD/Log*P*	ACD/Log*D* pH 7	ACD/Log*D* pH 1
(**78**) Desglymidodrine	R = H	0.378±0.279	−1.14	−3.42
(**79**) Midodrine	R = -$COCH_2NH_2$	−0.319±0.595	−0.83	−2.72

8.4.3.2. N-α-Acyloxyalkoxycarbonyl Derivatives Carbamates are of limited use in prodrug design due to their general resistance to enzymatic cleavage *in vivo*. The introduction of an enzymatically labile ester group in the carbamate structure could render them sensitive to esterase-catalyzed hydrolysis leading to activation. Thus, *N*-α-acyloxyalkoxycarbonyl derivatives (**80**) of primary and secondary

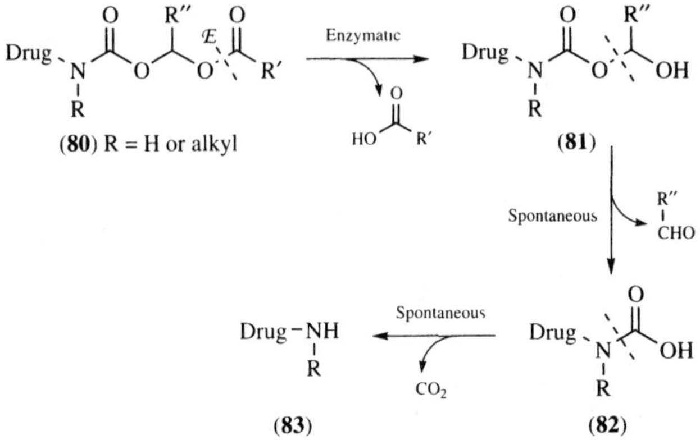

Drug–N(R)–C(=O)–O–CH(R″)–O–C(=O)–R′ (80) R = H or alkyl → Enzymatic (– HO–C(=O)–R′) → Drug–N(R)–C(=O)–O–CH(R″)–OH (81) → Spontaneous (– R″–CHO) → Drug–N(R)–C(=O)–OH (82) → Spontaneous (– CO_2) → Drug–NH–R (83)

Scheme 12

amines may be readily transformed, as shown in Scheme 12, to the parent amine (**83**) *in vivo*.[49,50] Esterase-catalyzed hydrolysis of the ester moiety in these derivatives leads to an unstable α-hydroxyalkoxycarbonyl intermediate (**81**) which spontaneously decomposes into the parent amine via a labile carbamic acid (**82**). These α-acyloxyalkyl carbamate derivatives are neutral and combine high stability in aqueous solution with high susceptibility to enzymatic reconversion to the active agent triggered by hydrolysis of the terminal ester functions; they may be promising reversible prodrugs for amino-containing compounds. These are used in the preparation of orally active aminomethyl-THF 1β-methylcarbapenems (**56** → **57**).

This approach has also been applied to peptides and peptidomimetics in order to improve their unfavorable physicochemical characteristics (e.g., size, charge, hydrogen-bonding potential), which prevent them from permeating biological barriers such as the intestinal mucosa, and by their lack of stability against enzymatic degradation.[51,52] Many of the structural features of a peptide, such as the *N*-terminal amino group, the *C*-terminal carboxyl group, and the side chain carboxyl, amino, and hydroxyl groups, which bestow upon the molecule affinity and specificity for its pharmacological receptor, severely restrict its ability to permeate biological barriers and render it as a substrate of proteases. Bioreversible cyclization of the peptide backbone is one of the most promising new approaches in the development of peptide prodrugs. Cyclization of the peptide backbone enhances the extent of intramolecular hydrogen bonding and reduces the potential for intermolecular hydrogen bonding to aqueous solvent. Linking the *N*-terminal amino group to the *C*-terminal carboxyl group via an α-acyloxyalkoxy promoiety, as in (**84**), is an interesting approach that has been shown to work on a number of model peptides (Scheme 13). These cyclic prodrugs were designed to be susceptible to esterase-catalyzed hydrolysis (slow step), leading to a cascade of chemical reactions resulting in the generation of the linear peptide. In pH 7.4 buffer at 37°C, the cyclic prodrugs (**84**) were shown to degrade quantitatively to their corresponding linear peptides (**87**). In human plasma, the rates of hydrolysis of cyclic prodrugs were significantly faster

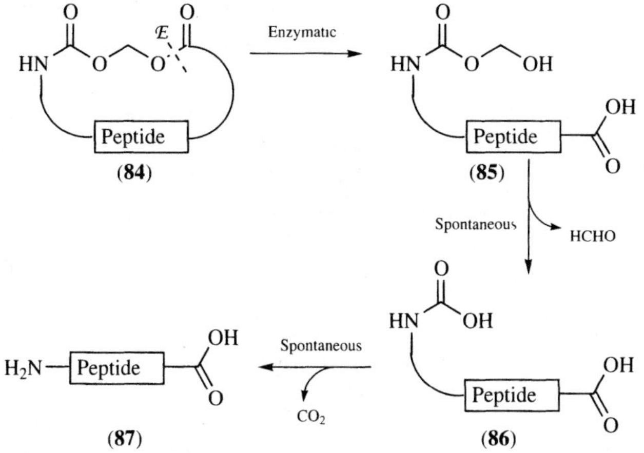

Scheme 13

than in buffer and were inhibited by paraoxon, a potent esterase inhibitor. In comparison to the linear peptides, the cyclic prodrugs were at least 70 times more permeable in cell culture models of the intestinal mucosa.

8.4.3.3. N-Mannich Bases *N*-Mannich bases have been used successfully to obtain prodrugs of amide- (see Section 8.4.2.1) as well as amine-containing drugs. Due to their rapid cleavage, with half-lives between 10 and 40 minutes at physiological pH and a pronounced decrease in their basicity of 3–4 pK_a units,[53] salicylamide *N*-Mannich bases (**90**) were evaluated as prodrug forms for primary and secondary amines (Scheme 14). In this case, the amide part of a Mannich base is the promoiety. To improve their stability *in vitro* and avoid stability-associated formulation problems, the hydroxyl group of the salicylamide *N*-Mannich bases (**90**)

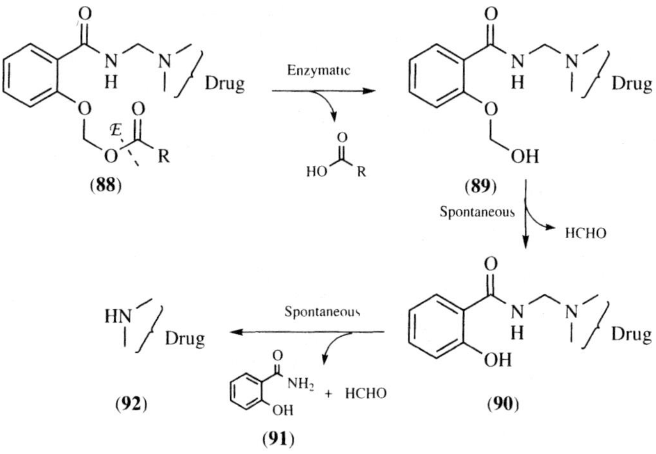

Scheme 14

can be blocked by *O*-acyloxymethylation. *O*-Acyloxymethylated derivatives (**88**) were much more stable in acidic and neutral aqueous solutions than the parent salicylamide *N*-Mannich base (**90**) and could be readily converted to the latter in the presence of human plasma by enzymatic hydrolysis. In addition to providing an *in vitro* stabilizing effect, the concept of *O*-acyloxymethylation makes it possible to obtain prodrug derivatives of a given amine drug with varying physicochemical properties of importance for drug delivery, such as lipophilicity and water solubility. This can simply be effected by the selection of an appropriate α-acyloxymethyl group.[54]

8.4.3.4. Azo Prodrugs Amines have been incorporated into an azo linkage to form prodrugs that can be activated through azo reduction. In fact, sulfa drugs were discovered because of prontosil (**93**), an inactive azo dye that was converted *in vivo* to the active sulfanilamide (**95**) (Scheme 15). Clinically useful balsalazide (**23**), olsalazine (**25**), and sulfasalazine (**26**) are azo prodrugs of mesalazine (**27**). They are converted *in vivo* by bacterial azo reductases in the gut to the active 5-aminosalicylic acid (5-ASA or mesalazine, **27**), which is responsible for their anti-inflammatory activity in the treatment of ulcerative colitis, as discussed earlier.

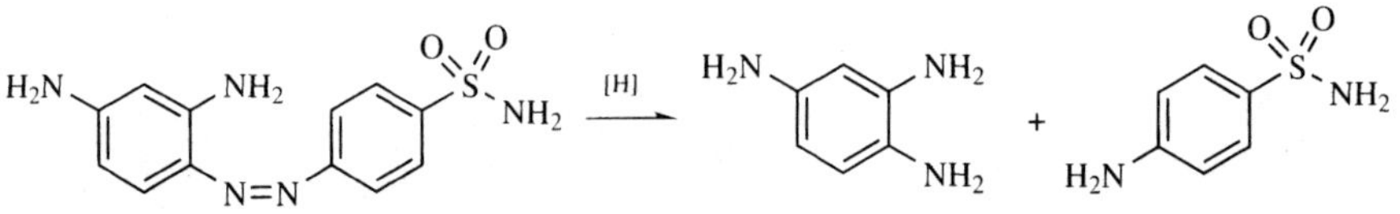

(**93**) Prontosil —[H]→ (**94**) + (**95**) Sulfanilamide

Scheme 15

8.4.3.5. Schiff Base Prodrugs Amines can form reversible Schiff bases with aldehydes and ketones. Although they are of limited use in small-molecule prodrugs, Schiff bases have been used to conjugate amine-containing drugs to polymers with carbonyl groups as macromolecular prodrugs for slow release and targeting. Doxorubicin was conjugated to polyethylene glycol (PEG) through a Schiff base linkage, and the resulting conjugate was found to release doxorubicin under the lysosomal acidic conditions *in vitro* and very slowly under physiological conditions. Moreover, the conjugate showed strong cytotoxic activity similar to that of free doxorubicin against lymphocytic leukemia cells *in vitro*.[55]

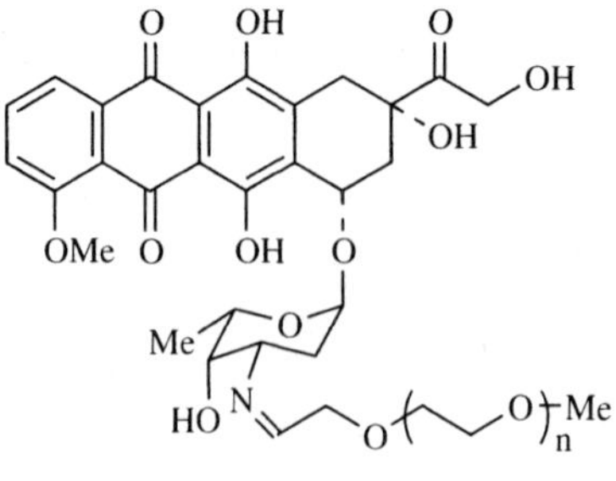

(**96**)

8.4.4. Prodrugs for Compounds Containing Carbonyl Groups

8.4.4.1. Schiff Bases and Oximes Schiff bases and oximes formed from ketones or aldehydes with amines or hydroxyl amines are chemically reversible under acidic or basic conditions. They could be used as prodrugs of compounds containing either an amine or carbonyl functionality.

Oximes of enones (**98**, **100**, and **101**) have been used as prodrugs of contraceptive norethindrone (**97**) and levonorgestrel (**99**). The oximes are highly bioavailable and are converted *in vivo* through chemical hydrolysis to their corresponding active drugs.[56,57]

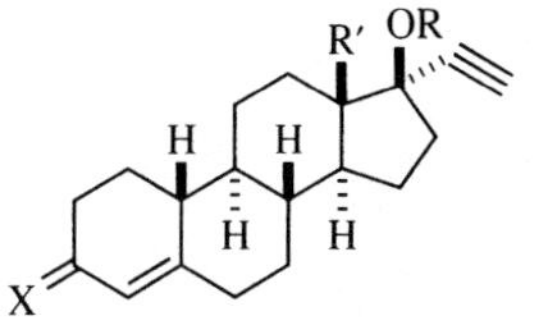

Compound	Substituents	ACD/Log*P*
(**97**) Norethindrone	R = H, R′ = Me, X = O	3.384±0.348
(**98**) Norethindrone-3-oxime	R = H, R′ = Me, X = N-OH	3.866±0.607
(**99**) Levonorgestrel	R = H, R′ = Et, X = O	3.916±0.348
(**100**) Levonorgestrel-3-oxime	R = H, R′ = Et, X = N-OH	4.398±0.607
(**101**) Norgestimate	R = Ac, R′ = Et, X = N-OH	5.002±0.611

A more recent application of oxime derivatives as prodrugs is the design of cascade prodrugs of dopamine agonists for the treatment of Parkinson's disease. As shown in Scheme 16, enones such as *S*-(-)-6-(*N*,*N*-di-*n*-propylamino)-3,4,5,6,7,8-hexahydro-2*H*-naphthalen-1-one (**103**) can be oxidized *in vivo* to catecholamines

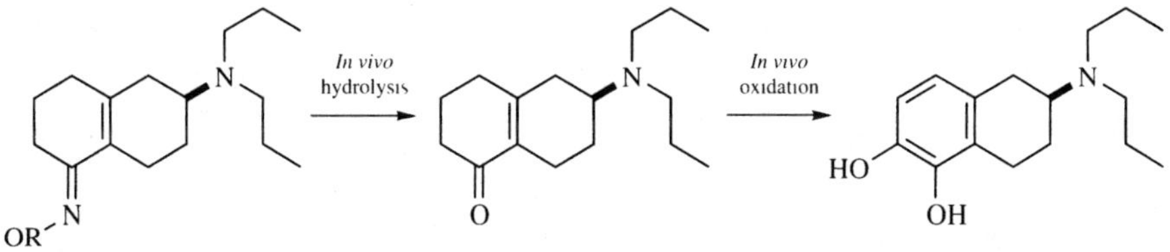

In vivo hydrolysis → *In vivo* oxidation →

(**102**) Oxime prodrug (**103**) (**104**) (−)-5,6-diOH-DPAT

Scheme 16

such as (−)-5,6-dihydroxy-2-(*N*,*N*-di-*n*-propylamino)tetralin ((−)-5,6-diOH-DPAT, **104**) which are known as "mixed dopamine D_1/D_2 agonists" with potential utility in the treatment of Parkinson's disease.[58] Upon oral administration of catecholamines, the phenol and catechol moieties are rapidly metabolized to an extent that limits the therapeutic usefulness of these compounds. Enones such as **103** are prodrugs of such catecholamines and have been shown to improve their bioavailability and extend the duration of action. Compound **103** was found to be efficacious *in vivo* in models for Parkinson's disease in the rat. To potentially further increase the usefulness of enone **103**, a number of oxime ethers and oxime esters (**102**) were prepared as potential cascade prodrugs.[59] It was found that the unsubstituted oxime and the acetyl-oxime induced a pronounced and long-lasting effect *in vivo*. The oxime derivatives were readily hydrolyzed under acidic and alkaline conditions. The fact that these oximes as well as **103** were inactive at the dopamine receptor, yet induced dopamine D_1 and D_2 receptor-related effects *in vivo*, suggested that they were acting as prodrugs and were being converted *in vivo* to the active species (**104**).

Oximes can be acylated to make prodrugs, as in the above example and in the case of FLM 5011, which is a strongly lipophilic, poorly water-soluble, lipoxygenase inhibitor. The water solubility was mproved by using the succinate monoester prodrug. The bioavailability of FLM 5011 in rabbits after oral administration was markedly increased by its prodrug.[60]

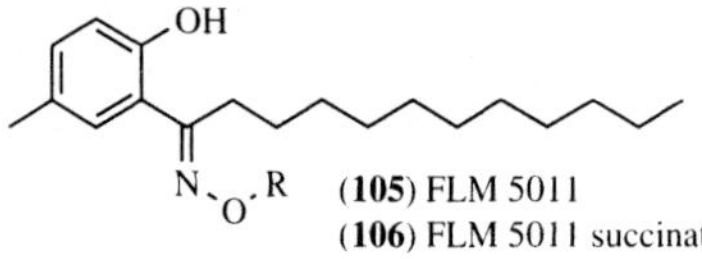

		ACD/Log*P*	ACD/Log*D* pH 7	ACD/Log*D* pH 1
(**105**) FLM 5011	R = H	7.705±0.553	7.70	7 70
(**106**) FLM 5011 succinate	R = $COCH_2CH_2COOH$	7.708±0.599	5.23	7.71

8.4.4.2. Enol Esters Enol esters are rather stable, bioreversible derivatives of ketones and may be useful as prodrugs of agents containing enolizable carbonyl groups. As shown in Scheme 17, 6′-acetylpapaverine enol esters (**107**), prepared by acylation of the appropriate Li enolate with the respective anhydride, were hydrolyzed to 6′-acetylpapaverine (**108**) by esterases present in rat and human plasma, rat liver, and brain tissue supernatants. The intermediate 6′-acetylpapaverine cyclizes rapidly to coralyne (**109**), which has antitumor activity but has

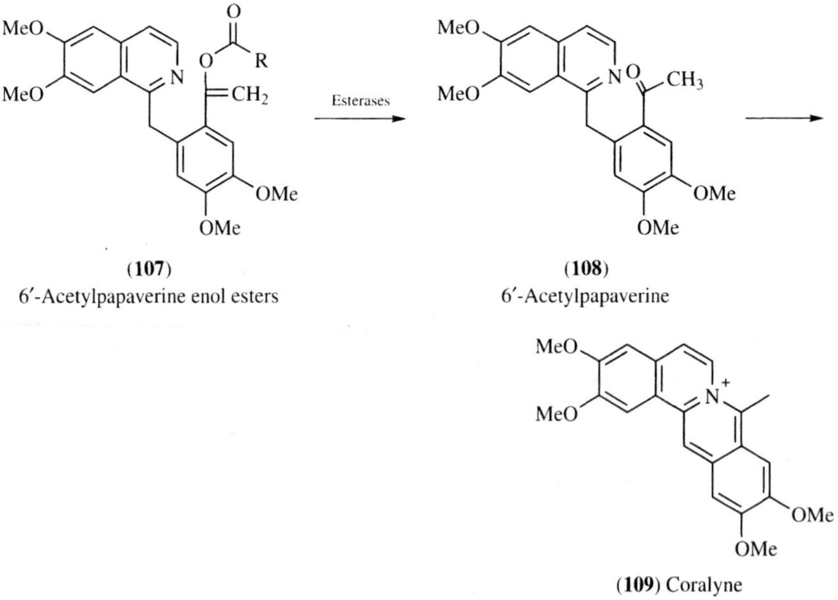

Esterases

(**107**)
6′-Acetylpapaverine enol esters

(**108**)
6′-Acetylpapaverine

(**109**) Coralyne

Scheme 17

difficulty passing through the blood-brain barrier due to the presence of the positive charge. 6′-Acetylpapaverine enol esters are neutral and stable in aqueous solution and are potential prodrugs for enhancing delivery of coralyne to brain tissues.[61]

8.4.4.3. Oxazolidines The kinetics of hydrolysis of several oxazolidines derived from Tris and various aldehydes and ketones were investigated to explore their suitability as prodrug forms for β-aminoalcohols such as (−) ephedrine (**110**) and for carbonyl-containing substances.[62-64]

Oxazolidines were easily and completely hydrolyzed at pH 1–11 at 37°C. The hydrolysis rates were subject to general acid-base catalysis by buffer substances and depended strongly on pH. Most oxazolidines showed sigmoidal pH-rate profiles with maximum rates at pH 7–7.5. At pH 7.4 and 37°C, the half-lives of hydrolysis for the various ephedrine oxazolidines (**111**) ranged from 5 seconds to 30 minutes (Scheme 18). The reaction rates in neutral and basic solutions decreased with

Oxazolidines (**111**) → (**110**) (−) ephedrine

R, R′	$t_{1/2}$ (s)
H, H	5
Et, H	18
Ph, H	300
2-HOPh, H	5
tBut, H	1800
Me, Me	280
Cyclohexylidene	360

Scheme 18

increasing steric effects of the substituents derived from the carbonyl component and decreased with increasing basicity of oxazolidines. Oxazolidines are weaker bases (pK_a 5.2–6.9) than the parent β-amino alcohol and are more lipophilic at physiological pH. Thus, oxalolidines can be considered as potentially useful prodrugs for drugs containing a β-amino alcohol moiety or carbonyl groups.[62,63] Molecular complexation with cyclodextrins might be able to enhance the stability of oxazolidine prodrugs to make them potentially more useful.[65,66]

The stability characteristics of various *N*-acylated oxazolidines were also studied in an attempt to develop approaches which may solve the stability problems associated with the use of oxazolidines as prodrug forms. The *N*-acylated oxazolidines, including a carbamate derivative, were in fact found to be highly stable in an aqueous solution, but they also proved to be resistant to hydrolysis by plasma enzymes. The latter limits the use of *N*-acylated oxazolidines in prodrug design.[64]

8.4.4.4. Thiazolidines Thiazolidines of some α,β-unsaturated 3-ketone steroids including progesterone, testosterone, and hydrocortisone (**112, 113**) were prepared from the reaction with cysteine alkyl esters and cysteamines as potential prodrugs.[67,68] The thiazolidines readily reverted to their parent steroidal ketones, thus meeting the requirements for a prodrug. Most of the thiazolidines were more lipophilic than their parent steroids, thereby imparting the desired change in the physicochemical properties to the derivatives of the steroids. Thus, they can function as bioreversible derivatives of the parent steroids, cysteines, and cysteamines. Cysteine derivatives are particularly attractive as promoieties due to the release of cysteine as the by-product upon activation. Both cysteines and cysteamines were also used as chemoprotective agents against side effects of chemotherapy and radiation therapy. Thus, thiazolidines of cysteines and cysteamines could be used as prodrugs for chemoprotection during chemotherapy and radiation therapy.

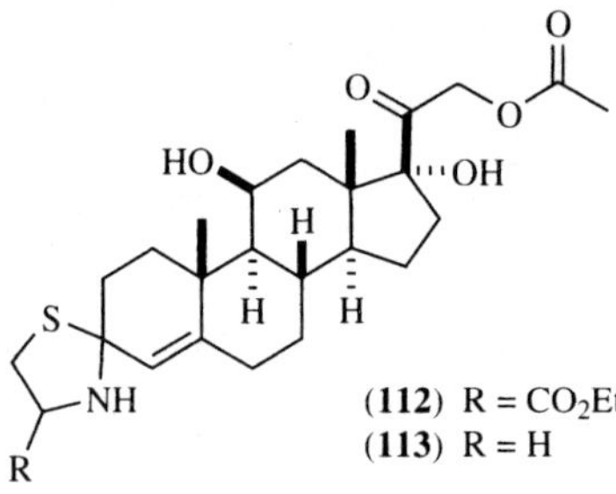

(**112**) R = CO_2Et
(**113**) R = H

8.5. DRUG RELEASE AND ACTIVATION MECHANISMS

Most prodrugs rely on enzymatic hydrolysis by esterases or proteases and, to a less extent, on chemical hydrolysis to achieve a one-step cleavage of the promoiety and the release of the original active drug (Figure 8.1Ai). These systems, as well as other one-step activation mechanisms (Figure 8.1Aii–iv) are simple and, in many cases, sufficient in achieving useful regeneration rates of the active agent. Otherwise, a cascade release/activation mechanism can be incorporated by taking advantage of autodegradation or intramolecular cyclization reactions to effect the release and activation of a prodrug. Some of the cascade strategies that have been employed in the design of prodrugs are shown in Figure 8.1B and will be briefly discussed here.

8.5.1. Cascade Release Facilitated by Linear Autodegradation Reactions

A number of examples are known for the release of an active drug facilitated by a linear autodegradation process. This can be achieved through chemically unstable intermediates such as α-hydroxy amines, amides, and esters. Many of the double prodrugs discussed earlier belong to this category.

Another interesting linear autodegradation process often used in prodrug design involves an electron "push and pull" mechanism through a conjugative aromatic ring that is linked to a good leaving group such as an ester in the *para* benzylic position. In such an approach, an electron-donating amino or hydroxy is masked as an electron-withdrawing nitro group or an amide or ester group in the prodrug.

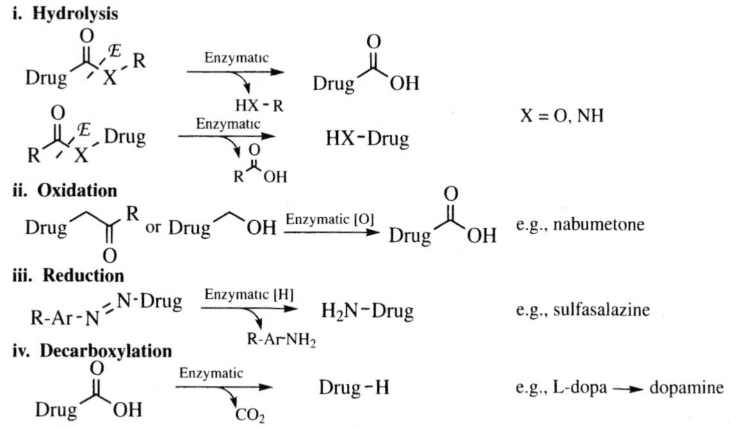

B. Cascade release/activation mechanisms initiated by an enzymatic triggering step

i. Linear releasing system

Enzymatic

HX-Drug

EWG = NO_2, N_3, RCONH, RCO_2

EDG = NH_2, NHOH, OH

+ HX-Drug

ii. Cyclization releasing system

Masked

+ HX-Drug

Masked nucleophiles

iii. Cyclization to form cyclic drug

Masked

HX-R′ or H_2O

X = O, NH

X = CH_2

Figure 8.1. Examples of drug-release/activation mechanisms.

Upon unmasking via reduction or hydrolysis, the resulting electron-donating group will be able to push electrons through the conjugative system to the *para* position, leading to the cleavage of the benzylic carbon-oxygen bond. The rate of this cleavage is not enzyme-dependent, but rather relies on the electron-pushing capability of the unmasked electron-donating group and the electron-pulling ability of the leaving group. The formation of the negatively charged species can serve as the activation mechanism of the drug. This approach has recently been used in our effects to develop a novel and superior class of nitroaryl phosphoramides as potential prodrugs for nitroreductase-mediated enzyme-prodrug therapy.[33]

Several nitroaryl phosphoramides were designed and synthesized, each with a strategically placed nitro group on the benzene ring in the *para* position to the benzylic carbon (Scheme 19). Compound **114** is a cyclophosphamide analog

116, **118**, and **121** could cross-link DNA strands, leading to cytotoxicity.

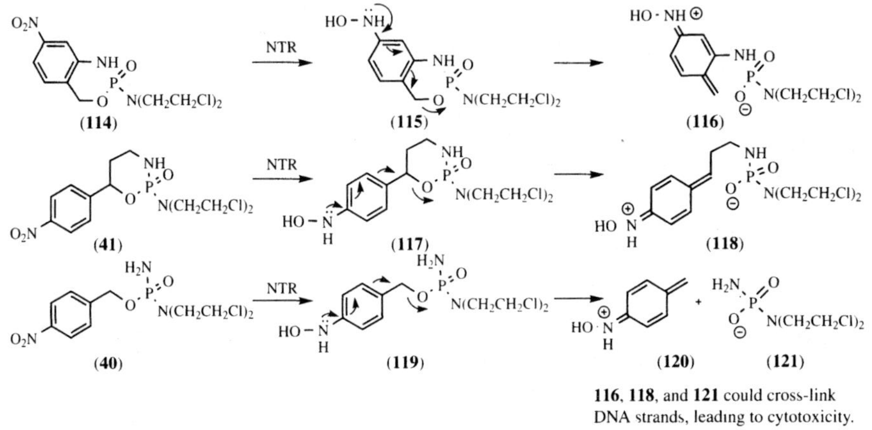

Scheme 19

with the cyclophosphamide ring fused with a benzene ring and a nitro group placed in the *para* position to the benzylic carbon. Compound **41** is a 4-nitrophenyl substituted cyclophosphamide analog, and **40** is an acyclic nitrobenzyl phosphoramide mustard (LH7). The nitro group is a strong electron-withdrawing group (Hammett σ_p electronic parameter $= 0.78$) and is converted to an electron-donating hydroxylamino group ($\sigma_p = -0.34$) upon nitroreductase reduction. This large difference in electronic effect ($\Delta\sigma_p = 1.12$) is exploited to effect the formation of the highly cytotoxic phosphoramide mustard or like reactive species. After reduction by nitroreductase (NTR), the resulting hydroxylamines **115**, **117**, and **119** relay their electrons to the *para* position and facilitate the cleavage of the benzylic C—O bond, producing the anionic cytotoxic species phosphoramide mustard **121** or like reactive species **116** and **118**. Structurally, the phosphoramide portion in **116** and **118** closely resembles phosphoramide mustard **121**, the reactive alkylating agent produced following the metabolic activation of cyclophosphamide in the liver, and could also be the ultimate cytotoxic alkylating agent.

Phosphoramide mustard is the proven cytotoxic metabolite of cyclophosphamide, a successful clinical anticancer prodrug that requires cytochrome P450 activation in the liver. These nitroaryl phosphoramides in combination with nitroreductase could effectively move the site of activation from liver in the case of cyclophosphamide into nitroreductase-expressing tumor tissues. All compounds were shown to be excellent substrates of *E. coli* nitroreductase, but with varying degrees of cytotoxicity against nitroreductase-expressing V79 and SKOV3 cells. Compounds *cis*-**41** and *trans*-**41**, the best of the cyclic series, were over 22,000× more cytotoxic in nitroreductase-expressing Chinese hamster V79 cells and **40**, the acyclic compound LH7, was 167,500× more cytotoxic in the same cell line, with an IC_{50} as low as 0.4 nM upon 72-hour drug exposure. This level of activity is about 100× more active and 27× more selective than CB1954. Even when the V79 cells were exposed to each test compound for 1 hour before the media were replaced with non-drug-containing fresh media, the IC_{50} was 10 nM, which was about 30× lower than that of CB1954 (**31**). The high selectivity of *cis*-**41**, *trans*-**41**, and **40** was reproduced in SKOV3 human ovarian carcinoma cells infected with an adenovirus expressing *E. coli* nitroreductase. Enzyme kinetic analysis indicates that compound **40** was a much better substrate of *E. coli* nitroreductase with a specificity constant 20× that of CB1954.

8.5.2. Cascade Release Facilitated by Intramolecular Cyclization Reactions

Intramolecular reactions are usually thermodynamically favored over intermolecular reactions because they have lower activation energy and therefore more stable transition states. The lower activation energy is attributed to a better entropic situation. When a reaction is performed between two different reactants, the two molecules need to collide in a specific orientation and the reaction leads to a decrease in the number of molecules, resulting in an increase in order and therefore a loss of entropic energy. In an intramolecular reaction, the two reaction centers, e.g., a nucleophile and an electrophile, are both present within the same molecule and are in a good position to interact and form the cyclic product. The positioning of the two reaction centers within the same molecule is very important for the reaction to take place. Steric factors resulting in less flexible molecules and better positioning of the two reaction centers would lead to increased reaction rates. Generally speaking, intramolecular reactions leading to the formation of 5- or 6-membered rings are much more favorable and occur at faster rates.

A series of alkylaminoalkyl carbamates of 4-hydroxyanisole (**122**) were evaluated as prodrugs of the melanocytotoxic phenol (**123**) that could be activated through intramolecular cyclization (Scheme 20).[69] The carbamates were relatively stable at low pH but released 4-hydroxyanisole cleanly in a nonenzymatic fashion at pH 7.4 at rates that were structure-dependent. A detailed study of the *N*-methyl-*N*-[2-(methylamino)ethyl]carbamate showed that generation of the parent phenol followed first-order kinetics with $t_{1/2} = 36.3$ minutes at pH 7.4, 37°C, and was accompanied by formation of *N*,*N*-dimethylimidazolidinone (**124**). In comparison, the related derivative with three methylene units between the two N atoms releases the phenol at a much slower rate with $t_{1/2} = 942$ minutes. These basic carbamates are examples of cyclization-activated prodrugs in which generation of the active drug is not linked to enzymatic cleavage but rather depends solely upon a predictable, intramolecular cyclization-elimination reaction.

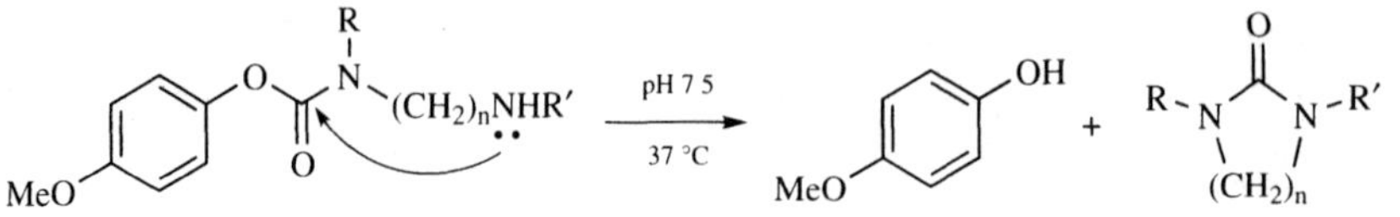

n	R, R′	$t_{1/2}$ (min)
2	Me, Me	36.3
2	Et, Et Me,	118
2	H	304
2	H, Me	335
2	H,H	724
3	H,H	910
3	Me,Me	942

Scheme 20

The terminal amino group in the above system could be masked as an amide, thus avoiding the stability problem encountered when using the basic carbamates as prodrugs. Unmasking of the amide could be catalyzed by a specific protease *in vivo* at certain target sites, thus achieving target specificity. Another alternative is to use *o*-nitroaromatic, as in compounds **125** (Scheme 21), which could be

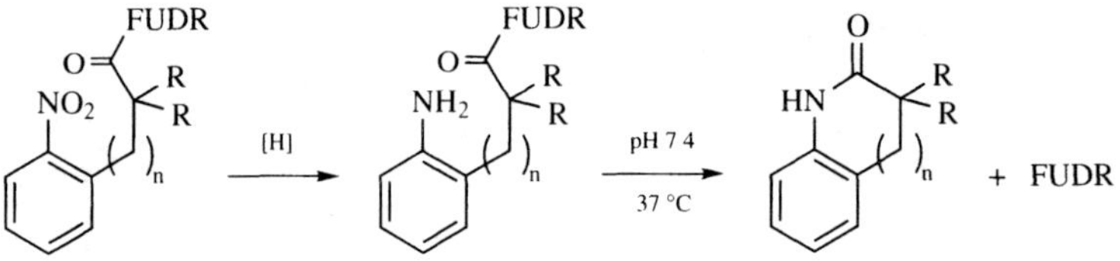

n	R	$t_{1/2}$ (min)
0	H	14
0	Me	~0
1	H	4
1	Me	2

Scheme 21

converted to a nucleophilic aromatic amine upon bioreduction in hypoxic tumor tissues or by other reductases delivered to targeted cells.[70–72] Kinetic analysis of the cyclization activation process indicates that the addition of two α methyl groups to the ester carbonyl would restrict the rotational freedom of the ground state molecule and promote the cyclization reaction. The nitro group can be reduced *in vivo* to a nucleophilic aromatic amine with a low pK_a (<4), which would be present in neutral nucleophilic form under most physiological conditions. At pH 7.4, 37°C, the amines **126** cyclized quickly to the lactam **127**, releasing the anticancer drug floxuridine (FUDR) (**128**).[70] For tumor targeting purposes, subsequent drug release after initial specific enzymatic activation should be very fast (preferably <1 minutes) in order to prevent the active drug from escaping the targeted site.

Reduction of quinone propionic esters or amides **129** bearing three Me groups in the so-called trialkyl lock positions (*o*-, **β**-, **β**-positions) or hydrolysis of the corresponding phenolic esters **130** has been shown to undergo spontaneous lactonization with the release of alcohol or amine, respectively (Scheme 22).[73–75] Several amides

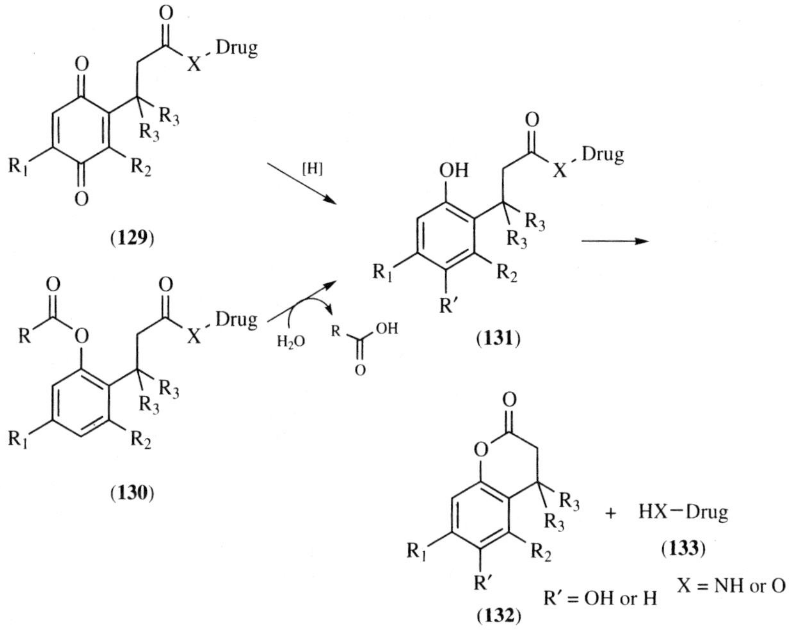

(**129**) (**130**) (**131**) (**132**) + HX−Drug (**133**)

R′ = OH or H X = NH or O

R′	R_1	R_2	R_3	$t_{1/2}$*
H	Me	Me	Me	65.4 s
H	H	H	Me	10.5 h
H	H	H	H	19.4 d

* with *p*-$MePhNH_2$ as a model amine, at pH 7.5, 30°C

Scheme 22

129 were synthesized and tested as model redox-sensitive cascade prodrugs of amines. The reduction of model amide prodrugs (**129**) generated hydroxy amide intermediates **131**, the lactonization of which resulted in amine release. The half-lives for appearance of the product lactone **132** from these intermediates ranged from 1.4 to 3.4 minutes at pH 7.4 and 37°C. With such rapid lactonization rates, it is believed that reduction would be the rate-limiting step in the two-step conversion of the prodrugs to amines.[75] Comparison of the observed rates of lactonization at pH 7.5 and 30°C for three hydroxy amides obtained from the hydrolysis of phenolic ester prodrugs (**130**) allowed an estimate of the extent of rate enhancement provided by the addition of a partial or total trimethyl lock for the hydroxy amide lactonization reaction under near-physiological conditions.[74] The half-life for the hydroxy amide with a full trimethyl lock was 65 seconds, a rate enhancement of 2.54×10^4 as compared to the corresponding hydroxy amide without the three methyl groups.

Still another intramolecular cyclization system is the coumarin-based prodrug system **134** that can be used for bioreversible derivatization of amine and alcohol drugs and the preparation of cyclic peptide prodrugs (Scheme 23).[76,77] This system takes advantage of the known facile lactonization of coumarinic acid and its

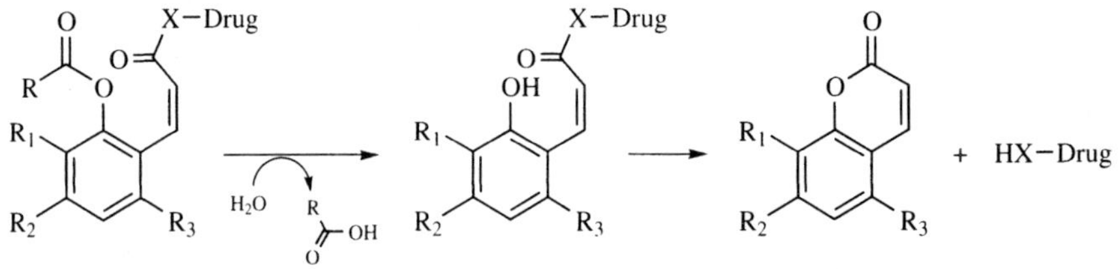

(**134**) (**135**) (**136**) + HX−Drug

R_1	R_2	R_3	$t_{1/2}$ (min)*
Me	H	Me	4.8
Me	Me	H	6.3
H	H	H	32.5

* with $PhCH_2NHMe$ as a model an at pH 7.4, 37 °C

Scheme 23

derivatives. Such a system can be used for the development of esterase- and phosphatase-sensitive prodrugs of amines and alcohols. Esterase-sensitive prodrugs of a number of model amines prepared by using this system readily released the amines upon incubation in the presence of porcine liver esterase, with $t_{1/2}$ values ranging from 100 seconds to 35 minutes.[77]

8.5.3. Cascade Activation Through Intramolecular Cyclization to Form Cyclic Drugs

Pilocarpine (**139**) is used as a topical miotic for controlling elevated intraocular pressure associated with glaucoma. The drug presents significant delivery problems due to its low ocular bioavailability (1–3% or less) and its short duration of action. The poor bioavailability was partly attributed to its poor permeability across the corneal membrane due to its low lipophilicity. Because of the low bioavailability, a large ophthalmic dose is required to enable an effective amount of pilocarpine to reach the inner eye receptors and reduce the intraocular pressure for a suitable duration. This in turn gives rise to concerns about systemic toxicity, since most of the applied drug is then available for systemic absorption from the nasolacrimal duct. The systemic absorption of pilocarpine may lead to undesired side effects, e.g., in those patients who display sensitivity to cholinergic agents. Upon instillation of pilocarpine into the eye, intraocular pressure is reduced for only about 3 hours. As a consequence, the frequency of administration is at an inconvenient three to six times per day.

To improve the ocular bioavailability and prolong the duration of action, various pilocarpic acid mono- and diesters were evaluated as prodrugs for pilocarpine. As shown in Scheme 24, the pilocarpic acid monoesters (**138**) undergo a quantitative

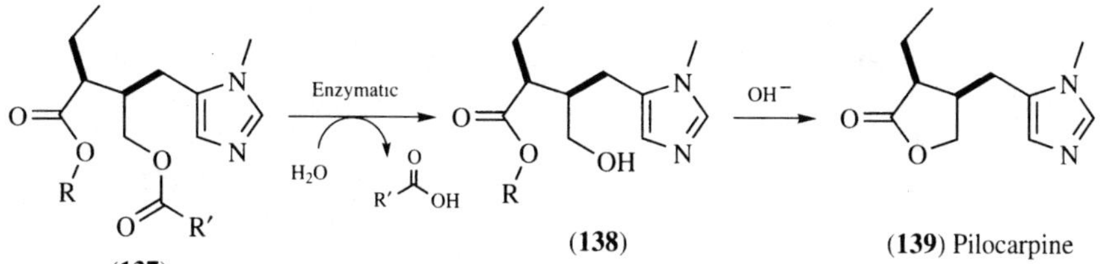

Enzymatic
H_2O
OH⁻
(**137**)
(**138**)
(**139**) Pilocarpine

Scheme 24

cyclization to pilocarpine (**139**) in aqueous solution, the rate of cyclization being a function of the polar and steric effects within the alcohol portion of the esters. At pH 7.4 and 37°C, half-lives ranging from 30 to 1105 minutes were observed for the various esters. A main drawback of these monoesters is their poor solution stability, but this problem was overcome by esterification of the free hydroxy group. A number of pilocarpic diesters (**137**) were highly stable in aqueous solution (shelf lives were estimated to be more than 5 years at 20°C) and, most significantly, were readily converted to pilocarpine under conditions simulating those occurring *in vivo* through a cascade process involving rapid enzymatic hydrolysis of the *O*-acyl bond followed by spontaneous lactonization of the intermediate pilocarpic acid monoester (**138**). Both the pilocarpic acid monoesters and, in particular, diesters enhanced the ocular bioavailability of pilocarpine and significantly prolonged the duration of its activity following topical instillation, as determined by a miosis study in rabbits.[78]

Derivatives of *N*-alkylbenzophenones and peptidoamino-benzophenones undergo hydrolysis, with subsequent intramolecular condensation that results in the formation of the 1,4-benzodiazepine hypnotics. A number of such compounds, e.g. rilmazafone (**140**), were suggested to have more beneficial pharmacological properties in comparison to standard benzodiazepines. Rilmazafone is a ring-opened derivative of 1,4-benzodiazepine (Scheme 25) and was developed in Japan

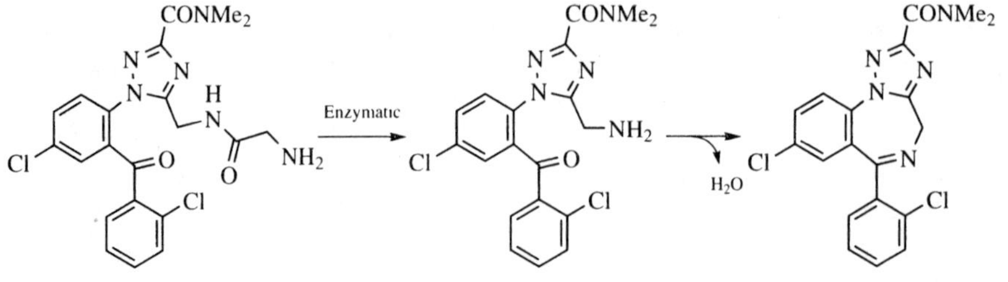

(140) Rilmazafone (141) (142)

Scheme 25

as an orally active sleep inducer. Rilmazafone is exclusively metabolized by aminopeptidases in the small intestine to the labile desglycylated metabolite **141** and then to its cyclic form **142**. The concentration of **142** in the systemic plasma (i.e., bioavailability) after oral administration of rilmazafone has been reported to be higher than that observed after administration of **142** due to the lower hepatic extraction of **141** than **142**.[79]

8.6. PRODRUGS AND INTELLECTUAL PROPERTY RIGHTS—TWO COURT CASES

The primary purpose for developing prodrugs is, of course, not to circumvent intellectual property rights but to obviate certain disadvantages that may have precluded an active agent from being used in clinical applications. Therefore, if undesirable properties of a drug molecule cannot be overcome by conventional changes in the pharmaceutical formulation or route of administration, the method of choice is to use one of the prodrug approaches discussed above. The parent drug usually came first and was followed by the prodrug. The prodrug is inactive or much less active, and it was the parent drug that would ultimately act in the human body.

If a prodrug is sufficiently distinct from the parent drug and if it possesses unexpected improved properties over the parent drug, the prodrug can be patented. Chapter 20 focuses on the intellectual property issues related to drug delivery. Here only a brief discussion involving prodrugs is presented. In many cases, the patent will be granted to the same inventor or company that developed the parent drug. Even if a third party patented the prodrug, the prodrug patent would, of course, not prevent the user of the parent drug from continuing to use it. However, an interesting question arises when a third party decides to manufacture, use, import, or sell the prodrug form of a patented drug: whether the owner of a patent covering the parent drug can object to the use of the prodrug by third parties (irrespective of whether the prodrug has been patented). The answer to this question may have to be determined on a case-by-case basis in the courts. The following two cases, though they do not fully address this question, do show the potential legal ramifications.

The first case relates to a decision by the British House of Lords on the previously discussed hetacillin (**71**).[80] The question was whether the British patent covering the antibiotic ampicillin was infringed by the importation and use of the antibiotic hetacillin in England. The House of Lords ruled that the prodrug hetacillin infringed the patent pertaining to ampicillin. It reasoned that when the prodrug came into contact with water in the gullet, it underwent a chemical reaction and became the active substance ampicillin. The court believed that there was no therapeutic or other added value associated with the use of hetacillin. It was decided that this was an infringement, despite the existence of more than insignificant structural diversity between the claims and the infringing product. It was important to note that the court believed the prodrug did not add any value to the known invention.[80]

The second case relates to the relationship between terfenadine (**143**) and its active metabolite, now known as fexofenadine (or Allegra, **144**) (Scheme 26). As

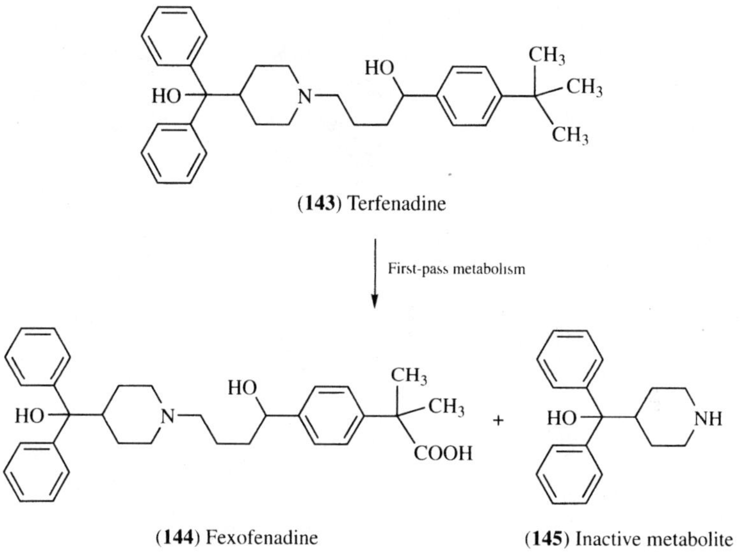

(**143**) Terfenadine

First-pass metabolism

(**144**) Fexofenadine + (**145**) Inactive metabolite

Scheme 26

a now discontinued "second-generation" antihistamine, terfenadine itself was active and developed as a histamine H_1 receptor antagonist; thus, it was not a prodrug in the strict sense. But this court case does have ramifications for prodrug design. Terfenadine was successfully marketed for a long period of time, without knowledge that terfenadine was, in fact, acting *in vivo* through its active metabolite, fexofenadine. On the basis of a mass balance study using ^{14}C-labeled terfenadine, oral absorption of terfenadine was estimated to be at least 70%. However, terfenadine is not normally detectable in plasma at levels >10 ng/ml; it undergoes extensive (99%) first-pass metabolism to two primary metabolites, an active acid metabolite, fexofenadine, and an inactive dealkylated metabolite. This led some to refer to terfenadine as a "prodrug" of fexofenadine. While the drug–active metabolite relationship was unknown, both to the patent owners and to the public, a new patent application was filed 7 years later covering the compound fexofenadine (formerly known as "MDL 16,455").[81,82] Therefore, the owner of both patents and the public realized only later that the compound fexofenadine was in fact the active drug all along.

After expiration of the terfenadine patent, the owner of the metabolite patent believed that the marketing of terfenadine-containing products infringed upon the substance and use claims of the later-filed metabolite patent. This led the owner of the metabolite patent to file infringement lawsuit in Germany, the United Kingdom, and the United States against generic companies that had launched terfenadine-containing drug products. In these cases, it was the alleged infringers, not the plaintiffs, that received sympathy from the courts. The infringement lawsuits were regarded as attempts to extend the monopoly of a lapsed patent.

It is clear from the above discussion that prodrugs can be patented if they are designed to add value to, not to circumvent, a known invention, are sufficiently distinct from the parent drug, and possess unexpected improved properties compared to the parent drug. But one should bear in mind the potential legal ramifications arising from working on prodrugs of the patented inventions of others.

REFERENCES

1. Bundgaard, H. *Drug Future* **1991**, *16*, 443–458.
2. Balant, L. P.; Doelker, E. In *Burger's Medicinal Chemistry and Drug Discovery. Volume 1: Principles and Practice*, 5th ed.; Wolff, M. E., Ed. John Wiley & Sons: New York, **1995**, pp. 949–982.
3. Larsen, C. S.; Ostergaard, J. In *Textbook of Drug Design and Discovery*, 3rd ed.; Krogsgaard-Larsen, P., Liljefors, T., Madsen, U., Eds. Taylor & Francis: London and New York, **2002**, pp. 410–458.
4. Lipinski, C. A.; Lombardo, F.; Dominy, B. W.; Feeney, P. J. *Adv. Drug Deliv. Rev.* **1997**, *23*, 3–25.
5. Ensink, J. M.; Vulto, A. G.; van Miert, A. S.; Tukker, J. J.; Winkel, M. B.; Fluitman, M. A. *Am. J. Vet. Res.* **1996**, *57*, 1021–1024.
6. Friedman, D. I.; Amidon, G. L. *Pharm. Res.* **1989**, *6*, 1043–1047.
7. Ribeiro, W.; Muscara, M. N.; Martins, A. R.; Moreno, H. J.; Mendes, G. B.; de Nucci, G. *Eur. J. Clin. Pharmacol.* **1996**, *50*, 399–405.
8. Todd, P. A.; Goal, K. L. *Drugs* **1992**, *43*, 346–383.
9. Menard, J.; Patchett, A. A. In *Drug Discovery and Design Advances in Protein Chemistry*, Vol. 56, Scolnick, E. M., Ed. Academic Press: London, Boston, and New York, **2001**, pp. 13–75,
10. Beauchamp, L. M.; Orr, G. F.; De Miranda, P.; Burnette, T.; Krenitsky, T. A. *Antiviral Chem. Chemother.* **1992**, *3*, 157–164.

11. Crooks, R. J.; Murray, A. *Antiviral Chem. Chemother.* **1994**, *5 (suppl.)*, 31–37.
12. Boyd, M. R.; Safrin, S.; Kern, E. R. *Antiviral Chem. Chemother.* **1993**, *4 (suppl.)*, 3–11.
13. Shinkai, I.; Ohta, Y. *Bioorg. Med. Chem.* **1996**, *4*, 1–2.
14. Pue, M. A.; Pratt, S. K.; Fairless, A. J.; Fowles, S.; Laroche, J.; Georgiou, P.; Prince, W. *J. Antimicrob. Chemother.* **1994**, *33*, 119–127.
15. Cirelli, R.; Herne, K.; McCrary, M.; Lee, P.; Tyring, S. K. *Antiviral Res.* **1996**, *29*, 141–151.
16. Han, H. K.; Oh, D. M.; Amidon, G. L. *Pharm. Res.* **1998**, *15*, 1382–1386.
17. Kane, J. M. *J. Clin. Psychiatry* **1984**, *45*, 5–12.
18. Tammara, V. K.; Narurkar, M. M.; Crider, A. M.; Khan, M. A. *Pharm. Res.* **1993**, *10*, 1191–1199.
19. Waller, D. G.; George, C. F. *Br. J. Clin. Pharmacol.* **1989**, *28*, 497–507.
20. Sinha, V. R.; Kumria, R. *Pharm. Res.* **2001**, *18*, 557–564.
21. Yang, L. B.; Chu, J. S.; Fix, J. A. *Int. J. Pharm.* **2002**, *235*, 1–15.
22. Chourasia, M. K.; Jain, S. K. *J. Pharm. Pharm. Sci.* **2003**, *6*, 33–66.
23. Repta, A. J.; Rawson, B. J.; Shaffer, R. D.; Sloan, K. B.; Bodor, N.; Higuchi, T. *J. Pharm. Sci.* **1975**, *64*, 392–396.
24. Kennedy, K. A.; Teicher, B. A.; Rockwell, S.; Sartorelli, A. C. *Biochem. Pharmacol.* **1980**, *29*, 1–8.
25. Siim, B. G.; Atwell, G. J.; Anderson, R. F.; Wardman, P.; Pullen, S. M.; Wilson, W. R.; Denny, W. A. *J. Med. Chem.* **1997**, *40*, 1381–1390.
26. Wilson, W. R. In *Cancer Biology and Medicine*, Vol. 3; Waring, M. J., Ponder, B. A. J., Eds. Kluwer Academic: Lancaster, **1992**, pp. 87–131.
27. Rauth, A. M.; Goldberg, Z.; Misra, V. *Oncol. Res.* **1997**, *9*, 339–349.
28. Ast, G. *Curr. Pharm. Design* **2003**, *9*, 455–466.
29. Eisenbrand, G.; Lauck-Birkel, S.; Tang, W. C. *Synthesis* **1996**, 1246–1258.
30. Ray, J. M.; Stetler-Stevenson, W. G. *Eur. Respir. J.* **1994**, *7*, 2062–2072.
31. DeFeo-Jones, D.; Garsky, V. M.; Wong, B. K.; Feng, D. M.; Bolyar, T.; Haskell, K.; Kiefer, D. M.; Leander, K.; McAvoy, E.; Lumma, P.; Wai, J.; Senderak, E. T.; Motzel, S. L.; Keenan, K.; Van Zwieten, M.; Lin, J. H.; Freidinger, R.; Huff, J.; Oliff, A.; Jones, R. E. *Nat. Med.* **2000**, *6*, 1248–1252.
32. Melton, R. G.; Knox, R. J., Eds. *Enzyme-Prodrug Strategies for Cancer Therapy.* Kluwer Academic/Plenum: New York, 1999.
33. Hu, L. Q.; Yu, C. Z.; Jiang, Y. Y.; Han, J. Y.; Li, Z. R.; Browne, P.; Race, P. R.; Knox, R. J.; Searle, P. F.; Hyde, E. I. *J. Med. Chem.* **2003**, *46*, 4818–4821.
34. Leamon, C. P.; Low, P. S. *Drug Discov. Today* **2001**, *6*, 44–51.
35. Vickers, S.; Duncan, C. A.; White S, D.; Breault, G. O.; Royds, R. B.; de Schepper, P. J.; Tempero, K. F. *Drug Metab. Dispos.* **1978**, *6*, 640–646.
36. Huang, H. P.; Ayres, J. W. *J. Pharm. Sci.* **1988**, *77*, 104–109.
37. Easthope, S. E.; Jarvis, B. *Drugs* **2002**, *62*, 1253–1287.
38. Lin, Y. I.; Bitha, P.; Li, Z.; Sakya, S. M.; Strohmeyer, T. W.; Lang, S. A.; Yang, Y. J.; Bhachech, N.; Weiss, W. J.; Petersen, P. J.; Jacobus, N. V.; Bush, K.; Testa, R. T. *Bioorg. Med. Chem. Lett.* **1997**, *7*, 1811–1816.
39. Mangan, F. R.; Flack, J. D.; Jackson, D. *Am. J. Med.* **1987**, *83*, 6–10.

40. Bundgaard, H.; Johansen, M. *Int. J. Pharm.* **1981**, *9*, 7–16.
41. Vej-Hansen, B.; Bundgaard, H. *Arch. Pharm. Chem. Sci. Ed.* **1979**, *7*, 341–353.
42. Durbin, A. K.; Rydon, A. N. *Chem. Commun.* **1970**, 1249–1250.
43. Schwartz, M. A.; Hayton, W. L. *J. Pharm. Sci.* **1972**, *61*, 906–909.
44. Bundgaard, H.; Nielsen, N. M. *Acta Pharm. Suecica* **1987**, *24*, 233–246.
45. Bundgaard, H.; Johansen, M. *Int. J. Pharm.* **1984**, *22*, 45–56.
46. Biel, J. H.; Somani, P.; Jones, P. H.; Minard, F. N.; Goldberg, L. I. *Biochem. Pharmacol.* **1974**, *23*, 748–750.
47. Cruz, D. N. *Expert Opin. Pharmacother.* **2000**, *1*, 835–840.
48. McClellan, K. J.; Wiseman, L. R.; Wilde, M. I. *Drug. Aging* **1998**, *12*, 76–86.
49. Gogate, U. S.; Repta, A. J.; Alexander, J. *Int. J. Pharm.* **1987**, *40*, 235–248.
50. Gogate, U. S.; Repta, A. J. *Int. J. Pharm.* **1987**, *40*, 249–255.
51. Pauletti, G. M.; Gangwar, S.; Siahaan, T. J.; Aube, J.; Borchardt, R. T. *Adv. Drug Deliv. Rev.* **1997**, *27*, 235–256.
52. Song, X. P.; Xu, C. R.; He, H. T.; Siahaan, T. J. *Bioorg. Chem.* **2002**, *30*, 285–301.
53. Johansen, M.; Bundgaard, H. *Int. J. Pharm.* **1980**, *7*, 119–127.
54. Bundgaard, H.; Klixbull, U.; Falch, E. *Int. J. Pharm.* **1986**, *29*, 19–28.
55. Ohya, Y.; Kuroda, H.; Hirai, K.; Ouchi, T. *J. Bioact. Compat. Pol.* **1995**, *10*, 51–66.
56. Juchem, M.; Pollow, K.; Elger, W.; Hoffmann, G.; Moebus, V. *Contraception* **1993**, *47*, 283–294.
57. Li, Q. G.; Huempel, M. *Eur. J. Drug Metab. Pharmacokinet.* **1992**, *17*, 281–291.
58. Venhuis, B. J.; Wikström, H. V.; Rodenhuis, N.; Sundell, S.; Wustrow, D.; Meltzer, L. T.; Wise, L. D.; Johnson, S. J.; Dijkstra, D. *J. Med. Chem.* **2002**, *45*, 2349–2351.
59. Venhuis, B. J.; Dijkstra, D.; Wustrow, D.; Meltzer, L. T.; Wise, L. D.; Johnson, S. J.; Wikstroem, H. V. *J. Med. Chem.* **2003**, *46*, 4136–4140.
60. Tscheuschner, C.; Neubert, R.; Fuerst, W.; Luecke, L.; Fries, G. *Pharmazie* **1993**, *48*, 681–684.
61. Repta, A. J.; Patel, J. P. *Int. J. Pharm.* **1982**, *10*, 29–42.
62. Bundgaard, H.; Johansen, M. *Int. J. Pharm.* **1982**, *10*, 165–175.
63. Johansen, M.; Bundgaard, H. *J. Pharm. Sci.* **1983**, *72*, 1294–1298.
64. Buur, A.; Bundgaard, H. *Arch. Pharm. Chem. Sci. Ed.* **1987**, *15*, 76–86.
65. Bakhtiar, R.; Hop, C. E. C. A.; Walker, R. B. *Rapid Commun. Mass Spec.* **1997**, *11*, 598–602.
66. Walker, R. B.; Dholakia, V. N.; Brasfield, K. L.; Bakhtiar, R. *Gen. Pharmacol.* **1998**, *30*, 725–731.
67. Bodor, N.; Sloan, K. B.; Little, R. J.; Selk, S. H.; Caldwell, L. *Int. J. Pharm.* **1982**, *10*, 307–321.
68. Bodor, N.; Sloan, K. B. *J. Pharm. Sci.* **1982**, *71*, 514–520.
69. Saari, W. S.; Schwering, J. E.; Lyle, P. A.; Smith, S. J.; Engelhardt, E. L. *J. Med. Chem.* **1990**, *33*, 97–101.
70. Liu, B.; Hu, L. Q. *Bioorg. Med. Chem.* **2003**, *11*, 3889–3899.
71. Hu, L.; Liu, B.; Hacking, D. R. *Bioorg. Med. Chem. Lett.* **2000**, *10*, 797–800.
72. Jiang, Y.; Zhao, J.; Hu, L. *Tetrahedron Lett.* **2002**, *43*, 4589–4592.

73. Carpino, L. A.; Triolo, S. A.; Berglund, R. A. *J. Org. Chem.* **1989**, *54*, 3303–3310.
74. Amsberry, K. L.; Borchardt, R. T. *J. Org. Chem.* **1990**, *55*, 5867–5877.
75. Amsberry, K. L.; Borchardt, R. T. *Pharm. Res.* **1991**, *8*, 323–330.
76. Shan, D. X.; Nicolaou, M. G.; Borchardt, R. T.; Wang, B. H. *J. Pharm. Sci.* **1997**, *86*, 765–767.
77. Liao, Y.; Wang, B. *Bioorg. Med. Chem. Lett.* **1999**, *9*, 1795–1800.
78. Bundgaard, H.; Falch, E.; Larsen, C.; Mosher, G. L.; Mikkelson, T. J. *J. Med. Chem.* **1985**, *28*, 979–981.
79. Koike, M.; Futaguchi, S.; Takahashi, S.; Sugeno, K. *Drug Metab. Dispos.* **1988**, *16*, 609–615.
80. House of Lords *Beecham Group v. Bristol Laboratories of 30 03* **1978**, R.P.C. 153.
81. House of Lords *Merrell Dow Pharmaceuticals Inc. v. H.N. Norton & Co. Ltd* **1996**, RPC 76.
82. *Marion Merrell Dow v. Geneva Pharmaceuticals, 877 F. Supp. 531* **1994**, (D. Colo.).

9

RECEPTOR-MEDIATED DRUG DELIVERY

CHRISTOPHER P. LEAMON

Endocyte, Inc., 1205 Kent Avenue, West Lafayette, IN 47906

PHILIP S. LOW

Purdue University, West Lafayette, IN 47907

Drug Delivery: Principles and Applications Edited by Binghe Wang, Teruna Siahaan, and Richard Soltero
ISBN 0-471-47489-4

9.1. INTRODUCTION

Most current drugs distribute nonspecifically and randomly throughout the body, entering both healthy and pathological cells with roughly equal efficiency. Not surprisingly, when normal cells are sensitive to such drugs, their health can be compromised, leading to side effects that can limit use of the therapeutic agents. In the case of drugs designed to promote only minor changes in cell behavior (e.g., aspirin), such side effects are usually acceptable. However, when the drug is designed to cause cell death or induce a significant change in cell behavior, toxicity to normal cells can undermine its use. The development of receptor-targeted therapeutic agents has been initiated primarily to limit the distribution of toxic drugs to only the pathological cells, thus minimizing collateral damage to normal cells. However, as will be noted below, receptor-mediated drug delivery can also enable otherwise membrane-impermeant drugs to enter target cells by receptor-mediated endocytosis, or it can induce desirable changes in cell behavior by activating a receptor's normal signaling pathway. Because endocytosis is intimately involved in each of the above merits of receptor-mediated drug delivery, we will begin this chapter by summarizing the basic characteristics of this process.

Endocytosis constitutes the pathway by which extracellular material is carried into a cell by membrane invagination and internalization.[1–3] Endocytosis occurs in virtually all eukaryotic cells,[4,5] and can assist in such diverse processes as hormone signaling and removal, vitamin and mineral uptake, extracellular solute uptake, pathogen removal, and even simple membrane turnover. In fact, endocytosis is so active in some cells that the entire plasma membrane is internalized and replaced in less than 30 minutes.[6]

Endocytosis can be divided into three subcategories. The first is commonly referred to as "phagocytosis," or the process of cellular "eating." Phagocytosis plays a major role in host defense mechanisms by mediating the ingestion and degradation of microorganisms. Phagocytosis is also essential for tissue remodeling/differentiation and elimination of cellular debris. In contrast to other forms of endocytosis, phagocytosis is generally carried out in higher eukaryotes by professional phagocytes, such as polymorphonuclear granulocytes, monocytes, and macrophages.

The second subcategory of endocytosis is called "fluid phase pinocytosis," which arises from entrapment of solutes by vesicles that invaginate from the cell surface. Importantly, the amount of material taken in by this route is proportional to a component's concentration in the extracellular environment. As such, pinocytosis is generally regarded as the means by which solutes enter cells nonspecifically.

The third subcategory of endocytosis is referred to as "receptor-mediated." Receptors belong to a special class of cell surface proteins that utilize the endocytosis machinery to carry exogenous ligands into cells. When a specific cell surface receptor is overexpressed on a pathological cell, receptor-mediated endocytosis can be exploited for targeted drug delivery.

The specific events that occur during receptor-mediated endocytosis are illustrated in Figure 9.1. Initially, exogenous ligands bind to externally oriented

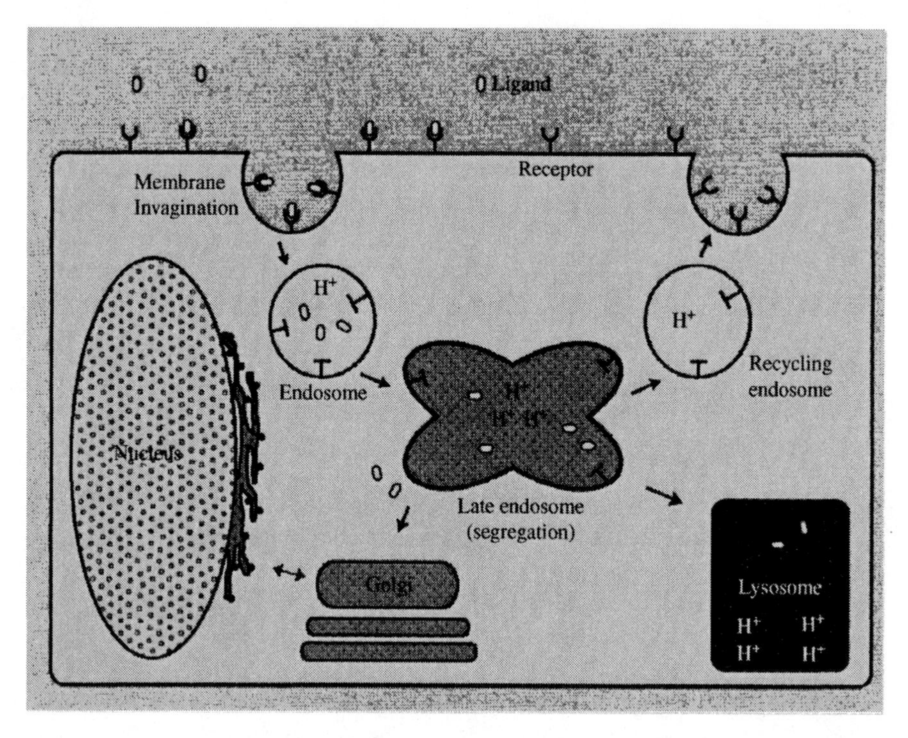

Figure 9.1. Receptor-mediated endocytosis. Exogenous ligands (yellow ovals) bind specifically to their cell surface receptors. The plasma membrane invaginates around the ligand-receptor complexes to form an intracellular vesicle (endosome). As the lumen of the maturing endosome acidifies, the receptor often releases the ligand. Eventually, the fates of both the ligand and the receptor are determined during a sorting process within late endosomal compartments.

receptors on the cell membrane. This is a highly specific event, i.e., analogous to a key (ligand) inserting into a lock (receptor). Ligand binding usually occurs within minutes, but the kinetics of this event are dictated by the rate of ligand diffusion and the intrinsic affinity of the ligand for its receptor. Immediately after binding, the plasma membrane surrounding the ligand-receptor complex begins to invaginate until a distinct internal vesicle, called an "early endosome," forms within the cell.[7] The pH of the vesicle lumen is then often lowered to $\sim$5 through the action of proton pumps, after which the ligand often dissociates from its receptor.

Endosomal vesicles often move to their intracellular destinations along tracks of microtubules in a random, salutatory motion.[8] They may eventually interact with the trans Golgi reticulum, where they are believed to fuse with membranous compartments prior to converting into late endosomes or multivesicular bodies. These latter compartments are capable of sorting the dissociated ligands from their empty receptors. At this juncture, there are four possible fates of the ligand and receptor:

- Both the ligand and its receptor can be directed to the lysosomes for destruction (e.g., various hormones).
- The ligand can be directed to a lysosome for destruction, while its receptor is recycled back to the plasma membrane to participate in another round of endocytosis (e.g., asialoglycoprotein).
- The ligand can be transferred into the cytosol, while its receptor is recycled back to the plasma membrane to participate in another round of endocytosis (e.g., folic acid).
- Both the ligand and its receptor can be recycled back to the plasma membrane (e.g., transferrin, folic acid).

Peculiarly, the fates of many receptor-ligand complexes can change from one of the above categories to another, depending on the percentage of occupancy of the cell surface receptor.[9] Thus, high receptor occupancy often causes a traditional recycling receptor to divert into a degradative pathway. When the same cell surface receptor is exploited for receptor-mediated drug delivery, the ligand-drug conjugate generally follows the intracellular itinerary of the free ligand. The only known exception to this rule arises when multiple ligands are attached to a single therapeutic particle (e.g., a liposome). Under these conditions, the natural endocytic pathway can be aborted, and the multivalent complex may be trafficked to lysosomes or some other unnatural destination.

9.2. SELECTION OF A RECEPTOR FOR DRUG DELIVERY

The choice of a receptor for receptor-mediated drug delivery is generally based on several criteria. First, the receptor should be present at high density on the pathological cell, but largely absent or inaccessible on normal cells. For tumor targeting applications, receptors expressed on the apical surfaces of epithelial cells often constitute good targets, since such receptors in normal epithelia are inaccessible to parenterally administered drugs; however, upon neoplastic transformation these sites become accessible as a result of loss of cell polarity (also note that 80% of human cancers derive from epithelial cells). A second criterion often considered in receptor selection concerns the heterogeneity in its expression on the pathological cells. Thus, receptors that are present at high levels on only a small percentage of pathological cells would be a poor target for drug delivery because the targeted drug would enter the diseased tissue unevenly. Third, the receptor should not be shed in measurable amounts into the circulation, thereby generating a decoy that would compete for ligand-drug conjugates. And except for applications relating to antibody-dependent enzyme-prodrug therapy (ADEPT) or immunotherapy, the receptor should internalize and recycle in order to permit maximal drug delivery into the pathological cells. Because receptor specificity and internalization/recycling can be so important, we will now elaborate on these two characteristics in greater detail.

9.2.1. Specificity

Perhaps the most significant advantage of receptor-mediated drug delivery lies in the researcher's ability to restrict drug deposition to tissues that express the ligand's receptor. Thus, the biodistribution of a ligand-drug conjugate should, in principle, follow the expression pattern of the ligand's receptor in the body. In our experience, this approximation is, in fact, realized if (1) the affinity of the receptor for the ligand is high, (2) the attached drug introduces no competing affinity of its own, and (3) the conjugate does not become trapped in nontargeted compartments.

The ligand-drug conjugate's specificity for its receptor can and should be evaluated *in vitro* before it is tested *in vivo*. Typically, such specificity can be established by showing that (1) a ligand-drug conjugate binds to and becomes internalized by receptor-positive cell lines, (2) association of the ligand-drug conjugate with these cells is blocked when an excess of free ligand is either pre- or coincubated with the cells, and (3) no measurable cell association occurs with either receptor-negative cell lines or cells from which the receptor has been cleaved.

In vivo specificity can similarly be evaluated by (1) comparing uptake of the ligand-drug conjugate in a known receptor-positive tissue (e.g., tumor) with its uptake in several receptor-negative tissues (e.g., lung, liver, heart), and (2) examining the competitive blockade of the ligand-drug conjugate's enrichment in target tissue upon pre- or coinjection of the animal with excess free ligand. While some nonspecific retention of conjugate in normal tissues cannot usually be avoided, in our experience nontargeted uptake can be minimized by constructing the conjugate such that its linker and therapeutic cargo exhibit little affinity for cell surfaces on their own. In general, the more hydrophilic a conjugate is, the less it will be plagued by nonspecific tissue adsorption.

9.2.2. Receptor Internalization/Recycling

As noted above, following ligand binding and endocytosis, some (but not all) receptors unload their ligands and recycle back to the cell surface, where they participate in another round of ligand binding and endocytosis. In order to maximize drug delivery, it would seem intuitive to try to identify such a receptor, since recycling receptors can continually deliver ligand-drug conjugates into their target cells. A simple calculation will serve to emphasize the importance of this consideration. Assume that a target cell expresses 500,000 molecules of receptor and that the net efficiency of receptor unloading, endosome escape, and release of free drug in the cytosol is only ~1%. If the receptor cannot recycle back to the cell surface, then ~5000 active drug molecules would enter an average aqueous cytosolic space of ~0.4 picoliters per cell,[10] leading to a cytosolic concentration of only 20 nM. In contrast, if the receptor recycles say, every 2 hours, then a controlled-release formulation of the same drug delivered over a week's time could establish a cytosolic concentration of 1.68 μM. Notably, many drugs are active in the micromolar but not in the nanomolar range.

9.3. DESIGN OF A LIGAND-DRUG CONJUGATE: LINKER CHEMISTRY

A typical structure for a ligand-drug conjugate is presented in Figure 9.2. While ligand, linker, and drug can all take on a diversity of sizes, shapes, and chemistries, a few fundamental principles can be followed to enhance therapeutic efficacy. We will begin by briefly outlining the desirable features to include in the design of a linker and then describe the preferred characteristics of both the ligand and the drug.

Because one's freedom to change the chemistry of either ligand or drug is frequently limited by the functional roles these components must perform, the investigator's greatest creativity is often required in designing a linker that endows the conjugate with the optimal properties. Thus, not only must the linker be equipped with appropriate groups to react with available functional moieties on both the ligand and the drug, but when improvements in water solubility, kidney excretion, serum protein binding, or other pharmacokinetic properties are required, the linker is often the only site where such modifications can be made.

The length of the linker can also be critical to drug delivery, since drug moieties positioned too close to a low molecular weight ligand can sterically reduce or even eliminate the affinity of the ligand for its receptor. Conversely, drug moieties separated too far from their targeting ligands by flexible spacers can often loop back and interact with the ligand, thereby also compromising the ligand's affinity for its receptor. Further, depending on the nature of the drug's activity, release of an intact unmodified drug may be critical to full expression of activity. Such processes have, in fact, evolved in nature to yield plant, fungal, and bacterial protein toxins of extraordinary potency, and similar release strategies have recently been shown to maximize the biological activities of ligand-targeted therapeutics (see below).

Interestingly, knowledge gained from the study of protein toxins has proven highly useful in the design of receptor-targeted drugs. Thus, it was learned early

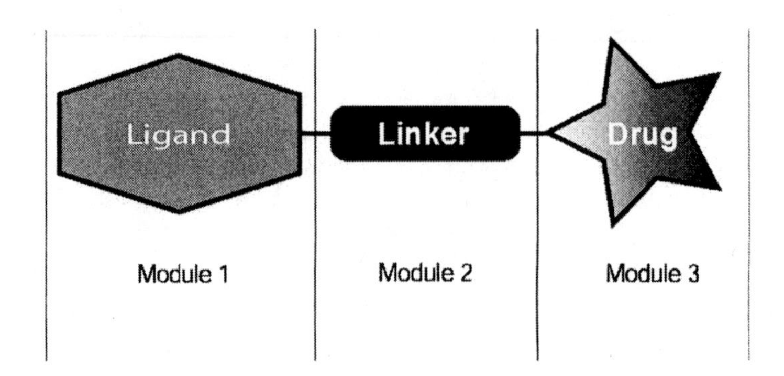

Figure 9.2. Structural design of a ligand-targeted drug conjugate.

in the characterization of protein toxins that replacement of the natural binding (B) chain with an alternative ligand would produce a powerful therapeutic agent with the exogenous ligand's cell-targeting specificity. These and related studies demonstrated that such toxins were constructed of independent binding and active domains, much like the drug conjugate presented in Figure 9.2. Second, it was shown that natural release of the toxic domain from the binding domain frequently involves intraendosomal disulfide bond reduction,[11–13] suggesting that disulfide bond reduction might also be exploited to release synthetic drugs from their ligands following entry into their target cells.[14–18]

Knowledge that endocytic vesicles rapidly become acidified (to ~pH 5)[19] has also prompted some to explore the use of acid-labile linkers in their designs of ligand-drug conjugates. As anticipated from the pH profiles of endocytic compartments, a drug attached to its ligand via an acid-sensitive linker can often be released shortly after formation of the endosome. Importantly, most of the progress in this area has come from studies of hydrazone-, acetal-, and ketal-based linkers. For example, Neville et al. reported that the potency of a IgG-ketal-diphtheria toxin conjugate is increased 50-fold over its noncleavable counterpart, and researchers at Wyeth-Ayerst have successfully demonstrated remarkable activity with an anti-CD33-hydrazone-calicheamicin construct in acute myelogenous leukemia patients.[20,21] Notably, the latter construct has recently been approved by the Food and Drug Administration (FDA) under the name Myelotarg, and it represents the first marketed antibody-targeted chemotherapeutic agent.[22]

A third strategy for enabling the release of a drug from its targeting ligand following endocytosis consists of insertion of a peptide linker whose sequence is recognized and cleaved by endosomal/lysosomal enzymes. Indeed, the peptide Gly-Phe-Leu-Gly has been used successfully to promote lysosome-specific release of a variety of drugs,[23,24] but its utility is probably limited to use with ligands that naturally target the destructive lysosomal compartment.

For any of the aforementioned release strategies, one must be concerned about the chemical nature of the liberated drug fragment. For instance, as illustrated above for the enzymatic technique, the released drug fragment will contain a portion of the cleaved peptide (Phe-Gly-drug or drug-Gly-Leu if the construct design were reversed). In some cases, this added chemical baggage may affect the drug's intrinsic activity or its ability to traverse the endosomal membrane. The same principle applies to all release strategies. In fact, the authors have experienced the inactivation of a potent microtubule-stabilizing drug following its hydroxy esterification with a thiopropionyl linker moiety (unpublished observations). Overall, we believe that the best release strategy consists of one that discharges the drug in its original unmodified form.

9.4. SELECTION OF LIGANDS

A variety of biological ligands have been used to deliver drugs to target cells. Table 9.1 lists some of the most common ligand-receptor systems exploited to date for the delivery of therapeutic molecules. Most of these have been coupled to functionally active peptides and proteins, and in some cases to small molecular

TABLE 9.1 Ligands Frequently Used for Receptor-Mediated Drug Delivery

Ligand	Drug Payload	References
Insulin	Drugs, enzymes	25, 26
Epidermal growth factor	Protein toxins	27, 28
Transferrin	Drugs, protein toxins, gene therapy vectors	29
Thyrotropin-releasing hormone	Protein toxins	30
Human chorionic gonadotropin	Protein toxins	31, 32
Leutinizing hormone	Protein toxins	33, 34
Interleukin-2	Protein toxins	35, 36
Mannose-6-phosphate	Enzymes	37, 38
Asialoglycoprotein	Drugs, protein toxins and gene therapy vectors	39
IgG	All pharmaceutical classes	40, 41
Vitamin B_{12}	Drugs, peptides	42–45
Folate	All pharmaceutical classes	See Section 9.5

weight chemotherapeutic drugs. Each receptor system has advantages and disadvantages. Unfortunately, limitations of space preclude an in-depth discussion of each system. Therefore, the reader is referred to the listed references when additional information is needed. However, for the purpose of completing this discussion on receptor-targeted drug delivery, we shall illustrate in detail how the folate-targeted drug delivery pathway has been exploited at both the academic and clinical levels.

9.4.1. Selection of Therapeutic Drug

While the nature of the pathology often dictates the choice of the therapeutic agent, wherever multiple selections exist, a few guidelines can be beneficial. First, because receptor-mediated delivery pathways are frequently of low capacity, higher activity will likely occur with those conjugates constructed with the more potent drugs. Second, since target cell penetration is mediated by an endocytic pathway, membrane permeability is often not a necessary property of the drug. In fact, the more hydrophobic drugs are frequently less desirable for ligand-targeted conjugates, since they can promote nonspecific adsorption to cell surfaces and the consequent unwanted toxicity to nontargeted cells. Finally, if specificity for the pathological cells is high, the toxicity characteristics of the free drug can be ignored, since targeted delivery can prevent uptake by the sensitive normal cells. Thus, drugs that have been discarded because of poor toxicity profiles can often be reconsidered for use in targeted drug therapies.

9.5. FOLATE-MEDIATED DRUG DELIVERY

As a more detailed example of receptor-targeted drug delivery, we have elected to elaborate on the delivery pathway that exploits the cell surface receptor for folic

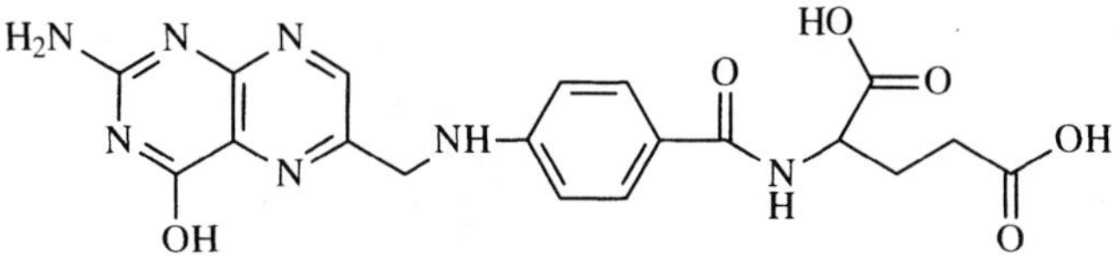

Figure 9.3. Structure of folic acid.

acid as a means of targeting drugs to receptor-expressing cells. Folates are low molecular weight vitamins required by all eukaryotic cells for 1-carbon metabolism and *de novo* nucleotide synthesis. Since animal cells lack key enzymes of the folate biosynthetic pathway, their survival and proliferation are dependent on their ability to acquire the vitamin from their diet. Thus, effective mechanisms for capturing exogenous folates are needed to sustain all higher forms of life. While most cells rely on a low-affinity ($K_D \sim 1$–$5\ \mu M$) membrane-spanning protein that transports reduced folates directly into the cell (termed the "reduced folate carrier"[46]), a few cells also express a high-affinity ($K_D \sim 100$ pM) receptor, generally referred to as the "folate receptor" (FR), that preferentially mediates the uptake of oxidized forms of the vitamin (e.g., folic acid) by receptor-mediated endocytosis.[47,48] As will be explained below, attachment of folic acid (Figure 9.3) via one of its carboxyl groups to a therapeutic or imaging agent allows targeting of the conjugate to cells that express FR,[49] with no measurable uptake by cells that express the reduced folate carrier.

9.5.1. Expression of FRs in Malignant Tissues

In 1991, a clinically valuable tumor marker was purified from ovarian cancers, and sequence analysis showed that it was the receptor for folic acid.[50] Subsequent to that finding, FRs were shown to be overexpressed on the cell surfaces of many different types of human cancers.[50–53] In general, the FR is up-regulated in malignant tissues of epithelial origin. As detailed in Table 9.2, FR expression has been detected at very high levels in >90% of ovarian and other gynecological cancers and at high to moderate levels in brain, lung, and breast carcinomas.[51,54–56] Cancers of the endometrium, kidney, and colon have also been found to frequently express FR.[50–59] Notably, the FR gene was recently mapped to region 11q13, a chromosomal locus that is amplified in >20% of tissue samples from breast and head/neck tumors.[60,61]

FR expression may also be dependent on the histological classification of a tissue or cancer. For example, using a recently developed quantitative assay for measuring functional FR in cells and tissues, it was found that normal ovarian tissue and the mucinous form of ovarian cancer express very low levels of FR (~1 pmol FR/mg protein; unpublished data, Endocyte, Inc., West Lafayette, IN). However, serous ovarian carcinomas express high amounts of FR, with the average expression level being 30 pmol FR/mg protein, or about 30-fold higher than that of the normal

TABLE 9.2 FR Levels in Various Human Carcinoma Tissues

	IHC[55]		RT-PCR[52]		Other Studies
Tissue	% Positive	*N*	% Positive	*N*	% Positive (*N*)
Ovarian	93%	56	100%	4	91% (34),[62] 90% (136),[58] 83% (40)[63]
Endometrial	91%	11	100%	3	64% (25)[63]
Breast	21%	53	80%	5	
Lung	33%	18	33%	3	50% (49)[64]
Adenocarcinoma	n.d.		n.d.		Adenocarcinoma 90% (10)[64]
Colorectal	22%	27	20%	5	
Kidney	50%	18	100%	4	67% (3)[63]
Prostate	0%	5	n.d.		
B-cell lymphoma	0%	21	n.d.		
Brain	n.d.		100%	8	
Mets.	80%	5	n.d.		
Head and neck	n.d.		50%	4	
Pancreas	13%	8	n.d.		
Bladder	0%	14	n.d.		

n.d., not determined.

ovary. Endometrioid and metastatic ovarian carcinomas also express FR, although not as much as the serous form. Interestingly, others have observed a strong correlation between FR expression and both the histological grade and stage of the tumor.[58] In general, highly dedifferentiated metastatic cancers express considerably more FR than their more localized, low-grade counterparts.

9.5.2. Expression of FRs in Normal Tissues

FRs have also been detected in normal tissues, particularly those involved in the retention and uptake of the vitamin. For example, the choroid plexus in the brain expresses high levels of FR. However, the receptor is primarily localized to the brain side of the blood-brain barrier, where it is inaccessible to blood-borne folates or folate-drug conjugates.[59,65,66] The FR has similarly been located on the apical membrane surface of the intestinal brush border,[67] but again, access to these docking sites requires either escape into the lumen of the intestine or oral ingestion of the folates. FR is also expressed at high levels in the proximal tubules of the kidney, where it is believed to capture folates (and small molecular weight folate-drug conjugates) prior to their urinary excretion and then return them back into circulation via a transcellular reabsorption process.[68–70] Recent reports further reveal FR expression on activated but not resting macrophages,[71,72] an observation that could find utility in the treatment of autoimmune diseases. And finally, FR has also been detected at low levels in a few other normal tissues,[52] but its relative expression level in these tissues is very low compared to that of many FR-positive cancers. This latter finding is evidenced by the fact that (1) the FR has been reliably employed as a tumor marker for many years,[50–59] and, more importantly, (2) intravenously administered folates or folate-drug conjugates accumulate predominantly in FR-expressing tumor and kidney tissue.[73–75]

In summary, FR is expressed on the apical membrane surface of some normal polarized epithelia, where it is largely inaccessible to folates and their drug conjugates present in the blood. FR is also expressed on the surfaces of many malignant cells, where it is fully accessible to parenterally administered folate-drug conjugates due to the collapse of membrane polarization associated with transformation. Since ~80% of human cancers arise from polarized epithelia, and since FR is further up-regulated in many of these malignancies, the FR has emerged as a useful target for receptor-directed therapies for cancer. The fact that FRs bind folate-drug conjugates with high affinity (10^{-9} M) and that the conjugates are subsequently transported nondestructively into the target cells only adds to the utility of this strategy for tumor-specific drug delivery.

9.5.3. Applications of Folate-Mediated Drug Delivery

Initial studies on folate conjugate targeting were conducted with radiolabeled and fluorescent proteins (drug surrogates) covalently attached to folic acid.[49] These conjugates were shown to bind and become internalized by FR-positive cells via a nondestructive, functionally active endocytic process.[49] When FR was subsequently identified as a major tumor-associated antigen,[50,76] much effort was quickly devoted to determining the types of attached cargo that might be easily targeted with folic acid. These studies have revealed that conjugates of radiopharmaceutical agents,[73,74,77–82] magnetic resonance imaging (MRI) contrast agents,[83] low molecular weight chemotherapeutic agents,[17,84] antisense oligonucleotides and ribozymes,[85–89] proteins and protein toxins,[49,90–94] immunotherapeutic agents,[95–98] liposomes with entrapped drugs,[99–103] drug-loaded nanoparticles,[104–106] and plasmids[107–115] can all be selectively delivered to FR-expressing cancer cells. Indeed, the major limitation associated with the above targeting efforts appears to be the intrinsic permeability of the tumor. Thus, where perfusion barriers do not limit access to FR-expressing tumor cells, folate conjugate binding, FR-mediated endocytosis, and intracellular drug release are readily achievable if the fundamental principles outlined above are followed.

In the remainder of this chapter, we shall illustrate the techniques of folate-targeted radiodiagnostic imaging, chemotherapy, and immunotherapy. However, the reader is encouraged to review the listed references if information on other folate conjugates is desired.

9.5.3.1. Tumor Targeting Through the FR: Radiodiagnostic Imaging The field of nuclear medicine has been revitalized with the advent of tissue-specific radiopharmaceutical targeting technologies. Ligands capable of concentrating at pathological sites have been derivatized with chelator-radionuclide complexes and then used as noninvasive probes for diagnostic imaging. Folate-targeted radiopharmaceuticals have been explored for the purpose of both (1) developing an imaging agent for the localization, sizing, and characterization of cancers and (2) obtaining "proof-of-principle" data regarding the ability of folic acid to deliver attached drugs to human tumors *in vivo*. Several animal models that contain tumors with FR levels similar to those found in common human carcinomas have been used to test uptake of folate-based radiopharmaceuticals in living organisms, including (1) nude mice with implanted human KB, MDA-231, or IGROV tumors, (2) C57BL/6 mice implanted with 24JK tumors, (3) Dupont's c-neu Oncomouse,[116] (4) DBA mice implanted with L1210A tumors, and (5) Balb/c mice implanted with syngeneic M109 tumors. This continually growing list of acceptable tumor models indicates that the location and nature of the tumor are relatively unimportant so long as the tumors express appreciable levels of FR.

In 1999, Phase I/II clinical studies were initiated by Endocyte, Inc., to evaluate ^{111}In-DTPA-folate.[75,82] Patients suspected of having ovarian cancer received a 5 mCi (2 mg) intravenous dose of the radiopharmaceutical, and whole body single photon emission computed tomographic (SPECT) images were taken 4 hours later to identify the location of the probe. Representative images from two enrolled patients are shown in Figure 9.4. The image displayed in Panel A shows that in a cancer-free patient, ^{111}In-DTPA-folate (a folate-drug conjugate) primarily concentrates in the FR-positive kidneys, while the remaining tissues of the body effectively clear the radiopharmaceutical by 4 hours post injection. However, in addition to the kidneys (and to some extent the liver), the folate-targeted radiopharmaceutical accumulates in the widely disseminated malignant tissue in ovarian cancer patients (Panel B). Notably, a similar distribution pattern has also been demonstrated in patients receiving a new ^{99m}Tc-based radiopharmaceutical called EC20 (Ref. 81 and unpublished clinical observations). Taken together, and following the treatment of more than 45 enrolled subjects, we have noted a pattern of (1) consistent uptake of folate conjugates into malignant masses (including $\sim$1 cm metastatic abdominal lesions), (2) absence of uptake into benign masses, (3) consistent uptake by the kidneys, and (4) little or no uptake by other normal tissues. In summary, these clinical results have provided valuable "proof-of-principle" data confirming that folate-mediated tumor targeting also occurs in humans. It further suggests that folate-targeted radiodiagnostic imaging agents may be useful for noninvasively identifying the loci of pathological FR-positive tissue within patients.

9.5.3.2. Folate-Targeted Chemotherapy The fundamentals of folate-cytotoxin therapy were first illustrated by the targeted killing of FR-positive cells *in vitro* using folate conjugates of numerous protein synthesis–inhibiting enzymes.[92] For example, folate conjugates of cell-impermeant, ribosome-inactivating proteins (e.g., momordin, saporin, gelonin, ricin A) were found to kill cultured FR-positive malignant cells without harming receptor-negative normal cells. The selectivity of this approach was confirmed by many important controls which demonstrated that (1) an excess of free folic acid quantitatively blocked folate-cytotoxin cell killing,

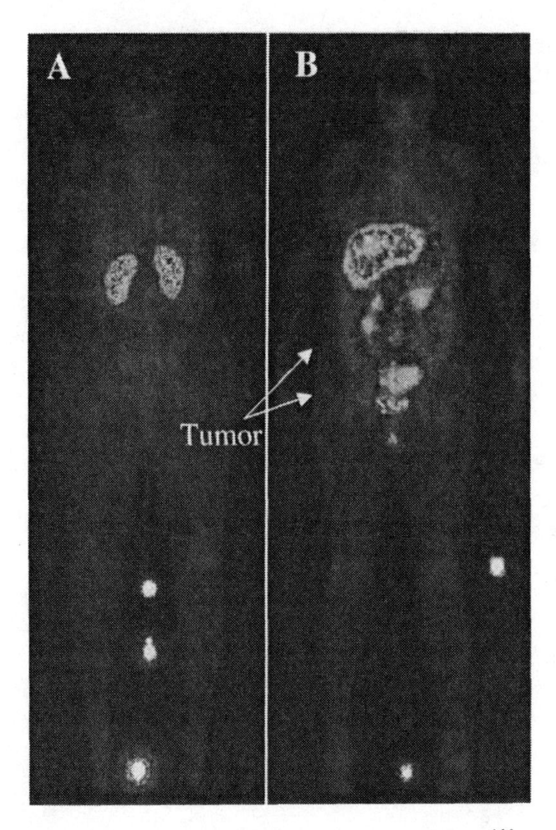

Figure 9.4. Anterior SPECT images of two patients receiving ^{111}In-DTPA-folate. (A) Image of a female patient without cancer. (B) Image of a female patient with stage IIIc ovarian carcinoma.

(2) the underivatized protein was not toxic to the target cells, and (3) pre-treatment of cells with phosphatidylinositol-specific phospholipase C, an enzyme that removes glycosylphosphatidylinositol-linked proteins (like the FR) from a cell's surface, effectively blocked folate-cytotoxin cell killing. Importantly, the same folate-cytotoxin conjugates were also shown to selectively kill only malignant cells when cocultured in the same dish with "normal," non-transformed cells.[94] Overall, high activity (IC_{50} values of $\sim$1 nM) was observed using a number of FR-expressing cell lines.

While it is tempting to speculate on the antitumor activity that may result from testing such folate-protein toxin conjugates *in vivo*, practical pharmaceutical considerations diminished the priority for their development. Instead, the focus shifted toward folate conjugates of conventional drug molecules. To date, both successes and failures with this approach have been published. In one study, researchers linked folate to a maytansinoid derivative (a natural product that blocks tubulin polymerization) via a disulfide bond. The conjugate was found to exhibit an IC_{50} of 50 pM for FR-positive KB cells but was nontoxic to FR-negative A375 cells, even though both cell lines were killed with equal potency (IC_{50} ~25 pM) by the free nontargeted DM1 maytansinoid.[17] However, subsequent studies with "nonreleasable" folate conjugates of paclitaxel or a nitroheterocyclic bis(haloethyl)phosphoramidate prodrug were much less impressive.[84,117] Unfortunately, none of the above drug conjugates was ever evaluated for activity against FR-positive tumors *in vivo*.

Endocyte Inc., has recently collected data on the activities of a number of folate-drug conjugates both *in vitro* and *in vivo*. For example, folate has been conjugated to a potent small molecular weight DNA alkylator through a disulfide linker, and the resultant conjugate (referred to as EC72) was found to promote target cell killing *in vitro* (IC_{50} ~ 3 nM; unpublished data). Although the toxicity of the EC72 conjugate could be quantitatively blocked by the presence of excess free folic acid (to demonstrate FR specificity), neither EC72 nor the underivatized drug could effectively kill 100% of the cancer cells *in vitro*, possibly because of an intrinsic resistance to the mechanism of the drug's activity. Nevertheless, despite this recognized limitation, the EC72 conjugate was evaluated in a pilot study using Balb/c mice bearing FR-positive M109 tumors. The study had two goals: (1) to determine if daily treatment with EC72 could prolong the lives of FR-positive tumor-bearing mice beyond that which the parent drug could do alone when tested under an identical dosing regimen and (2) to examine the pathological effects of EC72 treatment on normal tissues, including the FR-positive kidneys.

As shown in Figure 9.5, all control mice died 22 days following an intraperitoneal inoculation with M109 cells, while a 39% increase in life span was observed

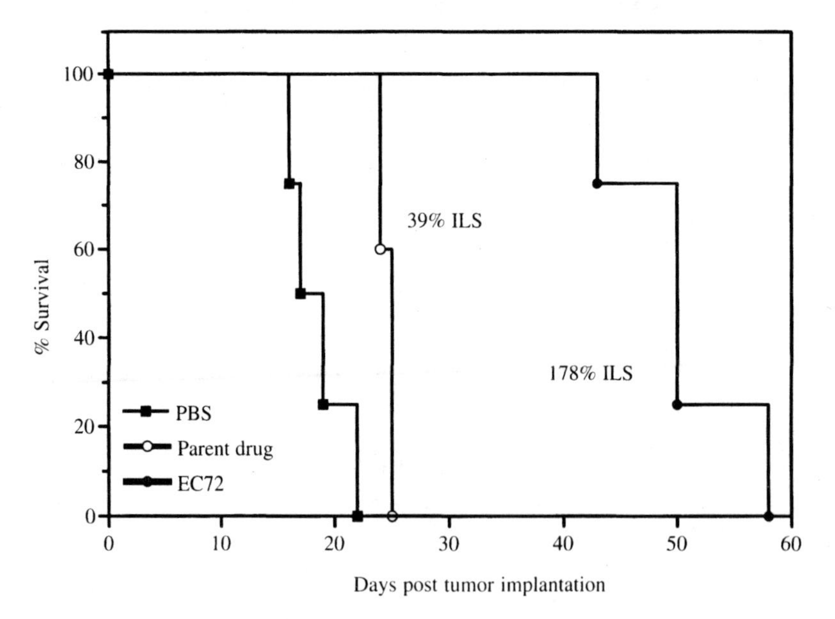

Figure 9.5. Survival of treated M109 tumor-bearing mice. Four days post inoculation, Balb/c mice bearing intraperitoneal M109 tumors were treated once daily with either the unmodified parent drug or the folate-derivatized drug. ILS: increased life span.

for the animals treated with the unmodified drug. More importantly, animals treated with the EC72 conjugate lived on average 178% longer.

Following euthanasia, major organs were collected from both the EC72 and parent drug–treated animals, and they were sent to a certified pathologist for examination. The nontargeted drug-treated animals were found to suffer from massive myelosuppression (which is a characteristic dose-limiting side effect of the parent drug's therapy), and all of the animals in this cohort died from obvious drug-related side effects. In dramatic contrast, animals treated with EC72 (folate-SS-drug) displayed no evidence of myelosuppression or kidney damage. Further, examination of blood collected from EC72-treated animals indicated normal blood urea nitrogen and creatinine levels following 30 consecutive daily injections with the conjugate. We concluded from these results that the use of folate-drug conjugates may be an effective form of chemotherapy that does not cause injury to normal tissues, including the FR-positive kidneys. While the latter conclusion was surprising, these observations do support the hypothesis that the FRs in the kidney proximal tubules function primarily to shuttle scavenged folates (or small folate-drug conjugates) back into systemic circulation rather than to deliver the conjugates into the kidney cells.[69,70]

9.5.3.3. Folate-Targeted Immunotherapy

9.5.3.3.1. Intracellular Delivery versus Cell-Surface Loading In contrast to most hormone receptors, not all FRs endocytose following ligand binding.[118] Rather, a substantial fraction of the occupied FRs remain on the cell surface as a means of storing folates for use at a later date. Since folate-drug conjugates interact with FR in the same manner as free folate,[49] it was of no surprise to learn that a fraction of these conjugates also remains extracellularly bound.[49,93] This dual destiny of occupied FR obviously allows for two distinct uses of the folate-targeting technology: (1) delivery of folate-drug conjugates into FR-expressing cells and (2) decoration of FR-expressing cell surfaces with folate-drug conjugates. This latter application has led to the advent of folate-targeted immunotherapy,[119] as outlined below.

Based on the ability of folate to position an attached drug on the surface of an FR-expressing cancer cell, it was recognized that a malignant cell could be converted from its normal immunologically invisible state (the condition that permits its proliferation *in vivo*) to a state where it is vividly recognized as foreign by the immune system (and consequently subject to immune-mediated elimination) simply by targeting a highly immunogenic antigen to its cell surface. This concept is presented in Figure 9.6. Different from targeted chemotherapy, this treatment begins with a series of subcutaneous inoculations of a hapten-based vaccine to stimulate production of anti-hapten antibodies in the patient. After induction of an adequate antibody titer, a folate-hapten conjugate is administered to enable "marking" of all FR-expressing tumor cells with the hapten. This process rapidly promotes the tumor cell's opsonization with the previously induced endogenous anti-hapten antibodies. Mechanistically, the folate-hapten conjugate forms a molecular "bridge" between the tumor cell and the endogenous circulating anti-hapten

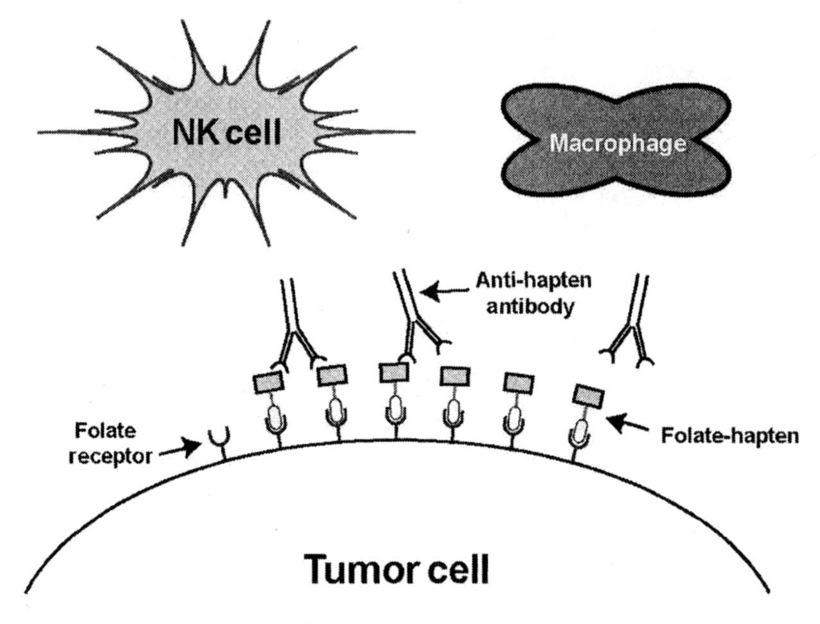

Figure 9.6. Folate-targeted immunotherapy strategy.

antibody. Ultimately, this marking process enables Fc receptor-bearing effector cells, such as natural killer (NK) cells and macrophages, to recognize and destroy the antibody-coated tumor cells.

Although we have found this folate-targeted immunotherapy to be effective in tumor-bearing rodent models, we have also noted that complete and reproducible cures are obtained only when the animals are concomitantly dosed with low levels of the cytokines, interleukin-2 and interferon-α.[119] The purpose of adding these cytokines is to stimulate those immune effector cells (like NK and macrophage cells) that are responsible for killing opsonized tumor cells. Importantly, the low cytokine dose levels applied in this technique were found not to produce effective antitumor responses in the absence of the folate-hapten conjugate. Although the exact effector mechanisms involved in this therapy have not yet been established, there is strong evidence to suggest that the cured animals develop independent cellular immunity against the tumor cells and that this cellular immunity prevents the animals from growing tumors when rechallenged with the same malignant cells, even without additional treatment.[119] These findings also support the notion that a long-term, perhaps T cell-specific immunity, develops during the hapten-mediated immunotherapy, and that a shift of the host's antitumor response from humoral to cellular immunity occurs. Because this therapy has generated cures in multiple tumor-bearing animal models, and since the complete therapeutic regimen (vaccination and drug therapy at dose levels approaching 100-fold of the human dose equivalent) did not produce test article–related toxicities in toxicology studies, the strategy was approved for clinical trial by the FDA for treatment of advanced ovarian and renal cell cancers beginning in 2003. Results of these clinical studies should provide much information on the strengths and weaknesses of folate-targeted cancer therapies in humans.

9.6. CONCLUSIONS

Nature has designed biological membranes to serve as formidable barriers against the unwanted entry of harmful agents into cells. Life, however, cannot exist without the cell's ability to selectively capture and internalize certain biosupportive molecules. Receptor-mediated endocytosis is but one mechanism by which cells retrieve required molecules from their environment, and it is also a powerful means for delivering normally impermeable molecules into target cells for medicinal purposes. The list of available ligands for use in this exploitive technique keeps growing, as does the potential for inventing alternative drug delivery strategies. It is important, however, to remember that not all ligand-receptor systems will function in the same manner. Many variations exist among cell types, including their level of receptor expression, internalization rate, ligand affinity, intracellular compartmentalization, recycling capabilities, etc. Thus, in spite of recent technological advances, many challenges remain for the future development of receptor-targeted therapies. The most significant issues will undoubtedly be related to the confirmation of target tissue specificity (i.e., receptor distribution among target and nontarget tissues) as well as the design of releasable linkers for certain ligand-linker-drug conjugates. Receptor-mediated delivery technology is undoubtedly influencing rational drug design, and it is our hope that the scientific and pharmaceutical communities will continue to invest in this exiting new field of medicine.

ACKNOWLEDGMENT

This work was supported in part by grants from Endocyte, Inc., and the NIH (CA89581).

REFERENCES

1. McPherson, P. S.; Kay, B. K.; Hussain, N. K. *Traffic* **2001**, *2*, 375–384.
2. Mukherjee, S.; Ghosh, R. N.; Maxfield, F. R. *Physiol. Rev.* **1997**, *77*, 759–803.
3. Schwartz, A. L. *Pediatr. Res.* **1995**, *38*, 835–843.
4. Stahl, P.; Schwartz, A. L. *J. Clin. Invest.* **1986**, *77*, 657–662.
5. Smythe, E.; Warren, G. *Eur. J. Biochem.* **1991**, *202*, 689–699.
6. Marsh, M.; Helenius, A. *J. Mol. Biol.* **1980**, *142*, 439–454.
7. Pastan, I.; Willingham, M. C. *The Pathway of Endocytosis*; Plenum Press: New York, 1985.

8. Willingham, M. C.; Pastan, I. *Cell* **1980**, *21*, 67–77.
9. Lai, W. H.; Cameron, P. H.; Wada, I.; Doherty, J. J., 2nd; Kay, D. G.; Posner, B. I.; Bergeron, J. J. *J. Cell Biol.* **1989**, *109*, 2741–2749.
10. Hanvey, J. C.; Peffer, N. J.; Bisi, J. E.; Thomson, S. A.; Cadilla, R.; Josey, J. A.; Ricca, D. J.; Hassman, C. F.; Bonham, M. A.; Au, K. G.; Carter, S. G.; Bruckenstein, D. A.; Boyd, A. L.; Noble, S. A.; Babiss, L. E. *Science* **1992**, *258*, 1481–1485.
11. Terada, K.; Manchikalapudi, P.; Noiva, R.; Jauregui, H. O.; Stockert, R. J.; Schilsky, M. L. *J. Biol. Chem.* **1995**, *270*, 20410–20416.
12. Scheiber, B.; Goldenberg, H. *Arch. Biochem. Biophys.* **1993**, *305*, 225–230.
13. Liu, Y.; Peterson, D. A.; Kimura, H.; Schubert, D. *J. Neurochem.* **1997**, *69*, 581–593.
14. McIntyre, G. D.; Scott, C. F.; Ritz, J.; Blattler, W. A.; Lambert, J. M. *Bioconjugate Chem.* **1994**, *5*, 88–97.
15. Leamon, C. P.; Pastan, I.; Low, P. S. *J. Biol. Chem.* **1993**, *268*, 24847–24854.
16. Leamon, C. P.; DePrince, R. B.; Hendren, R. W. *J. Drug Targeting* **1999**, *7*, 157–169.
17. Ladino, C. A.; Chari, R. V. J.; Bourret, L. A.; Kedersha, N. L.; Goldmacher, V. S. *Int. J. Cancer* **1997**, *73*, 859–864.
18. Liu, C.; Tadayoni, B. M.; Bourret, L. A.; Mattocks, K. M.; Derr, S. M.; Widdison, W. C.; Kedersha, N. L.; Ariniello, P. D.; Goldmacher, V. S.; Lambert, J. M.; Blattler, W. A.; Chari, R. V. *Proc. Natl. Acad. Sci. USA* **1996**, *93*, 8618–8623.
19. Lee, R. J.; Wang, S.; Low, P. S. *Biochim. Biophys. Acta* **1996**, *1312*, 237–242.
20. Neville, D. M.; Srinivasachar, K.; Stone, R.; Scharff, J. *J. Biol. Chem.* **1989**, *264*, 14653–14661.
21. Hamann, P. R.; Hinman, L. M.; Beyer, C. F.; Lindh, D.; Upeslacis, J.; Flowers, D. A.; Bernstein, I. *Bioconjug. Chem.* **2002**, *13*, 40–46.
22. Hamann, P. R.; Hinman, L. M.; Hollander, I.; Beyer, C. F.; Lindh, D.; Holcomb, R.; Hallett, W.; Tsou, H. R.; Upeslacis, J.; Shochat, D.; Mountain, A.; Flowers, D. A.; Bernstein, I. *Bioconjug. Chem.* **2002**, *13*, 47–58.
23. Rihova, B.; Srogl, J.; Jelinkova, M.; Hovorka, O.; Buresova, M.; Subr, V.; Ulbrich, K. *Ann. NY Acad. Sci.* **1997**, *831*, 57–71.
24. Seymour, L. W.; Ulbrich, K.; Wedge, S. R.; Hume, I. C.; Strohalm, J.; Duncan, R. *Br. J. Cancer* **1991**, *63*, 859–866.
25. Poznansky, M. J.; Singh, R.; Singh, B. *Science* **1984**, *223*, 1304–1306.
26. Poznansky, M. J.; Hutchison, S. K.; Davis, P. J. *Faseb J.* **1989**, *3*, 152–156.
27. Shaw, J. P.; Akiyoshi, D. E.; Arrigo, D. A.; Rhoad, A. E.; Sullivan, B.; Thomas, J.; Genbauffe, F. S.; Bacha, P.; Nichols, J. C. *J. Biol. Chem.* **1991**, *266*, 21118–21124.
28. Cawley, D. B.; Herschman, H. R.; Gilliland, D. G.; Collier, R. *J. Cell* **1980**, 22, 563–570.
29. Qian, Z. M.; Li, H.; Sun, H.; Ho, K. *Pharmacol. Rev.* **2002**, *54*, 561–587.
30. Bacha, P.; Murphy, J. R.; Reichlin, S. *J. Biol. Chem.* **1983**, *258*, 1565–1570.
31. Sakai, A.; Sakakibara, R.; Ishiguro, M. *J. Biochem.* **1989**, *105*, 275–280.
32. Sakai, A.; Sakakibara, R.; Ohwaki, K.; Ishiguro, M. *Chem. Pharm. Bull. (Tokyo)* **1991**, *39*, 2984–2989.
33. Singh, V.; Sairam, M. R.; Bhargavi, G. N., Akhras, R. G. *J. Biol. Chem.* **1989**, *264*, 3089–3095.
34. Singh, V.; Curtiss, R., 3rd. *Mol. Cell Biochem.* **1994**, *130*, 91–101.

35. Bacha, P.; Williams, D. P.; Waters, C.; Williams, J. M.; Murphy, J. R.; Strom, T. B. *J. Exp. Med.* **1988**, *167*, 612–622.
36. Shapiro, M. E.; Kirkman, R. L.; Kelley, V. R.; Bacha, P.; Nichols, J. C.; Strom, T. B. *Targeted Diagn. Ther.* **1992**, *7*, 383–393.
37. Karson, E. M.; Neufeld, E. F.; Sando, G. N. *Biochemistry* **1980**, *19*, 3856–3860.
38. Sato, Y.; Beutler, E. *J. Clin. Invest.* **1993**, *91*, 1909–1917.
39. Wu, J.; Nantz, M. H.; Zern, M. A. *Front. Biosci.* **2002**, *7*, d717–25.
40. Pennell, C. A.; Erickson, H. A. *Immunol. Res.* **2002**, *25*, 177–191.
41. Garnett, M. C. *Adv. Drug Deliv. Rev.* **2001**, *53*, 171–216.
42. Collins, D. A.; Hogenkamp, H. P.; Gebhard, M. W. *Mayo Clin. Proc.* **1999**, *74*, 687–691.
43. Swaan, P. W. *Pharmaceutical Res.* **1998**, *15*, 826–834.
44. Bauer, J. A.; Morrison, B. H.; Grane, R. W.; Jacobs, B. S.; Dabney, S.; Gamero, A. M.; Carnevale, K. A.; Smith, D. J.; Drazba, J.; Seetharam, B.; Lindner, D. J. *J. Natl. Cancer Inst.* **2002**, *94*, 1010–1019.
45. Russell-Jones, G. J.; Alpers, D. H. *Pharm. Biotechnol.* **1999**, *12*, 493–520.
46. Antony, A. C. *Blood* **1992**, *79*, 2807–2820.
47. Kamen, B. A.; Capdevila, A. *Proc. Natl. Acad. Sci. USA* **1986**, *83*, 5983–5987.
48. Antony, A. C. *Annu. Rev. Nutr.* **1996**, *16*, 501–521.
49. Leamon, C. P.; Low, P. S. *Proc. Natl. Acad. Sci. USA* **1991**, *88*, 5572–5576.
50. Coney, L. R.; A., T.; Carayannopoulos, L.; Frasca, V.; Kamen, B. A.; Colnaghi, M. I.; Zurawski, V. R. J. *Cancer Res.* **1991**, *51*, 6125–6132.
51. Weitman, S. D.; Lark, R. H.; Coney, L. R.; Fort, D. W.; Frasca, V.; Zurawski, V. R.; Kamen, B. A. *Cancer Res.* **1992**, *52*, 3396–3401.
52. Ross, J. F.; Chaudhuri, P. K.; Ratnam, M. *Cancer* **1994**, *73*, 2432–2443.
53. Weitman, S. D.; Frazier, K. M.; Kamen, B. A. *J. Neurol. Oncol.* **1994**, *21*, 107.
54. Boerman, O. C.; van Niekerk, C. C.; Makkink, K.; Hanselaar, T. G.; Kenemans, P.; Poels, L. G. *Int. J. Gynecol. Pathol.* **1991**, *10*, 15–25.
55. Garin-Chesa, P.; Campbell, I.; Saigo, P. E.; Lewis, J. L.; Old, L. J.; Rettig, W. J. *Am. J. Pathol.* **1993**, *142*, 557–567.
56. Mattes, M. J.; Major, P. P.; Goldenberg, D. M.; Dion, A. S.; Hutter, R. V. P.; Klein, K. M. *Cancer Res. Suppl.* **1990**, *50*, 880S.
57. Weitman, S. D.; Weiberg, A. G.; Coney, L. R.; Zurawski, V. R.; Jennings, D. S.; Kamen, B. A. *Cancer Res.* **1992**, *52*, 6708–6711.
58. Toffoli, G.; Cernigoi, C.; Russo, A.; Gallo, A.; Bagnoli, M.; Boiocchi, M. *Int. J. Cancer* **1997**, *74*, 193–198.
59. Holm, J.; Hansen, S. I.; Hoier-Madsen, M.; Bostad, L. *Biochem. J.* **1991**, *280*, 267–271.
60. Berenson, J. R.; Yang, J.; Mickel, R. A. *Oncogene* **1989**, *4*, 1111–1116.
61. Orr, R. B.; Kriesler, A. R.; Kamen, B. A. *J. Natl. Cancer Inst.* **1995**, *87*, 299–303.
62. Li, P. Y.; Vecchio, S. D.; Fonti, R.; Carriero, M. V.; Potena, M. I.; Botti, G.; Miotti, S.; Lastoria, S.; Menard, S.; Colnaghi, M. I.; Salvatore, M. *J. Nucl. Med.* **1996**, *37*, 665–672.
63. Veggian, R.; Fasolato, S.; Menard, S.; Minucci, D.; Pizzetti, P.; Regazzoni, M.; Tagliabue, E.; Colnaghi, M. I. *Tumori* **1989**, *75*, 510–513.

64. Franklin, W. A.; Waintrub, M.; Edwards, D.; Christensen, K.; Prendegrast, P.; Woods, J.; Bunn, P. A.; Kolhouse, J. F. *Int. J. Cancer* **1994**, *Suppl. 8*, 89–95.
65. Patrick, T. A.; Kranz, D. M.; van Dyke, T. A.; Roy, E. J. *J. Neurooncol.* **1997**, *32*, 111–123.
66. Kennedy, M. D.; Jallad, K. N.; Low, P. S.; Ben-Amotz, D. *Pharmaceut. Res.* **2003**,
67. Zimmerman, J. *Gastroenterology* **1990**, *99*, 964–972.
68. Morshed, K. M.; Ross, D. M.; McMartin, K. E. *J. Nutr.* **1997**, *127*, 1137–1147.
69. Birn, H.; Selhub, J.; Christensen, E. I. *Am. J. Physiol.* **1993**, *264*, C302–C310.
70. Birn, H.; Nielsen, S.; Christensen, E. I. *Am. J. Physiol.* **1997**, *272*, F70–F78.
71. Turk, M. J.; Breur, G. J.; Widmer, W. R.; Paulos, C. M.; Xu, L. C.; Grote, L. A.; Low, P. S. *Arthritis Rheum.* **2002**, *46*, 1947–1955.
72. Nakashima-Matsushita, N.; Homma, T.; Yu, S.; Matsuda, T.; Sunahara, N.; Nakamura, T.; Tsukano, M.; Ratnam, M.; Matsuyama, T. *Arthritis Rheum.* **1999**, *42*, 1609–1616.
73. Mathias, C. J.; Wang, S.; Lee, R. J.; Waters, D. J.; Low, P. S.; Green, M. A. *J. Nucl. Med.* **1996**, *37*, 1003–1008.
74. Mathias, C. J.; Wang, S.; Waters, D. J.; Turek, J. J.; Low, P. S.; Green, M. A. *J. Nucl. Med.* **1998**, *39*, 1579–1585.
75. Leamon, C. P.; Low, P. S. *Drug Discov. Today* **2001**, *6*, 44–51.
76. Campbell, I. G.; Jones, T. A.; Foulkes, W. D.; Trowsdale, J. *Cancer Res.* **1991**, *51*, 5329–5338.
77. Wang, S.; Lee, R. J.; Mathias, C. J.; Green, M. A.; Low, P. S. *Bioconjugate Chem.* **1996**, *7*, 56–62.
78. Ilgan, S.; Yang, D. J.; Higuchi, T.; Zareneyrizi, F.; Bayham, H.; Yu, D.; Kim, E. E.; Podoloff, D. A. *Cancer Biother. Radiopharm.* **1998**, *13*, 427–435.
79. Guo, W.; Hinkle, G. H.; Lee, R. J. *J. Nucl. Med.* **1999**, *40*, 1563–1569.
80. Linder, K. E.; Wedeking, P.; Ramalingam, K.; Nunn, A. D.; Tweedle, M. F. *Soc. Nucl. Med. Proc. 47th Annual Meeting* **2000**, *41*, 119P.
81. Leamon, C. P.; Parker, M. A.; Vlahov, I. R.; Xu, L. C.; Reddy, J. A.; Vetzel, M.; Douglas, N. *Bioconjugate Chem.* **2002**,
82. Wang, S.; Luo, J.; Lantrip, D. A.; Waters, D. J.; Mathias, C. J.; Green, M. A.; Fuchs, P. L.; Low, P. S. *Bioconjugate Chem.* **1997**, *8*, 673–679.
83. Konda, S. D.; Aref, M.; Brechbiel, M.; Wiener, E. C. *Invest. Radiol.* **2000**, *35*, 50–57.
84. Lee, J. W.; Lu, J. Y.; Low, P. S.; Fuchs, P. L. *Bioorg. Med. Chem.* **2002**, *10*, 2397–2414.
85. Citro, G.; Szczylik, C.; Ginobbi, P.; Zupi, G.; Calabretta, B. *Br. J. Cancer* **1994**, *69*, 463–467.
86. Li, S.; Huang, L. *J. Liposome Res.* **1997**, *7*, 63–75.
87. Li, S.; Huang, L. *J. Liposome Res.* **1998**, *8*, 239–250.
88. Li, S.; Deshmukh, H. M.; Huang, L. *Pharm. Res.* **1998**, *15*, 1540–1545.
89. Leopold, L. H.; Shore, S. K.; Newkirk, T. A.; Reddy, R. M.; Reddy, E. P. *Blood* **1995**, *85*, 2162–2170.
90. Ward, C. M.; Acheson, N.; Seymour, L. M. *J. Drug Target.* **2000**, *8*, 119–123.
91. Lu, J. Y.; Lowe, D. A.; Kennedy, M. D.; Low, P. S. *J. Drug. Target.* **1999**, *7*, 43–53.
92. Leamon, C. P.; Low, P. S. *J. Biol. Chem.* **1992**, *267*, 24966–24971.

93. Leamon, C. P.; Low, P. S. *Biochem. J.* **1993**, *291*, 855–860.
94. Leamon, C. P.; Low, P. S. *J. Drug Target.* **1994**, *2*, 101–112.
95. Kranz, D. M.; Patrick, T. A.; Brigle, K. E.; Spinella, M. J.; Roy, E. J. *Proc. Natl. Acad. Sci. USA* **1995**, *92*, 9057–9061.
96. Cho, B. K.; Roy, E. J.; Patrick, T. A.; Kranz, D. M. *Bioconjugate Chem.* **1997**, *8*, 338–346.
97. Kranz, D. M.; Manning, T. C.; Rund, L. A.; Cho, B. K.; Gruber, M. M.; Roy, E. J. *J. Controlled Release* **1998**, *53*, 77–84.
98. Rund, L. A.; Cho, B. K.; Manning, T. C.; Holler, P. D.; Roy, E. J.; Kranz, D. M. *Int. J. Cancer* **1999**, *83*, 141–149.
99. Lee, R. J.; Low, P. S. *J. Biol. Chem.* **1994**, *269*, 3198–3204.
100. Lee, R. J.; Low, P. S. *Biochim. Biophys. Acta* **1995**, *1233*, 134–144.
101. Vogel, K.; Wang, S.; Lee, R. J.; Chmielewski, J.; Low, P. S. *J. Am. Chem. Soc.* **1996**, *118*, 1581–1586.
102. Rui, Y.; Wang, S.; Low, P. S.; Thompson, D. H. *J. Am. Chem. Soc.* **1998**, *120*, 11213–11218.
103. Gabizon, A.; Horowitz, A. T.; Goren, D.; Tzemach, D.; Mandelbaum-Shavit, F.; Qazen, M. M.; Zalipsky, S. *Bioconjugate Chem.* **1999**, *10*, 289–298.
104. Zhang, Y.; Kohler, N.; Zhang, M. *Biomaterials* **2002**, *23*, 1553–1561.
105. Oyewumi, M. O.; Mumper, R. J. *Bioconjugate Chem.* **2002**, *13*, 1328–1335.
106. Oyewumi, M. O.; Mumper, R. J. *Int. J. Pharm.* **2003**, *251*, 85–97.
107. Gottschalk, S.; Cristiano, R. J.; Smith, L. C.; Woo, S. L. C. *Gene Ther.* **1994**, *1*, 185–191.
108. Mislick, K. A.; Baldeschwieler, J. D.; Kayyem, J. F.; Meade, T. J. *Bioconjugate Chem.* **1995**, *6*, 512–515.
109. Douglas, J. T.; Rogers, B. E.; Rosenfeld, M. E.; Michael, S. I.; Feng, M.; Curiel, D. T. *Nature Biotech.* **1996**, *14*, 1574–1578.
110. Leamon, C. P.; Weigl, D.; Hendren, R. W. *Bioconjugate Chem.* **1999**, *10*, 947–957.
111. Guo, W.; Lee, R. J. *PharmSci.* **1999**, *1*, article 19.
112. Reddy, J. A.; Dean, D.; Kennedy, M. D.; Low, P. S. *J. Pharm. Sci.* **1999**, *88*, 1112–1118.
113. Reddy, J. A.; Low, P. S. *J. Controlled Release* **2000**, *64*, 27–37.
114. Reddy, J. A.; Abburi, C.; Hofland, H.; Howard, S. J.; Vlahov, I.; Wils, P.; Leamon, C. P. *Gene Ther.* **2002**, *9*, 1542–1550.
115. Hofland, H. E.; Masson, C.; Iginla, S.; Osetinsky, I.; Reddy, J. A.; Leamon, C. P.; Scherman, D.; Bessodes, M.; Wils, P. *Mol. Ther.* **2002**, *5*, 739–744.
116. Liu, S.; Edwards, D. S.; Barrett, J. A. *Bioconjugate Chem.* **1997**, *8*, 621–636.
117. Steinberg, G.; Borch, R. F. *J. Med. Chem.* **2001**, *44*, 69–73.
118. Kamen, B. A.; Wang, M. T.; Streckfuss, A. J.; Peryea, X.; Anderson, R. G. W. *JBC* **1988**, *263*, 13602–13609.
119. Lu, Y.; Low, P. S. *Cancer Immunol. Immunother.* **2002**, *51*, 153–162.

10

ORAL PROTEIN AND PEPTIDE DRUG DELIVERY

RICK SOLTERO
Pharma Directions, Inc.

10.1. INTRODUCTION

Endogenous peptides play a pivotal role in almost all regulatory processes of the body functions and act with high specificity and potency. Their vast therapeutic potential has prompted biomedical scientists to greatly expand research into the identities of biologically active polypeptides and proteins and their activities.

Drug Delivery: Principles and Applications Edited by Binghe Wang, Teruna Siahaan, and Richard Soltero
ISBN 0-471-47489-4

Concurrently, pharmaceutical companies around the world have endeavored to develop processes for producing therapeutically active entities at commercial scales.

As a result of this research, the availability of biotechnology products to develop is no longer an issue. At the present time, over 350 peptide and protein drugs are in various stages of clinical development. Many other candidates will likely be identified as genomic studies successfully translate genetic data into knowledge about proteins and their functions.

The broad availability of structurally diverse peptides possessing a spectrum of pharmacological effects has not been matched, however, by their clinical development, mostly because of the poor delivery properties of peptide- and protein-based products. Today, the most common means for administering these protein drugs remains injection (i.e., intravenous, intramuscular, or subcutaneous administration). Patient compliance with drug administration regimens by any of these parenteral routes, is generally poor and severely restricts the therapeutic value of the drug, particularly for diseases such as diabetes.

Delivering therapeutically active proteins and peptides by any route other than the invasive injection methods has been a challenge and a goal for many decades.[1] Among the alternate routes that have been tried with varying degrees of success are the oral, buccal,[2] intranasal,[3] pulmonary,[4] transdermal,[5] ocular,[6] and rectal[7] approaches. The many benefits of oral administration of therapeutic agents make oral polypeptide delivery stand out from among these drug delivery alternatives. In addition to high levels of patient acceptance and long-term compliance, oral delivery also benefits the patient through avoidance of the pain and discomfort associated with injections; greater convenience; implementation of a more socially compatible dosing regimen; and the elimination of concerns about needle-related infections. In addition, a growing body of data suggests that for certain polypeptides, such as insulin, the oral delivery route is more physiological.[8,9] Thus, oral delivery would be an ideal route if appropriate oral dosage forms of therapeutic peptides and proteins were available.

The oral route of administration of these therapeutic agents, however, is among the most problematic delivery regimens. Drug delivery via the gastrointestinal (GI) tract requires relatively lengthy exposure to a multifaceted system that is designed to degrade nutrients and dietary materials into small molecules and to prevent the indiscriminate passage of macromolecules, as well as other large entities such as microbes or foreign antigens that may present dangers to the host. The various barriers, including physiological, chemical, and biochemical ones, for oral drug delivery are discussed in detail in Chapter 2 and therefore will not be presented in this chapter.

In spite of the obstacles to oral delivery, substantial evidence suggests that pharmaceutical polypeptides are absorbed through the intestinal mucosa, although in minute amounts.[10] Small amounts of polypeptide drugs can be absorbed by the action of specific peptide transporters in the intestinal mucosa cells.[11] This suggests that properly formulated proteins or peptide drugs may be administered by the oral route with retention of sufficient biological activity for their therapeutic use.

Designing and formulating a polypeptide drug for delivery through the GI tract requires a multitude of strategies.[12,13] The dosage form must initially stabilize the drug while making it easy to take orally.[14] It must then protect the polypeptide from the extreme acidity and action of pepsin in the stomach.[15] When the drug reaches the intestine, the formulation must incorporate some means for limiting drug degradation by the plethora of enzymes that are present in the intestinal lumen.[16] In addition, the polypeptide and/or its formulations must facilitate both aqueous solubility at near-neutral pH and lipid layer penetration in order for the protein to traverse the intestinal membrane and then the basal membrane for entry into the bloodstream.[17] To accomplish this, formulation excipients that promote absorption may be required.[18,19] Finally, when the modified polypeptide enters the systemic circulation, the structural modifications may add to the functionality of the drug, e.g., by extending its half-life in the circulation. However, any structural changes that may have been employed to enhance oral bioavailability must not interfere with receptor binding and uptake at the site of biological activity.

In this chapter, we summarize the general approaches that have been used to successfully achieve the formulation goals for oral delivery: minimize enzymatic degradation; enhance intestinal absorption; maximize blood level reproducibility; deliver drug through the gut wall; and produce a palatable and acceptable dosage form. Then insulin will be used as an example to show how oral bioavailability has been achieved through chemical modification.

10.2. OVERCOMING PHYSIOLOGICAL BARRIERS WITH FORMULATION APPROACHES

Oral delivery of polypeptides has been an ongoing challenge. Nearly every oral dosage form used for delivery of conventional small-molecule drugs has been used to explore oral delivery of polypeptides. Except for cases where the polypeptide has been chemically modified or where a proprietary absorption enhancer has been used, the results have been disappointing. There have been only a handful of human clinical trials where there has been adequate bioavailability and pharmacokinetics to suggest that commercializing an orally delivered polypeptide is feasible. Some of the approaches used to date are described in general terms in this section. For more in-depth discussions, readers should consult the numerous reviews that cover specific areas of research.[20–26]

10.2.1. Enzyme Inhibitors

Researchers have evaluated the use of protease inhibitors with the aim of slowing the rate of degradation of proteins and peptides in the GI tract. The hypothesis was that a slow rate of degradation would increase the amount of protein and peptide drug available for absorption.

For example, enzyme degradation of insulin is known to be mediated by the serine proteases trypsin, α-chymotrypsin, and thiol metalloproteinase insulin-degrading enzymes. The stability of insulin has been evaluated in the presence of excipients that inhibit these enzymes. Representative inhibitors of trypsin and α-chymotrypsin include pancreatic inhibitor and soybean trypsin inhibitor,[27] FK-448,[28] camostat mesylate,[29] and aprotinin.[30] Inhibitors of insulin-degrading enzyme include 1,10 phenanthroline, *p*-choromeribenzoate,[31] and bacitracin.[32] Ziv et al.[33] reported the use of a combination of an enhancer, sodium cholate, and a protease inhibitor to achieve a 10% increase in rat intestinal insulin absorption.

Another approach to enzyme inhibition is to manipulate the pH to inactivate local digestive enzymes. Lee et al. conducted studies demonstrating that oral absorption properties of salmon calcitonin can be modulated by changing intestinal pH. Reducing the intestinal pH in the GI tract, increased absorption of the intact peptide.[34] A sufficient amount of a pH-lowering buffer that lowers local intestinal pH to values below 4.5 can deactivate trypsin, chymotrypsin, and elastase.

10.2.2. Absorption Enhancers

Permeation enhancers improve the absorption of protein and peptides by increasing their paracellular and transcellular transports. An increase in paracellular transport is mediated by modulating the tight junctions of the cells, and an increase in transcellular transport is associated with an increase in the fluidity of the cell membrane. Chapter 2 describes in depth the various paracellular and transcellular transport pathways, which will not be discussed here.

Paracellular permeation enhancers include calcium chelators, bile salts, and fatty acids. Calcium chelators, such as ethylendiaminetetraacetic acid (EDTA),[5] act by inducing calcium depletion, thereby creating global changes in the cells, including disruption of actin filaments, disruption of adherent junctions, and diminished cell adhesion.[35] Zonula occludens toxin (ZOT) acts specifically on the actin filaments of tight junctions,[36] and chitosan triggers opening of the tight junction between the cells.[37]

Transcellular permeation enhancers include surfactants, medium chain fatty acids, nonionic surfactants,[38] sodium cholate and other bile salts,[39] and many other surfactants.

10.2.2.1. Nanoparticles Nanoparticles have been studied in recent decades as particulate carriers to deliver the protein and peptides drugs orally. This approach is supported by the literature, which states that particles in the nanosize range are absorbed intact by the intestinal epithelium, especially Peyer's patches, and travel to sites such as the liver, the spleen, and other tissues.[40–42] The proteins and peptides encapsulated in the nanoparticles are less sensitive to enzyme degradation through their association with polymers. The extended release of protein and peptide drug from the particles could have pharmacological and clinical significance.

Many researches have demonstrated that protein and peptide encapsulated in the nanoparticles have better absorption through the GI tract compared to their native counterpart.[43] The factors affecting uptake include the particle size of the particulate, the surface charge of the particles, the influence of surface ligands, and the dynamic nature of particle interactions in the gut. In one example, the interaction of nanoparticles consisting of hydrophobic polystyrene, bioadhesive chitosan, and (PLA-PEG) with two human intestinal cell lines was investigated and compared to the *in vivo* uptake in rats. After intraduodenal administration of chitosan nanoparticles in rats, particles were detected in both epithelial cells and Peyer's patches. Chitosan nanoparticle seemed to be taken up and transported by adsorptive transcytosis, while polystyrene nanoparticles uptake was probably mediated by nonadsorptive transcytosis.[44] In another example, insulin was encapsulated in nanosphere formulations using phase inversion nanoencapsulation. The encapsulated insulin was released from the nanospheres over a period of approximately 6 hours, was shown to be active orally, and had 11.4% of the efficacy of intraperitoneally delivered insulin.[45]

One problem in using nanoparticles for peptide and protein delivery is the erratic nature of nanoparticle absorption. For example, the proportion of intact particles reaching the systemic circulation was estimated to be generally below 5%.[46] Considering the generally low encapsulation efficiency of the protein in the particulates, the overall oral bioavailability of proteins and peptides is not significant.

10.2.2.2. Emulsions Liquid emulsions have been used to deliver proteins and peptide orally. Emulsions are thought to protect the drug from chemical and enzymatic breakdown in the intestinal lumen. Drug absorption enhancement is dependent on the type of emulsifying agent, particle size of the dispersed phase, pH, solubility of drug, type of lipid phase used, etc. Water-in-oil microemulsions have been shown to enhance oral bioavailability of proteins and peptides.[47,48] The lipid phase of microemulsions composed of medium chain fatty acid triglycerides increased the bioavailability of muramyl dipeptide analog.[49]

Many successful formulations are based on emulsion techniques. In these formulations, oil and water are mixed in such a way that small, uniformly shaped oil droplets are dispersed in the water phase (oil in water) or water droplets are dispersed in a continuous oil phase (water in oil). An emulsion appears as a cloudy suspension. When an oil-in-water emulsion has oil droplets so small as to produce a clear solution, then formulation is called a "microemulsion." The oil phase of the emulsion can provide lipophilic proteins with some protection from enzymatic digestion while the product is in the intestinal tract. Water-in-oil microemulsion formulations have been developed for oral insulin delivery.[50]

10.2.2.3. Micelle Formulations A micelle system can be either water-based or oil-based. The use of a micelle formulation for poorly water-soluble drugs for systemic delivery has been well recognized. In recent years, the effective development of self-emulsifying microemulsions or mixed micelle-based lipid formulations products, such as Sandimmun Neoral (cyclosporin), Norvir (ritonavir), and Fortovase (saquinavir), has substantially increased interest in the application of lipid-based micelle formulation to improve oral delivery of poorly water-soluble drugs as well as protein and peptide drugs.[51]

P. Modi patented a mixed micellar for administering insulin to the buccal mucosa using a metered dose inhaler.[52] The formulation includes a micellar proteinic pharmaceutical agent, lauryl sulphate, salicylate, edetate, and at least one absorption-enhancing compound such as lecithin, hyaluronic acid, glycolic acid, oleic acid, linolenic acid, or glycerin.

10.2.2.4. Liposomes Liposomes have also been studied as a way to deliver peptides and proteins orally. Chapter 19 discusses in detail the use of liposomes as drug delivery vehicles. Therefore, only issues relevant to peptide and protein drug delivery are discussed here. Liposomes are prone to the combined degrading effects of the acidic pH of the stomach, bile salts, and pancreatic lipase upon oral administration. Compared to the parenteral route, there have been fewer attempts to develop oral formulations to deliver the protein and peptides using a liposome system. In cases where the encapsulated agents within liposomes have increased bioavailability, it is not clear whether the liposome was absorbed intact or if the lipid caused permeation of the released agent at the site of absorption.[53] There are several reports on the intact liposomes uptake by cells in *in vitro* and *in situ* experiments;[54–56] the results are, however, not convincing for the oral delivery of proteins with a liposome system. Attempts have been made to improve the stability of liposomes either by incorporating polymers at the liposome surface[57,58] or by using GI-resistant lipids.[59]

10.2.3. Chemical Modifications

Chemically modified proteins and peptides offer some significant advantages over native proteins. Proteins are inherently unstable to digestive enzymes; however, chemical modifications can be made that inhibit enzyme attack. Such modifications include prodrug approaches[60] and permanent modifications.[61] Prodrugs, including peptide prodrug derivatization, are covered in Chapter 8 in detail and therefore will not be discussed in this chapter. Permanent modifications, if conducted in the appropriate functional group, may increase the stability and oral bioavailability of a peptide or protein without compromising its biological activity. Various permanent modification approaches have been studied, especially with proteins. These include glycosylation (reference), pegylation (reference), cross-linking (reference), and other polymer conjugation. So far, two of the most successful examples in protein drug delivery using chemical modification are calcitonin and insulin.

10.3. IMPROVING THE ORAL ACTIVITY OF CALCITONIN AND INSULIN THROUGH CHEMICAL MODIFICATION

Calcitonin is a 32-amino acid peptide hormone that participates in calcium and phosphorus metabolism. In mammals, the major source of calcitonin is from the parafollicular or C cells in the thyroid gland. Calcitonin contains a single disulfide bond, which causes the amino terminus to assume the shape of a ring. It is used to treat Paget's disease of bone. It also may be used to prevent continuing bone loss in women with postmenopausal osteoporosis and to treat hypercalcemia (too much calcium in the blood).

Chemical modification such as adding polyethylene glycol polymers can improve the *in vivo* characteristics of peptides. An example is CT-025, which is a diconjugate of recombinant salmon calcitonin (sCT).[62] The chemical modification of sCT results in an orally available diconjugate that retains hormonal activity while having increased amphiphilicity. Modification can occur at sites where an amine is available. On sCT, there are sites at Cys-1, Lys-11, and Lys 18, as shown in Figure 10.1.

Insulin is the major hormonal regulator of glucose metabolism. It is a 51-amino acid polypeptide that is produced and secreted by the β-cells of pancreatic islets.[64,65] All known vertebrate insulins are composed of two polypeptide chains that are linked to one another by two disulfide bonds, one between CysA7 and CysB7 and the other between CysA20 and CysB19. A third, intrachain disulfide bond connects CysA6 and A11. The amino acid sequence among vertebrate species is also highly conserved. One-third of the amino acid residues in the two insulin chains are strictly invariant, and more than one-third of the remaining amino acids are either invariant or highly conserved, suggesting unique structural requirements

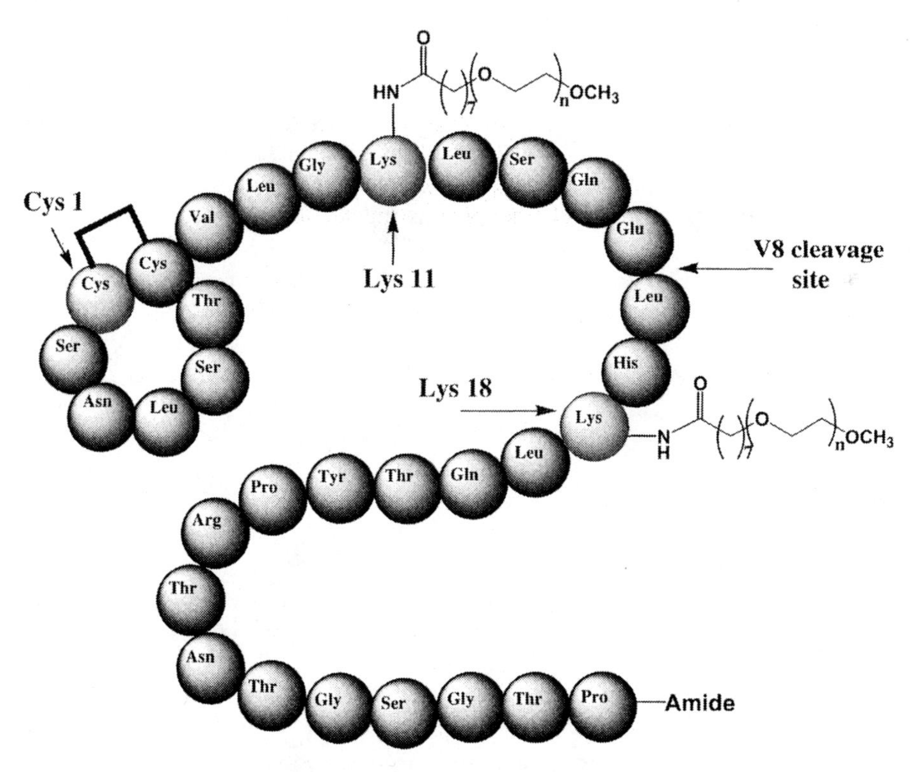

Figure 10.1. Chemical modification sites on sCT.[63]

for receptor recognition and biological activity. Moreover, there is evidence that receptor binding is associated with a structural change that must be accommodated by insulin analogs if biological activity is to be preserved.[66]

For over 70 years, approved insulin products have been given to diabetic patients exclusively by parenteral administration. Due to the challenges and obstacles associated with oral administration of peptide drugs, no oral dosage form for insulin has been marketed to date.[67] In general, studies of insulin's oral bioavailability have shown uptakes that range from near zero to a small percentage of the administered dose.[68–75]

As part of continuing efforts to design and develop oral delivery dosage forms for peptide drugs, researchers have extensively investigated the effects of structural manipulations and formulation changes on the oral availability of insulin, These efforts have resulted in the preparation of libraries of amphiphilic oligomers that, when conjugated in certain patterns and number to the drugs, may be used to enhance their pharmacological performance. Moreover, when the drug-oligomer conjugates are used in combination with common formulation strategies, oral bioavailability can be dramatically increased, especially relative to that of the unconjugated, native drug.

In previous reports,[76–78] researchers showed that conjugation of properly designed amphiphilic oligomers to insulin provides insulin analogs that maintain solubility in aqueous and lipophillic matrices, are resistant to enzymatic degradation, have improved physicochemical properties for absorption across intestinal mucosa, and are orally active. These results also suggested that a particular insulin modification, oligomer-conjugation at the epsilon-amino group of lysine-29 on the B chain of insulin, would be a suitable oral insulin candidate.[78] It has long been recognized that chemical modifications of proteins can improve bioavailability, delivery to systemic targets, and other pharmaceutical and pharmacological performance characteristics.[79]

Of the functional groups available for modification, the primary amino groups present at the N termini of the protein chain(s) and the ε-amino groups of lysine residues are frequently selected, since most polypeptides and proteins have at least one of these sites and a wide variety of chemical agents are available that exhibit specificity for these functional groups.

Insulin, for example, has three primary amino groups that are available for modification: the glycine and phenylalanine residues at the N termini of the A and B chains (GlyA1 and PheB1, respectively) and the ε-amino group of lysine-29 on the B chain (LysB29). Several groups have reported that modification of GlyA1 causes diminished receptor binding and a loss of biological activity.[7] Conversely, (MPEG) modification of PheB1 reduces the antigenicity of the protein, and PheB1 modification with glucosyl-PEG has been reported to suppress the association of insulin monomers/dimers into hexamers.[80,81]

Likewise, modification of either PheB1 or LysB29 has been shown to improve the pharmaceutical performance (e.g., increased plasma half-life, reduced immunogenicity and antigenicity, improved resistance to proteolysis, increased aqueous/organic solubilities) of the insulin conjugates thus produced while not significantly compromising receptor binding or biological activity.[82,83] A review by Brange and Volund provides detailed physicochemical and pharmacological characterization of numerous insulin analogs that have been developed over the past 70 years or are being investigated today.[84]

Fortunately, the pKa values of these three amino groups (<7.0, $\simeq 8.0$, and $\simeq 10.5$, respectively, for PheB1, GlyA1, and LysB29) are sufficiently different that some degree of selectivity can be achieved by controlling the reaction pH of the acylating agent. This observation was first reported by Lindsay and Shall, who showed that GlyA1 and PheB1 were preferentially monoacylated by *N*-hydroxysuccinimyl acetate at pH 6.9, whereas at pH 8.5, the yield of monoacylated PheB1 was decreased and monoacyl LysB29 insulin could be obtained.[85] More recently, this strategy has been enhanced and used to prepare insulin analogs in which LysB29 is derivatized with fatty acyl groups.[86]

To date, true site specificity of insulin modification has been achieved only through a multistep synthetic approach. First, two of the three reactive amino groups are masked using selectively labile protecting groups (e.g., Boc or dimethylmaleyl groups), and then the remaining unprotected amino group is modified with activated esters. Following acylation, the protecting groups are removed by treatment with acid; the more stable acyl amide is not affected. For example, Shah and Shen blocked the two α-amino groups of insulin by reaction with dimethylmaleic anhydride at a pH of about 7 before acylating the ε-amino group of LysB29 at pH 10; subsequently, the maleyl protecting groups were selectively removed by dialysis at pH 4.[87] The site-specifically modified-insulin conjugates thus obtained were further purified using chromatography.

The real test of a modified therapeutic protein is evaluations in human clinical trials. Several Phase I and Phase II trials have been conducted with modified insulin to show the safety and effectiveness of its glucose-lowering effects. Several Phase I studies were conducted to determine plasma glucose and insulin levels after administration of oral doses of a chemically modified oral insulin product in fasting, insulin-deprived adult patients with type 1 diabetes. The product was effective in preventing the expected rise in plasma glucose concentrations in these patients.[88,89] Another exploratory study of this product assessed the postprandial glucose-lowering effects in patients with type 2 diabetes. Single oral doses of the chemically modified insulin were as effective as subcutaneous regular insulin in controlling postprandial glycemia with respect to a number of parameters.[90]

10.4. CONCLUSIONS

Delivering protein and peptides by the oral route is extremely challenging. The very nature of the digestive system is designed to break down these polypeptides into amino acids prior to absorption. The techniques and formulations that have been developed to date must protect against the acidity of the stomach and the natural enzymatic processes of digestion. Once that protection is provided, the formulation or chemical modification must enhance the absorption of the polypeptide across the epithelial layer. These challenges are likely to be overcome only by using multiple approaches simultaneously.

REFERENCES

1. Wearley, L. L. *Crit. Rev. Ther. Drug Carrier Syst.* **1991**, *8*(4), 331–394.
2. Sayani, A. P.; Chien, Y. W. *Crit. Rev. Ther. Drug Carrier Syst.* **1996**, *13*(1–2), 85–184.
3. Torres-Lugo, M.; Peppas, N. A. *Biomaterials* **2000**, *21*(12), 1191–1196.
4. O'Hagan, D. T.; Illum, L. *Crit. Rev. Ther. Drug Carrier Syst.* **1990**, *7*(1), 35–97.
5. Banga, A. K.; Chien, Y. W., *Pharm. Res.* **1993**, *10*(5), 697–702.
6. Lee, Y. C.; Yalkowsky, S. H. *Int. J. Pharm.* **1999**, *185*(2), 199–204.
7. Burgess, D. J. In *Biotechnology and Pharmacy*; Pezzuto, J. M., Johnson, M. E., Manasse, H. R., Eds. Chapman and Hall: New York, **1993**, pp. 116–151.
8. Hoffman, A.; Ziv, E. *Drug Disiposition* **1996**, *33*, 285–301.
9. Gwinup, G.; Elias, A. N.; Domurat, E. S. *Gen. Pharm.* **1991**, *22*, 243–246.
10. Lee, Y. H.; Sinko, P. J. *Adv. Drug Deliv. Rev.* **2000**, *42*(3), 225–238.
11. Yang, C. Y.; Dantzig, A. H.; Pidgeon, C. *Pharm. Res.* **1999**, 16, 1331–1343.
12. Cleland, J. L.; Daugherty, A.; Mrsny, R. *Curr. Opin. Biotechnol.* **2001**, *12*(2), 212–219.
13. Lee, V. H. *Eur. J. Pharm Sci.* **2000**, *11* (Suppl 2), S41–S50.
14. Sayani, A. P.; Chien, Y. W. *Crit. Rev. Ther. Drug Carrier Syst.* **1996**, *13*(1–2), 85–184.
15. Creighton, T. E. *Proteins: Structures and Molecular Properties,* 2nd ed., W.H. Freeman and Company: New York, **1997**, pp. 429–431.
16. Yamamoto, A.; Taniguchi, T.; Fujita, T.; Murakami, M.; Muranishi, S. *Proc. Int. Symp. Controlled Release Bioact. Mater.* **1994**, *21*, 324–325.
17. Ziv, E.; Bendayan, M. *Microsc. Res. Tech.* **2000**, *49*(4), 346–352.
18. Cholewinski, M.; Lueckel, B.; Horn, H. *Pharm. Acta Helv.* **1996**, *71*(6), 405–419.
19. Baudys, M.; Mix, D.; Kim, S. W. *J. Controlled Rel.* **1996**, *39*, 145–151.
20. Catnach, S. M.; Fairclough, P. D.; Hammond, S. M. *Gut* **1994**, *35*, 441–444.
21. Davis, S. S. In *Advanced Delivery Systems for Peptides and Proteins—Pharmaceutical Considerations: Delivery Systems for Peptide Drugs*; Davis, S. S., Illum, L., Tomlinson, E., Eds. Plenum Press: New York, **1986**, pp. 1–21.
22. Humphrey, M. J.; Ringrose, P. S. *Drug Metab. Rev.* **1986**, *17*, 283–310.
23. Oliyai, R., Stella, V. J. *Annu. Rev. Pharmacol. Toxicol.* **1993**, *32*, 521–544.
24. Oliyai, R. *Adv. Drug Del. Rev.* **1996**, *19*, 275–286.
25. Kramer, W.; Wess, G.; Neckermann, G.; Schubert, G.; Fink, J.; Girbig, F.; Gutjahr, U.; Knowalewski, S.; Baringhaus, K.-H.; Boger, G.; Enhsen, A.; Falk, E.; Friedrich, M.; Glombik, H.; Hoffmann, A.; Pittius, C.; Urmann, M. *J. Biol. Chem.* **1994**, *269*(14), 10621–10627.
26. Lee, V. H. L.; Yamamoto, A. *Adv. Drug Del. Rev.* **1990**, 4, 171–207.
27. Laskowski, M. J.; Haessler, H. A.; et al. *Science* **1958**, *127*, 1115–1116.
28. Fujii, S., et al. *J. Pharm. Pharmacol.* **1985**, *37*, 545–549.
29. Tozaki, H., et al., *J. Pharm. Pharmacol.* **1997**, *49*, 164–168.

30. Yamamoto, A., et al. *Pharm. Res.* **1994**, *11*, 164–168.
31. Bai, J. P.; Chang, L. L. *J. Pharm. Pharmacol.* **1996**, *48*, 1078–1082.
32. Bai, J. P.; Chang, L. L. *Pharm. Res.* **1995**, *12*, 1171–1175.
33. Ziv, E., et al. *Biochem. Pharmacology* **1987**, *36*, 1035–1039.
34. Lee, Y. H.; Perry, B. A.; et al. *Pharm. Res.* **1999**, *16*(8), 1233–1239.
35. Citi, S. *J. Cell Biol.* **1992**, *117*, 169–178.
36. Fasano, A.; Uzzau, S. *J. Clin. Invest.* **1997**, *99*, 1158–1164.
37. Thanou, M.; Verheof, J. C.; Junginer, H. E. *Adv. Drug. Del. Rev.* **2001**, *50*, S91–S101.
38. Touitou, E.; Donbrow, M.; Rubinstein, A. *J. Pharm. Pharmacol.* **1980**, *32*, 108–110.
39. Yuichiro, N., et al. *J. Pharmacobio-Dyn.* **1989**, *12*, 736–743.
40. Florence, A. T. *Pharm. Res.* **1997**, *14*, 259–266.
41. Ermak, T. H. *Cell Tissue Res.* **1995**, *279*, 433–436.
42. Sakuma, S.; Hayashi, M.: Akashi, M. *Adv. Drug Deliv. Rev.* **2001**, *47*, 21–37.
43. Hussain, N.; Jaitley, V.: Florence, A. T. *Adv. Drug Deliv. Rev.* **2001**, *50*, 107–142.
44. Behrens, I.; Pena, A. I.; Alonso, M. J.; Kissel, T. *Pharm. Res.* **2002**, *19*, 1185–1193.
45. Carino, G. P.; Jacob, J. S.; Mathiowitz, E. *J. Controlled Release* **2000**, *65*, 261–269.
46. Allemann, E.; Leroux, J.; Gurny, R. *Adv. Drug Deliv. Rev.* **1998**, *34*, 171–189.
47. Sarciaux, J. M., et al. *Int. J. Pharm.* **1995**, *120*, 127.
48. Constantinides, P. P., et al. *Pharm. Res.* **1996**, *13*, 210.
49. Lyons, K. C., et al. *Int. J. Pharm.* **2000**, *199*, 17.
50. Bruinside, B. A.; Mattes, C. E.; McGuinness, C. M.; Rudnic, E. M.; Belendiuk, G. W. U.S. Patent **5,824,638**, issued October 20, 1998.
51. Christopher, J. H., et al. *Adv. Drug Del. Rev.*, **2001**, 50, S127–S147.
52. Modi, P. U.S. Patent **6,432,383**, issued October 13, 2002.
53. Arien, A., et al. *Life Sci.* **1993**, *53*, 1279–1290.
54. Patel, H. M., et al., *Biochim. Biophys. Acta* **1985**, *839*, 40–49.
55. Childers, N. K., et al. *Reg. Immunol.* **1990**, *3*, 8–16.
56. Tomizawai, H., et al. *Pharm. Res.*, **1993**, *10*, 549–552.
57. Iwanaga, K., et al. *J. Pharm. Sci.* **1999**, *88*, 248–252.
58. Chen, H., et al. *Pharm. Res.* **1996**, *13*, 1378–1383.
59. Sprott, G. D., et al. *FEMS Microbiol. Lett.* **1997**, *154*, 17–22.
60. Gangwar, S.; Pauletti, G. M.; Wang, B.; Siahaan, T.; Stella, V. J.; Borchardt, R. T. *Drug Discovery Today* **1997**, *2*, 148–155.
61. Wang, W.; Jiang, J.; Ballard, C. E.; Wang, B. *Curr. Pharm. Design* **1999**, *5*, 265–287.
62. Donaldson, S.; Radhakrishnan, B.; Soltero, R. "*Structure Determination of a Modified Amphiphilic Recomnbinant Salmon Calcitonin Conjugate and Related Conjuates*," presented at the AAPS National Meeting, Toronto, **November 2002**.
63. Reprinted with permission from Nobex Corporation.
64. Shoelson, S. E.; Halban, P. A. Insulin biosynthesis and chemistry. In *Joslin's Diabetes Mellitus*. 13th ed.; Kahn, C. R., Weir, G. C., Eds. Lippincott Williams & Wilkins: Philadelphia, **1994**, pp. 29–55.
65. Cutfield, J.; Cutfield, S.; Dodson, E.; et al. *Z. Physiol. Chem.* **1981**, *362*, 755–761.

66. Derewenda, U.; Derewenda, Z.; Dodson, E. J.; Dodson, G. O.; Bing, X.; Markussen, *J. Mol. Biol.* **1991**, *220*, 425–433.
67. Swenson, E. S.; Curatolo, W. J. *Adv. Drug Deliv. Rev.* **1992**, *8*, 39–92.
68. Danforth, E.; Moore, R. O. *Endocrinology* **1959**, *119*, 123.
69. Spangler, R. S. *Diabetes Care* **1991**, *13*(9), 911–922.
70. Touitou, E.; Donbrow, M.; Rubenstein, A. *J. Pharm. Pharmacol.* **1980**, *32*(2), 108–110.
71. Choudhari, K. B.; Labhesetwar, V.; Dorle, A. K. *J. Microencapsulation* **1994**, *11*(3), 319–325.
72. Ziv, E.; Kidron, M.; Raz, I; Krausz, M.; Blatt, Y.; Rotman, A.; Bar-On, H. *J. Pharm. Sci.* **1994**, *83*(6), 792–794.
73. Hosny, E. A.; Ghilzai, N. M.; Elmazar, M. M. *Pharm. Acta Helv.* **1997**, *2*(4), 203–207.
74. Damge, C.; Vranckx, H.; Balschmidt, P.; Couvreur, P. *J. Pharm. Sci.* **1997**, *86*(12), 1403–1409.
75. Musabayane, C. T.; Munjeri, O.; Bwititi, P.; Osim, E. E. *J. Endocrinol.* **2000**, *64*(1), 1–6.
76. Marschutz, M. K.; Bernkop-Schnurch, A. *Biomaterials* **2000**, *21*(14), 1499–1507.
77. Agarwal, V.; Reddy, I. K.; Khan, M. A. *Int. J. Pharm.* **2001**, *225*(1–2), 31–39.
78. Krishnan, B. R.; Rajagopalan, J.; Anderson, W. L.; Simpson, M.; Ackler, S.; Davis, C. M.; Ansari, A. M.; Harris, T. M.; Ekwuribe, N. *Proc. Intl. Symp. Control. Rel. Bioact. Mater.* **1998**, *25*, 124–125.
79. Smith, R. A. G.; Dewdney, J. M.; Fears, R.; Poste, G. *TIBTECH* **1993**, *11*, 397–403.
80. Uchio, T.; Baudyš, M.; Liu, F.; Kim, S. W. *Adv. Drug Deliv. Rev.* **1999**, *35*, 289–306.
81. Baudyš, M.; Uchio, T.; Mix, D.; Wilson, D.; Kim, S. W. *J. Pharm. Sci.* **1995**, *84*(1), 28–33.
82. Gliemann, J.; Gammeltoft, S. *Diabetologia* **1974**, *10*, 105–113.
83. Hashimoto, M.; Takada, K.; Kiso, Y.; Muranishi, S. *Pharm. Res.* **1989**, *6*, 171–176.
84. Brange, J.; Volund, A. *Adv. Drug Deliv. Rev.* **1999**, *35*, 307–335.
85. Lindsay, D. G.; Shall, S. *Biochem. J.* **1971**, *121*, 737–745.
86. Friesen, H. J. In *Chemistry, Structure and Function of Insulin and Related Hormones, Proceedings of the Second International Insulin Symposium, Aachen, Germany, Sept. 4–7, 1979*; Brandenburg, D., Wollmer, A., Eds. De Gruyter: Berlin, **1980**.
87. Shah, D.; Shen, W.-C. *J. Pharm. Sci.* **1996**, *85*(12), 1306–1311.
88. Clement, S.; Still, J. G.; et al. *Diabetes Tech. Ther.* **2002**, *4*(4), 459–466.
89. Clement, S.; Dandona, P.; et al. *Metabolism* **2004**, *53*(1), 54–58.
90. Kipnes, M.; Dandona, P.; et al. *Diabetes Care,* **2003**, *26*(2), 421–426.

11

METABOLIC ACTIVATION AND DRUG TARGETING

XIANGMING GUAN

Department of Pharmaceutical Sciences, College of Pharmacy, South Dakota State University, Brookings, SD 57007

Drug Delivery: Principles and Applications Edited by Binghe Wang, Teruna Siahaan, and Richard Soltero
ISBN 0-471-47489-4

11.1. INTRODUCTION

Drug targeting through metabolic activation provides a way to deliver a drug to the desired site of drug action. It has the advantage of increasing the drug's site selectivity and reducing systemic adverse effects. Since metabolic activation is an enzyme-mediated process, drug targeting through metabolic activation requires that the activating enzyme be unique to or at least highly enriched in the target site.[1] Successful examples have been reported in targeting a drug to the colon, kidney,[1] and liver.[2] In particular, drug targeting through metabolic activation has unique value in cancer chemotherapy, which will be the focus of this chapter.

Chemotherapy is one of the major treatments in cancer therapy. However, it is often associated with severe side effects due to the fact that anticancer drugs are primarily cytotoxic agents that not only kill cancer cells but also cause damage to normal cells, especially proliferating cells such as bone marrow and gut epithelia cells. As a result, the success of chemotherapy is often hampered by the severe systemic side effects of chemotherapeutic drugs. Consequently, increasing the selectivity of the chemotherapeutic agents has been an intense research effort in improving chemotherapeutic efficacy. There are three general approaches in increasing the selectivity: (1). identify agents that will be more selective in killing cancer cells than normal cells; (2). deliver the chemotherapeutic agent more selectively (ideally, specifically) to cancer cells; and (3). mask the chemotherapeutic agent in such a way that it will be released selectively (ideally, specifically) in cancer cells. Discussion of the first two approaches is beyond the scope of this chapter. In the third approach, the masked agent is called a "prodrug."

By definition, a prodrug is a compound that exhibits no desired biological activity and will be turned into the desired drug through (most often) an enzyme-mediated process. The prodrug approach has been applied in improving the physicochemical or pharmacokinetic properties of a drug, including solubility, bioavailability, and half-life, which will not be discussed here. This chapter will focus on prodrugs that are aimed at releasing cytotoxic anticancer drugs selectively at tumor sites. Different anticancer prodrug approaches and the biochemical processes based on which the prodrugs are designed will be presented.

11.2. TUMOR-TARGETED ANTICANCER PRODRUGS AND THEIR BIOCHEMICAL BASIS

One of the challenges in anticancer prodrug design is the identification of cancer-associated biochemical processes that can be utilized to release anticancer drugs from prodrugs. The obvious advantage of this approach is its high selectivity. Ideally, the cancer-associated biochemical processes do not or barely occur in normal cells. Two cancer-related features that have been extensively used for anticancer prodrug design are hypoxia and metastasis. Anticancer prodrugs based on the biochemical processes associated with hypoxia and metastasis have demonstrated high selectivity in killing cancer cells.[3–5] In addition, enzymes with elevated activities in cancer cells, such as β-glucuronidase, prostate specific antigen, and cytochrome P450, have also been exploited for anticancer prodrug design. Further, the enzyme selected to activate an anticancer prodrug can be enhanced or delivered to tumor sites through gene expression [gene-directed enzyme prodrug therapy (GDEPT)][6–8] or antibody-aided delivery [antibody-directed enzyme prodrug therapy (ADEPT)].[9–11] In both GDEPT and ADEPT, the therapy involves two steps. First, an enzyme is delivered to the tumor site, followed by the second step—prodrug administration. Therefore, GDEPT and ADEPT are also referred to as "two-step therapy." Therapy involving a prodrug alone is called "monotherapy."

11.2.1. Tumor-Activated Anticancer Prodrugs Based on Hypoxia

Hypoxia appears to be a common and unique property of cells in solid tumors and is a target for tumor-specific activation of anticancer prodrugs.[4] It is now well known that solid tumors often contain an inefficient microvascular system, and part of solid tumors exists under a hypoxic condition.[12] Hypoxia can be classified into two broad types: chronic and acute. Chronic hypoxia occurs in cells that are distant from their blood supply and suffer low oxygen tension permanently. Acute hypoxia results in cells experiencing temporary cessation of blood flow.

Hypoxia is a major problem in radiotherapy and chemotherapy.[13,14] The cytotoxic effects of radiation require the presence of oxygen. In chemotherapy, hypoxia reduces the distribution of chemotherapeutic agents. Further, hypoxic cells receive not only a reduced amount of oxygen but also a low supply of nutrients, which causes them to stop or reduce their rate of progress through the cell cycle. Since most anticancer agents are more effective against rapidly proliferating cells, the anticancer agent will be far less effective in eradicating hypoxic tissues.

Hypoxia of solid tumors has been exploited in designing hypoxia-selective anticancer drugs or prodrugs. In particular, the elevated activities of various reductive enzymes under a hypoxic condition have been extensively targeted for tumor-selective prodrugs. Another feature associated with hypoxia is low extracellular pH.[15] This is likely to be a result of an insufficient blood supply that leads to the accumulation of acidic metabolites. Anticancer prodrugs activated in a low-pH environment have also been reported.[16]

It is worth noting that because only a small proportion of the cells in a solid tumor are likely to be chronically hypoxic, hypoxia-selective prodrugs are not expected to be curative on their own. Other agents need to be used to eradicate aerobic cancer cells. However, because of acute hypoxia, a significant fraction of solid tumor cells may be killed, theoretically, by hypoxia-selective prodrugs when they are experiencing transient hypoxia. Another mechanism by which hypoxia-selective anticancer prodrugs kill aerobic cells is the diffusion effect. Active drug can diffuse and cause the killing of surrounding cells (bystander effect). However, this effect is relatively limited since most active drugs cannot travel far due to their high chemical reactivity.

Although anticancer prodrugs designed by targeting hypoxia-related enzymes have demonstrated high tumor selectivity, toxicity associated with this approach has been observed, specifically retinal toxicity. The inner retina is vascularized, but the outer retina is avascular and relies on diffusion of oxygen from the inner retinal vessels and from the choriocapillaris. The retinal toxicity of the 2-nitroimidazole alkylating agents CI-1010 and triapazamine has been linked to the physiological hypoxia in the retina.[17] Nevertheless, this toxicity does not appear to be related to some other hypoxia-selective anticancer prodrugs such as the quinone porfiromycin, the anthraquinone *N*-oxide AQ4N, and the nitrogen mustard prodrugs SN 23816 and SN 25341.[17]

11.2.1.1. Tumor-Activated Anticancer Prodrug Based On Reductive Enzymes and Low Oxygen Tension Most hypoxia-selective anticancer prodrugs are activated by reducing enzymes that are also present in aerobic cells. The enzymes that are involved in the reductive activation of anticancer prodrugs include NADPH-cytochrome P450 reductase (EC1.6.2.3), NADH-cytochrome b_5 reductase (EC 1.6.2.2), NADH:ubiquinone oxidoreductase (EcbEC 1.6.5.3), ferredoxin-NADP$^+$ reductase (EC 1.6.7.1), and NAD(P)H:quinone acceptor oxidoreductase (EC1.699.2), also called "DT-diaphorase."[18] These are flavoenzymes, and all catalyze a single-electron reduction except DT-diaphorase, which carries out two-electron reductions. The hypoxia selectivity of targeting the single-electron reductases arises from the fact that upon a single-electron reduction, a prodrug is converted to a single-electron reduction adduct which, in the absence of molecular oxygen, is further reduced to the active drug. However, in the presence of molecular oxygen, the one-electron adduct is oxidized back to the prodrug (Scheme 1). Therefore, formation of the active drug is restricted to hypoxic tissue. In the case of DT-diaphorase, a two-electron reductase, the selectivity is derived from the fact that the enzyme is found to exist at high levels throughout many human solid tumors such as thyroid,

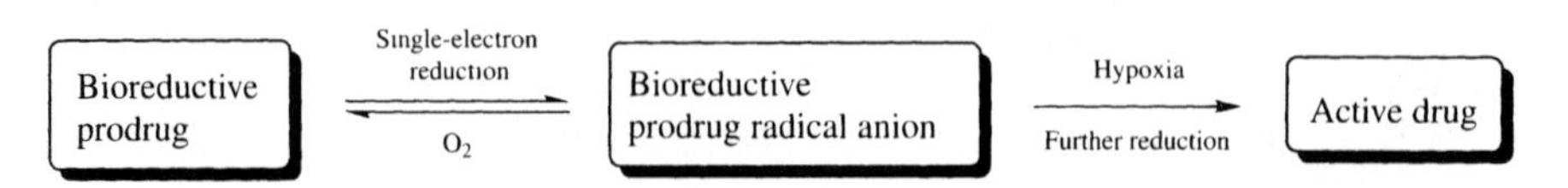

Scheme 1. Bioactivation of prodrugs by a single-electron reductase.

adrenal, breast, ovarian, colon, and non-small-cell lung cancer.[19] Further, bioreductive activation of a prodrug is not necessarily restricted to one enzyme. It appears that several different enzymes can participate to different extents with the various bioreductive agents.[20]

The reducible organic functional groups employed for these reductases include quinones, *N*-oxides, aromatic or heteroaromatic nitro groups, and sulfoxides. These compounds are also termed "bioreductive agents."

11.2.1.1.1. Hypoxia-Selective Quinone-Containing Prodrugs Quinones can be reduced by a variety of reductases including one-electron donating enzymes such as NADPH:cytochrome P450 reductase and xanthine oxidase[21–24] and by the two-electron donating enzyme DT-diaphorase. Hypoxia-selective quinone-containing prodrugs include some naturally occurring anticancer prodrugs, such as mitomycin, and compounds designed to be hypoxia-selective.

Mitomycin C (**1**, Scheme 2) is an anticancer agent that has been shown to be more cytotoxic to hypoxic tumor cells than to their aerobic counterparts.

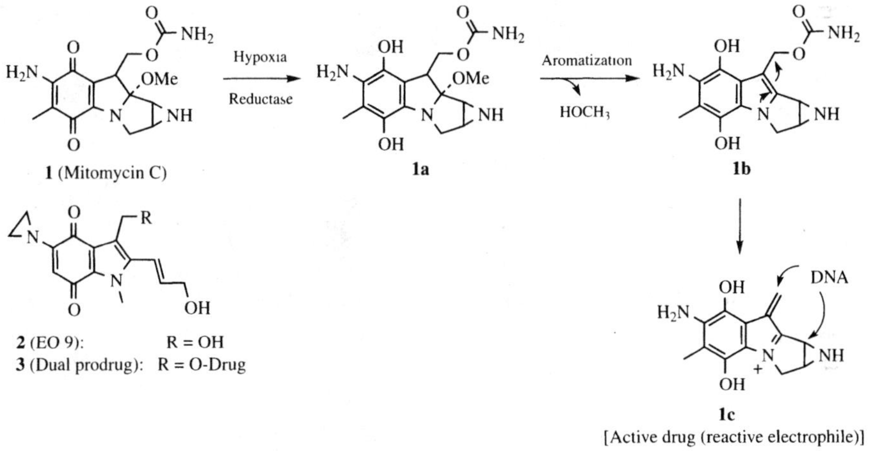

Scheme 2. Bioreductive activation of mitomycin C and its derivatives.

Mitomycin C is considered the prototypical bioreductive alkylating agent.[25] Although the agent was not designed as a prodrug, its mode of action is thought to involve a bioreductive activation to a species (**1c**, Scheme 2) that alkylates DNA.[26] Similar activation occurs with structurally related compounds such as indolquinone EO9 (**2**, Scheme 2). Notice that the aziridine ring in the structure provides an additional site that can be attacked by DNA. This group may account for the aerobic toxicity of the compound.

An elegant design of a dual-action prodrug by coupling EO9 with another anticancer drug (**3**, Scheme 2) has been proposed in which the dual-action prodrug, upon reduction, releases two active anticancer drugs.[27]

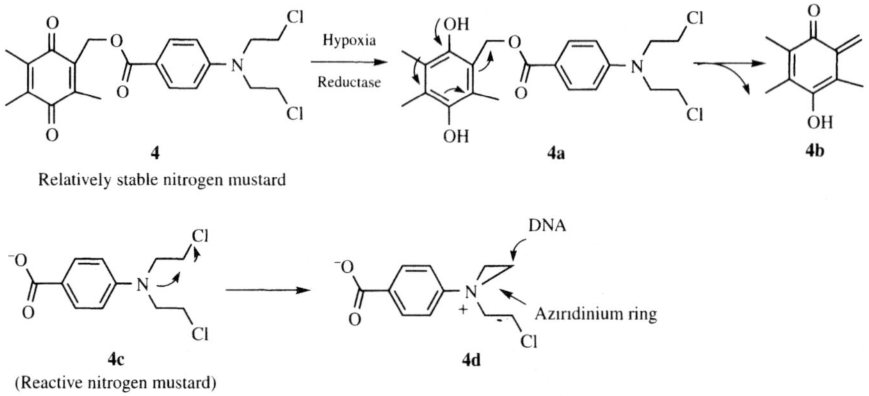

4
Relatively stable nitrogen mustard

Hypoxia
Reductase

4a

4b

4c
(Reactive nitrogen mustard)

4d

Azıridinium ring

Scheme 3. Bioreductive activation of prodrug **4**.

In a similar fashion, prodrug **4** (Scheme 3) is bioreductively activated in hypoxic cells to produce species **4b** and a nitrogen mustard, **4c**, both of which are capable of reacting with nucleophiles, e.g. DNA, and proteins.[4] Notice that the nitrogen mustard in prodrug **4** is much less reactive than the one in **4c** due to the reduced electron density of the aromatic ring by the electron-withdrawing effect of the ester functionality. Under physiological pH, the anionic nature of the carboxylic group increases the electron density that favors the formation of an aziridinium ring (**4d**, Scheme 3), a reactive electrophilic group that alkylates DNA.

11.2.1.1.2. Hypoxia-Selective Aromatic or Heteroaromatic Nitro-Containing Prodrugs An aromatic nitro (Ar-NO_2) group can be reduced to an aromatic amine through the intermediates of a nitroso group (Ar-NO) and a hydroxylamine (Ar-NH-OH) group. The reduction turns a strong electron-withdrawing group (EWG) (-NO_2) into an electron-donating group (EDG) (-NH_2). The change in electron density, as a result of the reduction, has been utilized in turning a chemically stable compound into a reactive electrophile. The systemic toxicity of nitrogen mustard anticancer agents is derived from its indiscriminate alkylation to cancer cells and normal cells. The mode of nitrogen mustard anticancer agents is illustrated in Scheme 3. The availability of the nitrogen lone pair electrons is crucial in determining the reactivity of the nitrogen mustard. During the formation of a reactive aziridinium ring (Scheme 3), the lone pair electrons aid the elimination of the chloride through a neighboring group participation effect (**4c** to **4d**, Scheme 3). The resulting aziridinium ring alkylates DNA, leading to cell killing. One approach in reducing the systemic toxicity of nitrogen mustard is the use of a prodrug in which the electron density (or the basicity) of the nitrogen is reduced but will be regained in tumor cells. This concept is illustrated in Scheme 3, in which an ester functional group is used to reduce the basicity of the nitrogen. In the following example, a strong electron-withdrawing nitro group is employed for the same purpose. Scheme 4 is a general scheme demonstrating the activation of aromatic nitro-containing

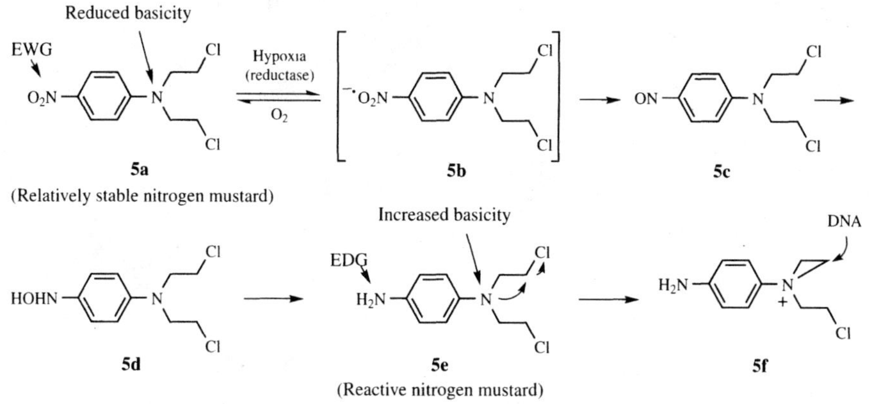

Scheme 4. Bioactivation of aromatic nitro-containing hypoxia-selective nitrogen mustard prodrugs.

nitrogen mustard prodrugs. Notice that the one-electron reduction of the nitro group leads to the nitro radical anion, which, in the presence of oxygen, is reversed back to the nitro group. Under a hypoxic condition, the reduction proceeds further to a nitroso compound, hydroxylamine, and amine. Figure 11.1 shows representative hypoxia-selective aromatic nitro-containing nitrogen mustard prodrugs.[17] One of the most effective bioreductive drugs against hypoxic cells in murine tumors is the 2-nitroimidazole alkylating agent CI-1010 (PD 144872), which is the *R*-enantiomer of the racemate RB 6145 [1-[(2-bromoethyl)amino]-3-(2-nitro-1*H*-imidazole-1-yl)-2-propanol hydrobromide] (**7**). However, further clinical evaluation of this compound was terminated due to its severe retinal toxicity, as mentioned earlier.[17]

Alternatively, an increase in electron density can be achieved through the conversion of an amide group (a much less EDG) to an amino group (a strong EDG), as illustrated by prodrug **9** (Scheme 5).[28] Upon reduction of the nitro group, the formed hydroxylamine (**9a**) rapidly cyclizes to release the alkylating agent **5e**.

A very elegant design using an aromatic nitro compound to release an active drug involves a reduction of the nitro group followed by a spontaneous fragmentation

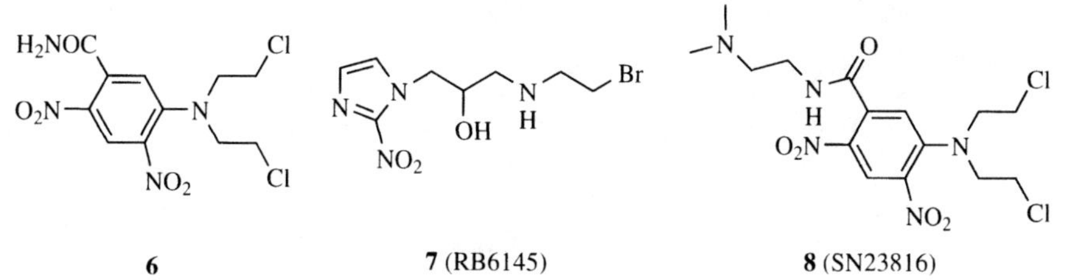

Figure 11.1. Representative aromatic nitro-containing hypoxia-selective nitrogen mustard prodrugs.

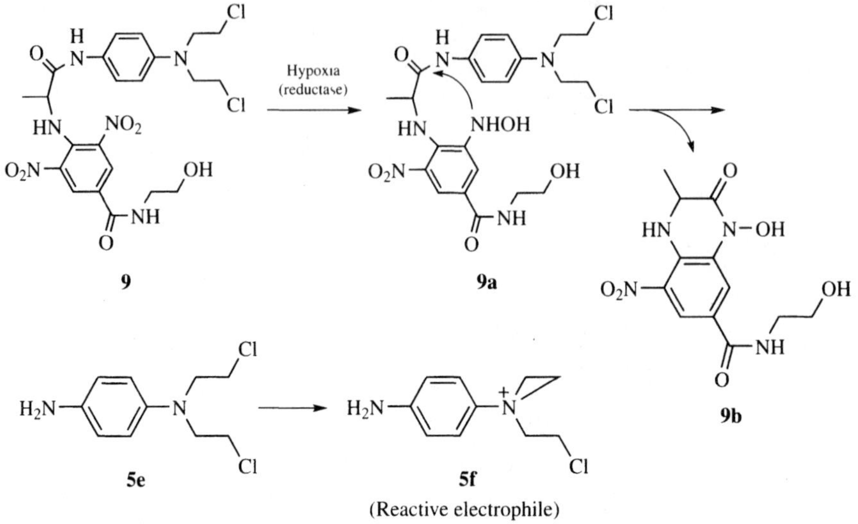

Scheme 5. Bioreductive activation of prodrug 9.

(1,6-elimination). Scheme 6 provides a general description of this approach and some of the representative prodrugs based on this design.[29–32] Most of the prodrugs produced by this design contain nitrobenzyl and carbamate functionalities. It has been reported that release of the active drug occurs after the formation of the hydroxylamine (**10a**, R′ = OH) rather than the amine (**10a**, R′ = H).[30]

In addition to nitrobenzyl carbamate prodrugs, some nitro heteroaromatic prodrugs have also been reported. Scheme 7 shows the general mode of active drug release.[33–36]

11.2.1.1.3. Hypoxia-Selective N-Oxide-Containing Prodrugs *N*-Oxide compounds can be reduced by a variety of reductases. Compound **20** is a prodrug of compound **20a** (Scheme 8),[37] which is a structural analog of the metal binding unit of bleomycin, an anticancer drug. The metal binding unit is key to the antitumor activity of bleomycin. It is believed that bleomycin forms a chelate with iron (Fe^{2+}). Five of the six coordination positions of Fe^{2+} are strongly coordinated to bleomycin. The sixth is available for coordination to oxygen. The chelate alters the redox potential of iron such that bound oxygen is reduced, converting the oxygen into a reactive radical species, the hydroxyl radical, that causes cell killing through DNA degradation.[26] By converting one of the nitrogens involved in metal chelating to *N*-oxide (prodrug **20**), the structure becomes incapable of metal chelating and therefore becomes nontoxic.

Tirapazamine (**21**, Scheme 9) is a benzotriazine di-*N*-oxide bioreductive anticancer prodrug. Tirapazamine is activated to a DNA-damaging oxidizing radical by cytochrome P450 reductase and other one-electron reductases in the absence of oxygen.[38,39] Tirapazamine has been demonstrated to be effective in killing

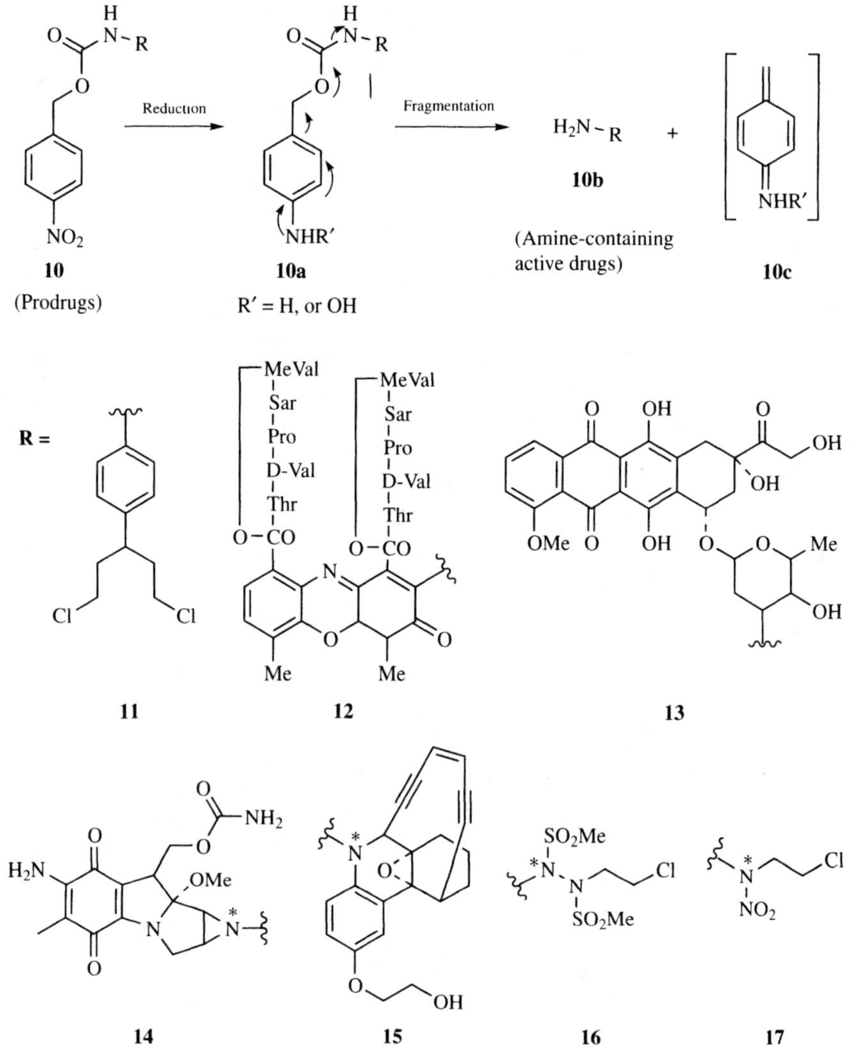

Scheme 6. Bioreductive activation through reduction of a nitro group followed by 1,6-elimination and representative related prodrugs. *The nitrogen is part of the carbamate structure.

hypoxic cells in murine tumors[38] and in sensitizing hypoxic cells to cisplatin.[40,41] Recent clinical studies have shown that tirapazamine enhances cisplatin activity against non-small-cell lung cancer.[42] However, like Cl-1010 (R enantiomer of **7**, Figure 11.1), tirapazamine has been shown to produce retinal toxicity.[17]

The *N*-oxide approach can also be used in reducing the basicity of the nitrogen in nitrogen mustards. Nitromin has been used with some success for the treatment of Yoshida ascites sarcomas in rats.[43] Recent studies showed that nitromin (**22**,

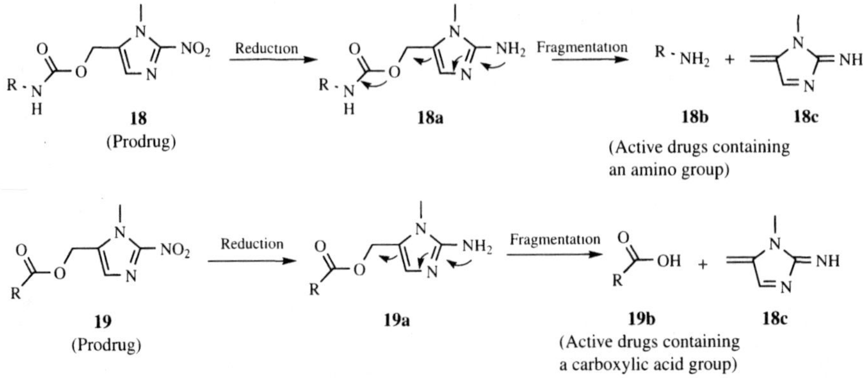

18 (Prodrug) Reduction 18a Fragmentation $R-NH_2$ + 18c
18b (Active drugs containing an amino group)

19 (Prodrug) Reduction 19a Fragmentation 19b + 18c
(Active drugs containing a carboxylic acid group)

Scheme 7. Bioreductive activation of heteroaromatic prodrugs via reduction of a nitro group followed by 1,5-elimination.

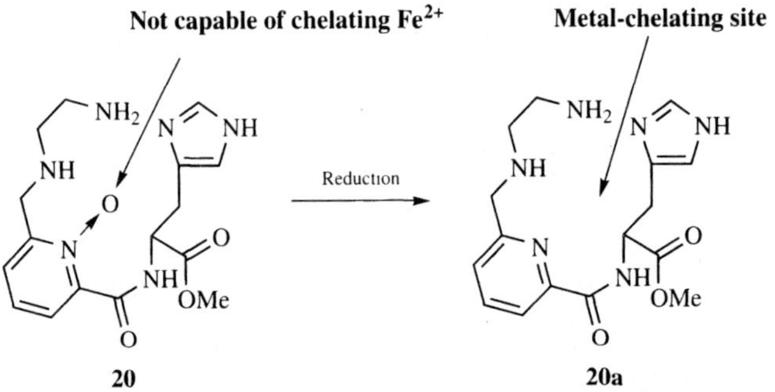

Not capable of chelating Fe^{2+} — Metal-chelating site

20 Reduction 20a

Scheme 8. Bioreductive activation of prodrug **20** to active drug **20a**.

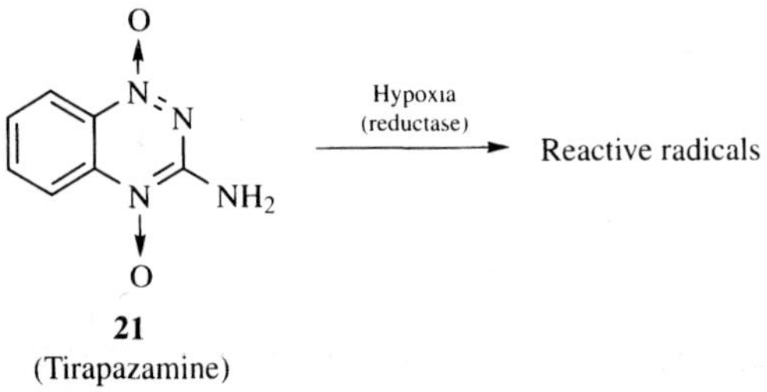

21 (Tirapazamine) Hypoxia (reductase) Reactive radicals

Scheme 9. Bioreductive activation of tirapazamine.

Scheme 10) serves as a bioreductive prodrug of a nitrogen mustard.[44,45] NADPH-dependent cytochrome P450 reductase has been shown to catalyze the reduction of nitromin to its active nitrogen mustard (**22a**).[45]

AQ4N (**23**) is a prodrug of the DNA binding agent and topoisomerase inhibitor AQ4 (**23a**). The less toxic prodrug is activated in hypoxic tissue through reduction of the *N*-oxide (Scheme 11).[46,47]

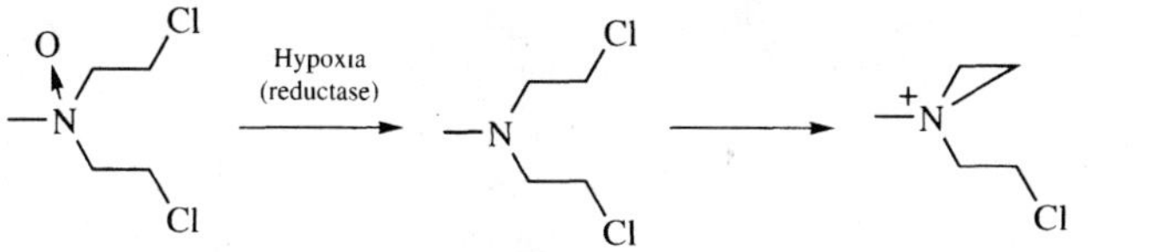

Hypoxia (reductase)

22 (Nitromin) **22a** **22b** (Reactive electrophile)

Scheme 10. Bioreductive activation of nitromin.

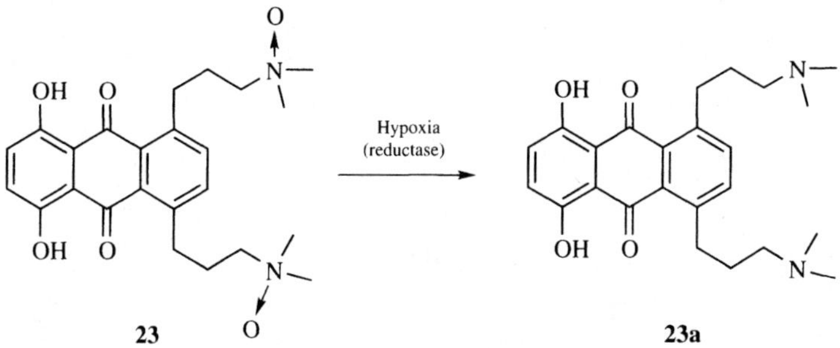

Hypoxia (reductase)

23 (AQ4N prodrug) **23a** (AQ4, DNA binding agent and topoisomerase inhibitor)

Scheme 11. Bioreductive activation of AQ4N.

11.2.1.1.4. Hypoxia-Selective S-Oxide-Containing Prodrugs The metabolic fate of sulfoxide (*S*-oxide) has been shown to be different under aerobic or anaerobic conditions. Under anaerobic conditions, sulfoxides (**24a**) can be reduced to sulfides in a reversible reaction (**24b**) (Scheme 12).[3] Under aerobic conditions, sulfoxides (**24a**) are oxidized to sulfones (**24**), a process that is not reversible (Scheme 12).

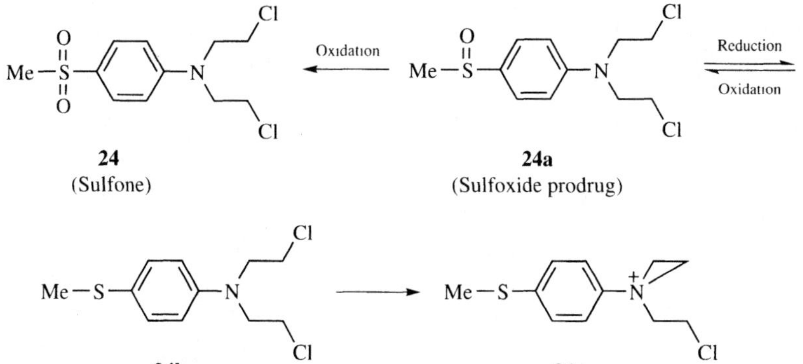

Oxidation Reduction Oxidation

24 (Sulfone) **24a** (Sulfoxide prodrug)

24b (Sulfide nitrogen mustard, active drug) **24c** (Reactive electrophile)

Scheme 12. Bioreductive activation of sulfoxide-containing nitrogen mustard prodrugs.

This difference has been exploited for the design of hypoxia-selective nitrogen mustard prodrugs. Since the sulfoxide group is an electron-withdrawing group, a sulfoxide nitrogen mustard prodrug is a less cytotoxic agent. This changes when, under a hypoxic condition, the sulfoxide group is converted into a sulfide, an electron-donating group that increases the electron density of the nitrogen, which, in turn, favors the formation of a reactive aziridinium ion (**24c**). The aziridinium ion then reacts with DNA, resulting in cell death. On the other hand, under aerobic conditions, a sulfoxide group is more likely to be oxidized to a corresponding sulfone, which is a stronger electron-withdrawing group and makes the formation of the reactive aziridinium ring more difficult.[3] A successful example of the hypoxia selectivity of a sulfoxide-containing anticancer prodrug was described by Kwon.[3] 1-[Bis(2-chloroethyl)amino-4-[{4-[bis(2-chloroethyl)amino]phenyl}sulfinyl]benzene, a diphenylsulfoxide-containing nitrogen mustard prodrug, was synthesized and found to exhibit high hypoxia selectivity.

11.2.1.2. Activation of Prodrugs Based on Hypoxic Irradiation Shibamoto et al. reported that upon irradiation under a hypoxic condition, [1-(2′-oxopropyl)-5-fluorouracil] (OFU001) (**25**, Scheme 13) was converted to the anticancer drug 5-fluorouracil (5-FU) (**25d**, Scheme 13).[48,49] Minimum conversion occurred when the

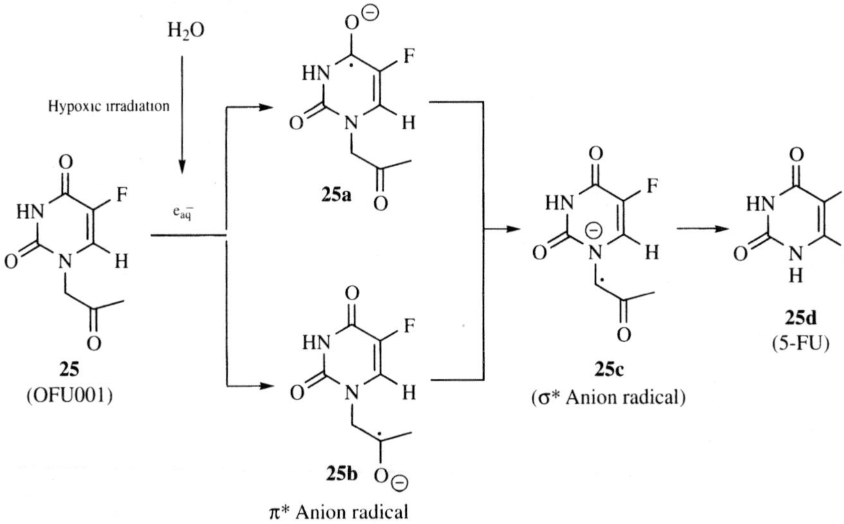

H_2O

Hypoxic irradiation

e_{aq}^-

25 (OFU001)

25a

25b π* Anion radical

25c (σ* Anion radical)

25d (5-FU)

Scheme 13. Proposed mechanism for conversion of OFU001 to 5-FU following hypoxic irradiation.

experiment was carried out under an aerobic condition. The mechanism for the conversion of OFU001 to 5-FU under a hypoxic condition is described in Scheme 13. Upon irradiation, hydrated electrons, which are active species derived from radiolysis of water, are thought to be incorporated into the compound to form a corresponding π* anion radical, which is thermally activated to a σ* anion radical with a weakened C-N bond. Subsequently, hydrolytic dissociation of the C-N bond occurs, releasing 5-FU (**25d**). Since hydrated electrons are rapidly deactivated by O_2 into superoxide anion radicals (O_2^-) under an aerobic condition, the conversion of OFU001 is therefore hypoxia specific. The 2′-oxoalkyl side chain (**25**, Scheme 13) in the structure is believed to be required for the activation.

11.2.1.3. Activation of Prodrugs Based on Low pH of Solid Tumors Lower extracellular pH has been observed in many solid tumors as a result of the limited blood flow.[15] The lower extracellular pH is believed to be caused by insufficient clearance of acidic metabolites from chronically hypoxic cells, a phenomenon that can lower the mean extracellular pH in tumors to below ca. 6.3—up to one pH unit lower than the intracellular pH, which is actively regulated. Prodrugs that are stable at physiological pH and can be activated by this lower extracellular pH in solid tumors have been described.[50] 2-Hexenopyranoside of aldophosphamide (**26**, Figure 11.2) is a prodrug of aldophosphamide. The prodrug was stable at physiological pH (7.4) but readily hydrolyzed to produce aldophosphamide under an acidic condition.

11.2.2. Tumor-Activated Prodrugs Based on Proteases

Tumor malignancy is often associated with an enhanced expression of proteolytic enzymes such as cysteine protease cathepsin B, matrix metalloproteinases such as collagenases and stromelysins, and serine proteases represented by plasminogen activator and plasmin.[51–53] These enzymes are thought to be critically involved in the events that lead to metastasis because they are capable of degrading the basement membrane and extracellular matrix around tumor tissue, allowing tumor cells to migrate and invade the surrounding stroma and endothelium. Although these enzymes are also produced by normal cells, their activity is normally tightly regulated by hormonal controls and by specific inhibitors.[54,55] Therefore, these enzymes are targeted as activating enzymes for anticancer prodrugs. In general, prodrugs targeting proteases contain two components—a peptide and a parent drug. The peptide chosen is usually a di-, tri-, or tetrapeptide which should give rise to a prodrug

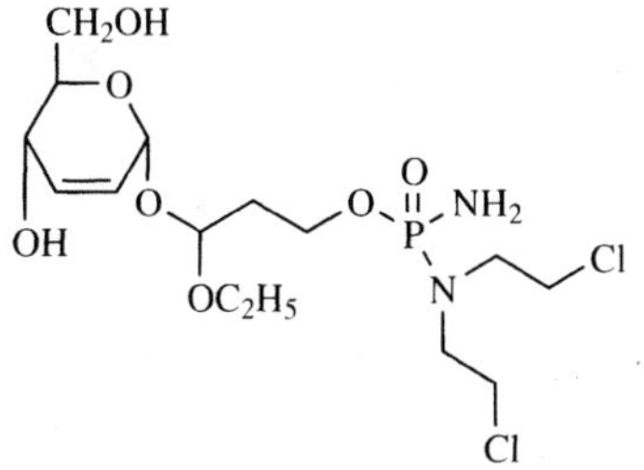

CH_2OH O OH O O O P NH_2 N Cl OC_2H_5 Cl

26

(2-Hexenopyranoside of aldophosphamide)

Figure 11.2. Structure of 2-hexenopyranoside of aldophosphamide.

resistant to the serum peptidase but susceptible to enzymes present around the tumor mass.

11.2.2.1. Prodrug Bond Linkage Between a Peptide and a Drug A peptide can be connected to a drug directly (Scheme 14) or indirectly through a linker termed a "spacer" (Schemes 15 and 16). The direct linkage of a peptide to a drug leads to a prodrug that can release either the parent drug or a drug that contains vestiges of the

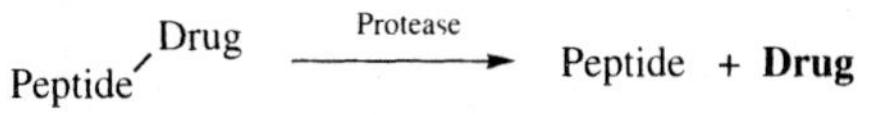

Peptide–Drug —Protease→ Peptide + **Drug**

Scheme 14. Prodrugs with a drug directly connected to a peptide.

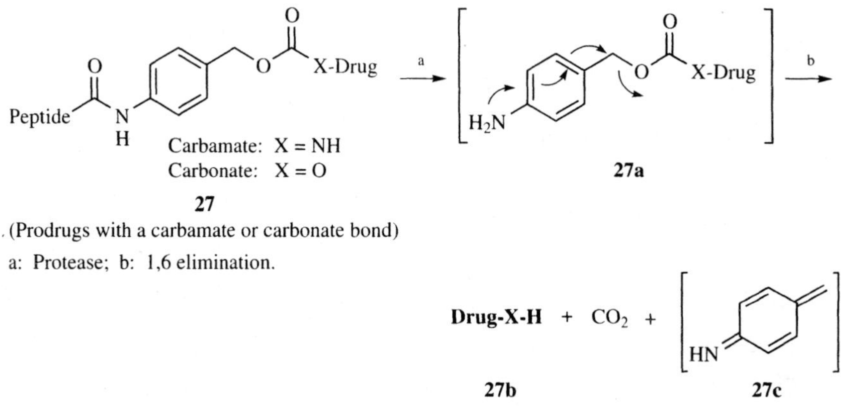

Peptide N H O O O X-Drug
Carbamate: X = NH
Carbonate: X = O
27
(Prodrugs with a carbamate or carbonate bond)
a: Protease; b: 1,6 elimination.

a → [H_2N O O X-Drug] b →
27a

Drug-X-H + CO_2 + [HN]
27b **27c**

Scheme 15. Hydrolytic activation of prodrugs with a *p*-aminobenzyl alcohol as a spacer that contains a carbamate or carbonate bond.

Peptide N H O O-Drug
28
(Prodrugs with a ether bond)
a: Protease; b: 1,6 elimination.

a → [H_2N O-Drug] b →
28a

Drug-OH + CO_2 + [HN]
28b **27c**

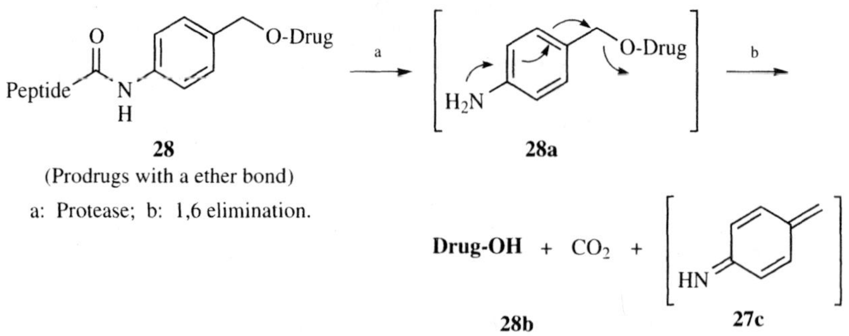

Scheme 16. Hydrolytic activation of prodrugs with a *p*-aminobenzyl alcohol as a spacer that contains an ether bond.

bound peptide.[56–58] In the latter case, the released drug may have impaired cytotoxic activity. An additional consideration for direct drug attachment to peptides is that the drug may reduce the hydrolysis rate of the peptide by the activating enzyme, resulting in a slow release of the active drug. This often occurs when a bulky drug is involved. The problem can be circumvented through an indirect linkage in which a self-immolative spacer is employed between the drug and the peptide. The spacer spatially separates the drug from the site of enzymatic cleavage so that the drug will not affect the hydrolysis of the peptide bond. Subsequent fragmentation of the spacer releases the active drug. Carl et al. developed one of the most commonly used spacers—a bifunctional *p*-aminobenzyl alcohol group, which is linked to the peptide through the amine moiety and to the drug through the alcohol moiety.[59] The alcohol moiety of the spacer can form a carbamate bond with an amino-containing drug or a carbonate bond with an alcohol-containing drug (Scheme 15). The formed prodrugs are activated upon protease-mediated cleavage of the amide bond followed by a 1,6-elimination that releases the drug, carbon dioxide, and remnants of the spacer (Scheme 15). Since the carbonate is easily susceptible to hydrolysis, only a few alcohol-containing drugs, such as paclitaxel, can be made stable enough prodrugs through a carbonate bond. Therefore, the chemistry of this drug attachment has generally been restricted to amine-containing drugs. For alcohol-containing drugs, a recently reported approach through an ether bond has proved to be more feasible than carbonate (Scheme 16). However, the ether bond did not undergo fragmentation as readily as the carbamate bond, and some of the prepared ether prodrugs did not undergo fragmentation to release the parent drug.[60]

It is noteworthy that the length of the spacer plays a significant role in the rate of enzymatic activation. This is especially true when *p*-aminobenzyl alcohol carbamate is used as a spacer and a bulky drug is involved. As shown in Scheme 15, after cleavage of the amide bond, the resulting amino group of the aromatic ring is an electron-donating group and initiates an electronic cascade that leads to the expulsion of the leaving group (1,6-elimination), which releases the free drug after elimination of carbon dioxide. The 1,6-elimination of the carbamate prodrug was found to be virtually instantaneous upon unmasking of the amine group.[61] Since the spacer rapidly eliminates after prodrug activation, the enzymatic activation itself determines the efficiency of drug release. This assumption is supported by the fact that prodrugs derived from bulky drugs often have a slower release rate.[61] To increase enzyme activation rates, de Groot et al. propose to increase the length of the spacer between the drug and the peptide to further keep the bulky drug away from the site of the enzymatic reaction.[61] This approach results in a significant increase in enzyme activation rates and will be discussed in Section 11.2.2.2.

Other spacers such as ethylene diamine, *o*-aminobenzyl alcohol, *p*-hydroxylbenzyl alcohol, and *o*-hydroxylbenzyl alcohol have also been utilized. The application of some of these spacers will be illustrated in the examples presented later.

It is worth mentioning that the use of a spacer is not limited to prodrugs involving a peptide. Spacers discussed above have also been extensively employed in other prodrugs as well.

11.2.2.2. Tumor-Selective Prodrugs Activated by Plasmin Plasmin is derived from inactive plasminogen by plasminogen activators (Pas) and is involved in extracellular matrix degradation. The levels of Pas are high in many types of malignant cells and human tumors such as malignant lung[62] and colon[63] tumors. Consequently, tumor-associated plasmin activity is highly localized.[53,62,64]

Plasmin is a protease with specificity for arginine or lysine as amino acids participating in bond cleavage.[65] Examination of the preferred sites for plasmin cleavage of the fibrinogen molecule (the physiological function of plasmin) shows that the preferred sites involve lysine linked to a hydrophobic amino acid. Thus, the choice of the peptidic sequence to be used has focused on a peptide having a hydrophobic amino acid linked to lysine.[66]

The idea of selecting plasmin to activate peptide-containing prodrugs was first proposed in 1980 by Carl et al.[65] A D-Val-Leu-Lys tripeptide connected to the amino functional group AT-125 (**29**, Figure 11.3) or phenylene diamine mustard (**30**, Figure 11.3) was prepared and found to generate the free drug upon treatment with plasmin. Selection of the D-configuration of the *N*-terminal amino acid (D-Val) prevents undesired proteolysis by serum peptidases or other ubiquitous enzymes. A five- to sevenfold increase in selectivity between Pas-producing cells in comparison with low-level Pas-containing cells was demonstrated.[65]

The first prodrugs of doxorubicin for plasmin activation were designed by Chakravarty et al.[67,68] Doxorubicin was directly linked to the tripeptide D-Val-Leu-Lys (**31**, Figure 11.4). The formed prodrug showed a sevenfold increase in selectivity for Pas-producing cells in comparison with low-Pas-containing cells, but the drug was very inefficiently released by plasmin, a phenomenon likely caused by steric hindrance of doxorubicin. To overcome this problem, de Groot et al. placed a spacer between doxorubicin and the peptide.[69] Compounds **32a-c** were prepared with a *p*-aminobenzyl alcohol as a self-immolative 1,6-elimination spacer to separate doxorubicin or its anthracycline derivatives from the tripeptide D-Ala-Phe-Lys. All prodrugs were stable in buffer and serum for 3 days and generated the parent drugs upon incubation with plasmin. Compound **32c** demonstrated the fastest plasmin cleavage rate. Upon incubation with seven human tumor cell lines, the prodrugs showed a marked decrease in cytotoxicity in comparison with

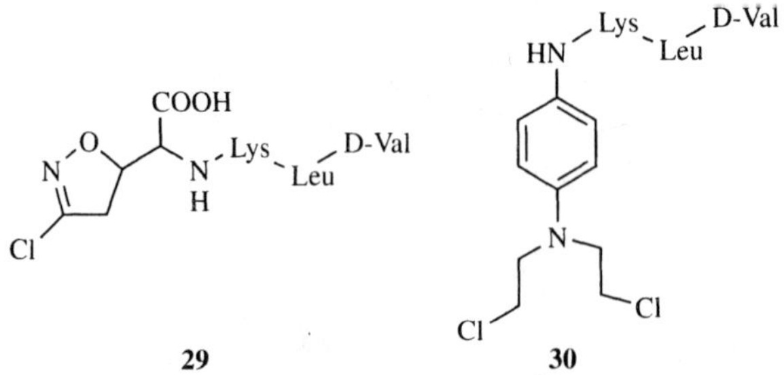

Figure 11.3. Anticancer prodrugs activated by plasmin.

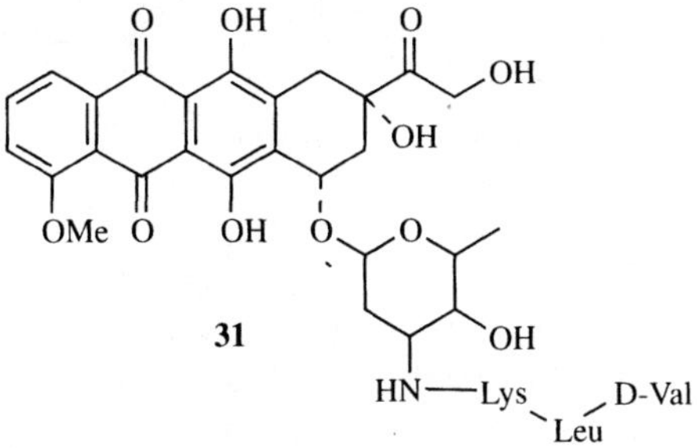

31

(A prodrug with a direct linkage of a peptide to doxorubicin)

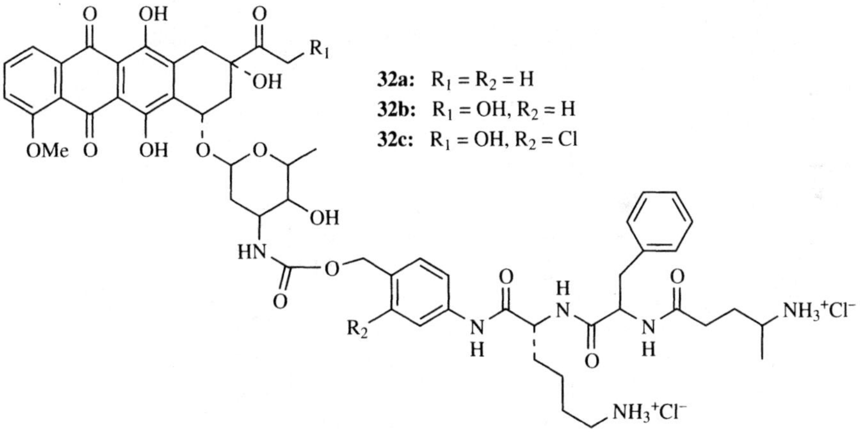

32a: $R_1 = R_2 = H$
32b: $R_1 = OH, R_2 = H$
32c: $R_1 = OH, R_2 = Cl$

(Prodrugs with one spacer between a peptide and doxorubicin or its derivatives)

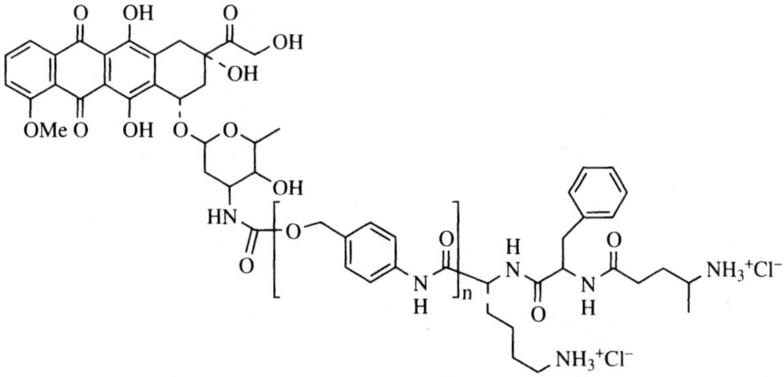

33: n=2, a prodrug with two spacers between a peptide and doxorubicin
34: n=3, a prodrug with three spacers between a peptide and doxorubicin

Figure 11.4. Prodrugs of doxorubicin and its derivatives with different numbers of spacers between the drug and a peptide.

the corresponding parent drugs. *In vitro* selectivity was demonstrated by incubation of the prodrugs with Pas-transfected MCF-7 cells in comparison with nontransfected cells. The prodrugs showed cytotoxicity similar to that of free doxorubicin only in the Pas-transfected cells, while exhibiting much less toxicity in the nontransfected cells.[69] The enzyme activation rate was even further increased when a longer spacer was placed between the drug and the peptide. De Groot et al. placed two *p*-aminobenzyl alcohol moieties between doxorubicin and the peptide (**33**, Figure 11.4) and found that the enzymatic release rate was increased by twofold.[61] The rate was further increased (ca. threefold) when three *p*-aminobenzyl alcohol moieties were incorporated (**34**, Figure 11.4).[61]

An elegant cyclization spacer was used in the prodrugs of *N*-nitrosourea.[70] Prodrugs **35a** and **35b** contain an ethylene diamine (**35a**) or monomethylated ethylene diamine (**35b**) spacer. Upon cleavage by plasmin, a cyclization reaction led to the formation of a pentacyclic urea derivative (imidazolidin-2-one), with concomitant expulsion of the reactive electrophile (**35g**, Scheme 17). Compared with *p*-aminobenzyl alcohol, which undergoes 1,6-elimination, cyclization with an ethylene diamine spacer occurs at a slower rate and often is a rate-determining factor in releasing the active drug.

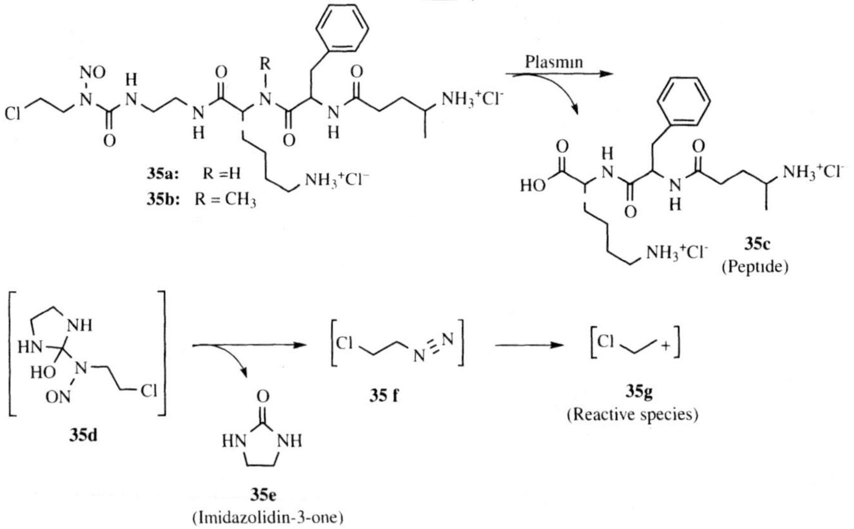

35a: R = H
35b: R = CH_3
Plasmin
35c (Peptide)
35d
35 f
35g (Reactive species)
35e (Imidazolidin-3-one)

Scheme 17. Hydrolytic bioactivation of prodrugs **35a** and **35b**.

De Groot et al. used the same approach to prepare prodrugs of paclitaxel for activation by plasmin.[69] However, most of the prepared prodrugs were either not stable or resistant to the hydrolysis by plasmin. An alternative approach using a *p*-aminobenzyl alcohol group as a spacer appeared to be successful. Prodrug **36** yielded free paclitaxel (**36a**) upon incubation with plasmin (Scheme 18). Further, prodrug **36** showed a dramatic decrease in cytotoxicity in seven human tumor cell lines in

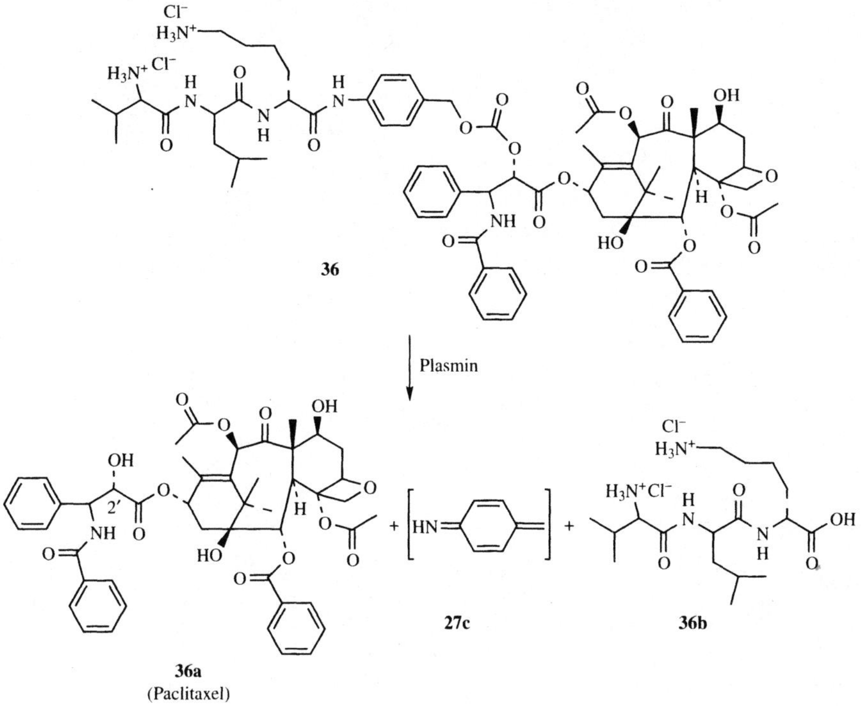

Scheme 18. Hydrolytic bioactivation of prodrug **36** by plasmin.

comparison with paclitaxel (**36a**). A similar approach to increase the enzymatic activation rate by increasing the space between paclitaxel and the peptide was also used.[61] Upon incubation with plasmin, the prodrug (**37**, Figure 11.5) with two *p*-aminobenzyl alcohol moieties released paclitaxel at a sixfold increased rate compared to the single-spacer-containing prodrug (**36**).[61]

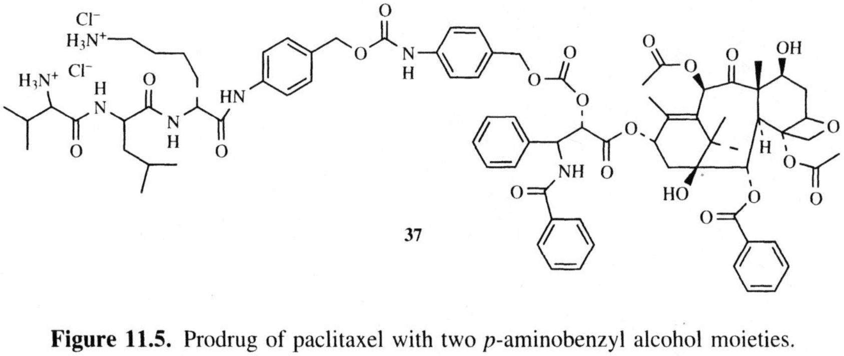

Figure 11.5. Prodrug of paclitaxel with two *p*-aminobenzyl alcohol moieties.

11.2.2.3. Tumor-Selective Prodrugs Activated by Cathepsin B Cathepsins are cysteine proteases. Cathepsins, especially cathepsin B, are overexpressed in tumors and may play an important role in cancer invasion.[71] Cathepsins may directly degrade extracellular matrix proteins or activate the Pas system.[71] Cathepsin B has been shown to be clinically relevant in cancer progression, with studies demonstrating that cytosolic enzyme levels were 11 times higher than those in benign breast tissue specimens.[72] Patients with high intratumor cathepsin B levels have a significantly worse prognosis than those with low levels.[72]

A number of prodrugs for cathepsin B activation were prepared (**38–43**, Figure 11.6). However, not much biological information is available on these compounds.

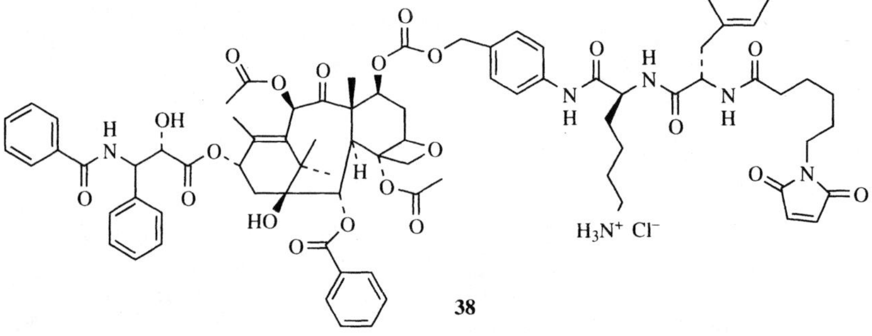

38

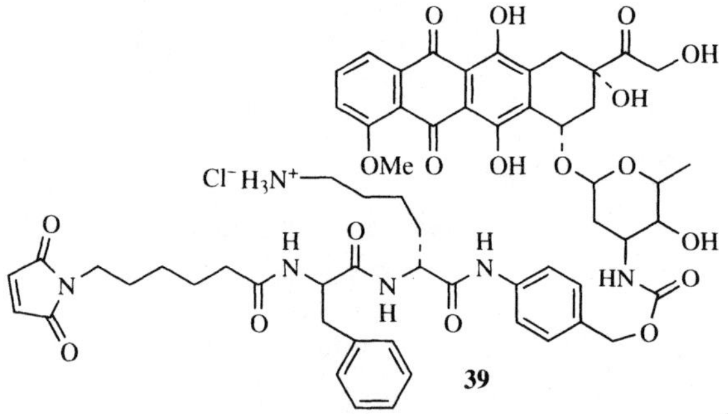

39

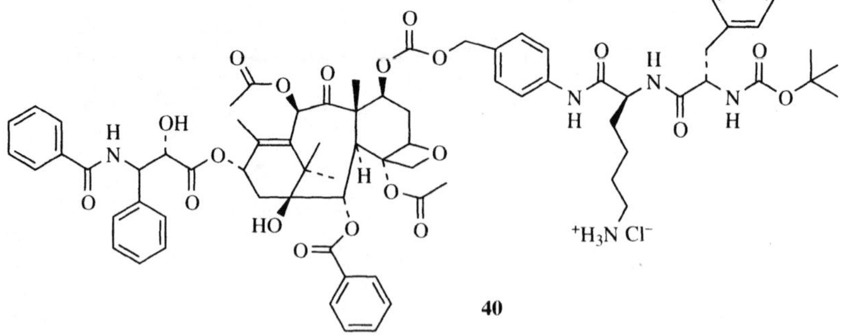

40

Figure 11.6. Representative prodrugs activated by cathepsin B.

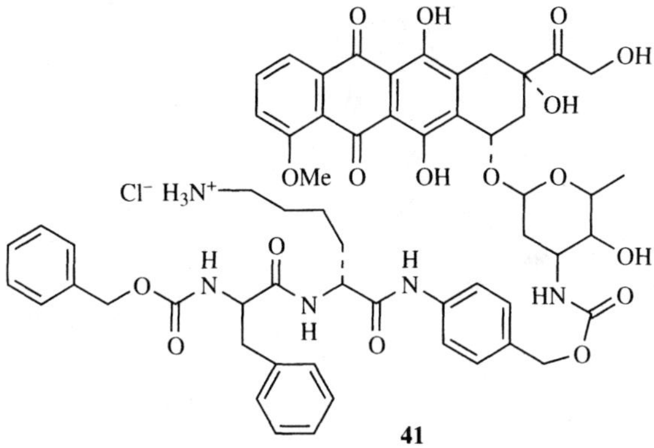

41

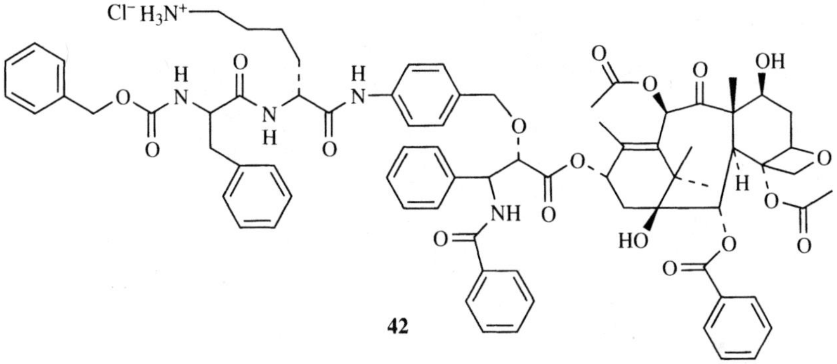

42

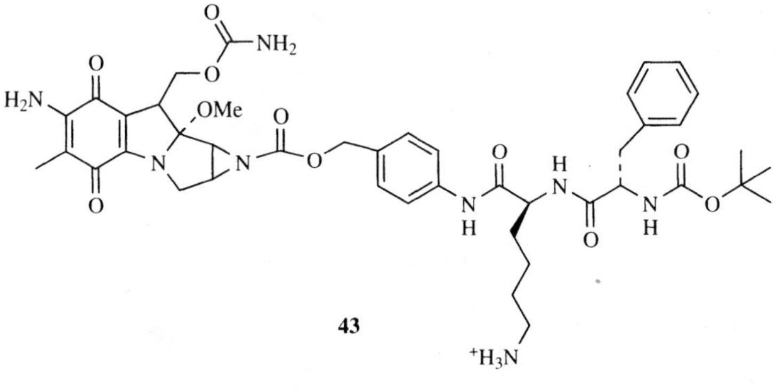

43

Figure 11.6. (*Continued*)

The half-lives of cathepsin B cleavage of doxorubicin prodrug **41** and mitomycin C prodrug **43** were much shorter than those of paclitaxel prodrugs **40** and **42**, indicating steric hindrance by the paclitaxel part of the prodrug.[4]

Recently, Toki et al. succeeded in developing cathepsin B-activated prodrugs of combretastatin A-4 (**44a**, Scheme 19) and etoposide (**45a**, Scheme 19).[60] Combretastatin A-4 is a promising antiangiogenic agent that inhibits the polymerization of tubulin.[73] Etoposide is a clinically approved topoisomerase inhibitor that has demonstrated utility in chemotherapeutic combinations for the treatment of

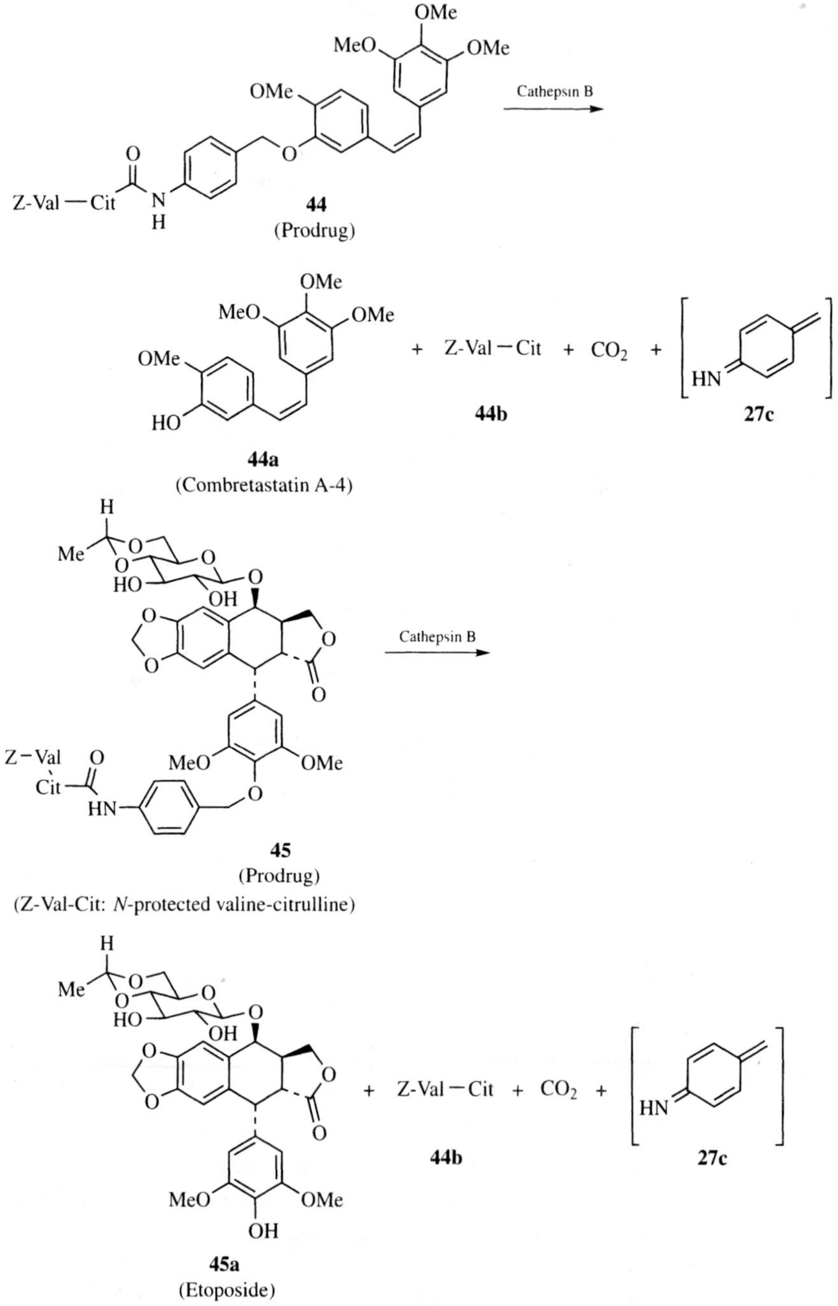

OMe
MeO OMe
OMe
Cathepsin B
O
Z-Val — Cit N O
H
44
(Prodrug)

OMe
MeO OMe
OMe + Z-Val — Cit + CO_2 +
HN
HO
44b **27c**

44a
(Combretastatin A-4)

H
Me O
O O
HO O
OH
O
O Cathepsin B
O
O
Z − Val O MeO OMe
Cit
HN O

45
(Prodrug)
(Z-Val-Cit: *N*-protected valine-citrulline)

H
Me O
O O
HO O
OH
O
O + Z-Val − Cit + CO_2 +
O HN
O
44b **27c**
MeO OMe
OH

45a
(Etoposide)

Scheme 19. Hydrolytic bioactivation of prodrugs by cathepsin B.

leukemia, lymphoma, germ cell tumors, small cell lung tumors, and several other carcinomas.[74] In a prodrug approach, combretastatin A-4 (**44a**) or etoposide (**45a**) was coupled to *Z*-valine-citrulline peptide, an *N*-protected valine-citrulline peptide, through an ether and amide bond using *p*-aminobenzyl alcohol as a spacer (Scheme 19). The formed prodrugs (**44** for **44a** and **45** for **45a**) were both substrates of cathepsin B and released the active drugs upon incubation with the enzyme, suggesting that these prodrugs can be activated by cathepsin B. It is noteworthy that this is the first example demonstrating 1,6-fragmentation with an ether bond instead of a carbamate bond when *p*-aminobenzyl alcohol is used as a spacer. The prodrug **44** was less potent than the parent drug **44a** by a factor of 13 in cell killing on L2987 human lung adenocarcinoma, while the prodrug **45** was 20 to 50 times less active than the parent drug **45a** in the cell lines L2987 (human lung adenocarcinoma), WM266/4 (human melanoma), and IGR-39 (human melanoma), confirming that the prodrugs were much less cytotoxic.[60] However, as mentioned earlier, the fragmentation of the ether bond did not proceed as readily as the fragmentation of the carbamate bond, which is illustrated by a prodrug derived from *N*-acetylnorephedrine (Scheme 20). *N*-Acetylnorephedrine underwent hydrolysis by cathepsin B but failed to release the drug *N*-acetylnorephedrine, suggesting that the alkoxyl

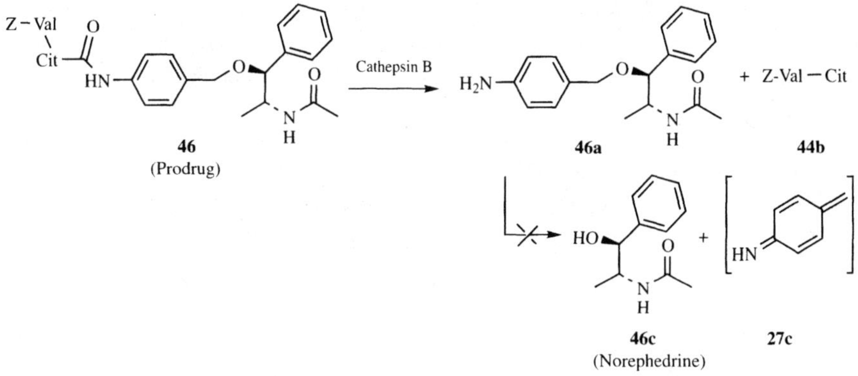

Scheme 20. The hydrolyzed prodrug (**46a**) fails to undergo 1,6-fragmentation to release norephedrine (**46c**).

ether bond was not as readily cleaved as those derived from phenoxyl ether-containing drugs, such as the one in etoposide or combretastatin. This phenomenon is consistent with the fact that a phenoxyl group is a better leaving group than an alkoxyl group.

11.2.3. Tumor-Activated Prodrugs Based on Enzymes with Elevated Activity at Tumor Sites

11.2.3.1. β-Glucuronidase β-Glucuronidase (EC 3.2.1.31) is an exoglycosidase that cleaves glucuronosyl-*O* bonds.[75] Glucuronidase is intracellularly located in lysosomes in many organs and body fluids such as macrophages, most blood cells, liver, spleen, kidney, intestine, lung, muscle, bile, intestinal juice, urine, and serum.[75] There is large interindividual variability in its activity and expression in the liver, kidney, and serum. The activity of the enzyme is relative high in some tissues, such as the liver. The rationale for selecting β-glucuronidase to activate anticancer prodrugs is based on the observation that the activity of the enzyme has been shown to be elevated in many tumors.[76,77] An advantage of targeting β-glucuronidase is that a glucuronide prodrug exhibits high hydrophilicity, which greatly reduces the distribution of the prodrug into cells, resulting in reduced systemic toxicity. However, the same feature also hampers the activation of a glucuronide prodrug at tumor sites. Since the enzyme is intracellularly located, a glucuronide prodrug needs to enter tumor cells in order to be activated. Bosslet et al. demonstrated that β-glucuronidase is liberated extracellularly in high local concentration in necrotic areas of human cancer.[77] Therefore, necrotic areas are the areas where glucuronide prodrugs can be activated. It is worth mentioning that, in addition to cancer, increased tissue or serum β-glucuronidase activity has been observed in other disease states, e.g., inflammatory joint disease, ichthyosiform dermatosis, some hepatic diseases and acquired immune deficiency syndrome (AIDS). For example, serum β-glucuronidase activity has been reported to be 16-fold higher in human immunodeficiency virus (HIV) infected patients than in healthy individuals.[78] The extracellular presence of β-glucuronidase in areas other than tumor sites leads to activation of β-glucuronidase-mediated prodrugs, resulting in side effects.

The β-glucuronidase-mediated release approach has been used in the prodrugs of many anticancer drugs. The prodrugs are formed by linking a drug to glucuronic acid through an anomeric ether bond directly or, more often, indirectly (Scheme 21). In the latter case, a spacer is used. In general, prodrugs with a spacer are

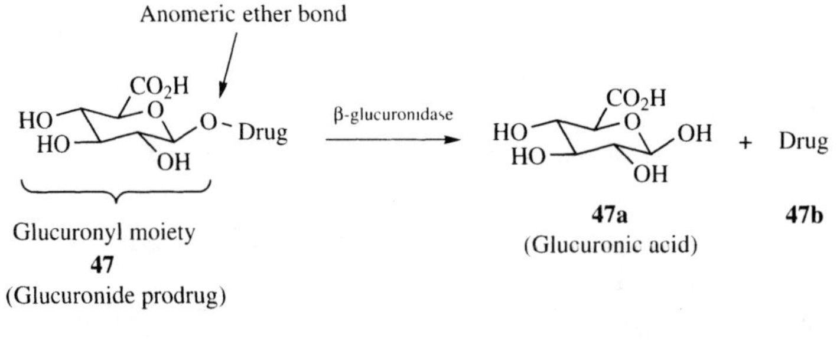

Anomeric ether bond

CO_2H HO HO O O~Drug OH —β-glucuronidase→ CO_2H HO HO O OH OH + Drug

Glucuronyl moiety
47
(Glucuronide prodrug)

47a
(Glucuronic acid)

47b

Scheme 21. Bioactivation of glucuronide prodrugs by β-glucuronidase.

cleaved more readily by the enzyme, especially when the parent drug molecules are bulky. Figure 11.7 shows the structures of doxorubicin prodrugs **48** and **49**. Both were much less toxic than the parent drug in a human ovarian cancer cell line (OVCAR-3), while prodrug **49**, with a spacer between the drug and the glucuronyl group, was activated much faster by β-glucuronidase.[79] Similarly,

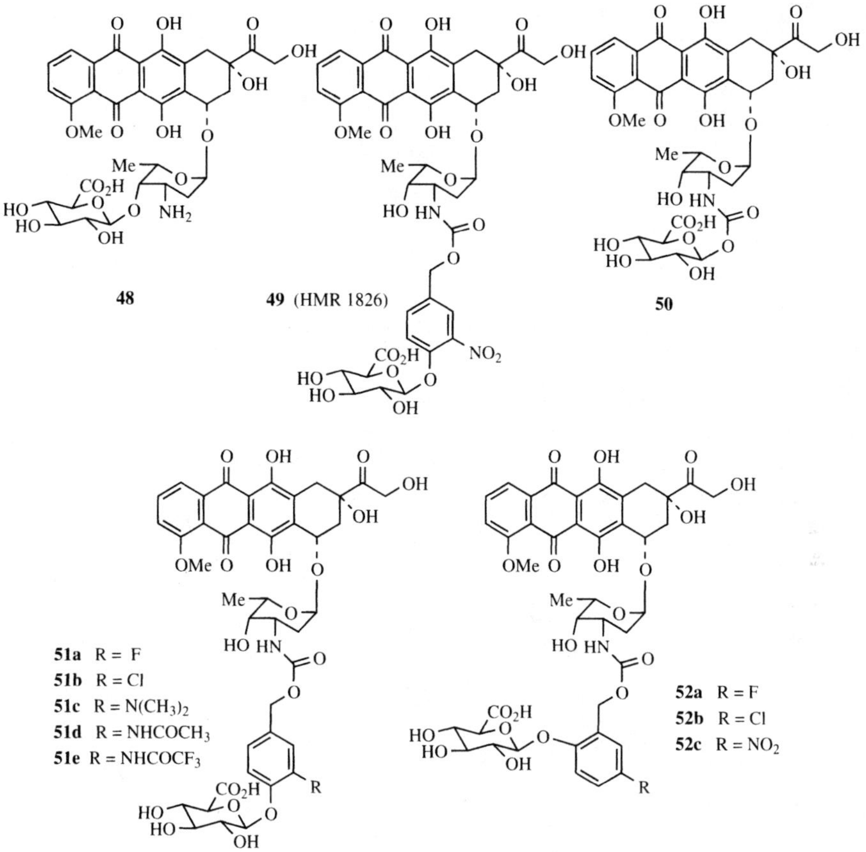

48 **49** (HMR 1826) **50**

51a R = F
51b R = Cl
51c R = $N(CH_3)_2$
51d R = $NHCOCH_3$
51e R = $NHCOCF_3$

52a R = F
52b R = Cl
52c R = NO_2

Figure 11.7. Representative prodrugs bioactivated by β-glucuronidase.

daunorubicin prodrug **50** showed very low toxicity against five human tumor cell lines, consistent with the observation that the compound appeared to be very stable against enzymatic cleavage by β-glucuronidase.[80] The spacers used most extensively are *p*- or *o*-hydroxybenzyl alcohol moieties.

Florent et al. prepared and evaluated a series of anthracycline glucuronide prodrugs with a *p*- or *o*-hydroxybenzyl alcohol as a spacer.[81] It was reported that doxorubicin glucuronide prodrugs **49, 51a**, and **51b** (Figure 11.7) appeared to be better substrates for β-glucuronidase than the corresponding prodrugs **52a–c** (Figure 11.7). Prodrugs **52a–c** also showed a slower elimination of the spacer.[81] Prodrug **49** was selected as the most appropriate prodrug for further evaluation, and it has been extensively investigated. An attempt to further increase the hydrophilicity of the prodrugs by introducing a hydrophilic group at the spacer moiety resulted in compounds **51c–e** (Figure 11.7).[82] Of these prodrugs, compound **51d** appeared most promising for further development due to its reduced cytotoxicity and fast hydrolysis by β-glucuronidase.

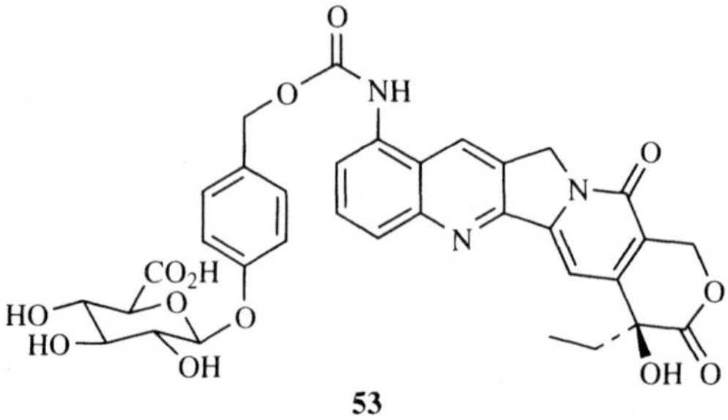

Figure 11.8. Structure of a glucuronide prodrug of 9-aminocamptothecin.

The camptothecins are a new class of very promising anticancer agents, several derivatives of which are in clinical use.[83] Prodrug **53** (Figure 11.8) of 9-aminocamptothecin showed 20- to 80-fold reduced toxicity in comparison with 9-aminocamptothecin.[84] The prodrug was readily cleaved by β-glucuronidase *in vitro*.

Schmidt et al. prepared and evaluated compound **54** as a prodrug of a phenol mustard (Scheme 22).[85] After removal of the glucuronic acid (**47a**) by β-glucuro-

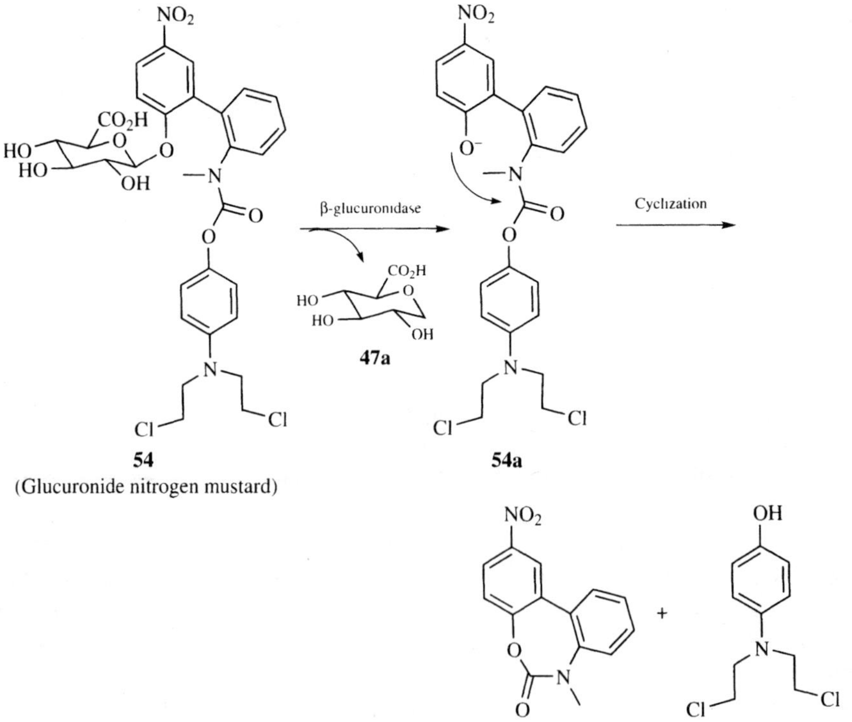

Scheme 22. Activation of glucuronide nitrogen mustard prodrug **54**.

nidase, the spacer was eliminated through cyclization liberating the phenol mustard **54c**. The prodrug was 80-fold less cytotoxic than the phenol mustard in LoVo cells. Chemically, the prodrug (**54**) was much more stable than the corresponding parent drug (**54c**) in phosphate buffer.

Another prodrug of **54c** is compound **55**. Prodrug **55**, which contains an aromatic and aliphatic bis-carbamate spacer, is activated by β-glucuronidase (Scheme 23).[86] A rapid cleavage of the glycosidic bond occurred ($t_{1/2} = 6.6$ minutes) with concomitant appearance of intermediate **55c**, of which the ethylene diamine spacer cyclized with a half-life of 2 hours. The cytotoxicity of **55** against LoVo cells was about 50 times less than that of the corresponding phenol mustard **54c**.

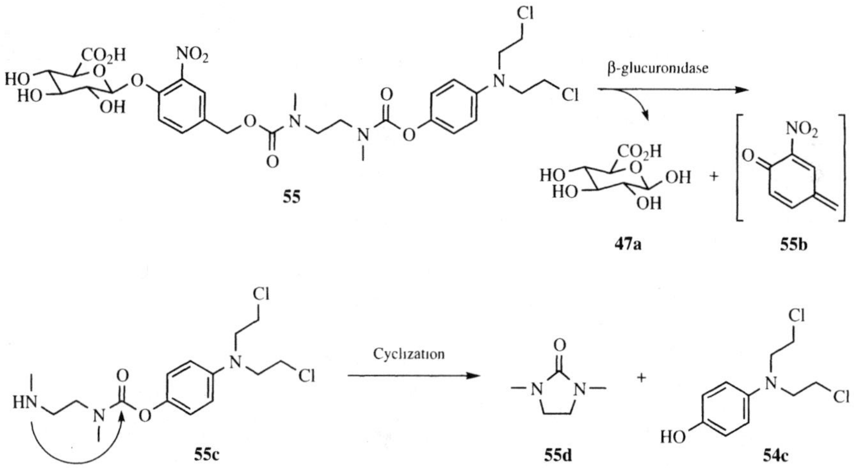

Scheme 23. Glucuronidase-mediated bioactivation of prodrug **55**.

A similar approach was also applied in paclitaxel prodrugs **56, 57a**, and **57b**.[87] After hydrolysis by β-glucuronidase, ring closure occurred, resulting in the release of paclitaxel (**36a**) and CO_2 (Schemes 24 and 25). The half-lives of β-glucuronidase-mediated activation were 2 hours and 45 minutes, respectively, for prodrugs **56** and **57b**. Both prodrugs were two orders of magnitude less cytotoxic than paclitaxel. Prodrug **57a** appeared to be as cytotoxic as paclitaxel, which was explained by the fact that the prodrug underwent spontaneous hydrolysis in buffer solution under physiological conditions.

Glucuronide prodrugs **58a, 58b**, and **59** of 5-fluorouracil exhibited half-lives of ~20 minutes when incubated with 25 μg/mL of β-glucuronidase and reduced (six to nine times less) cytotoxicity against LoVo cells in comparison with the parent drug 5-fluorouracil (Figure 11.9).[88]

11.2.3.2. Prostate-Specific Antigen (PSA) PSA is a serine protease that is present extracellularly in prostate cancers. PSA is inhibited in the bloodstream. As a result,

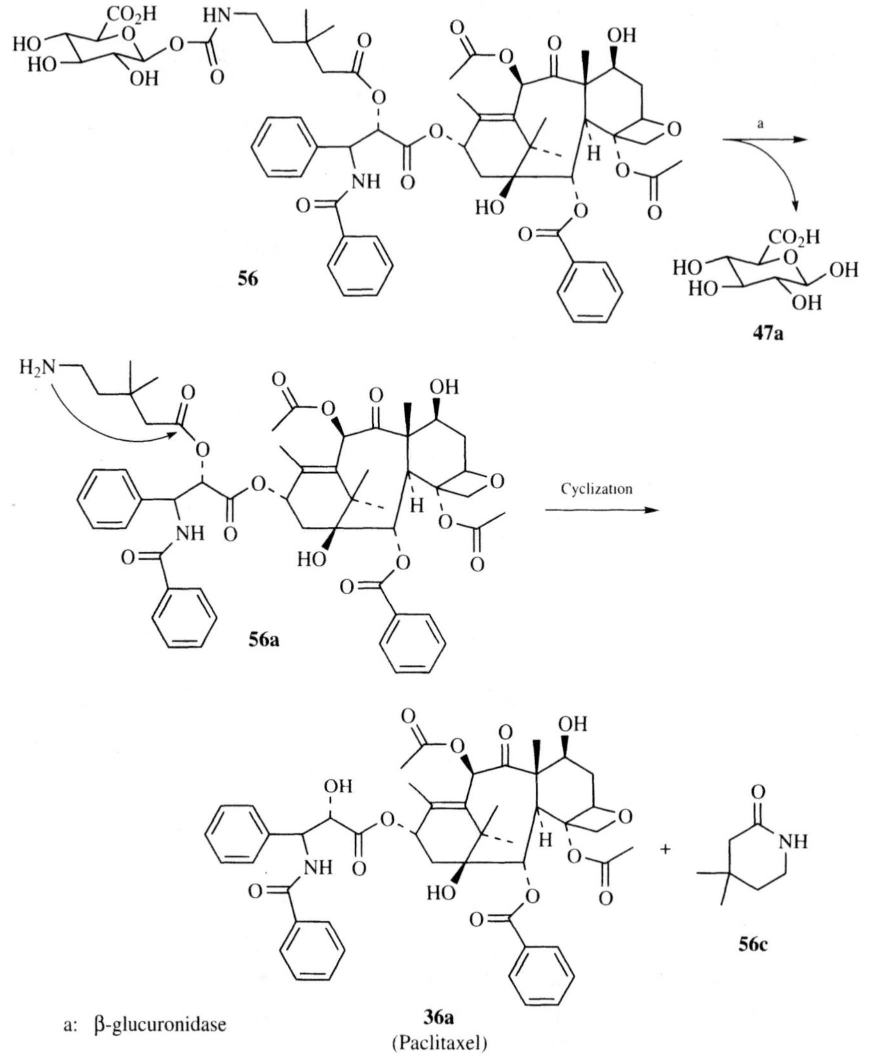

56

47a

56a

Cyclization

36a
(Paclitaxel)

56c

a: β-glucuronidase

Scheme 24. Bioactivation of prodrug **56** by β-glucuronidase.

active PSA is present only in prostate cancer.[4] A prodrug (**60**, Figure 11.10) of doxorubicin was formed by coupling the C-terminal carboxylic acid group of a heptapeptide to doxorubicin. The sequence His-Ser-Ser-Lys-Leu-Gln was selected because of its specificity and serum stability.[89] In compound **60**, an extra leucine residue was added after glutamine, which increased the distance between the peptide bond to be cleaved (leucine-glutamine) and doxorubicin. This increased distance reduced the steric hindrance that doxorubicin posed to the leucine-glutamine bond and facilitated the bond cleavage by PSA.[57] The prodrug **60** underwent cleavage by PSA at the leucine-glutamine bond to release not doxorubicin but

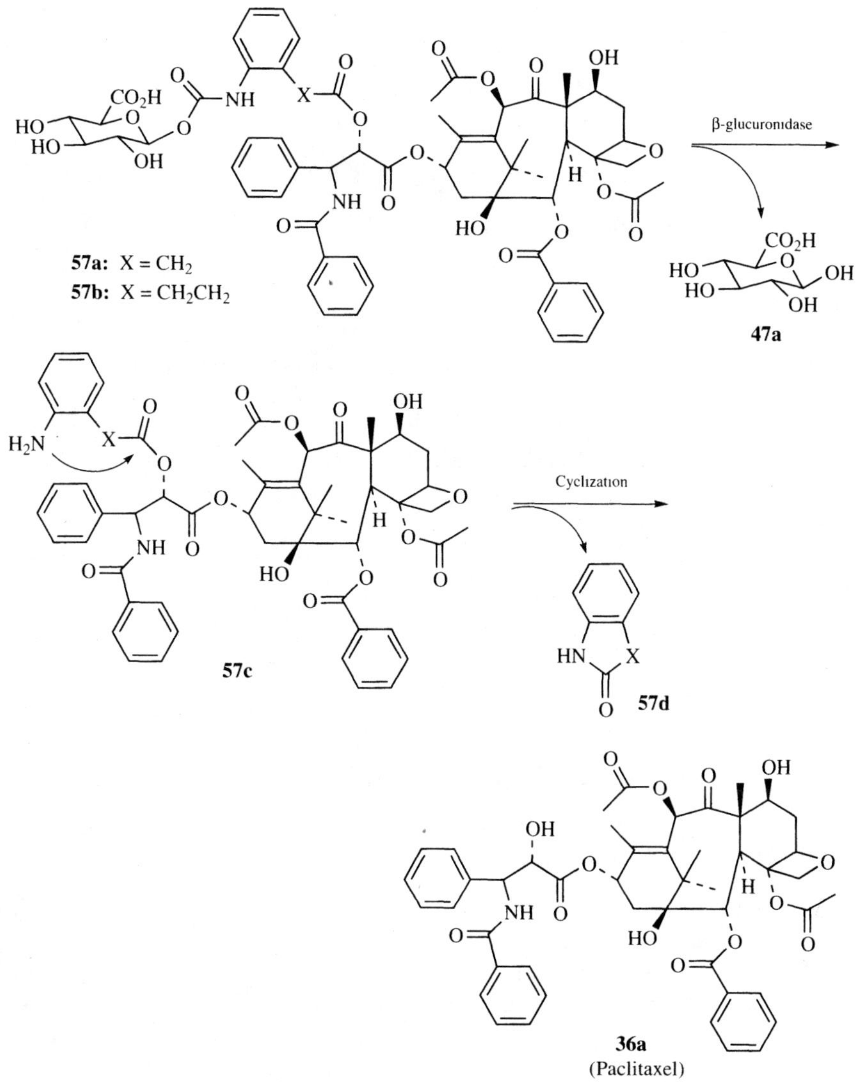

57a: X = CH_2
57b: X = CH_2CH_2

β-glucuronidase

47a

57c

Cyclization

57d

36a
(Paclitaxel)

Scheme 25. Bioactivation of prodrugs **57a** and **57b** by β-glucuronidase.

doxorubicin-leucine, which was an anticancer agent itself. *In vitro* selectivity was demonstrated by the fact that 70 nM of the prodrug killed 50% of the PSA-producing human prostate cancer cells (LNCaP cells), whereas doses as high as 1 microM had no cytotoxic effect on PSA-nonproducing TSU human prostate cancer cells.[57]

11.2.3.3. P450 as Activating Enzymes for Anticancer Prodrugs Cytochrome P450 (P450) comprises a superfamily of enzymes involved in the oxidation of a

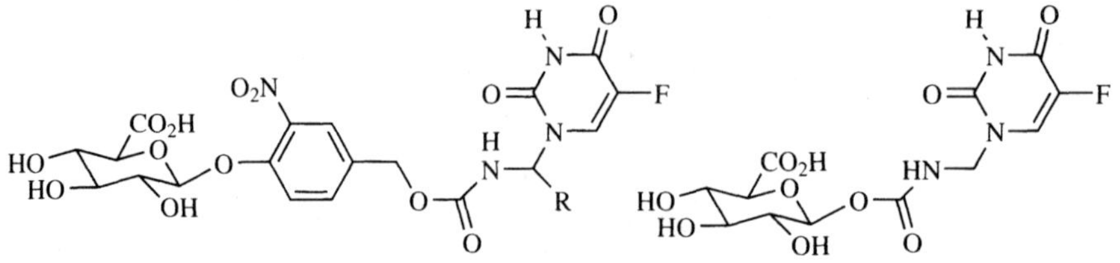

58a: R = H
58b R = COOH

59

Figure 11.9. Structures of glucuronide prodrugs **58a**, **58b**, and **59**.

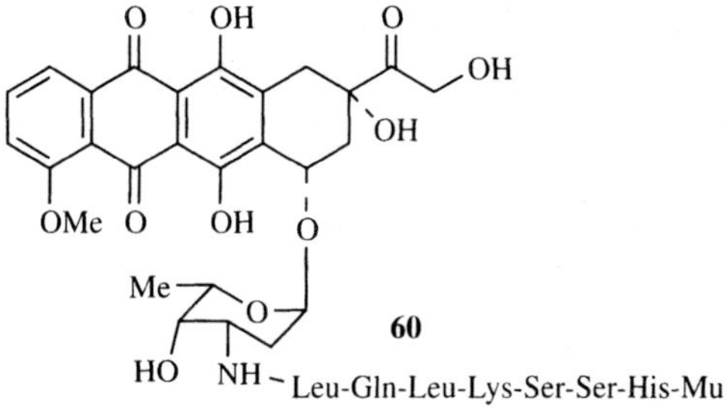

60

Mu = morpholinocarbonyl

Figure 11.10. Structure of a doxorubicin prodrug activated by PSA.

large number of exogenous and endogenous compounds.[90] Oxidation of compounds by P450 increases polarity and aids further metabolism or removal of the compound from the body. Therefore, P450 enzymes are viewed as the most important enzymes in removing exogenous compounds or toxic molecules from the body. P450 enzymes are present in normal tissues, with the highest levels in the liver consistent with the role of the liver as the detoxification organ. P450 enzymes are involved not only in inactivation of anticancer drugs but also in activation of several anticancer drugs such as cyclophosphamide[91] and its isomers ifosfamide and tegafur.[92] As expected, P450-activated anticancer prodrugs are not highly tumor-selective, since most of them are activated by liver P450. These liver P450-activated prodrugs will not be discussed here. However, efforts to increase tumor selectivity of P450-activated anticancer prodrugs have been made. Of particular interest is the approach used to deliver a P450 enzyme to tumor sites through gene-directed enzyme prodrug therapy (GDEPT).[93–96] Efforts to further increase tumor selectivity and decrease systemic toxicity by delivering a P450-activating enzyme to the tumor site via GDEPT and, at the same time, inhibiting liver P450 have also been reported.[97,98] A more detailed discussion of GDEPT is presented in Section 11.3.2.

11.3. ANTIBODY- AND GENE-DIRECTED ENZYME PRODRUG THERAPY (ADEPT AND GDEPT)

Antibody-directed enzyme prodrug therapy (ADEPT) involves two steps.[9–11] The first step is to deliver a chosen enzyme to the surface of cancer cells by a monoclonal antibody (mAb) followed by a second step—administration of a prodrug that is activated by the enzyme to release the active parent drug. GDEPT also involves two steps.[6–8] In the first step, a gene encoding a chosen enzyme is transported to cancer cells. After expression of the chosen enzyme, the prodrug that is activated by the enzyme is administered. In ADEPT, the enzyme is attached to the surface of cells, while in GDEPT, the enzyme can be expressed intracellularly or extracellularly.[99,100] The potential advantages of extracellular expression of the enzyme are twofold. First, it should produce an improved bystander effect because the active drug will be generated in the interstitial spaces within the tumor, rather than inside, as with an intracellularly expressed activating enzyme. Second, the prodrug does not need to enter cells to become activated; therefore, non-cell-permeable prodrugs can be used.

Enzymes used for ADEP or GDEPT approaches can be divided into three major classes: (I). enzymes of nonmammalian origin that have no mammalian homologs; (II). enzymes of nonmammalian origin with a mammalian homolog; and (III). enzymes of mammalian origin.[11] Examples of enzymes in class I include carboxypeptidase G2 (CPG2), β-lactamase, penicillin G amidase, and cytosine deaminase. The rationale for using such enzymes is that prodrugs can be designed to be nontoxic and not substrates for endogenous human enzymes. Since many of the nonmammalian enzymes are bacterial or are easily expressed in bacteria, they are available in large quantities. The main disadvantage is that they elicit immune responses in humans. The criterion for selecting enzymes in class II is that the activities of the endogenous counterparts of these enzymes should be low. Examples of class II enzymes include β-glucuronidase and nitroreductase. One advantage of employing class II enzymes is that enzymes with a higher catalytic rate can be selected. For example, bacterial β-glucuronidase exhibits a higher turnover rate than its human counterpart. Additionally, the exogenous enzyme could differ significantly enough from the endogenous counterpart that a prodrug can be designed to be cleaved only by the exogenous enzyme. One example is bacterial nitroreductase. As with the class I enzyme, class II enzyme also has limitations due to immunogenicity. The major advantage of using class III enzymes is that they are much less immunogenic than bacterial or fungal enzymes. The obvious disadvantage is that the designed prodrugs can also be activated by endogenous enzymes, resulting in nonspecific activation. Enzymes belonging to this class include alkaline phosphatase, carboxypeptidase A, β-glucuronidase, and cytochrome P450.

Clearly, in both ADEPT and GDEPT, the activity of the enzyme activating the prodrug is enhanced at tumor sites, leading to increased tumor selectivity of the prodrug. Compared with the prodrug therapy discussed earlier that is defined as "one-step prodrug therapy" or "prodrug monotherapy," ADEPT and GDEPT are "called two-step prodrug therapy."[4]

11.3.1. ADEPT

Monoclonal antibodies (mAbs) have been used to deliver chemotherapeutic drugs,[101,102] potent plant and bacterial toxins,[103] and radionuclides[104] to tumor sites. A number of mAb-based drugs have been approved clinically (Rituxan, Herceptin, and Panorex), and several others are now in advanced clinical trials. However, to date, no clinically approved mAb-based drugs are available for solid tumors. The major challenge in developing mAb-based drug therapy for solid tumors lies, in part, in the barriers to macromolecule penetration within the tumor masses and the heterogeneity of target antigen expression.[11] This has prompted research to find an alternative strategy that can dissociate the drug from the mAb delivery system after it reaches the target. ADEPT was a result of this effort. In the ADEPT approach, the activated anticancer drug is not linked to the antibody and can penetrate tumor masses through diffusion. Further, while the concentration of an anticancer drug is limited by the carrying capacity of the mAb-anticancer drug conjugate drug delivery system, it is expected to improve in the ADEPT since in ADEPT system, an active drug is not carried to the site, but rather is continually generated from the prodrug through enzymatic activation.

The general procedure for ADEPT is as follows. A chosen enzyme is conjugated to a tumor-specific mAb or a mAb fragment to form a mAb-enzyme conjugate. Formation of this conjugate can be achieved through chemical linkage of the mAb and the enzyme or through recombinant technology.[11] After administration, the conjugate will bind to the corresponding antigen, which is located on the surface of tumor cells. A clearance period is given to allow the non-antigen-bound mAb-enzyme conjugate to be removed from the body before a prodrug is administered. The administered prodrug will now be selectively activated by the enzyme attached to the surface of the cancer cells. One advantage of ADEPT is that since the enzyme is placed on the surface of cancer cells, the prodrug does not need to enter the cells for activation. Therefore, this approach will allow prodrugs that cannot enter cells to be activated. However, there are clear disadvantages of ADEPT, as discussed by de Groot and colleagues.[4] The main problem is the immunogenicity of the antibody-enzyme conjugate. Also, the mAb-enzyme conjugate does not always localize to the desired extent, and whole antibodies penetrate tumors poorly. Further, tumor cells express only a limited number of antigens. The clearance of unbound mAb-enzyme conjugate is also a problem. The unbound mAb-enzyme conjugate is often inadequately cleared, resulting in activation of the prodrug at nontumor sites and leading to systemic toxicity. Finally, antigen expression levels vary among individuals, which affects the efficacy of the prodrug. Nevertheless, ADEPT is an interesting approach in increasing the tumor selectivity of anticancer prodrugs. In the following discussion, carboxypeptidase G2 (CPG2) and β-lactamase will be used as examples to illustrate the application of ADEPT.

11.3.1.1. ADEPT with CPG2 CPG2 is a bacterial zinc-dependent metalloproteinase.[7] This enzyme is an exoprotease that specifically cleaves terminal glutamic acid amides.[11] Niculescu-Duvaz et al. describe an ADEPT system that uses a combination of mAb, CPG2, and nitrogen mustard prodrugs.[7] CPG2 is coupled to $F(ab')_2$ fragments of the mAb A5B7. A5B7 is an anti-human carcinoembryonic antigen (CEA) mouse mAb. The antibody enzyme conjugate [$F(ab')_2$- CPG2 conjugate] was investigated for its ability to activate the prodrug (2-chloroethyl)-(2-mesyloxygen)aminobenzoyl-L-glutamic acid (CMDA) and other *N*-L-glutamyl amide nitrogen mustard prodrugs. The general structural features of the prepared *N*-L-glutamyl amide nitrogen mustard prodrugs are shown in Scheme 26.

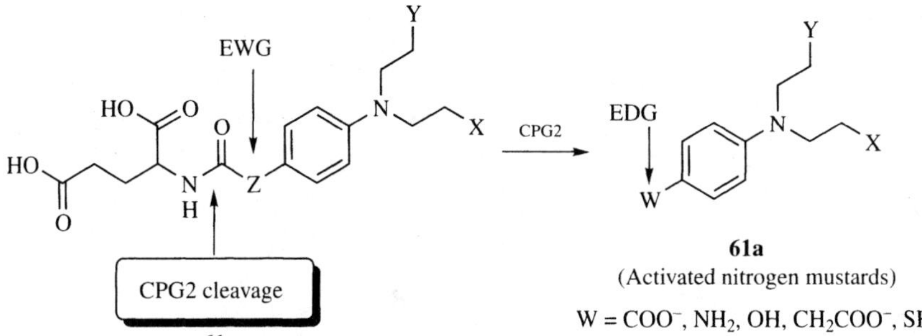

CPG2 cleavage

61

(*N*-L-glutamyl amide nitrogen mustard prodrugs)

X, Y = Cl, Br, I, OSO_2CH_3

Z = –, O, NH_2, CH_2, S

CMDA: X = Cl, Y = OSO_2CH_3, Z = -.

61a

(Activated nitrogen mustards)

W = COO^-, NH_2, OH, CH_2COO^-, SH.

Scheme 26. General structures of *N*-L-glutamyl amide nitrogen mustard prodrugs and activation of the prodrugs by CPG2.

The bond to be cleaved by CPG2 is the amide bond derived from L-glutamic acid. An additional advantage of including the L-glutamic acid moiety in the prodrug is the increased hydrophilicity of the molecule. As mentioned earlier, in ADEPT an activating enzyme is anchored on the outer membrane of tumor cells. It is advantageous if the prodrug is more hydrophilic than the parent drug since high hydrophilicity reduces the ability of the prodrug to cross the cell membrane. Therefore, the distribution of the prodrug into cells will be decreased, reducing the side effects to normal cells. The original work was conducted by Springer et al., who used nude mice implanted with chemoresistant choriocarcinoma xenografts as an animal model for the study of CMDA (Scheme 26).[105] It was demonstrated that 9 of 12 mice were long-time survivors (>300 days), whereas all control mice were dead by day 111. A clinical trial was also conducted in patients with advanced colorectal carcinoma of the lower intestinal tract.[9,106] A dose of 20,000 enzyme units/m^2 mAb-CPG2 conjugate was administered in the first step. This conjugate dose produced tumor CPG2 levels comparable to those found to be optimum in nude mice bearing xenografts. This treatment was followed 24–48 hours later by administration of a clearing agent, anti-CPG2 galactosylated Ab (220 mg/m^2), to help remove mAb-CPG2 conjugates that were not bound to the antigen. In the last step, prodrug CMDA was injected over 1–5 days up to a total dose of 1.2–10 g/m^2.

Oral cyclosporine was coadministered to suppress the host immune response. From eight evaluable patients, there were four partial responses and one mixed response. A more recent clinical trial involving patients with colorectal carcinoma expressing CEA using a similar approach demonstrates that the median tumor: plasma ratio of enzyme (CPG2) exceeded 10,000:1 at the time of prodrug administration. Enzyme concentrations in the tumor were sufficient to generate cytotoxic levels of the active drug.[107] A tumor response to the ADEPT was observed.[107] Thus the results obtained are encouraging and prove the feasibility of the ADEPT at the clinical level.

11.3.1.2. ADEPT with β-Lactamase Activation of β-lactam-based prodrugs is based on the well-established β-lactam chemistry. It was demonstrated that a molecule attached to the 3′ position of cephalosphorins was eliminated through a 1,4-fragmentation reaction.[11] Scheme 27 shows a general scheme of this reaction,

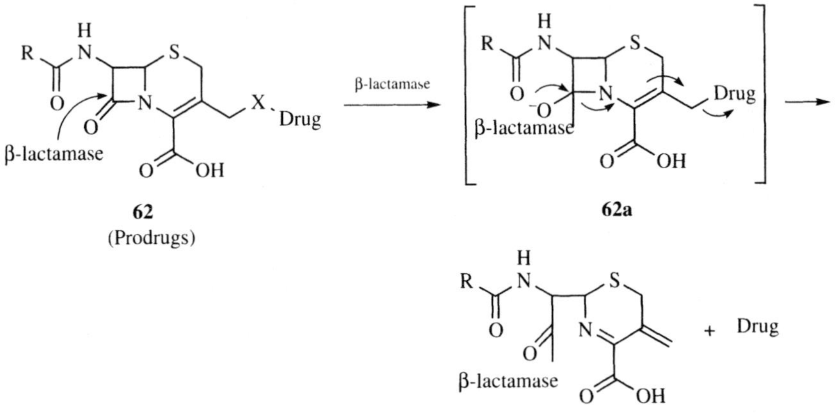

Scheme 27. Bioactivation of prodrugs by β-lactamase.

and Figure 11.11 shows the structures of the prodrugs of a vinca derivative (**63**), phenylenediamine mustard (**64**), doxorubicin (**65**), melphalan (**66**), paclitaxel (**67**), and mitomycin (**68**) prepared based on this chemistry.[11] The first report of *in vivo* activity in a mAb-lactamase system used the β-lactamase from *E. cloacae* and a cephalosphorin-vinca alkaloid prodrug LY266070 (**63**, Figure 11.11).[108] Nude mice implanted with human colorectal carcinoma were used for the investigation. The tumor inhibitory effects of the ADEPT with mAb-lactamase conjugate and LY266070 were superior to those produced by prodrug alone. The effects were also superior to those obtained when the vinca alkaloid (Figure 11.11) was attached directly to the mAb. Long-term regression of established tumors was observed in several dosing regimens, even in animals having tumors as large as 700 mm^3 at the initiation of therapy.[108]

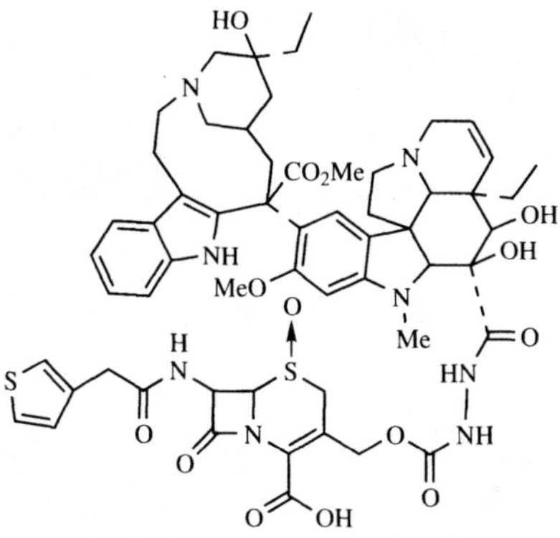

63

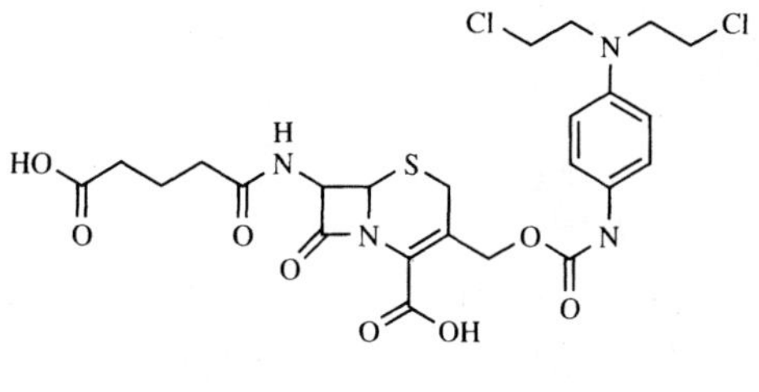

64

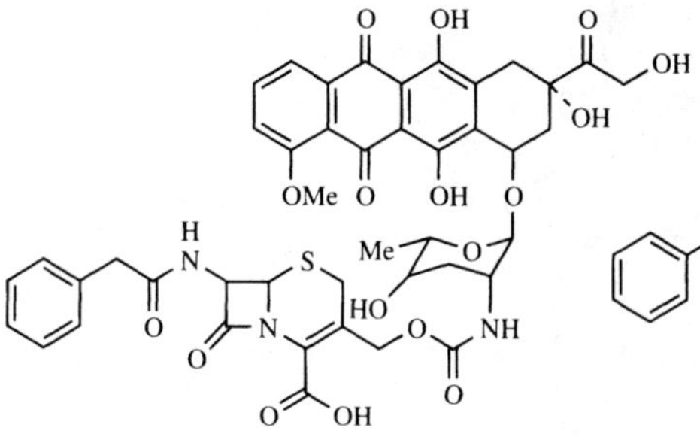

65

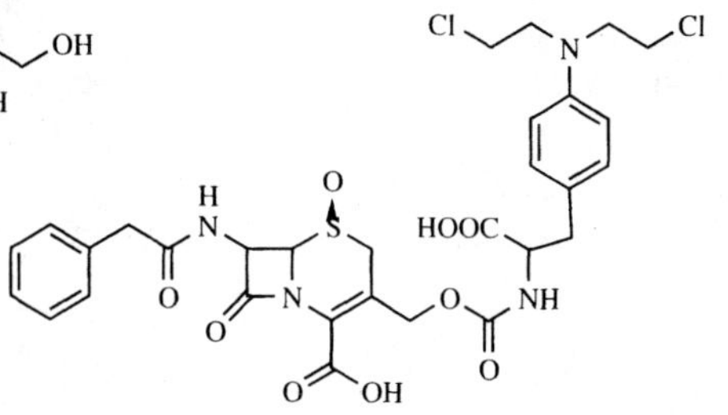

66

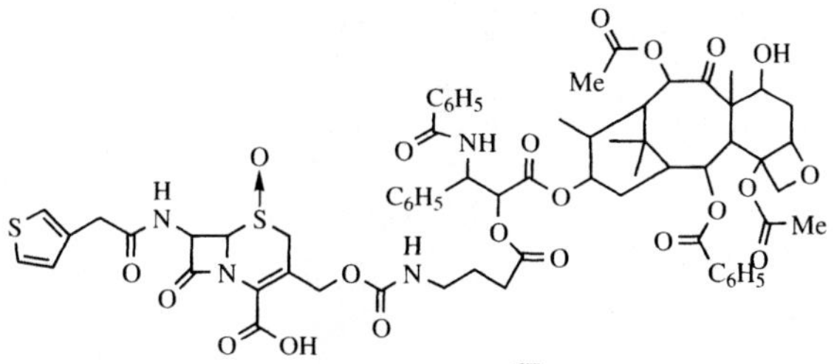

67

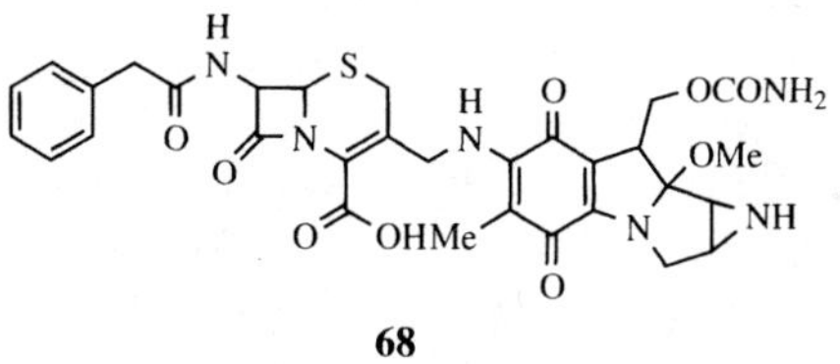

68

Figure 11.11. Structures of representative prodrugs that are activated by β-lactamase.

In a related study, Kerr et al. reported that ADEPT treatment with a combination of a mAb-lactamase conjugate and a prodrug of cephalosphorin-phenylenediamine mustard (**64**) in nude mice bearing subcutaneous 3677 human tumor xenografts produced regression in all the treated mice at doses that caused no apparent toxicity.[109] At day 120 post tumor implant, four of five mice who had received this treatment remained tumor free. Significant antitumor effects were even seen in mice that had large (800 mm^3) tumors before the first prodrug treatment.

11.3.2. GDEPT

Unlike ADEPT, the major problem with GDEPT is delivery of the gene to the tumor—the same problem associated with cancer gene therapy. Almost all clinical trials currently being performed for cancer gene therapy rely on direct intratumor injection of the vector. This is not only a major limitation for uniform distribution of the delivered gene throughout the primary tumor, but it also fails to address the problem of delivery to sites of metastatic spread, which may be too numerous, invisible, or inaccessible for direct injection. Efforts to overcome this problem include engineering adenoviral vectors to alter their tropism for infection,[110] use of cationic liposomes with surface ligands that are specific for tumors,[111] and use of live attenuated *Salmonella* that can preferentially accumulate in tumors by an as yet unknown mechanism.[112,113] Discussion of these efforts is beyond the scope of this chapter.

A closely related approach to GDEPT is virus-directed enzyme prodrug therapy (VDEPT).[114] The only difference between these approaches is that GDEPT involves both viral and nonviral vectors. For a more extensive discussion of GDEPT, readers are referred to earlier reviews.[6–8] In this chapter, GDEPT will be introduced through two representative examples.

11.3.2.1. Hypoxia-Selective GDEPT A very promising approach to deliver the gene to solid tumors was recently described by Liu and colleagues.[115] This approach takes advantage of the hypoxic/necrotic condition of solid tumors. The bacterial genus *Clostridium* comprises a large and heterogeneous group of gram-positive, spore-forming bacteria that become vegetative and grow only in the absence (or low levels) of oxygen. Therefore, strains of these bacteria have been suggested as tools to selectively deliver the gene vector to the hypoxic and necrotic region of solid tumors.[116–118] Liu and coworkers described, for the first time, the successful transformation of *C. sporogenes,* a clostridial strain, with the *Escherichia coli* cytosine deaminase (CD) gene.[115] They showed that intravenous injection of spores of *C. sporogenes* containing an expression plasmid for *E. coli* cytosine deaminase into tumor-bearing mice produced tumor-specific expression of cytosine deaminase protein, an enzyme that converts the nontoxic prodrug 5-fluorocytosine (5-FC) (**69**) to the anticancer drug 5-fluorouracil (5-FU) (**25d**, Scheme 28).

More importantly, significant antitumor efficacy of systemically injected 5-FC was observed following IV injection of these recombinant spores. The antitumor efficacy of the prodrug 5-FC following a single IV injection of the recombinant

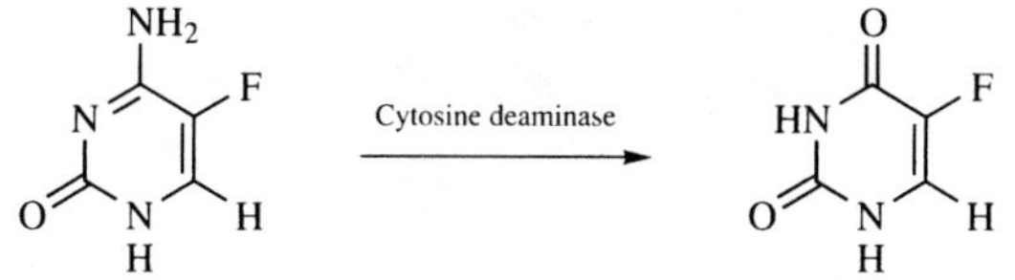

69
5-Fluorocytosine (5-FC)

Cytosine deaminase

25d
5-Fluorouracil (5-FU)

Scheme 28. Bioactivation of 5-FC by cytosine deaminase.

spores is equivalent to or greater than that produced by the maximum tolerated dose of the active drug 5-FU given by the same schedule. A major advantage of the clostridial delivery system is that not only is it tumor specific, it is also safe in humans. As cited by Liu et al.,[115] Mose et al. injected themselves with spores of *C. sporogenes* (a strain later named *C. oncolyticum*) and experienced a mild fever as the only side effect. The safety of *C. oncolyticum* was substantiated in clinical trials with cancer patients who had a variety of solid tumors[119] and also in more recent trials in noncancer patients with a *C. beijerinckii* strain.[120] An additional advantage is that injection of spores did not appear to elicit an immune response.[115]

Another example of hypoxia-selective GDEPT is prodrug **70** (CB 1954, Scheme 29). CB 1954 was originally synthesized over 30 years ago. It exhibits dramatic and

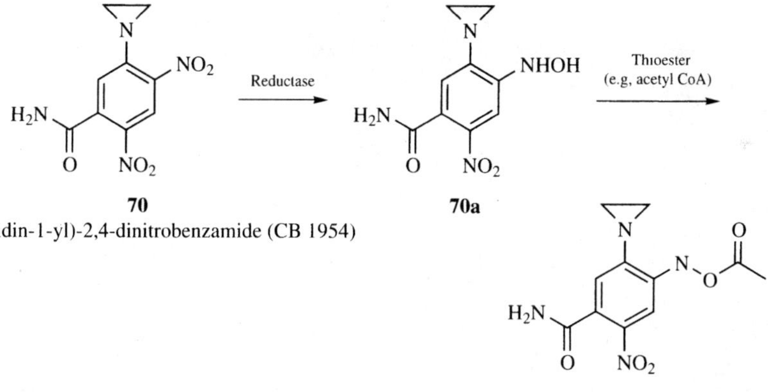

70
5-(Aziridin-1-yl)-2,4-dinitrobenzamide (CB 1954)

Reductase

70a

Thioester
(e.g, acetyl CoA)

70b
(DNA cross-linking agent)

Scheme 29. Bioactivation of CB 1954 by reductases.

highly specific antitumor activity against rat Walker 256 carcinoma cells.[121] The antitumor effect is due to efficient drug activation by rat DT-diaphorase.[122] CB 1954 entered clinical trials in 1970s, but little antitumor activity was observed, as human DT-diaphorase is much less active in the reduction of CB 1954 than the rat enzyme. Recent studies revealed that an amino acid difference at residue 104 between the human and rat enzymes is responsible for the catalytic difference to CB 1954.[123] A nitroreductase (NTR) gene isolated from *E. coli* has been demonstrated to activate the prodrug CB 1954 to its toxic form approximately 90-fold more rapidly than rat DT-diaphorase, suggesting the possibility of using CB 1954 with NTR in ADEPT[124,125] and GDEPT.[126–130]

Shibata et al. successfully transfected human HT 1080 tumor cells with the *E. coli* NTR gene.[131] The transfected human tumor cells expressed *E. coli* NTR in a time- and concentration-dependent manner under a hypoxic condition while expressing only trace levels under an aerobic condition, indicating that the enzyme's expression is induced by hypoxia. No NTR was observed with wild-type HT1080 cells. The expression of NTR conferred increased sensitivity of human tumor cells to CB 1954 both *in vitro* and *in vivo*. The IC_{50} value obtained with the transfected cells was reduced by 40- to 50-fold compared to the IC_{50} value with the wild cells in an *in vitro* experiment under a hypoxic condition. Significantly, no sensitivity difference was observed between the transfected and wild HT 1080 cells under an aerobic condition, consistent with the notion that NTR is induced under hypoxic conditions. A significant tumor growth delay was also observed in mice implanted with transfected clones of HT 1080 cells under a hypoxic condition. Similar to the *in vitro* result, no sensitivity difference to CB 1954 was observed when the *in vivo* experiment was conducted under an aerobic condition.

Other examples of GDEPT include using CPG2 to activate nitrogen mustard,[99,132] DT-diaphorase, and NADPH-cytochrome *c* (P450) reductase to activate mitomycin,[133] CYP2B1 to activate cyclophosphamide and ifosfamide,[94] and pyrimidine nucleoside phosphorylase (PyNPase) to activate 5′-DFUR to 5-FU.[134]

11.4. SUMMARY

The search for tumor-selective prodrugs has been a long and ongoing effort. Numerous approaches have been developed based on the exploitation of the biochemical differences between cancer and normal cells. The differences can be further amplified by the use of ADEPT and GDEPT. Among the various approaches used, anticancer prodrugs based on tumor hypoxia and metastasis appear to be the most extensively explored, and some of these prodrugs have proceeded to clinical trials. ADEPT and GDEPT provide the advantage of increased tumor selectivity. The high immune response and cost are the major disadvantage of these two approaches. However, the advances in genetic technology have prompted very active research in GDEPT which is expected to play a more significant role in tumor-selective anticancer prodrug development. Overall, tumor-selective anticancer prodrugs have been shown to be an effective way to improve therapeutic efficacy and reduce systemic side effects.

REFERENCES

1. Han, H. K.; Amidon, G. L. *AAPS PharmSci.* **2000**, *2*, E6.
2. Bebernitz, G. R.; Dain, J. G.; Deems, R. O.; Otero, D. A.; Simpson, W. R.; Strohschein, R. J. *J. Med. Chem.* **2001**, *44*, 512–523 and reference cited therein.
3. Kwon, C. H. *Arch. Pharm. Res.* **1999**, *6*, 533–541.
4. de Groot F. M.; Damen E. W.; Scheeren, H. W. *Curr. Med. Chem.* **2001a**, *8*, 1093–1122.
5. Denny, W. A. *Eur. J. Med. Chem.* **2001**, *36*, 577–595.
6. Deonarain, M. P.; Spooner, R. A.; Epenetos, A. A. *Gene Ther.* **1995**, *2*, 235–244.
7. Niculescu-Duvaz, I.; Cooper, R. G.; Stribbling, S. M.; Heyes, J. A.; Metcalfe, J. A.; Springer, C. J. *Curr. Opin. Mol. Ther.* **1999**, *1*, 480–486.
8. Aghi, M.; Hochberg, F.; Breakefield, X. O. *J. Gene Med.* **2000**, *2*, 148–164.
9. Bagshawe, K. D.; Sharma, S. K.; Springer, C. J.; Rogers, G. T. *Ann. Oncol.* **1994**, *5*, 879–891.
10. Melton, R. G.; Sherwood, R. F. *J. Natl. Cancer Inst.* **1996**, *88*, 153–165.
11. Senter, P. D.; Springer, C. J. *Adv. Drug Deliv. Rev.* **2001**, *53*, 247–264.
12. Vaupel, P.; Kallinowski, F.; Okunieff, P. *Cancer Res.* **1989**, *49*, 6449–6465.
13. Adams, G. E.; Hasan, N. M.; Joiner, M. C. *Radiother. Oncol.* **1997**, *44*, 101–109.
14. Brown, J. M. *Cancer Res.* **1999**, *59*, 5863–5870.
15. Tannock, I. F.; Rotin, D. *Cancer Res.* **1989**, *49*, 4373–4384.
16. Prezioso, J. A.; Hughey, R. P.; Wang, N.; Damodaran, K. M.; Bloomer, W. D. *Int. J. Cancer* **1994**, *56*, 874–879.
17. Lee, A. E.; Wilson, W. R. *Toxicol. Appl. Pharmacol.* **2000**, *163*, 50–59.
18. Gutierrez, P. L. *Free Radic. Biol. Med.* **2000**, *29*, 263–275.
19. Faig, M.; Bianchet, M. A.; Winski, S.; Hargreaves, R.; Moody, C. J.; Hudnott, A. R.; Ross, D.; Amzel, L. M. *Structure (Camb)* **2001**, *9*, 659–667.
20. Workman, P. *Int. J. Radiat. Oncol. Biol. Phys.* **1992**, *22*, 631–637.
21. Powis, G. *Pharmacol. Ther.* **1987**, *35*, 57–162.
22. Bachur, N. R.; Gordon, S. L.; Gee, M. V.; Kon, H. *Proc. Natl. Acad. Sci. USA* **1979**, *76*, 954–957.
23. Pan, S. S.; Andrews, P. A.; Glover, C. J.; Bachur, N. R. *J. Biol. Chem.* **1984**, *259*, 959–966.
24. Workman, P.; Walton, M. I. In Eds. Adams, G.E., Breccia, A., Fielden, E. M., Wardman, P., *Selective Activation of Drugs by Redox Processes*; Plenum Press; New York, **1990**, pp. 173–191.
25. Sartorelli, A. C. *Cancer Res.* **1988**, *48*, 775–778.
26. Callery, P. S.; Gannett, P. M. In *Cancer and Cancer Chemotherapy*; Williams, D. A., Lemke, T. L., Eds. Lippincott Williams & Wilkins: Philadelphia, 2002, pp. 924–949.
27. Jaffar, M.; Naylor, M. A.; Robertson, N.; Lockyer, S. D.; Phillips, R. M.; Everett, S. A.; Adams, G. E.; Stratford, I. J. *Anticancer Drug Des.* **1998**, *13*, 105–123.
28. Sykes, B. M.; Atwell, G. J.; Hogg, A.; Wilson, W. R.; O'Connor, C. J.; Denny, W. A. *J. Med. Chem.* **1999**, *42*, 346–355.
29. Mauger, A. B.; Burke, P. J.; Somani, H. H.; Friedlos, F.; Knox, R. J. *J. Med. Chem.* **1994**, *37*, 3452–3458

30. Hay, M. P.; Wilson, W. R.; Denny, W. A. *Bioorg. Med. Chem. Lett.* **1999a**, *9*, 3417–3422.
31. Shyam, K.; Penketh, P. G.; Shapiro, M.; Belcourt, M. F.; Loomis, R. H.; Rockwell, S.; Sartorelli, A. C. *J. Med. Chem.* **1999**, *42*, 941–946.
32. Reynolds, R. C.; Tiwari, A.; Harwell, J. E.; Gordon, D. G.; Garrett, B. D.; Gilbert, K. S.; Schmid, S. M.; Waud, W. R.; Struck, R. F. *J. Med. Chem.* **2000**, *43*, 1484–1488.
33. Everett, S. A.; Naylor, M. A.; Patel, K. B.; Stratford, M. R.; Wardman, P. *Bioorg. Med. Chem. Lett.* **1999**, *9*, 1267–1272.
34. Parveen, I.; Naughton, D. P.; Whish, W. J.; Threadgill, M. D. *Bioorg. Med. Chem. Lett.* **1999**, *9*, 2031–2036.
35. Hay, M. P.; Sykes, B. M.; Denny, W. A.; Wilson, W. R. A. *Bioorg. Med. Chem. Lett.* **1999b**, *9*, 2237–2242.
36. Hay, M. P.; Wilson, W.; Denny, W. A. *Tetrahedron* **2000**, *56*, 645–657.
37. Highfield, J. A.; Metha, L. K.; Parrick, J.; Candeias, L. P.; Wardman, P. *J. Chem. Soc. Perkin Trans. I*, **1999**, 2343–2351.
38. Brown, J. M.; Lemmon, M. J. *Cancer Res.* **1990**, *50*, 7745–7749.
39. Patterson, A. V.; Saunders, M. P.; Chinje, E. C.; Patterson, L. H.; Stratford, I. J. *Anticancer Drug Des.* **1998**, *13*, 541–573.
40. Dorie, M. J.; Brown, J. M. *Cancer Res.* **1993**, *53*, 4633–4636.
41. Siemann, D. W.; Hinchman, C. A. *Radiother. Oncol.* **1998**, *47*, 215–220.
42. von Pawel, J.; von Roemeling, R. *Proc. Am. Soc. Clin. Oncol.* **1998**, *17*, 454a.
43. Aiko, I.; Owari, S.; Torigoe, M. *J. Pharm. Soc. (Japan)* **1952**, *72*, 1297–1300.
44. Connors, T. A. In *Structure-Activity Relationships of Antitumor Agents*; Reinhoudt, D. N., Connors, T. A., Pinedo, H. M., van den Poll, K. W., Eds. Martinus Nijhoff: the Hague, **1983**, pp. 47–57.
45. White, I. N.; Suzanger, M.; Mattocks, A. R.; Bailey, E.; Farmer, P. B.; Connors, T. A. *Carcinogenesis* **1989**, *10*, 2113–2118.
46. Wilson, W. R.; Denny, W. A.; Pullen, S. M.; Thompson, K. M.; Li, A. E.; Patterson, L. H.; Lee H. H. *Br. J. Cancer Suppl.* **1996**, *27*, S43–S47.
47. Raleigh, S. M.; Wanogho, E.; Burke, M. D.; McKeown, S. R.; Patterson, L. H. *Int. J. Radiat. Oncol. Biol. Phys.* **1998**, *42*, 763–767.
48. Shibamoto, Y.; Zhou, L.; Hatta, H.; Mori, M.; Nishimoto, S. *Jpn. J. Cancer Res.* **2000**, 91, 433–438.
49. Shibamoto, Y.; Zhou, L.; Hatta, H.; Mori, M.; Nishimoto, S. I. *Int. J. Radiat. Oncol. Biol. Phys.* **2001**, *49*, 407–413.
50. Tietze, L. F.; Neumann, M.; Mollers, T.; Fischer, R.; Glusenkamp, K. H.; Rajewsky, M. F.; Jahde, E. *Cancer Res.* **1989**, *49*, 4179–4184.
51. Liotta, L. A.; Rao, C. N.; Wewer, U. M. *Annu. Rev. Biochem.* **1986**, *55*, 1037–1057.
52. Mignatti, P.; Rifkin, D. B. *Physiol. Rev.* **1993**, *73*, 161–195.
53. Vassalli, J. D.; Pepper, M. S. *Nature* **1994**, *370*, 14–15.
54. Naylor, M. S.; Stamp, G. W.; Davies, B. D.; Balkwill, F. R. *Int. J. Cancer* **1994**, *58*, 50–56.
55. Uria, J. A.; Ferrando, A. A.; Velasco, G.; Freije, J. M.; Lopez-Otin, C. *Cancer Res.* **1994**, *54*, 2091–2094.
56. Putnam, D. A.; Shiah, J. G.; Kopecek, J. *Biochem. Pharmacol.* **1996**, *52*, 957–962.

57. Denmeade, S. R.; Nagy, A.; Gao, J.; Lilja, H.; Schally, A. V.; Isaacs, J. T. *Cancer Res.* **1998**, *58*, 2537–2540.

58. Harada, M.; Sakakibara, H.; Yano, T.; Suzuki, T.; Okuno, S. *J. Control. Release* **2000**, *69*, 399–412.

59. Carl, P. L.; Chakravarty, P. K.; Katzenellenbogen, J. A. *J. Med. Chem.* **1981**, *24*, 479–480.

60. Toki, B. E.; Cerveny, C. G.; Wahl, A. F.; Senter, P. D. *J. Org. Chem.* **2002**, *67*, 1866–1872.

61. de Groot, F. M.; Loos, W. J.; Koekkoek, R.; van Berkom, L. W.; Busscher, G. F.; Seelen, A. E.; Albrecht, C.; de Bruijn, P.; Scheeren, H. W. *J. Org. Chem.* **2001b**, 66, 8815–8830.

62. Markus, G.; Takita, H.; Camiolo, S. M.; Corasanti, J. G.; Evers, J. L.; Hobika, G. H. *Cancer Res.* **1980**, *40*, 841–848.

63. Corasanti, J. G.; Celik, C.; Camiolo, S. M.; Mittelman, A.; Evers, J. L.; Barbasch, A.; Hobika, G. H.; Markus, G. *J. Natl. Cancer Inst.* **1980**, *65*, 345–351.

64. Campo, E.; Munoz, J.; Miquel, R.; Palacin, A.; Cardesa, A.; Sloane, B. F.; Emmert-Buck, M. R. *Am. J. Pathol.* **1994**, *145*, 301–309.

65. Carl, P. L.; Chakravarty, P. K.; Katzenellenbogen, J. A.; Weber, M. J. *Proc. Natl. Acad. Sci. USA* **1980**, *77*, 2224–2228.

66. Cavallaro, G.; Pitarresi, G.; Licciardi, M.; Giammona, G. *Bioconjug. Chem.* **2001**, *12*, 143–151.

67. Chakravarty, P. K.; Carl, P. L.; Weber, M. J.; Katzenellenbogen, J. A. *J. Med. Chem.* **1983a**, *26*, 633–638.

68. Chakravarty, P. K.; Carl, P. L.; Weber, M. J.; Katzenellenbogen, J. A. *J. Med. Chem.* **1983b**, *26*, 638–644.

69. de Groot, F. M.; de Bart, A. C.; Verheijen, J. H.; Scheeren, H. W. *J. Med. Chem.* **1999**, *42*, 5277–5283.

70. Eisenbrand, G.; Lauck-Birkel, S.; Tang, W. C. **1996**, 1246–1258.

71. Elliott, E.; Sloane, B. F. *Persp. Drug Disc. Design* **1996**, *6*, 12.

72. Thomssen, C.; Schmitt, M.; Goretzki, L.; Oppelt, P.; Pache, L.; Dettmar, P.; Janicke, F.; Graeff, H. *Clin. Cancer Res.* **1995**, *1*, 741–746.

73. Horsman, M. R.; Murata, R.; Breidahl, T.; Nielsen, F. U.; Maxwell, R. J.; Stodkiled-Jorgensen, H.; Overgaard, J. *Adv. Exp. Med. Biol.* **2000**, *476*, 311–323.

74. Hande, K. R. *Eur. J. Cancer* **1998**, *34*, 1514–1521.

75. Sperker, B.; Backman, J. T.; Kroemer, H. K. *Clin. Pharmacokinet.* **1997**, *33*, 18–31.

76. Bosslet, K.; Czech, J.; Hoffmann, D. *Tumor Targeting* **1995**, *1*, 45–50.

77. Bosslet, K.; Straub, R.; Blumrich, M.; Czech, J.; Gerken, M.; Sperker, B.; Kroemer, H. K.; Gesson, J. P.; Koch, M.; Monneret, C. *Cancer Res.* **1998**, *58*, 1195–1201.

78. Saha, A. K.; Glew, R. H.; Kotler, D. P.; Omene, J. A. *Clin. Chim. Acta* **1991**, *199*, 311–316.

79. Haisma, H. J.; van Muijen, M.; Pinedo, H. M.; Boven, E. *Cell Biophys.* **1994**, *24–25*, 185–192.

80. Leenders, R. G. G.; Scheeren, H. W.; Houba, P. H. J.; Boven, E.; Haisma, H. J. *Bioorg. Med. Chem. Lett.* **1995**, *5*, 2975–2980.

81. Florent, J. C.; Dong, X.; Gaudel, G.; Mitaku, S.; Monneret, C.; Gesson, J. P.; Jacquesy, J. C.; Mondon, M.; Renoux, B.; Andrianomenjanahary, S.; Michel, S.; Koch, M.; Tillequin, F.; Gerken, M.; Czech, J.; Straub, R.; Bosslet, K. *J. Med. Chem.* **1998**, *41*, 3572–3581.

82. Desbène, S.; Van, H. D.; Michel, S.; Koch, M.; Tillequin, F.; Fournier, G.; Farjaudon, N.; Monneret, C. *Anticancer Drug Des.* **1998**, *13*, 955–968.

83. Pantazis, P.; Giovanella, B. C.; Rothenberg, M. L. *The Camptothecins: From Discovery to the Patient.* New York Academy of Sciences: New York, **1996**.

84. Leu, Y. L.; Roffler, S. R.; Chern, J. W. *J. Med. Chem.* **1999**, *42*, 3623–3628.

85. Schmidt, F.; Florent, J. C.; Monneret, C.; Straub, R.; Czech, J.; Gerken, M.; Bosslet, K. *Bioorg. Med. Chem. Lett.* **1997**, *7*, 1071–1076.

86. Lougerstay-Madec, R.; Florent, J. C.; Monneret, C.; Nemati, F.; Poupon, M. F. *Anticancer Drug Des.* **1998**, *13*, 995–1007.

87. de Bont, D. B.; Leenders, R. G.; Haisma, H. J.; van der Meulen-Muileman, I.; Scheeren, H. W. *Bioorg. Med. Chem.* **1997**, *5*, 405–414.

88. Madec-Lougerstay, R.; Florent, J. C.; Monneret, C. *J. Chem. Soc., Perkin Trans. I*, **1999**, 1369–1375.

89. Denmeade, S. R.; Lou, W.; Lovgren, J.; Malm, J.; Lilja, H.; Isaacs, J. T. *Cancer Res.* **1997**, *57*, 4924–4930.

90. Guengerich, In de Montellano, P. R. Ed. Plenum Press: New York, 19 pp. 473–535.

91. Chang, T. K.; Weber, G. F.; Crespi, C. L.; Waxman, D. J. *Cancer Res.* **1993**, *53*, 5629–5637.

92. Komatsu, T.; Yamazaki, H.; Shimada, N.; Nakajima, M.; Yokoi, T. *Drug Metab. Dispos.* **2000**, *28*, 1457–1463.

93. Wei, M. X.; Tamiya, T.; Chase, M.; Boviatsis, E. J.; Chang, T. K.; Kowall, N. W.; Hochberg, F. H.; Waxman, D. J.; Breakefield, X. O.; Chiocca, E. A. *Hum. Gene Ther.* **1994**, *5*, 969–978.

94. Chen, L.; Waxman, D. J. *Cancer Res.* **1995**, *55*, 581–589.

95. Chen, L.; Waxman, D. J.; Chen, D.; Kufe, D. W. *Cancer Res.* **1996**, *56*, 1331–1340.

96. Jounaidi, Y.; Hecht, J. E.; Waxman, D. J. *Cancer Res.* **1998**, *58*, 4391–4401.

97. Huang, Z.; Waxman, D. J. *Cancer Gene Ther.* **2001**, *8*, 450–458.

98. Huang, Z.; Raychowdhury, M. K.; Waxman, D. J. *Cancer Gene Ther.* **2000**, *7*, 1034–1042.

99. Marais, R.; Spooner, R. A.; Light, Y.; Martin, J.; Springer, C. J. *Cancer Res.* **1996**, *56*, 4735–4742.

100. Marais, R.; Spooner, R. A.; Stribbling, S. M.; Light, Y.; Martin, J.; Springer, C. J. *Nat. Biotechnol.* **1997**, *15*, 1373–1377.

101. Dubowchik, G. M.; Walker, M. A. *Pharmacol. Ther.* **1999**, *83*, 67–123.

102. Chari, R. V. *Adv. Drug Deliv. Rev.* **1998**, *31*, 89–104

103. Brinkmann, U. *In Vivo* **2000**, *14*, 21–27.

104. Illidge, T. M.; Johnson, P. W. *Br. J. Haematol.* **2000**, *108*, 679–688.

105. Springer, C. J.; Bagshawe, K. D.; Sharma, S. K.; Searle, F.; Boden, J. A.; Antoniw, P.; Burke, P. J.; Rogers, G. T.; Sherwood, R. F.; Melton, R. G. *Eur. J. Cancer* **1991**, *27*, 1361–1366.

106. Bagshawe, K. D.; Sharma, S. K.; Springer, C. J.; Antoniw, P.; Boden, J. A.; Rogers, G. T.; Burke, P. J.; Melton, R. G.; Sherwood, R. F. *Dis. Markers* **1991**, *9*, 233–238.

107. Napier, M. P.; Sharma, S. K.; Springer, C. J.; Bagshawe, K. D.; Green, A. J.; Martin, J.; Stribbling, S. M.; Cushen, N.; O'Malley, D.; Begent, R. H. *Clin. Cancer Res.* **2000**, *6*, 765–772.

108. Meyer, D. L.; Jungheim, L. N.; Law, K. L.; Mikolajczyk, S. D.; Shepherd, T. A.; Mackensen, D. G.; Briggs, S. L.; Starling, J. J. *Cancer Res.* **1993**, *53*, 3956–3963.

109. Kerr, D. E.; Schreiber, G. J.; Vrudhula, V. M.; Svensson, H. P.; Hellstrom, I.; Hellstrom, K. E.; Senter, P. D. *Cancer Res.* **1995**, *55*, 3558–3563.

110. Krasnykh, V.; Dmitriev, I.; Navarro, J. G.; Belousova, N.; Kashentseva, E.; Xiang, J.; Douglas, J. T.; Curiel, D. T. *Cancer Res.* **2000**, *60*, 6784–6787.

111. Xu, L.; Pirollo, K. F.; Tang, W. H.; Rait, A.; Chang, E. H. *Hum. Gene Ther.* **1999**, *10*, 2941–2952.

112. Pawelek, J. M.; Low, K. B.; Bermudes, D. *Cancer Res.* **1997**, *57*, 4537–4544.

113. Low, K. B.; Ittensohn, M.; Le, T.; Platt, J.; Sodi, S.; Amoss, M.; Ash, O.; Carmichael, E.; Chakraborty, A.; Fischer, J.; Lin, S. L.; Luo, X.; Miller, S. I.; Zheng, L.; King, I.; Pawelek, J. M.; Bermudes, D. *Nat. Biotechnol.* **1999**, *17*, 37–41.

114. Kirn, D.; Niculescu-Duvaz, I.; Hallden, G.; Springer, C. J. *Trends Mol. Med.* **2002**, *8*(4 Suppl):S68–S73.

115. Liu, S. C.; Minton, N. P.; Giaccia, A. J.; Brown, J. M. *Gene Ther.* **2002**, 9, 291–296.

116. Minton, N. P.; Mauchline, M. L.; Lemmon, M. J.; Brehm, J. K.; Fox, M.; Michael, N. P.; Giaccia, A.; Brown, J. M. *FEMS Microbiol. Rev.* **1995**, *17*, 357–364.

117. Fox, M. E.; Lemmon, M. J.; Mauchline, M. L.; Davis, T. O.; Giaccia, A. J.; Minton, N. P.; Brown, J. M. *Gene Ther.* **1996**, *3*, 173–178.

118. Lemmon, M. J.; van Zijl, P.; Fox, M. E.; Mauchline, M. L.; Giaccia, A. J.; Minton, N. P.; Brown, J. M. *Gene Ther.* **1997**, *4*, 791–796.

119. Heppner, F.; Mose, J.; Ascher, P. W.; Walter, G.; *13th Int. Cong. Chemother.* **1983**, *226*, 38–45.

120. Fabricius, E. M.; Schneeweiss, U.; Schau, H.; Schmidt, W.; Benedix, A. *Res. Microbiol.* **1993**, *144*, 741–753.

121. Cobb, L. M.; Connors, T. A.; Elson, L. A.; Khan, A. H.; Mitchley, B. C.; Ross, W. C.; Whisson, M. E. *Biochem. Pharmacol.* **1969**, *18*, 1519–1527.

122. Knox, R. J.; Boland, M. P.; Friedlos, F.; Coles, B.; Southan, C.; Roberts, J. J. *Biochem. Pharmacol.* **1988**, *37*, 4671–4677.

123. Chen, S.; Knox, R.; Wu, K.; Deng, P. S.; Zhou, D.; Bianchet, M. A.; Amzel, L. M. *J. Biol. Chem.* **1997**, *272*, 1437–1439.

124. Anlezark, G. M.; Melton, R. G.; Sherwood, R. F.; Coles, B.; Friedlos, F.; Knox, R. J. *Biochem. Pharmacol.* **1992**, *44*, 2289–2295.

125. Knox, R. J.; Friedlos, F.; Sherwood, R. F.; Melton, R. G.; Anlezark, G. M. *Biochem. Pharmacol.* **1992**, *44*, 2297–2301.

126. Bridgewater, J. A.; Springer, C. J.; Knox, R. J.; Minton, N. P.; Michael, N. P.; Collins, M. K. *Eur. J. Cancer* **1995**, *31A*, 2362–2370.

127. Clark, A. J.; Iwobi, M.; Cui, W.; Crompton, M.; Harold, G.; Hobbs, S.; Kamalati, T.; Knox, R.; Neil, C.; Yull, F.; Gusterson, B. *Gene Ther.* **1997**, *4*, 101–110.

128. Drabek, D.; Guy, J.; Craig, R.; Grosveld, F. *Gene Ther.* **1997**, *4*, 93–100.

129. McNeish, I. A.; Green, N. K.; Gilligan, M. G.; Ford, M. J.; Mautner. V.; Young, L. S.; Kerr, D. J.; Searle, P. F. *Gene Ther.* **1998**, *5*, 1061–1069.

130. Connors, T. A. *Gene Ther.* **1995**, *2*, 702–709.

131. Shibata, T.; Giaccia, A. J.; Brown, J. M. *Neoplasia* **2002**, *4*, 40–48.

132. Friedlos, F.; Davies, L.; Scanlon, I.; Ogilvie, L. M.; Martin, J.; Stribbling, S. M.; Spooner, R. A.; Niculescu-Duvaz, I.; Marais, R.; Springer, C. J. *Cancer Res.* **2002**, *62*, 1724–1729.

133. Baumann, R. P.; Hodnick, W. F.; Seow, H. A.; Belcourt, M. F.; Rockwell, S.; Sherman, D. H.; Sartorelli, A. C. *Cancer Res.* **2001**, *61*, 7770–7776.

134. Nagata, T.; Nakamori, M.; Iwahashi, M.; Yamaue, H. *Eur. J. Cancer* **2002**, *38*, 712–717.

12

ULTRASOUND-MEDIATED DRUG DELIVERY

Ka-yun Ng*

Department of Pharmaceutical Sciences, School of Pharmacy, University of Colorado Health Sciences Center, Denver, CO 80262

Terry O. Matsunaga

Imarx Therapeutics Inc., Tucson, AZ 85719

*To whom correspondence and reprint requests should be addressed at the Department of Pharmaceutical Sciences, University of Colorado Health Sciences Center, Campus Box C-238, 4200 East Ninth Avenue, Denver, CO 80262. Phone: 303-315-6997; Fax: 303-315-0274; e-mail: Lawrence.Ng@UCHSC.edu.

Drug Delivery: Principles and Applications Edited by Binghe Wang, Teruna Siahaan, and Richard Soltero
ISBN 0-471-47489-4

12.1. INTRODUCTION

Ultrasound is best known for its imaging capability in diagnostic medicine. However, there have been considerable efforts recently to develop therapeutic uses for it. The purpose of this review is to summarize some of the recent advances made in the area of therapeutic ultrasound as they relate to drug delivery. The review will be divided into three sections. In the first section, we will briefly discuss the physical principles of ultrasound, as well as provide an overview of the history of the development of ultrasound in medicine. This will be followed by an in-depth analysis of ultrasound applications in drug delivery, with special emphases placed on the mechanistic effects of ultrasound in enhancing the delivery and efficacy of the following therapeutic drug classes: chemotherapeutic, thrombolytic, protein-, and gene-based drugs. Finally, because recent experimental evidence suggests that ultrasound contrast agents can be used as exogenous cavitation nuclei for enhancement of drug and gene delivery, we will conclude our review with an in-depth examination of this new class of agents and their roles in drug delivery.

In selecting studies for this review, only scientific journals that are published in English were used. As such, some of the interesting and important ultrasound work performed in Germany, Japan, and Russia were not included. We have also emphasized findings from experiments in which the acoustical parameters are in ranges of medical ultrasound as defined in NCRP Report No. 74.[1] These refer to frequencies in the megahertz (MHz) range (0.5 to 10 MHz) and average intensity between that of diagnostic and surgical applications of ultrasound. However, deviation from this norm may occur from time to time in the review in order to illustrate some recent and exciting findings on the use of low-frequency ultrasound (in the range of

20 kHz) for drug delivery purposes. It should also be noted that this review is not written with a specific route of drug administration in mind. Rather, many drug administration routes are selected for discussion in order to demonstrate the broad application of ultrasound for drug delivery purposes.

12.2. BACKGROUND

12.2.1. Basic Physics of Sound

Sound is our experience of the propagation of pressure waves through some physical elastic medium, such as air or liquid. The pressure waves are generated from mechanical vibrations, which become vibrating pressure waves, transferring energy to the medium and to the objects that the wave contacts. Because of this transfer of energy, the intensity of the sound energy is progressively decreased during its propagation. The transfer and absorption of sound energy by the objects in the path of sound propagation also causes a temperature increase. As can be seen later, this rise in temperature during medical application of ultrasound is referred to as the "thermal effect" of ultrasound.[2] Prediction of actual temperature increase produced by ultrasound is often difficult. It requires the knowledge of not only the acoustic absorption coefficients but the conduction and convection properties of the tissues involved.[2] Since discussion of these topics will be too complicated for this review, readers are referred to other technical resources for in depth review of this subject.[3,4]

"Ultrasound" refers to sound waves with frequencies above 20 kHz, which are beyond the upper limit of the human audible range.[1,5–9] It is commonly classified as a nonionizing radiation along with microwaves and radiowaves. Unlike microwaves and radiowaves, which are electromagnetic radiations, ultrasound is completely mechanical in nature. However, like a beam of electromagnetic radiation, ultrasound can be reflected, refracted, diffracted, and focused. Ultrasound is generated by a transducer, which converts electrical energy to acoustical energy or vice versa. The phenomenon of conversion of mechanical acoustical energy to electricity or vice versa is commonly referred to as "piezoelectricity."[10] In this regard, nearly all transducers use piezoelectric materials.[11] These materials can be either natural crystalline solid, such as quartz, or manufactured piezoceramics, such as barium titanate and lead zirconate titanate. To produce ultrasound, a suitable voltage is applied to the transducer. When the frequency of the input voltage reaches the resonance frequency for thickness vibrations of the piezoelectric material, the piezoelectric material responds by undergoing vibrations. The vibrations are then transmitted to the environment as ultrasonic pressure waves. As noted, this phenomenon is entirely reversible. Thus, the same piezoelectric material can be used as both a transmitter and a receiver of ultrasonic waves. This property forms the basis of future ultrasound applications such as SONAR (sound navigation and ranging) and other diagnostic techniques.[10]

12.2.2. History of the Development of Ultrasound in Medicine

Since ultrasound in general follows the principles delineated in acoustics, its development naturally reflects the developments in acoustics. For the purpose of this review, only some of the major developments in acoustics and ultrasonics will be highlighted. For in-depth review of this area, readers are referred to a website developed by Dr. Joseph Woo,[12] which contains some of the best information on the history of the study of acoustics and biomedical application of ultrasound.

Galileo Galilei (1564–1642) is often said to have started the modern studies of acoustics. But several ancient philosophers played important role in laying the groundwork for future study of acoustics. As early as the sixth century BC, Pythagoras had studied the properties of vibrating strings. The study was so popular that it led to the development of a tuning system that bears his name. Two centuries later, Aristotle correctly assumed that a sound wave resonates in air through motion of the air. Soon after that, a Roman architect named Vitruvius (first century BC) correctly determined the mechanism for the movement of sound waves that contributed significantly to the acoustic design of theatres. And in the sixth century AD, Boethius, a Roman philosopher, suggested that the human perception of pitch is related to the physical property of frequency. Galilei further elevated the study of vibrations by correlating the relationship between pitch and frequency using strict scientific standards. Then in 1877, the field of acoustics was catapulted to new heights with the publication of the famous treatise "The Theory of Sound" in England by Lord Rayleigh, in which he delineated the fundamental physics of sound vibrations (waves), transmission, and reflections.

Perhaps the most important breakthrough in ultrasound technique came in 1880 when the piezoelectric effects in certain crystals were discovered by the French scientist Pierre Curie and his brother Jacques Curie.[10] They observed that an electric potential was produced when mechanical pressure was exerted on a quartz crystal such as Rochelle salt; conversely, the application of an electric charge produced deformation in the crystal, causing it to vibrate. This work forms the basis of echo-sounding technique and lays the foundation for future ultrasound applications such as SONAR and other diagnostic techniques. Basic to this technique is a "pulse-echo" operation in which short pulses of ultrasound are emitted from one or more transmitting transducers. Upon impinging on an object, pulses are reflected as echo waves and detected by receiving transducers. The first large-scale application of SONAR was in antisubmarine warfare. In the late 1940s and mid-1950s, pulse-echo methods (similar to those used in SONAR) were introduced and applied to diagnostic medicine. This effort came, in part, from the desire of physicians to reduce dependence on other ionizing methods, which were viewed as being unsafe for imaging purposes, such as X-ray.

Interestingly, applications of ultrasound as a method of therapy predated diagnostic ultrasound, utilizing its heating and disruptive effects on animal tissues. The destructive ability of high-intensity ultrasound had been recognized in the 1920s from the time of Langevin (an eminent French physicist) when he noted destruction of schools of fishes in the sea and pain induced in the hand when placed

TABLE 12.1 Exposure Intensity Levels in Medical Applications

Category	Typical Intensity Range
Surgical	>10 W/cm^2
Therapeutic	0.5–3 W/cm^2
Diagnostic	0.0001–0.5 W/cm^2

Source: Reproduced from Ref. 1.

in a water tank insonated with high-intensity ultrasound. In the 1930s and 1940s, ultrasound was introduced as a method of therapy in Europe, wherein irradiation with ultrasound was used to generate heat. Thus, tissue deep in the body can be warmed by this means. In the 1940s, ultrasound had slowly evolved to become a neurosurgical tool, being used to destroy part of the basal ganglia in patients with parkinsonism. Ultrasonics was also extensively used in physical and rehabilitation therapy. The earliest reported use of ultrasound as a physical therapy in the United States was at the University of Colorado in 1953 by Jerome Gersten for treatment of patients with rheumatoid arthritis. For the past 50 years or so, ultrasound has been accepted as a useful tool in physical therapy. It should be noted that, while the intensity of ultrasound used for therapeutic (such as physical therapy) and diagnostic purposes is considered noninvasive, the intensity levels needed for such applications differ.[1] In the case of diagnostic ultrasound, intensities in the range of 0.0001 to 0.5 watts (W)/cm^2 are usually used (see Table 12.1). This is substantially less than the intensity needed for therapeutic (0.5 to 3 W/cm^2) or surgically based ultrasound (>10 W/cm^2) applications.

12.2.3. Therapeutic Applications of Ultrasound in Drug Delivery

The use of ultrasound to promote drug delivery was first reported by Fellinger and Schmid,[13] who developed a successful treatment for polyarthritis by using ultrasound to drive hydrocortisone ointment into the inflamed tissues. The technique of driving molecules of drug across the percutaneous barrier to the target area using ultrasonic perturbation is termed "sonophoresis" or "phonophoresis." Since then, a wide variety of drug/ultrasound combinations have been implemented for sonophoresis.[14] Most recently, ultrasound application has been used to promote delivery of high molecular weight proteins through intact skin.[15–17] In addition to sonophoresis research, ultrasound has been shown to enhance the effects of several therapeutic drug classes, including chemotherapeutic, thrombolytic, and gene-based drugs. Furthermore, ultrasound contrast agents, which were originally developed for diagnostic ultrasound, have been shown to augment the delivery and effectiveness of certain drugs. These ultrasound contrast agents can also be used as drug carriers for responsive and targeted drug delivery in the presence of ultrasound insonation.[18,19] In the following sections, we will offer an in-depth analysis of ultrasound applications in drug delivery of chemotherapeutic, thrombolytic, protein-, and gene-based drugs, and some of the recent applications of ultrasound contrast agents in drug delivery.

12.3. ULTRASOUND APPLICATIONS IN DRUG DELIVERY

12.3.1. Chemotherapeutic Drugs

12.3.1.1. Ultrasound in Cancer Therapy The use of ultrasound to enhance cancer therapy has been the subject of numerous biological and clinical investigations. In most of these studies, ultrasound has been used as an agent to induce hyperthermia for either direct treatment of small and localized cancerous tumors[20–27] or as adjuvant therapy to increase the efficacy of radiotherapy[28–30] and chemotherapy.[31–41] In the first application, acoustic intensities of thousands of W/cm^2 and temperatures in excess of 98°C are often used to coagulate cancerous tissues.[20,21,42] This form of hyperthermia is often referred to as "thermal therapy" because of its extreme temperature and energy.[43] Alternatively, lower ultrasound intensities (0.2 to several W/cm^2) produce a mild increase in temperature (41 to 45°C), and enhance the cytotoxicity of radiation therapy[28–30] and chemotherapy.[31,33,36,40,41] The enhancing effect of ultrasound on radiation therapy has been linked to the radiosensitization effect of hyperthermia, which increases radiation damage and prevents subsequent repair.[28] However, the precise mechanism for ultrasound-enhanced chemotoxicity is still the subject of debate.

12.3.1.2. Mechanisms of Ultrasound-Enhanced Chemotoxicity The mechanisms proposed to account for ultrasound-enhanced chemotoxicity include (1) (nonthermal) and (2) thermal mechanisms. The nonthermal effects of ultrasound may refer to conditions wherein ultrasound enhances chemotoxicity at or below normal physiological temperatures.[32–35,37–39] In experiments carried out at above physiological temperatures (i.e., >37°C), it can also refer to chemotoxicity enhancement that cannot be accounted for by the effects of heat alone (see Table 12.2).[31,33,36,40,41] The nonthermal effects of ultrasound on a biological system can occur via one or a combination of three mechanisms: (1) cavitation, (2) radiation pressure, and (3) acoustic microstreaming.[1,7,44] "Cavitation" refers to activity associated with small gaseous bubbles. Radiation pressure is time-averaged pressure elevations at any point in a medium traversed by ultrasound. They can act on and cause translational displacement of any discrete body by a steady force referred to as "radiation force." When radiation force acts on the elements of a homogeneous medium, such as a fluid, it causes the formation of a steady circulatory flow or acoustic streaming. This streaming can be both large and small in scale. In small-scale acoustic streaming (microstreaming), micro-scale eddying (on the order of a micron) in the vicinity of a small vibrating object occurs. Microstreaming is readily set up near gas bubbles; thus it is also closely related to cavitation. The

TABLE 12.2 Effect of Ultrasound Exposure on the Activity of Chemotherapy

Chemotherapeutic Agents: Enhanced Activity (+) or No Enhancement (−)	Ultrasound Exposure Conditions	Ref.
Nitrogen mustard (+)	2 MHz continuous wave (CW) ultrasound; 10 W/cm^2; temperature: 26–42.5°C	31
Doxorubicin (+); daunomycin (?)	1.92 MHz CW ultrasound; 0.66 W/cm^2; temperature: 35°C	32
Doxorubicin(+); amphotericin B (+); cisplatin (−); etoposide (?)	2.025 MHz CW ultrasound; 0.5–2 W/cm^2; temperature: 37–43°C	33
Doxorubicin (+)	2.6 MHz CW ultrasound; 2.3 W/cm^2; temperature: 37°C	34
Doxorubicin (+); diaziquone (+)	1.765 MHz tone burst and pulsed ultrasound; 0.25 W/cm^2; temperature: 37°C	35
Doxorubicin (+)	1.733 MHz CW ultrasound; 0.5–2 W/cm^2; temperature: 41–43°C	36
Doxorubicin (+); diaziquone (+); cisplatin (−); mitomycin C (−)	1.62 MHz tone burst and pulsed ultrasound; 0.5 W/cm^2; temperature: 37°C	37, 38
Ara-C (+)	48 kHz CW ultrasound; 0.3 W/cm^2; temperature: 37°C	39

nonthermal effects of ultrasound have been shown to influence cell surfaces,[1,7,44] and thereby might promote cellular uptake of cytotoxic drugs and enhance chemotoxicity. However, under the conditions of the chemotoxicity studies mentioned above, all three nonthermal mechanisms could occur. Thus, it is conceivable that each or a combination of these three nonthermal mechanisms of ultrasound might play an interesting and important role in chemotoxicity enhancement.

Although the nonthermal effects of ultrasound can enhance the cytotoxicity of chemotherapeutic drugs, not all drugs appear to benefit from such an approach (Table 12.2). Drugs rendered more cytotoxic include amphotericin B, daunorubicin, doxorubicin, nitrogen mustards (BCNU), and diaziquone. In contrast, the activity of cisplatin and mitomycin C are not enhanced by ultrasound. Compared to the drugs made more cytotoxic by ultrasound, mitomycin C and cisplatin are considered quite hydrophilic. As such, ultrasound tends to enhance the chemotoxicity of hydrophobic drugs. However, recent study indicated that low-frequency ultrasound (kHz range) could enhance the cytotoxicity of the hydrophilic chemotherapeutic drug, cytosine arabinoside (see Table 12.2).[39] This observation is consistent with recent findings indicating that low-frequency ultrasound could enhance intracellular accumulation of small molecular and high molecular weight hydrophilic drugs.[45–47] This enhancing effect is believed to be caused primarily by ultrasound-induced cavitation, which is the predominant nonthermal mechanism at the low-frequency range of ultrasound.[1,7,44] Further, the effect exhibits good correlation with acoustic energy exposure (the product of acoustic pressure and exposure time), where an increase in energy exposure causes temporal and reversible disruption of cell membrane for enhanced intracellular drug delivery.[45–47] However, since ultrasound-induced cavitation during low-frequency ultrasound treatment is known to lead to substantial cell damage, its use may be limited to *in vitro* or *ex vivo* gene transfer or other applications where the experiments are concerned only with the fate of the remaining viable cells.

The foregoing alludes to the importance of a nonthermal mechanism for ultrasound-enhanced drug uptake by cells. However, data in favor of a thermal effect for enhanced cellular drug uptake have recently been obtained in our laboratory (unpublished results). Utilizing a model lipophilic drug, rhodamine 123 (R123), in a 9L glioma cell culture system, we have shown that high-frequency (1 MHz) ultrasound treatment at below physiological temperature (i.e., 27°C) did not increase cellular uptake of R123 (see Figure 12.1). However, if the temperature was allowed to increase, increases in cellular accumulation of R123 were observed. This occurred regardless of whether the hyperthermic conditions were induced by ultrasound or nonultrasound sources. These data support the notion that a thermal rather than a nonthermal mechanism may mediate the ultrasound enhancement of cellular uptake. Accordingly, drugs whose entry into cells are limited by membrane permeability might find their permeabilities and hence cytotoxicities enhanced at elevated temperatures. Indeed, an example of such behavior exists for adriamycin or doxorubicin.[48] Aside from increasing membrane permeability to drugs, the ability of ultrasound to heat biological tissue also increases the cytotoxicity of certain

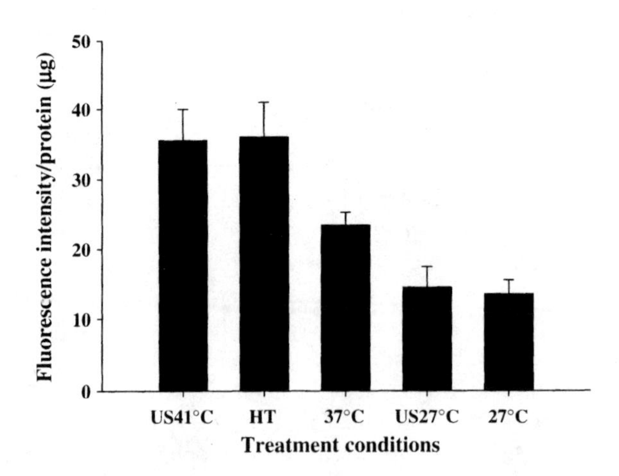

Figure 12.1. Effect of 41°C (HT), continuous wave (CW) ultrasound at 41°C (US41°C: 1 MHz at 0.4 W/cm^2), CW ultrasound at 27°C (US27°C: 1 MHz at 0.4 W/cm^2) on cellular uptake of R123 by 9L glioma cells *in vitro*. Cellular uptake of R123 was expressed as fluorescence intensity per μg of protein. For the study, ultrasound exposure was continued for 20 minutes. As controls, another groups of cells were treated at either 37°C or 27°C.

drugs, such as nitrosoureas, whose cytotoxic reactions follow an Arrhenius-type law, i.e., the rate constant governing the reaction increases more or less exponentially with temperature.[48]

12.3.1.3. Ultrasound as Adjuvant to Overcome the Clinical Effects of Multidrug Resistance Multidrug resistance (MDR) represents a major obstacle to successful chemotherapy of metastatic diseases.[49] One of the best-understood mechanisms of MDR is the overexpression of P-glycoprotein (P-gp).[50] This 170 kDa plasma membrane protein belongs to a larger family of ATP-binding cassette proteins and confers resistance to tumor cells by extruding many structurally and functionally unrelated hydrophobic anticancer drugs using the energy of ATP hydrolysis.[51]

High-frequency ultrasound (MHz) ranging from 0.5 W/cm^2 to several watts could increase treatment temperatures and enhance the activity of several hydrophobic drugs (Table 12.1). It would appear that ultrasound-induced heat could have potential in increasing cellular uptake of P-gp substrates and thereby reducing the clinical effects of MDR. We recently conducted studies to examine this possibility by investigating how mild heat (41°C) produced by a nonultrasound heat source (HT) or 1 MHz ultrasound (ultrasound-induced mild heat (USMH)) could affect (1) the cellular uptake of P-gp substrates, R123 and doxorubicin and (2) the cytotoxicity of doxorubicin in the parent and P-gp-overexpressing variants of two human cancer cell lines, MV522 and KB.[40] Our results indicated that both experimental conditions led to increased cellular uptake of R123 and doxorubicin in the parent and P-gp-overexpressing variants of the two cell lines (Figure 12.2). In parallel, the cytotoxicity of doxorubicin, was also enhanced. Importantly, the enhancement of uptake of R123 and doxorubicin, and of the cytotoxicity of doxorubicin produced by HT or USMH, was far better than that produced by the P-gp modulator, verapamil (Figure 12.3).

Because heating has been shown to induce a reduction in P-gp expression, leading to inhibition of drug efflux and enhancement of intracellular drug uptake, we recently examined the expression and activity of P-gp in both heat-treated and untreated cells. Both the expression level[52] and the activity of P-gp were unaffected by the heat treatment.[53] These results imply that a mechanism other than modulation of P-gp activity might be responsible for the enhanced drug accumulation by USMH. Our recent data suggest that the site of action of USMH might be the cell membrane and that USMH selectively enhances the permeability of hydrophobic but not hydrophilic drugs.[53] If USMH were to increase membrane permeability to hydrophobic drugs, it is possible that the enhanced drug entry rate might temporarily overwhelm P-gp and lead to more cellular drug accumulation.

It is reckoned that if the enhanced drug uptake could be maintained in the P-gp-expressing cells longer, an augmented binding of the anticancer drug to its target and, therefore, increased cytotoxicity might be obtained. To that end, we tested the effects of combining USMH, which enhances cellular drug uptake, with PSC 833, which modulates P-gp activity and hence promotes cellular drug retention, on cytotoxicity of doxorubicin in the parent and MDR cells.[41] The results of this study are shown in Figure 12.4. As shown in this figure, cytotoxicity of anticancer drugs

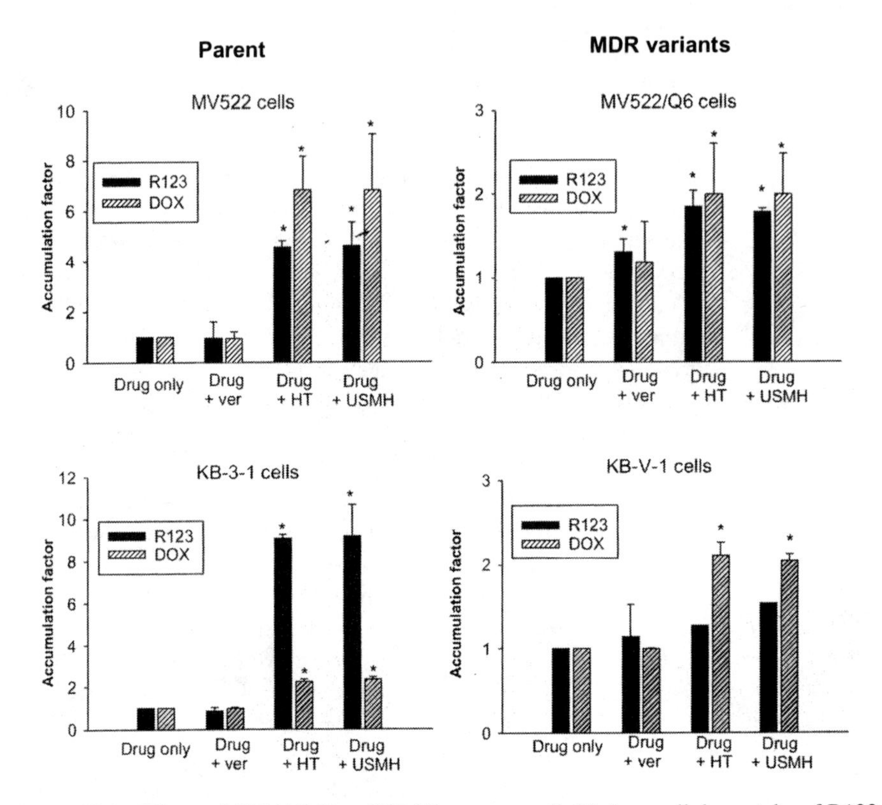

Figure 12.2. Effects of HT (41°C) or USMH or verapamil (Ver) on cellular uptake of R123 or doxorubicin (DOX) by the parent and MDR variants of human MV522 (human metastatic lung carcinoma) and KB (human epidermoid carcinoma) cell lines. Cellular uptake data are presented as accumulation factors. The accumulation factor for control cells is defined as equal to 1. The accumulation factor for treated cells is defined as the ratio of cellular R123/DOX accumulation in the presence of HT, USMH, or Ver to cellular accumulation in the absence of HT, USMH, or Ver. All experiments were carried out in triplicate. $^{*}P < 0.05$ compared to control values.

in P-gp-expressing cells can be significantly enhanced by the combined use of USMH and P-gp-modulating agent, where USMH increases uptake and the P-gp inhibitor reduces efflux of the cytcotoxic agent from the cell, resulting in significantly increased exposure and efficacy. These findings have significant implications for combined therapy using PSC 833 with cytotoxic anticancer drugs. For instance, PSC 833 has been found to alter the pharmacokinetic profile of the concomitant anticancer drugs, which leads to unexpected toxicity.[54] As a result, dosage reductions of anticancer drugs are often needed to avert unexpected toxicity. In the context of the present study, this reduction in dosage may be compensated for by the USMH, which specifically increases cellular uptake of the cytotoxic agent into the tumor.

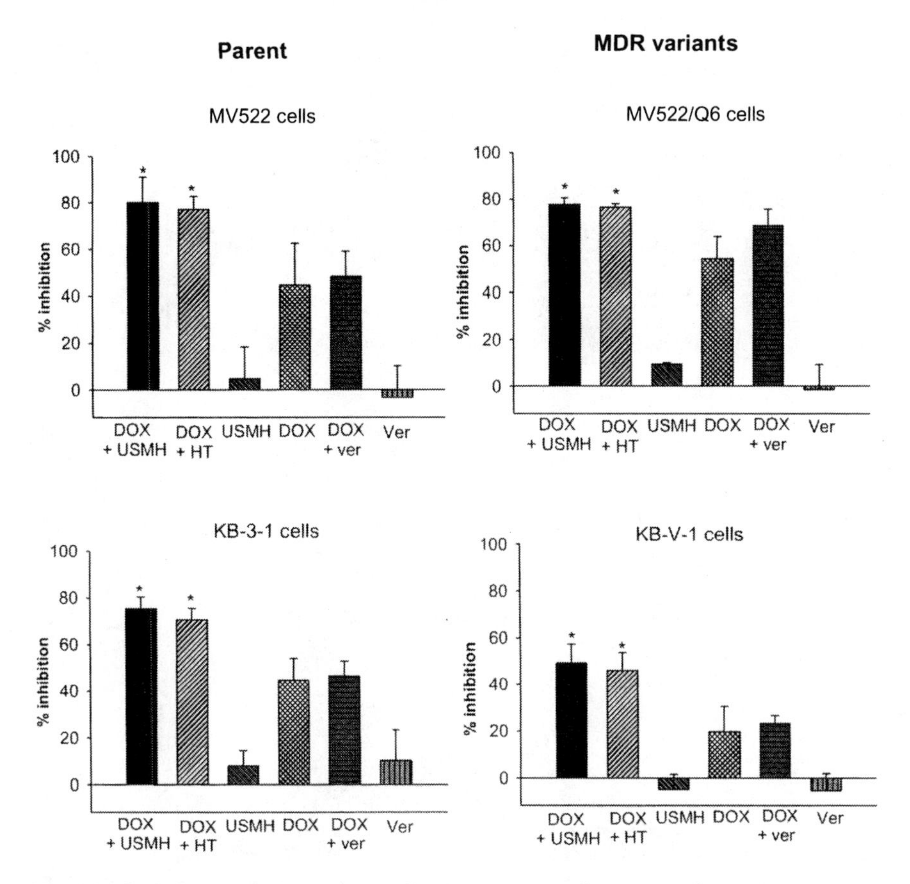

Figure 12.3. Effects of HT (41°C) or USMH or verapamil (Ver) on cytotoxicity enhancement of doxorubicin (DOX) in the parent or MDR variants of human MV522 and KB cell lines. Data are expressed as % inhibition calculated by the following formula: % inhibition = [1 − (counts of viable drug-exposed cells/counts of viable non-drug-exposed cells)] × 100. Cell viability was determined by hemocytometry technique after trypan blue staining. All experiments were carried out in triplicates. $^{*}P < 0.05$ compared to control values.

12.3.2. Thrombolytic Drugs

12.3.2.1. Thrombosis and Its Clinical Implications As a normal hemostatic response to limit hemorrhage from microscopic or macroscopic vascular injury, the body undergoes a process termed "local thrombosis." Specific proenzymes and proteins, platelets, and calcium participate in this process. The end result is the formation of insoluble fibrin, which mechanically blocks the flow of blood through ruptured vessels.[55] Thrombosis is usually counterbalanced by physiological anticoagulation and thrombolysis. Under normal conditions, the thrombus is confined to the area of vessel injury and rarely obstructs flow to critical areas. However,

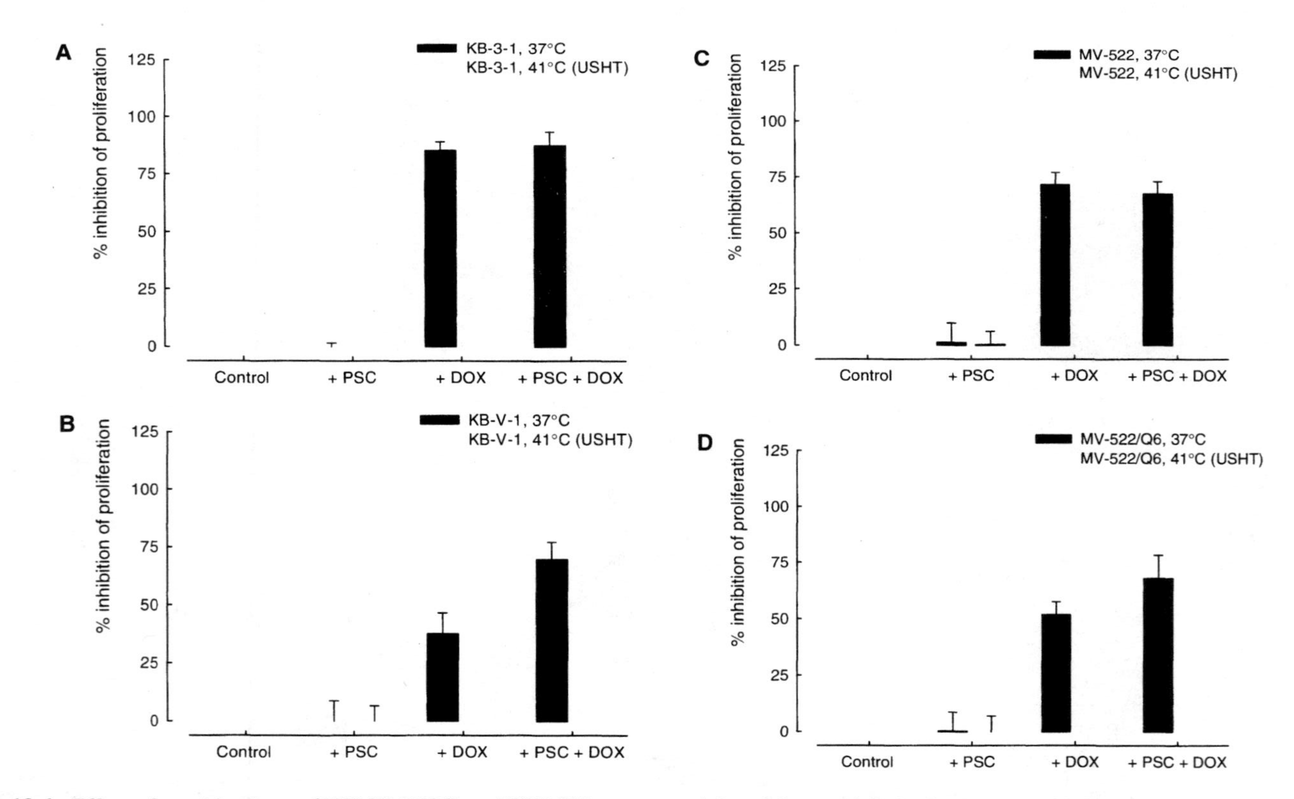

Figure 12.4. Effect of combined use of USMH (41°C) and PSC 833 on cytotoxicity of doxorubicin in the parent and MDR variant of MV522 and KB lines. Data were expressed as % inhibition calculated by the following formula: % inhibition = [1 − (counts of viable drug-exposed cells/counts of viable non-drug-exposed cells)] × 100. All experiments were carried out in triplicate (mean ± standard deviation).

under pathological conditions, the thrombus can dislodge from the area of vessel injury and enters the blood flow. The thrombus can then lodge in a critical vessel, and can obliterate valves and other structures that are essential to normal hemodynamic function. Abnormal thrombosis can occur in any vessel at any location in the body. The principal clinical syndromes that result are acute myocardial infarction, deep vein thrombosis, pulmonary embolism, acute non-hemorrhagic stroke, and acute peripheral arterial occlusion.[55]

12.3.2.2. Limitations of Present Treatment Modalities for Thrombosis The success of treatment of abnormal thrombosis depends on how rapidly one can restore blood flow to the obstructed vessels, because tissue death is closely related to the duration of ischemia.[55] Although recanalization of acutely thrombosed arteries can be achieved by mechanical interventions, such as balloon angioplasty,[56] this technique is not considered feasible for conditions such as stroke. Reperfusion by such techniques have serious limitations, including the requirement for specialized facilities and highly trained personnel, and these procedures often result in more complications in older patients.[57–62] Consequently, drug-induced thrombolysis has surfaced as an alternative for treatment of pathological thrombosis.[63] Thrombolytic agents available today are serine proteases, such as tissue type plasminogen activator (t-PA), urokinase, and streptokinase, that work by converting plasminogen to the natural thrombolytic agent, plasmin. Plasmin lyses thrombotic vascular occlusions by degrading fibrinogen and fibrin contained in a blood clot. Thrombolytic agents, however, also have limited success in recanalizing thrombotically occluded arteries.[64,65] For instance, in ischemic stroke treatment, thrombolytic therapy with t-PA demonstrated a small but significant improvement in neurological outcome in selected patients. As such, there is a need to enhance the effectiveness of thrombolytic agents by shortening the time to reperfusion.

12.3.2.3. Ultrasound and Its Role in the Treatment of Thrombosis Among the first to study the application of ultrasound to enhance the effectiveness of thrombolytic agents was Kudo,[66] who showed that, when used together with recombinant t-PA, ultrasound can accelerate recanalization of thrombotically occluded arteries *in vivo*. This observation was later confirmed by other investigators in various *in vitro* studies[67–76] and in animal models of arterial[9,77] and small-vessel thrombosis.[78,79] Consistent findings among these studies are: (1) the rate of thrombolytic enhancement was directly related to the temporal average intensity of the field; (2) the ultrasound-induced enhancement of thrombolysis was greater at higher thrombolytic concentrations; (3) a lower concentration of thrombolytic agent in the presence of ultrasound could induce more thrombolysis than a higher concentration in the absence of ultrasound; (4) enhancement was due primarily to accelerated enzymatic (thrombolytic) action rather than mechanical disruption of the clot;[80] and (5) enhancement was not limited to t-PA-specific drugs; the activity of other thrombolytic drugs, such as urokinase and streptokinase, was also increased.[80]

12.3.2.4. Mechanisms of Ultrasound-Enhanced Thrombolysis Ultrasound is known to have several biological effects, depending on the emission characteristics. At higher energy levels, ultrasound alone has a thrombolytic effect. This effect is already used for clinical purposes in interventional therapy using ultrasound catheters. At high intensities, ultrasound delivered via a transducer causes mechanical fragmentation through direct mechanical contact. In contrast, the use of lower-intensity ultrasound to accelerate enzymatic thrombolysis without mechanical disruption of the thrombus offers a conceptually different approach to thrombolytic therapy. The mechanisms by which ultrasound enhances enzymatic thrombolysis are multiple and relate primarily to transport (refer to Francis[80] for a recent review). Using 1 MHz ultrasound at 4 W/cm^2, Francis et al.[81] demonstrated that ultrasound significantly increases the rate of uptake and penetration of t-PA into the clots. Ultrasound also increases thrombolysis by pressure-mediated perfusion of thrombolytic agents into the matrix. For instance, Siddiqi et al.[82] examined the effects of ultrasound on fluid permeation through purified fibrin gels and found that at 2 W/cm^2, 1 MHz ultrasound increased flow through the gel. Flow through the gel was dependent on the ultrasound intensity (hence acoustic pressure) and the resistance of the fibrin gel (a property determined by the fibrin fiber structure). In the context of this study, ultrasound promotes thrombolysis by altering the fiber structure. Indeed, when the ultrastructure of fibrin gel after ultrasound treatment was examined using scanning electron microscopy, significant changes in structure were observed.[83] Such alterations could modify flow resistance as well as create additional binding sites for thrombolytic enzymes, thereby improving thrombolytic efficacy.

In order to study if inertial cavitation might be the mechanism by which ultrasound enhanced the dissolution of a blood clot when the clot was exposed to a thrombolytic agent, Everbach and Francis[84] examined the dissolution of radiolabeled fibrin clots exposed to 1 MHz ultrasound in a rotating sample holder, a condition which enhances ultrasound-induced nonthermal effects such as inertial cavitation.[44] To suppress acoustic cavitation, the exposure tank was positioned in a hyperbaric chamber capable of pneumatic pressurization to 10 atmospheres. In these studies, static pressure could only reduce the acceleration of thrombolysis by one-half, suggesting that other processes in addition to inertial cavitation might be responsible for the enhanced effect of ultrasound on thrombolysis. Hence, much work is needed to shed further light on the mechanism of this phenomenon.

12.3.2.5. Lower-Frequency Ultrasound in Ultrasound-Enhanced Thrombolysis The pioneering work by Kudo[66] used relatively low-frequency ultrasound (200 kHz or 0.2 MHz). However, most investigators used higher frequencies (MHz range) with subcavitation intensity. Suchkova et al.[85,86] and others[76,87] recently reemphasized the advantage of lower-frequency and low-intensity ultrasound (40 kHz; <1 W/cm^2) to optimize ultrasound-enhanced thrombolysis. Suchkova et al.[85,86] suggested that, at frequencies exceeding 1 MHz, the intensities necessary to achieve significant enhancement of thrombolysis might lead to substantial and harmful tissue heating. Lower frequencies should result in lower absorption of ultrasound

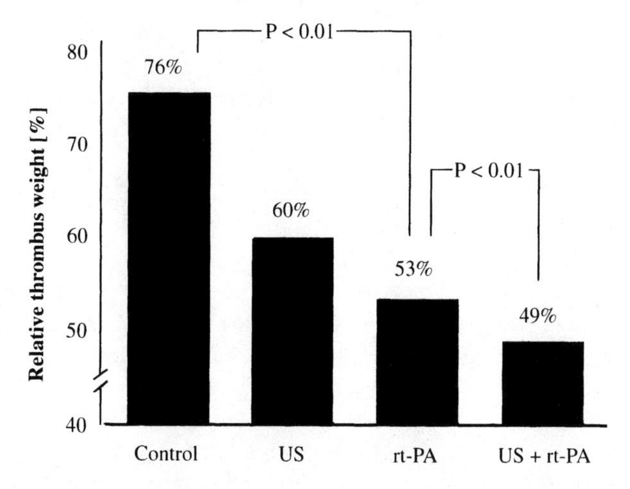

Figure 12.5. Thrombolytic activity: spontaneously after 1 hr in a control group, and after 1 hr treatment with transcranial US (33 kHz, CW, 0.5 W/cm^2) insonation only, after 1 hr treatment with rt-Pa only, and 1 hr treatment with combined US + rt-PA (from Ref. 90).

by tissues and cause less heating. In addition, with lower-frequency ultrasound, the acoustics field can be broader and more uniform, with greater depth of penetration than occurs at MHz frequencies.

Because high frequency ultrasound is transmitted with a high attenuation through the skull,[88] Daffertshofer et al.[76,89,90] have recently examined the potential use of low frequency ultrasound (20–300 kHz) for accelerating enzymatic thrombolysis as a potential therapy for ischemic stroke treatment. As a first step, they studied ultrasound transmission through a sample of different postmortem skulls. Their findings indicated very low attenuation in these skulls. It was also demonstrated that transmission of low-intensity (0.5 W/cm^2) and low-frequency ultrasound through the skull was sufficient to accelerate t-PA-mediated thrombolysis (Figure 12.5). These conditions were also found to be safe. No evidence for breakdown of the blood-brain barrier was observed in postmortem ultrasound-treated rats. Collectively, these data suggest that low-frequency ultrasound may reach the brain and intracranial vessels with thermally acceptable levels for enhancement of systemic intracranial t-PA-mediated thrombolysis.

12.3.3. Proteins

12.3.3.1. Limitations of Transdermal Drug Delivery Recent completion of the Human Genome Project[91] and advances in molecular biology have enabled the discovery of many novel drugs that are either peptides or proteins. Because peptides and proteins are metabolically labile and undergo extensive enzymatic degradation, they cannot be administered to patients by the traditional oral route. Currently, most peptide and protein drugs are given to patients by injection. However, injection is invasive and can instill psychological fears in patients, which decreases patient compliance. Thus, pharmaceutical scientists have looked at alternative routes such as the transdermal route for effective systemic delivery of these compounds.

Transdermal drug delivery offers several advantages over traditional drug delivery systems such as oral delivery and injections.[92] These include avoidance of first-pass metabolism, elimination of the pain associated with injection, and an opportunity for sustained release of drugs. The transdermal route of administration can be especially beneficial in the delivery of chronic injectable medications where patient compliance is low. However, transdermal transport of molecules is low because human skin is an effective and selective barrier to chemical permeation.[93] Experiments have traced the major barrier to drug permeation to the stratum corneum,[94] the outermost layer of the skin. The stratum corneum is composed of dead, flattened corneocytes surrounded by lipid bilayers. This arrangement resembles the "brick and mortar" structure of a wall where the dead corneocytes comprise the "brick" embedded in a "mortar" composed of multiple lipid bilayers of ceramides, fatty acids, cholesterol, and cholesterol esters (reviewed in Barry and William[95]). Indeed the low permeability of the stratum corneum is the main reason only a handful of low molecular weight drugs are administered by this route today.[92] A possible solution to this problem is to increase drug flux by reducing this barrier's hindrance using physicochemical forces such as chemical penetration enhancers or electric fields.[96] Penetration enhancers swell and disorder stratum corneum, facilitating passive diffusion of drugs. However, penetration enhancers, whether they are simple solvents like ethanol or complex compounds such as cyclic alkylamides, have significant limitations.[97] For instance, simple solvents tend to leave the transdermal drug delivery system and be absorbed rapidly into the body, and complex compounds may produce questionable metabolic products. Most importantly, penetration enhancers cause skin irritation and generally do little to enhance the delivery of hydrophilic high molecular weight proteins. An alternative approach is to use an electric field (iontophoresis) to drive drug molecules across the skin.[92,96] But this approach works only for charged molecules, and thus is ineffective for delivering drugs that carry either no charge or relatively low electric charges, such as proteins. In addition, iontophoresis is known to induce minor skin irritation. Thus, a relatively safe approach that works for a wider range of drugs will need to be developed for effective transdermal delivery of drugs such as proteins.

12.3.3.2. Ultrasound-Enhanced Transdermal Delivery of Proteins Ultrasound has been used for enhancing transdermal transport under a variety of conditions, using different combinations of frequency, intensity, and exposure time.[98] This phenomenon is referred to as "phonophoresis" or "sonophoresis." Most of the earliest transdermal drug delivery research used high-frequency ultrasound (MHz) to promote delivery of small molecular weight hydrophobic anti-inflammatory drugs into the skin for physical therapy.[98] Given the heightened need to develop delivery strategy for peptides and proteins, several research groups have looked at the use of therapeutic ultrasound (i.e., high frequency in the MHz range) to promote transdermal delivery of proteins. However, measurable enhancement of protein transport has been reported in only a few cases.[15,16] Previous work has demonstrated that application of ultrasound at therapeutic frequencies (MHz range) induces growth and oscillation of air pockets present in the corneocytes of the stratum corneum.[99] These oscillations disorganize the stratum corenum lipid bilayer, thereby enhancing transdermal transport. Because cavitational effects are inversely proportional to ultrasound frequency,[100] Robert Langer and his group at the Massachusetts Institute of Technology reasoned that application of ultrasound at frequencies lower than those corresponding to therapeutic ultrasound might induce sufficient bilayer disorganization so that proteins might be able to diffuse across the skin. Indeed, their seminal work indicated that application of low-intensity (mW/cm^2 range) and low-frequency ultrasound (20 KHz) increased the *in vitro* human skin permeability of various proteins such as insulin (molecular weight ~6000), interferon-γ (~17,000), and erythropoeitin (~48,000) by several orders of magnitude.[17] This enhancement effect was found to vary nearly exponentially with ultrasound intensity, a result the investigators attributed to the highly nonlinear dependence of cavitation on ultrasound intensity.[101] The observations suggest that ultrasound intensity might potentially be used to control transdermal protein delivery. Their data also indicated that low-intensity and low-frequency ultrasound did not cause any permanent loss of the barrier properties of the skin; most of the skin barrier properties could be recovered very quickly (less than a day) after sonophoresis. The results suggest that sonophoresis delivers intensity-dependent protein doses across the skin, which provides a platform for future noninvasive delivery of proteins across the skin. As the investigators correctly alluded to in their study, the success of such an endeavor would require extensive investigation on the physiological and immunological effects of ultrasound exposure and optimal selection of ultrasound parameters, such as frequency, pulse length, and intensity. Because recent studies indicated that coupling get might influence ultrasound-mediated transdermal drug delivery,[99,102,103] optimal selection of an ultrasound-coupling medium will be important for effective transdermal delivery of proteins. Finally, the first ultrasound experiments used laboratory-scale sonicators that are cumbersome. Thus, any realistic commercialization of this innovative concept of drug delivery will depend on the development of pocket-size sonicators. Realizing this potential problem, engineers at the Pennsylvania State University have miniaturized the system by arraying tiny cymbal-shaped transducers on a 1.5-in^2 patch.[104–106] Using this lightweight ultrasonic device, engineer Nadine Smith and her group recently reported that they delivered therapeutic levels of insulin to rats.[104] These results provide hope that a portable ultrasound system will one day be developed for transdermal drug delivery.

12.3.3.3. Mechanism of Enhancement of Transdermal Drug Delivery by Low-Frequency Ultrasound The biophysical modes of ultrasonic action on biological system can be classified into two categories: (1) thermal mechanism and (2) nonthermal mechanism.[2,107] The thermal effects of ultrasound, which are directly related to its intensity, result from transfer of energy from the vibrating pressure waves to the objects that the wave contacts. In transdermal application, this results in absorption of energy by the skin, causing a rise in skin temperature. Although literature supports the observation that increasing temperature leads to an increment in skin permeability,[99] recent studies indicate that thermal effects play an insignificant role in enhanced transdermal drug transport by low-frequency ultrasound.[108,109] For instance, low-frequency ultrasound (20 kHz at 15 W/cm^2 for 2 hours) caused a 20°C rise in temperature that resulted in a 35-fold increase in permeability of mannitol across *in vitro* porcine skin. In contrast, when the skin was heated (in the absence of ultrasound) in a manner similar to the thermal profile produced by sonophoresis, the permeability of mannitol increased by only 25%.[109] These data indicate that the key mechanism responsible for the observed skin permeability is related to the nonthermal effects of ultrasound. As mentioned before, the nonthermal effects of ultrasound on biological system can occur via one or a combination of three mechanisms: (1) cavitation, (2) radiation pressure, and (3) acoustic microstreaming.[2,107] Among these nonthermal effects, acoustic cavitation is believed to be the most important. Acoustic cavitation is defined as the acoustically induced activity of gas-filled cavities (involving nucleation, oscillation, and collapse).[107] Two types of cavitational activities are known: (1) noninertial or stable cavitation and (2) inertial or transient cavitation.[107] In noninertial cavitation, the radius of the bubbles oscillates about an equilibrium value without collapsing over a considerable number of acoustic cycles. In contrast, the bubbles in inertial cavitation grow rapidly within one or two acoustic cycles before collapsing violently during a single compression half cycle. This violent event may generate defects (pores or holes) in the skin that explain the enhanced skin permeability. Based on indirect experimental observations, such as the appearance of "holes" or irregularities in the skin tissue after low-frequency sonophoresis,[16,110,111] one could argue for transient cavitation as the predominant mechanism for enhanced skin permeability. Indeed, a recent study has provided evidences for such a claim.[112] In addition, the critical cavitation site has been identified and found to locate on, or in the vicinity of, the skin membrane.[112] In an effort to determine the dependence of transport pathways (defects that occur between/within the lipid lamellar organization in the stratum corneum) during low-frequency sonophoresis on ultrasound parameters, Tezel et al.[108] exposed *in vitro* porcine skin to low-frequency ultrasound over a range of frequencies (20–100 kHz) and energy densities (0–2000 J/cm^2: product of ultrasound intensity, I, and application time, t). Then the porous pathway model developed by Tang et al.[113] was used to study the dependence of average pore size, porosity, and tortuosity on ultrasound parameters. The data show that low-frequency ultrasound increases skin permeability by increasing skin porosity (up to 1700-fold) and/or decreasing tortuosity rather than by increasing the size of the pores. These findings provide quantitative guidelines for estimating the efficacy of sonophoresis, which should prove useful for future transdermal drug delivery research.

12.3.4. Gene-Based Drugs

12.3.4.1. Historical Perspective of Gene Therapy During the past several decades, the promising concept of gene therapy has emerged. This approach has potential implications for the treatment of over 4000 human inherited diseases associated with dysfunctional, nonfunctional, or missing gene-coded proteins.[114] The term "gene therapy" was originally coined to describe a proposed treatment of genetic disorders by directing treatment to the site of the defect itself—the mutant gene—rather than to secondary or pleiotropic effects of mutant gene products.[115] Technically, gene therapy consists of replacing, at least in a functional sense, a defective gene with its normal counterpart. Inevitably, the technical hurdle of gene therapy is gene delivery: that is, the introduction of genes into the appropriate cells of the patient. Thus, development of novel methods for rapid and efficient delivery of therapeutic genes of choice into mammalian cells has formed the basis of numerous gene therapy studies recently. Delivery of nucleic acids into mammalian cells can be made either via virus vector or by physical and chemical methods.[115] Viral vectors tend to achieve the highest efficiency; however, substantial concerns remain over their clinical safety and long-term efficacy.[116] Genes can also be delivered as plasmid DNA to cells directly by a number of methods. These techniques are categorized into two general groups: (1) naked DNA delivery by a physical method, such as electroporation, gene gun, high-pressure injection and ultrasound,[117,118] and (2) delivery mediated by a chemical carrier such as a cationic polymer and lipid.[117,118] Although the efficiency of gene delivery using nonviral techniques is relatively low (at least 10-fold less than that of viral vectors), the advantages of using plasmid DNA include (1) easy manipulation and ability to deliver large inserted sequences; (2) can be produced stably and cheaply to a high level of purity with minimal risk of replication or incorporation; and (3) plasmid DNA is weakly immunogenic and can therefore be administered repeatedly to immunocompetent subjects. As a result, a significant number of gene therapy researchers are attempting to increase the efficiency of plasmid DNA transfer into cells.

12.3.4.2. Ultrasound-Mediated Gene Delivery and Its Mechanism Ultrasound generated by a needle-tip sonicator (commonly used in the laboratory for cleaning and cell disruption) has been shown to enhance gene transfer into both mammalian and plant cells.[119–121] In these studies, relatively high peak ultrasonic power levels ($\sim$100 W) and low frequencies (20–50 kHz) were employed, leading to the possibility that acoustic cavitation played a major role in gene transfer in these studies. Miller et al.[44] demonstrated that the primary biological effect of acoustic cavitation is cell lysis. Nevertheless, the previous studies suggest that sublethal damage, such as transient permeabilization of the cell membrane, may also occur during cavitation, which leads to the uptake of molecules into the cells. This phenomenon, called "sonoporation," has been confirmed in numerous other studies. For example, ultrasonic shock waves (ultrasonic waves with very high pressure amplitude and hence high intensity), which favor production of cavitation, have been shown to increase cellular uptake of both small (adriamycin and fluorescein)[122] and macromolecular (fluorescein-labeled dextrans, ribosome-inactivating proteins gelonin and saporin)[105,123,124] molecules. The use of ultrasonic shock waves has been evaluated recently for enhancement of gene transfection. *In vitro* studies demonstrated that DNA plasmids entered cultured cells during ultrasound shock wave exposure, and that some of the cells survived to express the plasmid gene product.[125–130] However, sonoporation was accompanied by substantial cell lysis because of the very high ultrasound intensity used in these studies, and the transfection efficiency produced by such a method is very low (0.08% to 0.5%). In this regard, several studies have evaluated the use of therapeutic levels of ultrasound to enhance gene transfection. For example, Tata et al.[131] studied the ability of low-intensity (0.33 W/cm^2) and high-frequency ultrasound to enhance green fluorescent protein (GFP) reporter gene expression and obtained 20% transfection at about a 100 Hz pulse rate and a 120-second exposure in different cell types. Utilizing 1 MHz ultrasound at a spatial average peak positive pressure (SAPP) of 0.41 Mpa (~10 W/cm^2), Greenleaf et al.[132] obtained even better transfection reaching about 50% of the living cells. This transfection efficiency was comparable to that of other high-performance techniques of transfection, such as lipofection.[132] Because ultrasound is strictly a physical method, it is not bound by limitations currently affecting lipofection. For instance, the efficacy of plasmid DNA uptake into cells by lipofection depends on the net charge of the cationic lipids and DNA complexes.[133,134] This implies that an optimal cationic lipid to DNA ratio must be determined for each DNA type and concentration. In addition, the formation of cationic lipid/DNA complexes is highly dependent on other parameters such as the pH and concentration of electrolytes.[133,134] As a result, the use of cationic lipids in lipofection must be optimized for each experimental setting, and inconsistent results may occur. One possible drawback of using ultrasound to transfect plasmid DNA into cells is the relatively large amounts of plasmid DNA needed to obtain a competitive transfection rate.[132] However, plasmid DNA is comparatively inexpensive, and the possible site-specific *in vivo* applications of this method generally outweigh this small disadvantage.

It is clear that ultrasound can provide enhanced transfer of DNA plasmids into cells, but it will be important to extend these findings to an *in vivo* setting. Indeed, several studies have shown that ultrasound could enhance reporter gene expressions *in vivo*.[135–137] The main mechanism by which ultrasound enhances gene transfer is thought to be acoustic cavitation, which can affect transient nonlethal perforations in the plasmalemma and possibly the nuclear membranes.[138] Because the ultrasound contrast agent lowers the threshold for cavitation by ultrasound energy (see the next section), it is expected that its use during ultrasound application will further promote pore formation in cell membranes, thus facilitating the entry of plasmids into cells and their release from endosomes. Indeed, numerous studies that examined the combination of microbubbles with ultrasound have reported significant increase in gene transfection.

Most recently, low-intensity ultrasound has been used to enhance gene transfection by liposome.[139] In this study, ultrasound exposure (1 MHz, 0.4 W/cm^2) for 60 seconds enhanced transfection of naked or liposome-complexed luciferase reporter plasmid into cultured porcine vascular smooth muscle and endothelial cells. These results with liposome-complexed reporter plasmid are similar to those of Unger et al., who showed that relatively low levels of ultrasound energy (0.5 W/cm^2) enhanced gene expression.[140] In these two studies, ultrasound exposure did not begin until 30 minutes[139] or 60 minutes[140] into the transfection period. Accordingly, it may be hypothesized that ultrasound might enhance gene transfection through its ability to accelerate DNA escape from endosomes. Other site(s) of action of ultrasound might include effects on plasmid DNA entry, intracellular trafficking, lysosomal degradation, nuclear translocation, RNA transcription, or protein translation. Although not proven, these plausible mechanisms might be responsible for the recently published data on enhanced expression of a plasmid DNA-cationized gelation complex by ultrasound in murine muscle.[141]

12.4. ULTRASOUND CONTRAST AGENTS

12.4.1. Ultrasound Contrast Agents as Imaging Agents

Over the past decade, the development of stable microbubbles has extended to the pharmaceutical arena as ultrasound contrast agents. Knowledge of microbubbles or gas-water interfaces as efficient backscatterers of sound waves has existed for many decades. Even before it was conceived that they could be used for clinical ultrasound image enhancement, the U.S. Navy was interested in bubble technology because sound or sonar was an efficient means for detecting ships and submarines—the reason being that submarines could be made of stealth components but could not completely hide their propulsion systems, which would form a trail of bubbles or cavitation nuclei, detectable by ultrasound.

The first demonstration of the clinical application of microbubbles came in 1968 when Gramiak and Shah[142] demonstrated that introduction of agitated saline containing gas bubbles into the aortic root could elicit echo-contrast by diagnostic ultrasound. Since microbubbles produced by agitation are both large and unstable,[143] they are not considered viable contrast agents. Subsequently, a variety of techniques were developed to form microbubbles of the right size (1–10 μM) to traverse the capillaries and, at the same time, stable enough to provide enhancement over multiple cardiac cycles. Currently, microbubble technology uses a biocompatible gas such as air,[142] nitrogen,[144] perfluorocarbons,[145] or sulfur hexafluoride[146] in a biomaterial coating composed of components such as albumin,[147] phospholipids,[148] surfactants,[149] biopolymers,[150] or sugars (galactose).[151] In addition, perfluorocarbon emulsions,[152]which can be converted from the liquid phase to the gas phase upon reaching their boiling point in the body (e.g., perfluoropentane: bp = 29°C), have been utilized as ultrasound contrast agents as well.

The efficiency of ultrasound backscatter owes its sensitivity to the fact that the reflectivity of sound exhibits a r^6 dependence with respect to microbubble diameter. This was first theorized at the turn of the century by Lord Rayleigh and is appropriately referred to as "Rayleigh scattering." Thus, a 2 μm microbubble exhibits 64 times as much backscatter efficiency as a 1 μm microbubble. The objective of clinically relevant microbubble technology requires formulation of bubbles that are large enough to maximize the backscatter potential of the gas-water interface but can still freely course through the capillary bed. In addition, the microbubbles must be stable enough to minimize both aggregation and coalescence and, most importantly, maintain stability in the circulatory system.

The answer to the first problem, i.e., providing a microbubble that is small enough to freely pass through capillary beds yet large enough to maximize backscatter, was solved in a variety of ways, some of which were fortuitous. It turns out that the size of the microbubbles is highly dependent upon the surface tension of the shell and gas and is related to the pressure exerted on the bubble by the LaPlace equation:

$$P = 2\sigma/r$$

where

$$P = \text{pressure}$$
$$\sigma = \text{surface tension}$$
$$r = \text{bubble radius}$$

Thus, the use of gases that possess very low surface tension (e.g., perfluorocarbons) should reduce the pressure against bubble collapse such that the radius of the microbubbles formed would be small. Indeed, perfluorcarbon microbubbles prepared from perfluoropropane or perfluorobutane form bubble diameters on the order of 1 to 2 μm, which are ideal for passing through capillary beds. In addition, because of their extremely low solubility in aqueous media, the perfluorocarbon microbubbles would not dissolve in the venous blood and are therefore less prone to collapse.

With respect to aggregation and coalescence, techniques similar to liposomal technology have been employed to maintain microbubble stability. These techniques employ electrostatic charges on the membrane surface to promote electrostatic repulsion, which hinders aggregation. Typically, membrane lipids that carry charges, such as phosphatidic acid, are incorporated into the lipid-coated microbubbles to help microbubbles to stay apart. To minimize RES uptake and hence maintain the stability of the microbubbles in the circulatory system, researchers have turned to the use of stealth technology. Stealth or pegylation technology was first used in liposomal research, where drug-filled liposomes were pegylated to extend their vascular circulating half-life. An example of this was the study by Northfelt et al.,[153] who used membrane-bound polyethyleneglycol (PEG) to prevent RES recognition in their anticancer agent Doxil. Pegylation is now used in a currently marketed microbubble diagnostic agent, Definity, sold by Bristol Myers-Squibb.

Diagnostic applications of microbubbles have been primarily addressed in the area of echocardiography, where contrasting agents are mainly used to opacify the left ventricular chamber and aid in the delineation of wall motion.[154] In echocardiography, assessment of perfusion in the myocardial tissue has achieved somewhat limited success. Despite the ability of microbubbles to fill the myocardium, the large amount of contrast in the ventricular chamber causes dynamic range problems, making imaging of the smaller amount of bubbles in the myocardium very difficult. Although some researchers have been successful in qualitatively and semi-quantitatively determining myocardial perfusion, the technique is far from routine, with sophisticated pulsing regimes (phase inversion imaging and second harmonic imaging) and image subtraction often being necessary to image the myocardium.[155,156] Currently, microbubble-based ultrasound contrast agents have been approved only for echocardiography. However, applications for radiologic indications such as renal and liver perfusion are currently being pursued through the regulatory (Food and Drug Administration) process and should soon be approved. In addition, there have been efforts to develop microbubble-based, ultrasound-enhanced imaging of tumors. The rationale for such development is that the vascularity (or angiogenesis) around the tumor is greater than that of surrounding tissues and therefore can be contrasted to normal tissue for proper imaging.

12.4.2. Targeted Ultrasound Contrast Agents

Now that it has been demonstrated that stable microbubble technology can be utilized as ultrasound contrast agents, recent research efforts have turned to the development of microbubble technology selective identification of vascular landmarks. Earlier works by Lindner et al.[157–159] led to the observation that microbubbles could adhere to leukocytes in regions of inflammation. It was further demonstrated that after attachment to neutrophils and monocytes, the microbubbles were phagocytosed intact. Despite viscoelastic damping secondary to encapsulation, the microbubbles remained acoustically reflective. These results led to the conclusion that microbubbles could potentially image inflammation sites through passive targeting.

The use of targeting ligands to aid in site-selective targeting of microbubbles for ultrasound diagnostics was soon to follow. Work in the laboratories of Lanza and coworkers at Washington University in St. Louis introduced the development of a three-step approach to targeting ultrasound contrast agents selectively to thrombi.[160] Briefly, ligands, which consist of conjugating biotin to antifibrin monoclonal antibody that binds to fibrin clots, are administered first. This is followed by the addition of avidin, which binds to the biotin moiety of the ligands and cross-links the ligands on the fibrin clots. Finally, biotinlyated perfluorocarbon emulsions are introduced, which bind to the available sites on the avidin, thereby completing the ligand-avidin-constrast "sandwich" on the target surface. Using this system, Lanza and coworkers demonstrated enhanced ultrasound imaging of arterial thrombus in a canine model.

A variation of this targeting theme was demonstrated by Lindner and coworkers,[161] who hypothesized that tissue retention of microbubbles targeted to the endothelial cell adhesion molecule P-selectin would provide a means to assess inflammation with ultrasound imaging. In this technology, phospholipid microbubbles targeted to P-selectin (MBp) were made by conjugating the monoclonal antibodies against murine P-selectin to the lipid shell using a small spacer polymer. Compared with control microbubbles or isotype antibody, the P-selectin and perfluorobutane-filled microbubbles were assessed for binding to cremasteric venules using fluorescent labeling and visualization by intravital microscopy. The results

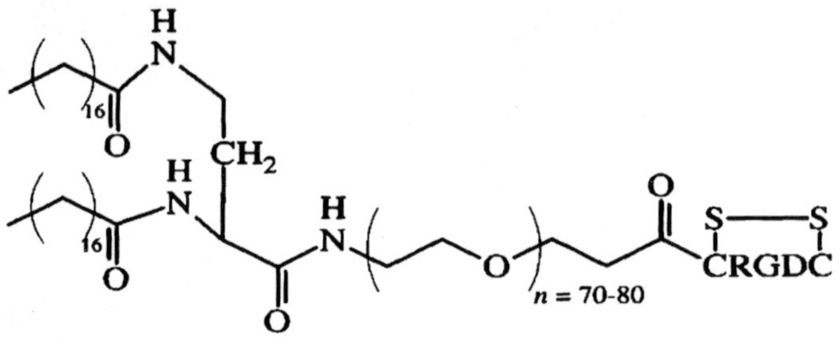

Figure 12.6. Bioconjugate ligand for targeting to GPIIbIIIa receptor of thrombi.

indicated that in the P-selection up-regulated [tumor necrosis factor (TNF-α)–stimulated] wild-type mice, microbubble binding of the MBp-conjugated microbubbles was significantly higher than in the P-selectin-deficient mice. Furthermore, in the P-selectin up-regulated mice, the P-selectin-conjugated microbubbles exhibited greater binding than the control or isotype antibody-conjugated microbubbles. Thus, it was demonstrated that targeted microbubbles would provide robust targeted imaging of the thrombus using perfluorocarbon gas microbubbles.

Schumann et al.[162] utilized another approach whereby a Arg-Gly-Asp-based (RGD) peptide ligand (tethered through a PEG moiety to a pseudo-isosteric lipid-simulating bioconjugate analog, Figure 12.6) that targets GPIIbIIIa receptor of thrombi was inserted into the lipid membrane of a microbubble for microvascular imaging of thrombi in the mouse cremasteric arterioles and venules. A pictorial design for the targeted microbubble is presented in Figure 12.7. Results indicated a clear linear relationship between bound microbubble and thrombus size, providing evidence for site targeting of microbubbles. In addition, flow stress studies were

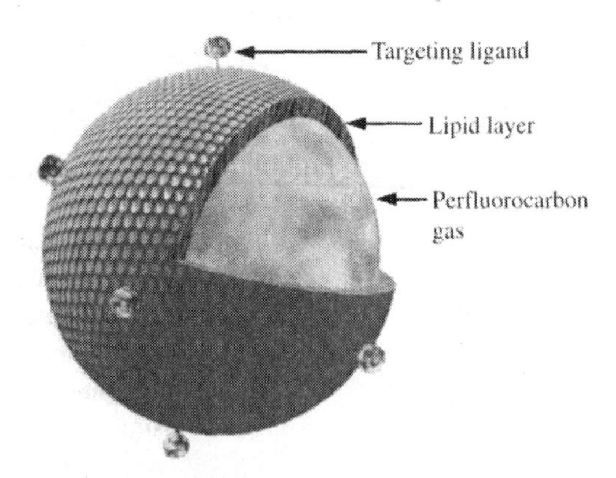

Figure 12.7. Pictorial design of a bioconjugate ligand on the surface of a microbubble for targeting to the GPIIbIIIa receptor of thrombi.

conducted which demonstrated that binding of these microbubbles could bind and adhere to clots under stress flow conditions similar to those of larger vessels.

12.4.3. Ultrasound Contrast Agents and Their Roles in Thrombolysis

Apart from diagnostic imaging, ultrasound-mediated manipulation of microbubbles may soon have therapeutic potential as well. As noted previously, the efficiency of thrombolytic agents was increased by nonthermal ultrasound energy such as acoustic cavitation. It has recently been shown that ultrasound contrast agents can significantly lower the acoustic cavitation production threshold.[163] This lowering of the cavitational threshold forms the basis for the induction of local shock waves (due to destruction of the microbubbles) that could be used as a therapeutic modality for lysing thrombi or blood clots. This therapeutic intervention, which uses sound waves to lyse thrombi, has been named "sonothrombolysis" or "sonolysis." The earliest study that reported the use of ultrasound contrast agents for sonolysis was by Tachibana and Tachibana.[164] In this study, the authors determined if the presence of albumin microbubbles (Albunex) used for diagnostic echo contrast could accelerate thrombolysis caused by urokinase. Their study indicated that the presence of the ultrasound contrast agent and ultrasound increased thrombolysis by urokinase. The data thus suggested that the ultrasound contrast agent or microbubbles could be used as an adjuvant to enhance ultrasound-induced thrombolysis. This observation was later confirmed by other investigators in various *in vitro*[72,73,165–167] and *in vivo*[87,113,166] studies. Because targeted microbubbles can detect areas of vascular pathology including stroke, deep vein thrombosis, myocardial infarct, and peripheral arterial disease, current efforts have focused on targeting the microbubble to thrombi or clots followed by the use of focused or unfocused ultrasound for disintegration/dissolution of the clots.

12.4.4. Ultrasound Contrast Agents and Their Roles in Gene Delivery

Over approximately the past 5 years, interest has been spurred in the area of ultrasound-mediated, microbubble-enhanced gene delivery for reliable and less toxic gene therapy. The use of microbubbles to carry a drug payload may be somewhat limiting, as the amount of drug to be carried on a microbubble surface (drugs cannot be carried in a gaseous interior matrix) may not provide sufficient capabilities to deliver a therapeutic load. Because gene-based drugs usually require significantly smaller amounts to generate therapeutic effects, this issue of insufficient drug loading might not be as limiting as was previously thought. In addition, adhering genes to the surface of the microbubbles may actually reduce the enzymatic degradation often seen with genes flowing freely in the vascular milieu[168] and should promote the overall efficacy of gene therapy.

Genes are amenable to adherence to the surface of lipid or proteinaceous microbubbles by virtue of their electrostatic (charge-charge) interactions. This has been achieved through the use of cationic lipids inserted into the membranes of lipid microbubbles as well as the use of cationic charges from albumin on the surface of proteinaceous microbubbles. In the laboratories of Dr. Evan Unger at ImaRx Therapeutics, avid DNA adherence to the microbubbles was demonstrated after the cationic lipids 1,2-dimyristyloxypropyl-3-dimethyl-hydroxy ethyl ammonium bromide (DMRIE) and dioleoylphosphatidylethanolamine (DOPE) were inserted into lipid microbubble membranes.[168] Similarly, utilizing perfluorocarbon-exposed and sonicated dextrose albumin microbubbles, Dr. Thomas Porter of the University of Nebraska demonstrated that microbubbles could bind oligonucleotides for the purpose of drug delivery.[154] Successful adherence of genetic materials using cationic gelatin-coated, gas-filled polymeric microspheres was also achieved in the laboratories of Dr. Fuminori Moriyasu.[169] Thus, it was demonstrated in many different formulations by many different laboratories that genes or genetic material could be bound to microbubble surfaces.

Once it was demonstrated that DNA could be bound to the microbubble surface, the next step was to demonstrate that microbubbles and ultrasound could be utilized as a potential gene delivery vehicle. Initial *in vitro* work with microbubbles was reported by Lawrie et al.[138] from the University of Sheffield. Using both a pGL3 luciferase plasmid and a second plasmid construct whereby the luciferase gene is driven by the Rous sarcoma virus promoter (pRSVLUC), Lawrie et al. studied the effect of insonation [1 MHz and a mechanical index (MI) of 1.1 and a 6% duty cycle for 3 hours] of vascular smooth muscle cells on gene expression. Following insonation in the presence of albumin-coated microbubbles (Optison), results indicated that there was up to a 300-fold increase in transgene expression compared to naked DNA alone. This expression was maximized to 3000-fold with the addition of a polyamine transfection reagent, TransIT-LT1. Since the increase was found using both constructs, the researchers concluded that the effect of cavitation was due to microbubbles and ultrasound and probably was not due to ultrasound-mediated free radical generation. Although the results were compelling, the range of the data occasionally varied by over an order of magnitude, which made the conclusion less robust. Nonetheless, there appeared to be a significant increase in transfection secondary to microbubbles and ultrasound.

Having demonstrated that *in vitro* transfection could be achieved, a logical next step was to determine if *in vivo* expression of ultrasound-mediated gene delivery could be accomplished with microbubbles. To address this issue, Dr. Mani Vannan of the University of Michigan studied ultrasound-mediated, microbubble-enhanced gene delivery of the chloramphenicolacyltransferase gene (CAT) to myocardial tissue in a dog model.[170] Upon injection of cationic microbubbles with adherent CAT into the animal, myocardial tissue was exposed to insonation at 1 MHz ultrasound using a diagnostic ultrasound machine (HP Sonos 550). The results indicated that high levels of CAT protein were observed in the myocardium within regions of the focal zone of ultrasound insonation but not in the myocardial tissue devoid of ultrasound insonation. These results demonstrated that ultrasound and microbubbles could indeed increase delivery of the gene, with subsequent expression of the gene.

Most recently, interesting work in a mouse model has been conducted in the laboratories of Dr. Martin J.K. Blomley of Imperial College, Hammersmith Hospital,[171] wherein the use of Optison and ultrasound to aid plasmid DNA delivery following direct injection into mouse skeletal muscle was investigated. The investigators used 1 MHz ultrasound at 3 W/cm^2 and a 20% duty cycle for 60 seconds. Results indicated that Optison increased transgene expression in both the absence and presence of ultrasound. Interestingly, the presence of Optison appeared to markedly decrease muscle damage that was seen with naked plasmid alone. Of additional interest was the observation that transgene expression was age-related, as expression was less efficient in older mice (6 months old) than in younger mice (4 weeks old). However, the addition of ultrasound increased the transfection in the older mice but not in the young mice. This result, if further validated, could imply that cellular age and integrity may have a profound effect on microbubble and ultrasound-mediated gene delivery.

Clearly, the use of microbubbles and ultrasound for gene delivery is in its infancy, and many of the results reported in this section have yet to be validated. Although the exact mechanism by which microbubbles and ultrasound treatment enhance gene transfer is currently unknown, it is speculated that as the microbubbles enter the region of insonation, the microbubbles cavitate, causing local release of DNA. Cavitation also likely causes a local shock wave that improves cellular uptake of DNA. Nonetheless, the promising results generated to date, coupled with the fact that microbubbles and ultrasound may offer higher therapeutic to toxicity windows, lead to the conclusion that this avenue of research could result in compelling new clinical paradigms for the future of gene delivery.

12.5. CONCLUSION

Until recently, medical applications of ultrasound have mostly been associated with diagnostic medicine. However, the application of ultrasound for therapeutic purposes has recently gained impetus because of the various advances made in the use of ultrasound to aid drug delivery. Low-intensity (<10 W/cm^2) and high- or low-frequency (0.5 to 10 MHz) ultrasound can enhance the delivery and effectiveness of several therapeutic drug classes including chemotherapeutic, thrombolytic, protein-, and DNA-based drugs. Interestingly, ultrasound contrast agents, a recent invention of diagnostic ultrasound, appear to improve the ability of ultrasound to enhance the activity of thrombolytic and DNA-based drugs. Ultrasound contrast agents can significantly lower the threshold for cavitation, a process that has been suggested to play a partial but very important role in enhancement of drug effects. As such, ultrasound contrast agents hold great promise in the future of ultrasound-mediated drug delivery. In summary, the recent successes in ultrasound-related drug delivery research position ultrasound as a valuable therapeutic tool for drug delivery in the future.

ACKNOWLEDGMENT

This work was supported by Grant CA79708 from the NIH awarded to K. Ng.

REFERENCES

1. *NCRP Report No. 74. Biological Effects of Ultrasound: Mechanisms and Clinical Applications.* National Council on Radiation Protection and Measurements: Bethesda, MD, **1983**.
2. Barnett, S. B.; Rott, H. D.; ter Haar, G. R.; Ziskin, M. C.; Maeda, K. *Ultrasound Med. Biol.* **1997**, *23*, 805–812.
3. Bamber, J. C. In *Ultrasound in Medicine*; Duck, F. A., Baker, A. C., Starritt, H. C., Eds. Institute of Physics Publishing: Bristol and Philadelphia, **1998**, chapter 4.
4. Hand, J. W. In *Ultrasound in Medicine*; Duck, F. A., Baker, A. C., Starritt, H. C., Eds. Institute of Physics Publishing: Bristol and Philadelphia, **1998**, chapter 8.
5. Fry, F. J. *Ultrasound: Its Application in Medicine and Biology*; Elsevier Scientific: New York, **1978**.
6. Frizzell, L. A. In *Therapeutic Heat and Cold*; Lehmann, J. F., Ed. Williams & Wilkins: Baltimore, **1982**, chapter 8.
7. Williams, A. R. *Ultrasound: Biological Effects and Potential Hazards.* Academic Press: San Francisco, **1983**.
8. Asher, R. C. *Ultrasonic Sensors.* Institute of Physics Publishing: Bristol and Philadelphia, **1997**.
9. Riggs, P. N.; Francis, C. W.; Bartos, S. R.; Penney, D. P. *Cardiovasc. Surg.* **1997**, *5*, 201–207.
10. Mason, W. P. *J. Acoust. Soc. Am.* **1998**, *70*, 1561–1566.
11. Jaffe, B.; Cook, W. R; Jaffe, H. *Piezoelectric Ceramics.* Academic Press: New York, 1971.
12. Woo, J. http://www.ob-ultrasound.net/history.html
13. Fellinger, K.; Schmid, J. *Klinik und Therapie des Chronischen Gelenkreumatismus.* Maudrich: Vienna, **1954**, pp. 549–554.
14. Tyle, P.; Agrawala, P. *Pharm. Res.* **1989**, *6*, 355–361.
15. Newman, J. T.; Nellermoe, M. D.; Carnett, J. L. *J. Am. Podiatr. Med. Assoc.* **1992**, *82*, 432–435.
16. Tachibana, K. *Pharm. Res.* **1992**, *9*, 952–954.
17. Mitragotri, S.; Blankschtein, D.; Langer, R. *Science* **1995**, *269*, 850–853.
18. Unger, E. C.; McCreery, T. P.; Sweitzer, R. H.; Caldwell, V. E.; Wu, Y. *Invest. Radiol.* **1998**, *33*, 886–892.
19. Frinking, P. J.; Bouakaz, A.; de Jong, N.; Ten Cate, F. J.; Keating, S. *Ultrasonics* **1998**, *36*, 709–712.
20. Chapelon, J. Y.; Margonari, J.; Vernier, F.; Gorry, F.; Ecochard, R.; Gelet, A. *Cancer Res.* **1992**, *52*, 6353–6357.
21. Madersbacher, S.; Kratzik, C.; Susani, M.; Marberger, M. *Urologe A* **1995**, *34*, 98–104.
22. Madersbacher, S.; Pedevilla, M.; Vingers, L.; Susani, M.; Marberger, M. *Cancer Res.* **1995**, *55*, 3346–3351.
23. Nakamura, K.; Baba, S.; Fukazawa, R.; Homma, Y.; Kawabe, K.; Aso, Y.; Tozaki, H. *Int. J. Urol.* **1995**, *2*, 176–180.
24. Mulligan, E. D.; Lynch, T. H.; Mulvin, D.; Greene, D.; Smith, J. M.; Fitzpatrick, J. M. *Br. J. Urol.* **1997**, *79*, 177–180.

25. Chen, W.; Wang, Z.; Wu, F.; Bai, J.; Zhu, H.; Zou, J.; Li, K.; Xie, F.; Wang, Z. *Zhonghua Zhong Liu Za Zhi* **2002**, *24*, 278–281.
26. Chen, W.; Wang, Z.; Wu, F.; Zhu, H.; Zou, J.; Bai, J.; Li, K.; Xie, F. *Zhonghua Zhong Liu Za Zhi* **2002**, *24*, 612–615.
27. Ogan, K.; Cadeddu, J. A. *J. Endourol.* **2002**, *16*, 635–643.
28. Overgaard, J. *Int. J. Radiat. Oncol. Biol. Phys.* **1989**, *16*, 535–549.
29. Sneed, P. K.; Stauffer, P. R.; Gutin, P. H.; Phillips, T. L.; Suen, S.; Weaver, K. A.; Lamb, S. A.; Ham, B.; Prados, M. D.; Larson, D. A.; et al. *Neurosurgery* **1991**, *28*, 206–315.
30. Sneed, P. K.; Phillips, T. L. *Oncology (Huntingt)* **1991**, *5*, 99–108; discussion 109–110, 112.
31. Kremkau, F. W.; Kaufmann, J. S.; Walker, M. M.; Burch, P. G.; Spurr, C. L. *Cancer* **1976**, *37*, 1643–1647.
32. Yumita, N.; Okumura, K.; Nishigaki, R.; Umemura, K.; Umemura, S. *Jpn. J. Hyperthermic. Oncol.* **1987**, *3*, 175–182.
33. Saad, A. H.; Hahn, G. M. *Cancer Res.* **1989**, *49*, 5931–5934.
34. Loverock, P.; ter Haar, G.; Ormerod, M. G.; Imrie, P. R. *Br. J. Radiol.* **1990**, *63*, 542–546.
35. Harrison, G. H.; Balcer-Kubiczek, E. K.; Eddy, H. A. *Int. J. Radiat. Biol.* **1991**, *59*, 1453–1466.
36. Saad, A. H.; Hahn, G. M. *Ultrasound Med. Biol.* **1992**, *18*, 715–723.
37. Harrison, G. H.; Balcer-Kubiczek, E. K.; Gutierrez, P. L. *Ultrasound Med. Biol.* **1996**, *22*, 355–362.
38. Harrison, G. H.; Balcer-Kubiczek, E. K.; Gutierrez, P. L. *Radiat. Res.* **1996**, *145*, 98–101.
39. Tachibana, K.; Uchida, T.; Tamura, K.; Eguchi, H.; Yamashita, N.; Ogawa, K. *Cancer Lett.* **2000**, *149*, 189–194.
40. Liu, Y.; Cho, C. W.; Yan, X.; Henthorn, T. K.; Lillehei, K. O.; Cobb, W. N.; Ng, K. Y. *Pharm. Res.* **2001**, *18*, 1255–1261.
41. Liu, Y.; Lillehei, K.; Cobb, W. N.; Christians, U.; Ng, K. Y. *Biochem. Biophys. Res. Commun.* **2001**, *289*, 62–68.
42. Cheng, S. Q.; Zhou, X. D.; Tang, Z. Y.; Yu, Y.; Wang, H. Z.; Bao, S. S.; Qian, D. C. *J. Cancer Res. Clin. Oncol.* **1997**, *123*, 219–223.
43. Skinner, M. G.; Iizuka, M. N.; Kolios, M. C.; Sherar, M. D. *Phys. Med. Biol.* **1998**, *43*, 3535–3547.
44. Miller, M. W.; Miller, D. L.; Brayman, A. A. *Ultrasound Med. Biol.* **1996**, *22*, 1131–1154.
45. Liu, J.; Lewis, T. N.; Prausnitz, M. R. *Pharm. Res.* **1998**, *15*, 918–924.
46. Keyhani, K.; Guzman, H. R.; Parsons, A.; Lewis, T. N.; Prausnitz, M. R. *Pharm. Res.* **2001**, *18*, 1514–1520.
47. Guzman, H. R.; Nguyen, D. X.; McNamara, A. J.; Prausnitz, M. R. *J. Pharm. Sci.* **2002**, *91*, 1693–1701.
48. Hahn, G. M. *Hyperthermia and Cancer.* Plenum Press: New York, **1982**.
49. Ling, V. *Cancer Chemother. Pharmacol.* **1997**, *40 Suppl*, S3–S8.
50. Gottesman, M. M.; Pastan, I. *Annu. Rev. Biochem.* **1993**, *62*, 385–427.
51. Ambudkar, S. V.; Dey, S.; Hrycyna, C. A.; Ramachandra, M.; Pastan, I.; Gottesman, M. M. *Annu. Rev. Pharmacol. Toxicol.* **1999**, *39*, 361–398.

52. Ng, K. Y.; Cho, C. W.; Henthorn, T. K.; Tanguay, R. T. *J. Pharm. Sci.* **2004**, *93*, 896–907.

53. Cho, C. W.; Liu, Y.; Cobb, W. N.; Henthorn, T. K.; Lillehei, K.; Christians, U.; Ng, K. Y. *Pharm. Res.* **2002**, *19*, 1123–1129.

54. Sikic, B. I.; Fisher, G. A.; Lum, B. L.; Halsey, J.; Beketic-Oreskovic, L.; Chen, G. *Cancer Chemother. Pharmacol.* **1997**, *40 Suppl*, S13–S19.

55. Braunwald, E. *Harrison's 15th Edition Principles of Internal Medicine*. McGraw-Hill Professional: New York, **2001**.

56. Detre, K.; Holubkov, R.; Kelsey, S.; Cowley, M.; Kent, K.; Williams, D.; Myler, R.; Faxon, D.; Holmes, D., Jr.; Bourassa, M.; et al. *N. Engl. J. Med.* **1988**, *318*, 265–270.

57. Cowley, M. J.; Dorros, G.; Kelsey, S. F.; Van Raden, M.; Detre, K. M. *Am. J. Cardiol.* **1984**, *53*, 12C–16C.

58. Holmes, D. R., Jr.; Vlietstra, R. E.; Smith, H. C.; Vetrovec, G. W.; Kent, K. M.; Cowley, M. J.; Faxon, D. P.; Gruentzig, A. R.; Kelsey, S. F.; Detre, K. M.; et al. *Am. J. Cardiol.* **1984**, *53*, 77C–81C.

59. Ellis, S. G.; Roubin, G. S.; King, S. B. 3rd; Douglas, J. S., Jr; Shaw, R. E.; Stertzer, S. H.; Myler, R. K. *J. Am. Coll. Cardiol.* **1988**, *11*, 211–216.

60. Serruys, P. W.; Luijten, H. E.; Beatt, K. J.; Geuskens, R.; de Feyter, P. J.; van den Brand, M.; Reiber, J. H.; ten Katen, H. J.; van Es, G. A.; Hugenholtz, P. G. *Circulation* **1988**, *77*, 361–371.

61. Detre, K. M.; Holmes, D. R., Jr; Holubkov, R.; Cowley, M. J.; Bourassa, M. G.; Faxon, D. P.; Dorros, G. R.; Bentivoglio, L. G.; Kent, K. M.; Myler, R. K. *Circulation* **1990**, *82*, 739–750.

62. Fischman, D. L.; Leon, M. B.; Baim, D. S.; Schatz, R. A.; Savage, M. P.; Penn, I.; Detre, K.; Veltri, L.; Ricci, D.; Nobuyoshi, M.; et al. *N. Engl. J. Med.* **1994**, *331*, 496–501.

63. Collins, R.; Peto, R.; Baigent, C.; Sleight, P. *N. Engl. J. Med.* **1997**, *336*, 847–860.

64. Lincoff, A. M.; Topol, E. J. *Circulation* **1993**, *88*, 1361–1374.

65. Every, N. R.; Parsons, L. S.; Hlatky, M.; Martin, J. S.; Weaver, W. D. *N. Engl. J. Med.* **1996**, *335*, 1253–1260.

66. Kudo, S. *Tokyo Jikeikai Med. J.* **1989**, *104*, 1005–1012.

67. Lauer, C. G.; Burge, R.; Tang, D. B.; Bass, B. G.; Gomez, E. R.; Alving, B. M. *Circulation* **1992**, *86*, 1257–1264.

68. Francis, C. W.; Onundarson, P. T.; Carstensen, E. L.; Blinc, A.; Meltzer, R. S.; Schwarz, K.; Marder, V. J. *J. Clin. Invest.* **1992**, *90*, 2063–2068.

69. Blinc, A.; Francis, C. W.; Trudnowski, J. L.; Carstensen, E. L. *Blood* **1993**, *81*, 2636–2643.

70. Luo, H.; Steffen, W.; Cercek, B.; Arunasalam, S.; Maurer, G.; Siegel, R. J. *Am. Heart J.* **1993**, *125*, 1564–1569.

71. Harpaz, D.; Chen, X.; Francis, C. W.; Marder, V. J.; Meltzer, R. S. *J. Am. Coll. Cardiol.* **1993**, *21*, 1507–1511.

72. Olsson, S. B.; Johansson, B.; Nilsson, A. M.; Olsson, C.; Roijer, A. *Ultrasound Med. Biol.* **1994**, *20*, 375–382.

73. Harpaz, D.; Chen, X.; Francis, C. W.; Meltzer, R. S. *Am. Heart J.* **1994**, *127*, 1211–1219.

74. Akiyama, M.; Ishibashi, T.; Yamada, T.; Furuhata, H. *Neurosurgery* **1998**, *43*, 828–832; discussion 832–833.

75. Sakharov, D. V.; Hekkenberg, R. T.; Rijken, D. C. *Thromb. Res.* **2000**, *100*, 333–340.
76. Behrens, S.; Spengos, K.; Daffertshofer, M.; Wirth, S.; Hennerici, M. *Echocardiography* **2001**, *18*, 259–263.
77. Kornowski, R.; Meltzer, R. S.; Chernine, A.; Vered, Z.; Battler, A. *Circulation* **1994**, *89*, 339–344.
78. Kashyap, A.; Blinc, A.; Marder, V. J.; Penney, D. P.; Francis, C. W. *Thromb. Res.* **1994**, *76*, 475–485.
79. Larsson, J.; Carlson, J.; Olsson, S. B. *Br. J. Ophthalmol.* **1998**, *82*, 1438–1440.
80. Francis, C. W. *Echocardiography* **2001**, *18*, 239–246.
81. Francis, C. W.; Blinc, A.; Lee, S.; Cox, C. *Ultrasound Med. Biol.* **1995**, *21*, 419–424.
82. Siddiqi, F.; Blinc, A.; Braaten, J.; Francis, C. W. *Thromb. Haemost.* **1995**, *73*, 495–498.
83. Braaten, J. V.; Goss, R. A.; Francis, C. W. *Thromb. Haemost.* **1997**, *78*, 1063–1068.
84. Everbach, E. C.; Francis, C. W. *Ultrasound Med. Biol.* **2000**, *26*, 1153–1160.
85. Suchkova, V.; Siddiqi, F. N.; Carstensen, E. L.; Dalecki, D.; Child, S.; Francis, C. W. *Circulation* **1998**, *98*, 1030–1035.
86. Suchkova, V. N.; Baggs, R. B.; Francis, C. W. *Circulation* **2000**, *101*, 2296–2301.
87. Siegel, R. J.; Atar, S.; Fishbein, M. C.; Brasch, A. V.; Peterson, T. M.; Nagai, T.; Pal, D.; Nishioka, T.; Chae, J. S.; Birnbaum, Y.; Zanelli, C.; Luo, H. *Echocardiography* **2001**, *18*, 247–257.
88. Fry, F. J.; Barger, J. E. *J. Acoust. Soc. Am.* **1978**, *63*, 1576–1590.
89. Behrens, S.; Spengos, K.; Daffertshofer, M.; Schroeck, H.; Dempfle, C. E.; Hennerici, M. *Ultrasound Med. Biol.* **2001**, *27*, 1683–1689.
90. Daffertshofer, M.; Fatar, M. *Eur. J. Ultrasound* **2002**, *16*, 121–130.
91. The human genome 2001. *Nature* **2001**, *409*, 813–958.
92. Naik, A.; Kalia, Y. N.; Guy, R. H. *PSTT* **2000**, *9*, 318–326.
93. Barry, B. W. *Dermatological Formulations: Percutaneous Absorption*. Marcel Dekker: New York, **1983**.
94. Scheuplein, R. J.; Blank, I. H. *Physiol. Rev.* **1971**, *51*, 702–747.
95. Barry, B. M.; William, A. C. In *Encyclopedia of Pharmaceutical Technology, Vol. 11*; Swarbrick, J.; Boylan, J. C.; Eds. Marcel Dekker: New York, **1995**, pp. 449–493.
96. Barry, B. W. *Eur. J. Pharm. Sci.* **2001**, *14*, 101–114.
97. Asbill, C. S.; Michniak, B. B. *PSTT* **2000**, *3*, 36–41.
98. Machet, L.; Boucaud, A. *Int. J. Pharm.* **2002**, *243*, 1–15.
99. Mitragotri, S.; Edwards, D. A.; Blankschtein, D.; Langer, R. *J. Pharm. Sci.* **1995**, *84*, 697–706.
100. Gaertner, W. *J. Acoust. Soc. Am.* **1954**, *26*, 977.
101. Apfel, R. E. *IEEE Trans. Ultrason. Ferroelectr. Freq. Control* **1986**, *UFFC-33*, 139.
102. Tachibana, K.; Tachibana, S. *Anesthesiology* **1993**, *78*, 1091–1096.
103. Zhang, I.; Shung, K. K.; Edwards, D. A. *J. Pharm. Sci.* **1996**, *85*, 1312–1316.
104. Lee, S.; Smith, N. B. *Proc. IEEE* **2002**, Ultrason Symp, Munich, Germany, **2002**.
105. Maione, E.; Shung, K. K.; Meyer, R. J., Jr; Hughes, J. W.; Newnham, R. E.; Smith, N. B. *IEEE Trans. Ultrason. Ferroelectr. Freq. Control* **2002**, *49*, 1430–1436.

106. Smith, N. B.; Lee, S.; Maione, E.; Roy, R. B.; McElligott, S.; Shung, K. K. *Ultrasound Med. Biol.* **2003**, *29*, 319–325.

107. Nyborg, W. L. *Ultrasound Med. Biol.* **2001**, *27*, 301–333.

108. Tezel, A.; Sens, A.; Mitragotri, S. *Pharm. Res.* **2002**, *19*, 1841–1846.

109. Merino, G.; Kalia, Y. N.; Delgado-Charro, M. B.; Potts, R. O.; Guy, R. H. *J. Control Release* **2003**, *88*, 85–94.

110. Yamashita, N.; Tachibana, K.; Ogawa, K.; Tsujita, N.; Tomita, A. *Anat. Rec.* **1997**, *247*, 455–461.

111. Wu, J.; Chappelow, J.; Yang, J.; Weimann, L. *Ultrasound Med. Biol.* **1998**, *24*, 705–710.

112. Tang, H.; Wang, C. C.; Blankschtein, D.; Langer, R. *Pharm. Res.* **2002**, *19*, 1160–1169.

113. Tang, H.; Mitragotri, S.; Blankschtein, D.; Langer, R. *J. Pharm. Sci.* **2001**, *90*, 545–568.

114. Perez-Iratxeta, C.; Bork, P.; Andrade, M. A. *Nat. Genet.* **2002**, *31*, 316–319.

115. Friedmann, T. *Science* **1989**, *244*, 1275–1281.

116. Somia, N.; Verma, I. M. *Nat. Rev. Genet.* **2000**, *1*, 91–99.

117. Niidome, T.; Huang, L. *Gene Ther.* **2002**, *9*, 1647–1652.

118. Lu, Q. L.; Bou-Gharios, G.; Partridge, T. A. *Gene Ther.* **2003**, *10*, 131–142.

119. Fechheimer, M.; Boylan, J. F.; Parker, S.; Sisken, J. E.; Patel, G. L.; Zimmer, S. G. *Proc. Natl. Acad. Sci. USA* **1987**, *84*, 8463–8467.

120. Fechheimer, M.; Taylor, D. L. *Methods Cell Biol.* **1987**, *28*, 179–190.

121. Joersbo, M.; Brunstedt, J. *J. Virol. Methods* **1990**, *29*, 63–69.

122. Holmes, R. P.; Yeaman, L. D.; Taylor, R. G.; McCullough, D. L. *J. Urol.* **1992**, *147*, 733–737.

123. Gambihler, S.; Delius, M.; Ellwart, J. W. *J. Membr. Biol.* **1994**, *141*, 267–275.

124. Delius, M.; Adams, G. *Cancer Res.* **1999**, *59*, 5227–5232.

125. Huber, P. E.; Jenne, J.; Debus, J.; Wannenmacher, M. F.; Pfisterer, P. *Ultrasound Med. Biol.* **1999**, *25*, 1451–1457.

126. Delius, M.; Hofschneider, P. H.; Lauer, U.; Messmer, K. *Lancet* **1995**, *345*, 1377.

127. Lauer, U.; Burgelt, E.; Squire, Z.; Messmer, K.; Hofschneider, P. H.; Gregor, M.; Delius, M. *Gene Ther.* **1997**, *4*, 710–715.

128. Miller, D. L.; Bao, S.; Gies, R. A.; Thrall, B. D. *Ultrasound Med. Biol.* **1999**, *25*, 1425–1430.

129. Bao, S.; Thrall, B. D.; Miller, D. L. *Ultrasound Med. Biol.* **1997**, *23*, 953–959.

130. Bao, S.; Thrall, B. D.; Gies, R. A.; Miller, D. L. *Cancer Res.* **1998**, *58*, 219–221.

131. Tata, D. B.; Dunn, F.; Tindall, D. J. *Biochem. Biophys. Res. Commun.* **1997**, *234*, 64–67.

132. Greenleaf, W. J.; Bolander, M. E.; Sarkar, G.; Goldring, M. B.; Greenleaf, J. F. *Ultrasound Med. Biol.* **1998**, *24*, 587–595.

133. Harrison, G. S.; Wang, Y.; Tomczak, J.; Hogan, C.; Shpall, E. J.; Curiel, T. J.; Felgner, P. L. *Biotechniques* **1995**, *19*, 816–823.

134. Zabner, J.; Fasbender, A. J.; Moninger, T.; Poellinger, K. A.; Welsh, M. J. *J. Biol. Chem.* **1995**, *270*, 18997–19007.

135. Huber, P. E.; Pfisterer, P. *Gene Ther.* **2000**, *7*, 1516–1525.

136. Anwer, K.; Kao, G.; Proctor, B.; Anscombe, I.; Florack, V.; Earls, R.; Wilson, E.; McCreery, T.; Unger, E.; Rolland, A.; Sullivan, S. M. *Gene Ther.* **2000**, *7*, 1833–1839.

137. Amabile, P. G.; Waugh, J. M.; Lewis, T. N.; Elkins, C. J.; Janas, W.; Dake, M. D. *J. Am. Coll. Cardiol.* **2001**, *37*, 1975–1980.
138. Lawrie, A.; Brisken, A. F.; Francis, S. E.; Cumberland, D. C.; Crossman, D. C.; Newman, C. M. *Gene Ther.* **2000**, *7*, 2023–2027.
139. Lawrie, A.; Brisken, A. F.; Francis, S. E.; Tayler, D. I.; Chamberlain, J.; Crossman, D. C.; Cumberland, D. C.; Newman, C. M. *Circulation* **1999**, *99*, 2617–2620.
140. Unger, E. C.; McCreery, T. P.; Sweitzer, R. H. *Invest. Radiol.* **1997**, *32*, 723–727.
141. Aoyama, T.; Hosseinkhani, H.; Yamamoto, S.; Ogawa, O.; Tabata, Y. *J. Control Release* **2002**, *80*, 345–356.
142. Gramiak, R.; Shah, P. M. *Invest. Radiol.* **1968**, *3*, 356–366.
143. Forsberg, F.; Merton, D. A.; Liu, J. B.; Needleman, L.; Goldberg, B. B. *Ultrasonics* **1998**, *36*, 695–701.
144. Unger, E. C.; Lund, P. J.; Shen, D. K.; Fritz, T. A.; Yellowhair, D.; New, T. E. *Radiology* **1992**, *185*, 453–456.
145. Unger, E. C.; Fritz, T. A.; Matsunaga, T.; Ramaswami, V.; Yellowhair, D.; Wu, G. Therapeutic Drug Delivery Systems, U.S. Patent # 5,580, 575.
146. Schneider, M.; Arditi, M.; Barrau, M. B.; Brochot, J.; Broillet, A.; Ventrone, R.; Yan, F. *Invest. Radiol.* **1995**, *30*, 451–457.
147. Feinstein, S. B.; Ten Cate, F. J.; Zwehl, W.; Ong, K.; Maurer, G.; Tei, C.; Shah, P. M.; Meerbaum, S.; Corday, E. *J. Am. Coll. Cardiol.* **1984**, *3*, 14–20.
148. Unger, E. C.; Lund, P. *Radiology* **1991**, *181*, 225.
149. Singhal, S.; Moser, C. C. *Langmuir* **1993**, *9*, 2426–2429.
150. Narayan, P.; Wheatley, M. A. *Polym. Eng. Sci.* **1999**, *30*, 2242–2255.
151. Schlief, R. In Goldberg, B.B., Ed. *Ultrasound Contrast Agents*. Martin Dunitz: London, pp. **1997**, 75–82.
152. Forsberg, F.; Liu, J. B.; Merton, D. A.; Rawool, N. M.; Goldberg, B. B. *J. Ultrasound Med.* **1995**, *14*, 949–957.
153. Northfelt, D. W.; Martin, F. J.; Working, P.; Volberding, P. A.; Russell, J.; Newman, M.; Amantea, M. A.; Kaplan, L. D. *J. Clin. Pharmacol.* **1996**, *36*, 55–63.
154. Mulvagh, S. L.; DeMaria, A. N.; Feinstein, S. B.; Burns, P. N.; Kaul, S.; Miller, J. G.; Monaghan, M.; Porter, T. R.; Shaw, L. J.; Villanueva, F. S. *J. Am. Soc. Echocardiogr.* **2000**, *13*, 331–342.
155. Wei, K.; Kaul, S. *Curr. Opin. Cardiol.* **1997**, *12*, 539–546.
156. Lindner, J. R.; Villanueva, F. S.; Dent, J. M.; Wei, K.; Sklenar, J.; Kaul, S. *Am. Heart J.* **2000**, *139*, 231–240.
157. Lindner, J. R.; Ismail, S.; Spotnitz, W. D.; Skyba, D. M.; Jayaweera, A. R.; Kaul, S. *Circulation* **1998**, *98*, 2187–2194.
158. Lindner, J. R.; Coggins, M. P.; Kaul, S.; Klibanov, A. L.; Brandenburger, G. H.; Ley, K. *Circulation* **2000**, *101*, 668–675.
159. Lindner, J. R.; Dayton, P. A.; Coggins, M. P.; Ley, K.; Song, J.; Ferrara, K.; Kaul, S. *Circulation* **2000**, *102*, 531–538.
160. Lanza, G. M.; Wallace, K. D.; Fischer, S. E.; Christy, D. H.; Scott, M. J.; Trousil, R. L.; Cacheris, W. P.; Miller, J. G.; Gaffney, P. J.; Wickline, S. A. *Ultrasound Med. Biol.* **1997**, *23*, 863–870.

161. Lindner, J. R.; Song, J.; Christiansen, J.; Klibanov, A. L.; Xu, F.; Ley, K. *Circulation* **2001**, *104*, 2107–2112.

162. Schumann, P. A.; Christiansen, J. P.; Quigley, R. M.; McCreery, T. P.; Sweitzer, R. H.; Unger, E. C.; Lindner, J. R.; Matsunaga, T. O. *Invest. Radiol.* **2002**, *37*, 587–593.

163. Apfel, R. E.; Holland, C. K. *Ultrasound Med. Biol.* **1991**, *17*, 179–185.

164. Tachibana, K.; Tachibana, S. *Circulation* **1995**, *92*, 1148–1150.

165. Porter, T. R.; LeVeen, R. F.; Fox, R.; Kricsfeld, A.; Xie, F. *Am. Heart J.* **1996**, *132*, 964–968.

166. Nishioka, T.; Luo, H.; Fishbein, M. C.; Cercek, B.; Forrester, J. S.; Kim, C. J.; Berglund, H.; Siegel, R. J. *J. Am. Coll. Cardiol.* **1997**, *30*, 561–568.

167. Mizushige, K.; Kondo, I.; Ohmori, K.; Hirao, K.; Matsuo, H. *Ultrasound Med. Biol.* **1999**, *25*, 1431–1437.

168. Unger, E. C. In: *Interventional Cardiology.* Jerusalem, Isreal, **1999** (abstract).

169. Matsumura, T.; Moriyasu, F.; Tabata, Y. *J. Ultrasound Med.* **1998**, *16*, S28 (abstract).

170. Vannan, M.; McCreery, T.; Li, P.; Han, Z.; Unger, E.; Kuersten, B.; Nabel, E.; Rajagopalan, S. *J. Am. Soc. Echocardiogr.* **2002**, *15*, 214–218.

171. Lu, Q. L.; Liang, H. D.; Partridge, T.; Blomley, M. J. *Gene Ther.* **2003**, *10*, 396–405.

13

POLYCATIONIC PEPTIDES AND PROTEINS IN DRUG DELIVERY: FOCUS ON NONCLASSICAL TRANSPORT

LISA A. KUELTZO

Department of Pharmaceutical Sciences, University of Colorado Health Sciences Center, Denver, CO 80262

C. RUSSELL MIDDAUGH

Department of Pharmaceutical Chemistry, University of Kansas, Lawrence, KS 66047

Drug Delivery: Principles and Applications Edited by Binghe Wang, Teruna Siahaan, and Richard Soltero
ISBN 0-471-47489-4

13.1. INTRODUCTION

The use of polycationic moieties in drug delivery has been a subject of considerable interest for the past several decades. One of the earliest demonstrations of the potential value of polycations in drug delivery was a study by Tan in 1977, which demonstrated increased uptake of the SV40 virus when infections were carried out in the presence of polycations.[1] Other early efforts involved direct cationization of proteins to enhance cellular uptake.[2,3] Simultaneously, extensive investigations of the ability of a variety of polycationic vehicles to enhance the uptake of DNA were made.[4] These vehicles include cationic liposomes,[5,6] poly-L-lysine,[7] polyetheleneimine,[8] amino-dendrimers,[9] and a variety of other natural and synthetic polycations. Only in recent years, however, has the surprisingly diverse potential of polycation-based delivery systems been realized. This revelation has come primarily from the discovery and utilization of nonclassical transport-based pathways.

"Nonclassical transport" denotes a previously unrecognized method(s) of entry into and/or exit from cells. Proteins, peptides, and other polymers that demonstrate nonclassical transport abilities are thought to pass through cellular membranes in an efficient and energy-independent fashion, utilizing neither classic endocytotic nor Golgi-based exocytotic pathways.[10,11] More than 20 proteins have been found to possess nonclassical secretory properties,[10] and it seems likely that there exist many more with this ability. These include interleukin 1β,[12] fibroblast growth factors 1 and 2,[13] and chick ciliary neurotrophic factor.[14] More importantly, in the context of drug delivery, some of these proteins [human immuno deficiency virus (HIV) Tat,[15] the herpesvirus protein VP22,[16] and homeodomain proteins[17,18]], as well as a large number of cationic peptides,[19,20] also demonstrate nonclassical import activity. The discovery of this pathway has led to numerous successful attempts to utilize these vectors as delivery vehicles. By covalently (and in some cases noncovalently) attaching macromolecular cargoes to these vehicles, it has proven possible to tow the cargoes through cell membranes with no loss of biological activity of the conveyed molecule. Additionally, it has been found that many of these proteins and peptides are able to localize to the nucleus, even with a cargo attached.

The discovery of nonclassical delivery vectors opens up an entirely new approach to drug delivery, one that appears not to be limited by the intrinsic disadvantages of endocytotic-based methods. These disadvantages include the loss of bioavailability produced by degradation of therapeutics in lysosomes[21] (for which endosomes are precursors), as well as limitations in the tissues that can be targeted, since endocytosis is typically triggered by specific protein-receptor interactions on the cell surface. Additionally, the vectors frequently offer the possibility of nuclear localization, a transport pathway that is normally strictly limited by the size of the unactivated nuclear pore.[22,23] This latter aspect is of particular importance for gene therapy applications, which usually require nuclear localization for activity.

In this chapter we present an overview of current research in the field of nonclassical transport-based delivery. Examples of major vectors are described in brief, focusing on applications as well as physical properties such as the vehicles' unusual polycationic character. Methods of complexation of cargoes to vectors are also described. In addition, although no consensus has yet been reached regarding the details of the nonclassical transport mechanism, a variety of hypotheses and models are described in detail and evaluated in the context of applicability to general mechanisms of nonclassical transport.

13.2. CURRENT VECTORS

13.2.1. VP22

The protein VP22 is found in the amorphous tegument layer of herpes simplex virus 1 (HSV-1).[24] A relatively small (36 kDa), highly basic protein, it has been shown to undergo covalent modification (phosphorylation[25] and nucleotidylation[26]) and interacts with cellular components such as microtubules and chromatin.[16,27–29] Although the exact function of VP22 in the HSV-1 assembly and infection pathways is not yet known, a perhaps related function of the protein has been demonstrated, which has sparked considerable interest: VP22 demonstrates both nonclassical import and export activity. As originally demonstrated by Elliott and colleagues, the protein enters and exits cells in an energy-independent manner, although it lacks either a known N-terminal secretion signal or a nuclear localization sequence (NLS).[16] This latter absence is surprising because the protein is also found in the nucleus of the cell after import. (It is possible that the highly basic C-terminal region of the protein may in fact serve as an NLS, but this has yet to be directly demonstrated.)

Since its initial discovery, VP22 has successfully delivered cargoes in both expression-based systems (see Section 13.3.1)[16,30–33] and by direct application to cells in culture.[16] The cargoes can be attached to either the N or C terminus of the protein, and delivery has been achieved in both actively dividing and terminally differentiated cell lines.[30,33] In addition to direct macromolecule delivery, the VP22 system has been successfully employed in the enhancement of suicide gene therapy systems,[55,56] viral gene therapy,[57–59] and viral vaccines[52] (Table 13.1). A recent

TABLE 13.1 Examples of Nonclassical Transport Vectors and Their Applications

	VP22	Tat	FGFs	Penetratin	Cationic Cell-Penetrating Peptides*
Proteins (direct delivery)	Y[a] [16]	*Y*[b] [15,34,35]	Y [36–38]	Y[c] [39]	Y [40,41]
Peptides	Y [16]	Y [42]	ND	*Y* [43–45]	Y [20,46]
Oligonucleotides	Y [47]	Y [48]	ND	Y [18,49]	ND
PNAs	ND	Y [50]	ND	*Y* [45,51]	Y [40]
Viral delivery	ND	ND	ND	ND	ND
Vaccine enhancement[d]	Y [52]	ND	ND	ND	ND
DNA	ND	Y [53,54]	ND	ND	ND
Plasmid expression enhancement[d]	*Y* [55–60]	ND	ND	ND	ND
Particulates (>40 nm)	ND	Y [61,62]	ND	ND	ND

*Excluding Tat and penetratin.
[a]Y = yes; ND = not determined.
[b]Italics indicate successful use *in vivo*.
[c]See text for additional information (p. 284).
[d]Enhancement: the vector was not used for delivery, but rather was incorporated into a previously developed system (e.g., viral or nonviral) to enhance delivery/spread of a protein cargo.
FGF: fibroblast growth factor;
PNA: peptide nucleic acid.

study by Cheng and colleagues shows an increased antitumor effect for a replicon particle vaccine against human papillomavirus containing DNA that encoded a VP22-antigen fusion protein compared to a vaccine encoding the antigen alone.[52] Despite numerous delivery successes both *in vitro* and *in vivo*, however, the mechanism of VP22-dependent transport remains unknown.

Although the three-dimensional structure of VP22 is not known, spectroscopic studies of a transport-competent core protein (residues 159–301) have provided provocative information. The core protein is highly temperature labile. Under physiological conditions of temperature and pH, it appears to adopt a molten globule (MG)-like conformation.[63] (In MG states, the majority of tertiary structure is lost, while the majority of secondary structure remains intact.) Deletion studies have also been conducted to pinpoint the transport-competent region(s) of the protein. Initial studies found that deletion of the C-terminal region of the protein (residues 267–301) eliminated transport activity.[16] Further studies by Aints and colleagues, however, found that while the C-terminal region does facilitate VP22 transport, there is a core region (residues 81–195) that also contributes significantly to transport.[64] These studies suggest that the key to VP22's unique transport activity lies in some element of its structural characteristics. In addition to cellular components, VP22 has been shown to interact with polyanions such as heparin,[63] sucrose octasulfate (Kueltzo and Middaugh, unpublished data), and oligonucleotides.[47,63] The protein also interacts with lipid-like fluorescent probes[63] and negatively charged liposomes[65] in a temperature-dependent manner. The potential role of these interactions in the mechanism of VP22 transport is discussed below.

13.2.2. Fibroblast Growth Factors

The fibroblast growth factor (FGF) protein family is both large (23 known members) and diverse.[66] FGFs exhibit a variety of biological functions involving the stimulation of cell proliferation,[67] including functions such as general mitogenic activity (FGFs 1, 2, and 9)[68,69] and stimulation of dermal proliferation (FGFs 7 and 10).[70] The most intriguing FGF property, in the context of this work, is the nonclassical secretory activity possessed by a number of FGFs (FGFs 1, 2, 9, 16, and 20).[66,67,69] This activity has been investigated in greatest detail for FGF-1 and -2. A number of studies have suggested that cellular components such as the cytoskeleton (specifically F-actin),[71] molecular chaperones,[72] elements of the heat shock system,[73,74] and even a multiprotein secretory complex[75,76] may be involved in the FGF-1 and -2 secretion pathways.

Unlike the majority of potential nonclassical transport vectors, FGF import is primarily receptor mediated, involving both specific cell surface receptors[77] and low-affinity proteoglycan binding sites.[67,68,78] The differential distribution of receptor subtypes in tissues is thought to provide a conventional targeting system for FGF-based vectors.[66] Internalization of FGF fusion proteins, however, has also been observed, including such cargo proteins as *Pseudomonas* exotoxin,[36,79,80] diphtheria toxin,[37] and the constant region of human immunoglobulin G (IgG).[38] These studies suggest that the FGFs may possess true nonclassical import activity in addition to their demonstrated secretory activity. Evidence that this import is abolished in the presence of polyanions suggests that cell surface polyanions may play a role in the FGF import pathway[37] or that partially unfolded states of the protein may be necessary for transport (polyanions lock some FGFs into their native states; see Section 13.4.1.2). This is not surprising, since many of the FGF proteins strongly bind polyanions,[67,81,82] and cell surface proteoglycans play a significant role in other FGF-related processes such as formation of the FGF/FGF receptor complex.[77,83]

Unlike VP22, members of the FGF family contain primarily β sheet secondary structure.[66] The similarities between the proteins exist in their gross properties (low molecular weight, net positive charge under neutral pH conditions) as well as their thermal stabilities. Like VP22, many of the FGFs (at least FGF-1, -2, -7, and -10) are highly thermally labile[82,84] under physiological conditions, with FGF-1 well established as existing in an MG-like conformation under physiological conditions. The interaction of FGF-1 with negatively charged lipid bilayers is apparently dependent on this MG-like state, permitting partial penetration of the bilayer when the MG conformation is induced.[85] FGF-1 and -2 are also believed to localize to the nucleus of cells, utilizing apparent NLSs,[86–89] although this has been disputed.

13.2.3. HIV-1 Tat

The Tat protein was one of the first nonclassical transport vectors to be recognized. A nuclear transcription activating protein of the HIV-1 retrovirus,[90] Tat was initially studied in an effort to elucidate its role in the HIV-1 infection process. As Frankel, Pabo, Green, and Lowenstein discovered independently in the late 1980s, however, Tat also possesses potent nonclassical transport activity.[91,92] Studies of an 86-residue form of Tat have shown that the protein possesses both nonclassical secretion[93] and import activity,[91,92] as well as postinternalization nuclear localization.[94] Current research in the Tat transport field is concentrated on the Tat protein transduction domain (PTD), a highly basic core peptide (residues 48–57, Table 13.2) that appears to be the functional unit for Tat-based transport.[35,95]

Although both the Tat protein[15] and the Tat PTD[34,96] have been shown to transport proteins into cells, the majority of delivery studies have been conducted with the Tat PTD, presumably due to the simplicity of the smaller peptide-based system. The Tat PTD has successfully transported a wide variety of macromolecular and particulate cargoes (Table 13.1). The conjugation methods employed include covalent linkage[61] as well as direct fusion of the Tat PTD to the cargo molecule by peptide synthesis[42] or plasmid expression.[35] The Tat system has been successfully employed both *in vitro* and *in vivo*. Somewhat surprisingly, Tat-protein conjugates have even been observed to cross the blood-brain barrier.[35,97]

Despite the data available from numerous investigations of the Tat delivery system, the mechanism of import remains elusive. One potential component of the transport process is the cell-surface proteoglycan heparan sulfate (HS). Although an essential interaction with HS has been observed for the full-length Tat protein import,[98,99] the requirement for a similar interaction with Tat PTD import remains ambiguous.[100,101] This suggests that the Tat PTD and the full-length protein may utilize different import pathways. This possibility is supported by a study by Silhol and colleagues showing that the Tat PTD does not inhibit internalization of the full-length protein.[100] A second potential component in the system is an apparent lack of dependence of import on rigid secondary or tertiary structure. This argument is supported by evidence that (1) partial unfolding of the cargo protein is required for translocation of full-length Tat fusion constructs,[102,103] and that (2) the amphipathicity of the Tat peptide (suggested by modeling studies[104]) is not required for import. This latter point is supported by observations that partial D-amino acid analogs of the Tat PTD also possess highly efficient translocation activity.[20]

13.2.4. Homeoproteins: Penetratin

The nonclassical transport activity of the homeoprotein family of proteins was first demonstrated by Perez and colleagues during their studies of the *Drosophila* antennapedia homeoprotein (Antp).[105] A class of transactivating factors, the homeoproteins interact with DNA through a small (~60aa), highly conserved helical region called the "homeodomain."[106] It was the homeodomain of Antp that was originally thought to enable the nonclassical internalization and nuclear localization observed for this protein.[105,107] Additional studies, however, localized the activity to a 16-amino acid polycationic peptide (residues 43–58; Table 13.2), now designated Penetratin.[49,107,108] (Interestingly, although the Penetratin peptide now serves as the model for nonclassical homeodomain-derived transport, it has been suggested that an extended sequence actually internalizes more efficiently than the minimal peptide vector.[108]) Since the discovery of the Penetratin peptide, similar activity has been observed in other members of the homeoprotein family, including Hoxa-5,[109] Engrailed,[110] and Islet-1.[39] The Engrailed protein also exhibits nonclassical secretory activity,[111] which may be true for the many members of the homeoprotein family that lack secretion signal peptides.[105] This suggests that the potential for nonclassical import may exist throughout the homeoprotein family, especially in light of the high sequence conservation of the transport-competent homeodomain region.

Similar to the Tat PTD, Penetratin is a highly basic peptide possessing helical character, although once again, it has been shown that the intrinsic secondary structure of the peptide is not required for internalization.[106,107] As shown in Table 13.1, Penetratin has been successfully employed to deliver protein and DNA-based macromolecules both *in vivo* and *in vitro*.[112] Unlike other nonclassical transport-based vectors, however, the delivery capabilities of Penetratin appear to be limited by cargo size (100 residues for peptides and 55 base pairs for oligonucleotides).[112] Two exceptions to this rule are the translocation of the parent Antp protein as well as a study by Kilk and colleagues that demonstrates transport of a ~63 kDa streptavidin-biotin complex.[39] It is therefore not entirely clear that cargo size limitation is indeed a property of Penetratin delivery, nor is it known if this limitation will apply to other homeoprotein-derived vectors. Unlike other vectors, although the precise transport pathway has not been conclusively established for Penetratin, a specific internalization mechanism has been postulated, involving the formation of an inverted hexagonal phase in lipid bilayers[107] (see Section 13.4.2.1).

13.2.5. Cell-Penetrating Peptides

The number of peptides demonstrating nonclassical import activity has risen dramatically in recent years (see the reviews by Lindgren et al.,[19] Wadia and Dowdy,[113] and others[114–117]). This now extensive group of potential delivery vectors can be divided into two general groups: (1) cationic peptides and (2) amphipathic helices, although some miscellaneous peptides are also included in the latter group. In order to focus on the central subject of this chapter, polycation-based delivery, we choose to include in this discussion only the former group of cationic peptides. For information on the latter group, the reader is referred to selected amphipathic peptides in Table 13.2 and references therein.

The cationic peptide family includes both synthetic and naturally derived peptides, all of which have been successfully employed for both *in vitro* and *in vivo* delivery of a variety of macromolecules (Table 13.1). The prototypic example of a naturally derived cationic peptide is the Tat PTD, described above. Other examples include the pVEC peptide, which has been successfully used for both protein and oligonucleotide delivery,[40] as well as a series of small, RNA-binding peptides discovered by Futaki and colleagues.[20] The Penetratin vector is also included in this group. The inclusion of Tat and Penetratin in the cationic peptide group arises from the intrinsic definition of these peptides. *In general, the cationic delivery peptides*

TABLE 13.2 Select Cell-Penetrating Peptides

Vector	Source	Sequence	Type	NL[a]	Ref.
Tat	HIV-1 Tat	**RKKRRQRRR**	Cationic	Y	35
Penetratin	*Drosophila* antennapedia homeodomain	**R**QIKIWFQN**RRMKWKK**	Cationic	Y	108
Polyarginines	Synthetic	$(\mathbf{R})_n$ (n = 6–8, optimally)	Cationic	Y	20, 46
*p*VEC	Murine vascular endothelium cadherin	LLIL**RRRIRK**QAHAHS**K**	Cationic	NA	40
Rev (34–50)	HIV 1 Rev	T**R**QA**RR**N**RRRR**-W**R**E**R**Q**R**GC	Cationic	Y	20
Pep-1	Synthetic	**K**ETWWETWW-TEWSQP**KKKRK**V	Linked cationic/ hydrophobic	Y	118
E[rns] peptide	CSFV[b]	**R**QGAA**R**VTSWLG**R**QL-**R**IAG**KR**LEG**R**S**K**	Amphipathic	Y	119
PreS2-TLM	Hepatitis B virus surface protein	PLSSIFS**R**IGDP	Amphipathic	N	120
Transportan	Synthetic (galanin, mastoparan)	GWTLNSAGYLLG-**K**INL**K**ALAALA**KK**IL	Amphipathic	Y	121
Oehlke peptide	Synthetic	**K**LAL**K**LAL**K**AL**K**-AAL**K**LA	Amphipathic	N	122

[a]NL: nuclear localization.
[b]CSFV: classical swine fever virus.

are short (< 20 residues), often rich in arginine, and appear to have no secondary structure requirement for nonclassical transport.[20] As seen in Table 13.2, both Tat and Penetratin fit the primary sequence definition; the evidence that the helicity of the peptides is not required for transport was indicated previously. The recognition of these properties has also led to the construction of synthetic vectors, most notably simple polyarginine peptides. It is not surprising, based on our knowledge of the natural vectors, that synthetic polyarginine vectors as originally developed by Futaki and others demonstrate nonclassical transport.[20] What is surprising, however, is that the efficiency of this transport equals, and in some cases exceeds,[123] the transport activity of the Tat PTD, often a benchmark in the field. Interestingly, the length of the polyarginine region appears to relate directly to transport efficiency.[20] Continued investigation of synthetic vectors has shown that transport is achievable with polylysine-based peptides as well, although transport is usually less efficient than that of analogous polyarginine vectors.[123] Together, these studies reaffirm the requirement for cationic residues, especially arginine, in the transport of these peptides and suggest a specific role for a high degree of positive charge in the transport activity of these vectors (see Section 13.4.1.1).

13.3. APPLICATIONS: VECTOR CONSTRUCTION

The vectors currently under investigation that employ nonclassical transport represent a diverse group of entities. This diversity has required that a variety of protocols be developed for both the conjugation of the vectors to cargo molecules and the presentation of the vector-cargo complex to the target cells. In this section, an overview of the methods most commonly employed for vector production is presented. For more detailed information on the actual protocols, the reader is referred to the specific references mentioned herein.

13.3.1. Plasmid Expression

In this approach, the vector cargo construct is encoded within a plasmid as a fusion protein or peptide, depending on the exact vector/cargo combination. This method has been used primarily with the Tat PTD and the VP22 delivery systems, although the exact methods differ. In the Tat system, the resulting fusion construct is normally expressed and highly purified.[41] The purified protein is then introduced to cells in culture or, as shown to be highly effective by Schwarze and colleagues, can be directly injected into a host.[35] Alternatively, the VP22 system commonly involves transfecting cells with the construct-encoding plasmid itself and mixing these cells with target cells.[55] The expressed protein is then observed to spread from the primary expressing cell to the surrounding cells. It should be noted, however, that this spread has been postulated to be an artifact of the imaging protocols used in these studies (see Section 13.4.4). An additional use of plasmid expression involves incorporation of a vector-target construct into a conventional gene therapy delivery vehicle to enhance the delivery of the expressed therapeutic.[52] Again, this has been used predominantly with the VP22 delivery system, most commonly in viral gene therapy delivery vehicles,[57,58] although this approach could quite readily be used with the other vectors described above. It does, however, then involve the new problem of transfection with the encoding DNA itself, something typically accomplished with cationic vehicles such as lipids or polyethyleneimine with only moderate to low efficiency.[124]

13.3.2. Direct Synthesis

A number of methods have been developed to create vector-cargo conjugates by direct chemical synthesis. These methods are more widely applicable than plasmid expression methods, since they do not rely on any inherent character of the delivery vector or cargo (e.g., a requirement that both be composed of L-amino acids). The delivery protocol for the majority of these constructs involves direct application of the complex to cells in culture, which has been employed with significant success.

Although, as mentioned above, amino acid character is not always required for direct synthetic methods, it is necessary for the first method in the group, peptide synthesis. In this protocol, the constructs are synthesized using commercially available synthesis techniques and instruments, with or without a spacer amino acid between the delivery and cargo sequence. Direct synthesis has been used with the Tat PTD,[42] the Penetratin peptide,[43,44] and the pVEC peptide.[40] A second method that has been used extensively is disulfide bond linkage. This involves introducing a free thiol group into both the cargo and vector molecule, either through chemical modification or, as shown in Figure 13.1, by taking advantage of intrinsic thiol functional groups (e.g., cysteine residues). The free thiols then spontaneously oxidize to form a covalent disulfide bond that links the molecules.[112] The ability to introduce disulfide bonds into a variety of cargoes, as demonstrated

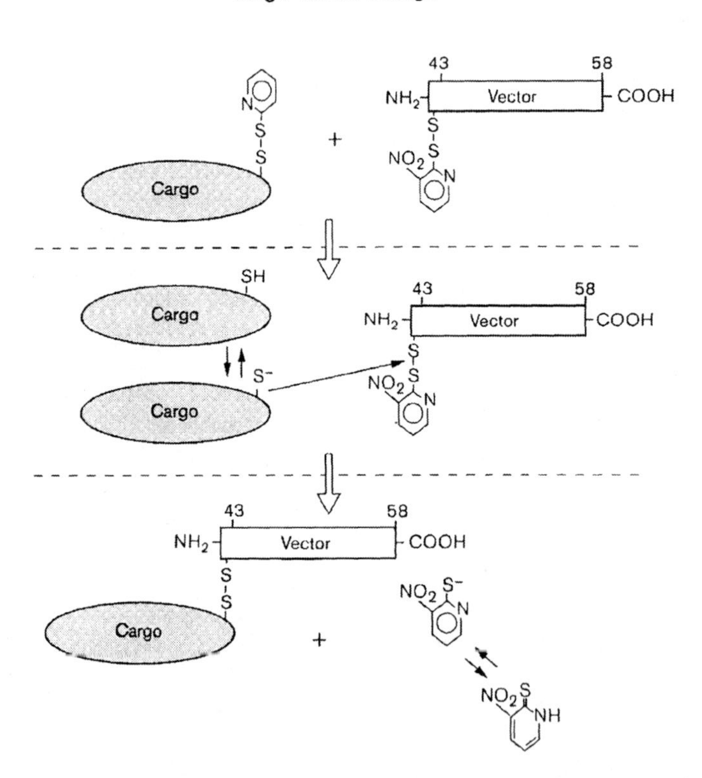

Figure 13.1. Disulfide linkage. In this protocol, the vector and cargo are independently synthesized, with the thiol groups of the cysteine residues protected by a pyridinium group. After synthesis, the pyridinium is removed, in the case of peptides simply by addition of a reducing agent, and the now free and reactive thiols quickly and efficiently oxidize to form a disulfide bond, linking vector to cargo molecule. (Reprinted from *Current Opinion in Neurobiology*, Vol. 6, Alain Prochiantz, *Getting hydrophilic compounds into cells: lessons from homeopeptides*, 629–634, © 1996, with permission from Elsevier.)

by Pooga and colleagues in a fusion of both transportan and Penetratin vectors to peptide nucleic acids,[51] demonstrates the applicability of this system to a wide variety of potential vector-cargo systems.

Other methods of direct conjugation have also been developed. Biotinylated delivery vehicles have been conjugated to streptavidin-labeled cargoes.[40,125] The Tat PTD has been covalently linked to polyethylene glycol and subsequently incorporated into liposomes, enhancing their delivery.[61] A technology recently developed by Eom and colleagues involves a novel tandem ligation protocol, permitting linkage of multiple peptide segments via chemical ligation and resulting in both linear and branched peptides.[126] As the delivery vectors and cargoes being investigated grow more diverse, it is probable that additional direct conjugation methods will be developed.

13.4. MECHANISM OF ACTION: CURRENT THEORIES

A number of ideas have been proposed to explain the mechanism by which cell-penetrating peptides and other nonclassical delivery vehicles induce transport across lipid bilayers. Some common properties among the vectors are apparent, suggesting that the exact mechanism of entry may become evident in the near future. In this section, these common properties of the vectors are described and the major mechanistic explanations that have been proposed are outlined. It should be emphasized, however, that at this time no general consensus on an exact mechanism has been reached.

13.4.1. Common Properties

13.4.1.1. Cationic Character: The Role of Guanidinium The most striking similarity among the nonclassical vectors is without a doubt the high prevalence and density of positive charges. Although most obvious in the cell-penetrating peptide group, areas of high positive charge density are also observed with all known protein vectors (e.g., VP22, FGF-1). This suggests the existence of a common internalization mechanism involving the cationic character of the vectors. This is further supported by competition for internalization of some cationic vectors by others[101] and reduction in efficiency among the arginine vectors as the extent of positive charge is reduced.[20] Closer examination of the cationic residues present in the vectors has shown a strong predilection for arginine residues, leading to the proposal by some groups that, in addition to its positive charge, some unique property of the arginine side chain may be a key factor in translocation.[101,113]

A number of studies have been conducted in an attempt to differentiate between a specific role for the guanidino moiety of the arginine side chain and a general requirement for cationic amino acids (which could also be satisfied by lysine or histidine residues). Examinations of polyarginine transport efficiency show either similar[41] or enhanced activity when compared to analogous polylysine,[123] polyhistidine, and poly-L-ornithine peptides.[127] Additionally, no significant differences in activity are observed in a comparison of import of polyarginine, homoarginine, and δ-guanidino α-amino butyric acid (δGABA),[127] a series that differs only in the number of side chain methylene groups. These data support the claim that transport efficiency is related to some unique property of arginine side chains, although it is clear that positive charge itself must play a significant role.

Although the precise molecular significance of the guanidino group in nonclassical transport is not yet known, the guanidino moiety has been found to form a number of specific interactions that could play a significant role. Most obviously, the delocalized π-system in the positively charged region of the side chain distinguishes it from that of lysine. For example, arginine can self-associate in solution in a face-to-face manner.[128] Perhaps more significantly, arginine-aromatic ring interactions are often observed due to the formation of discrete cation-π interactions (see Section 13.4.2.2).[129–131]

13.4.1.2. Polyanion Binding It is reasonable to assume from the abundance of cationic residues present in the majority of nonclassical transport-based vectors that these vectors should interact, at least to some extent, with the variety of polyanions found both at the cell surface and within the cell. In fact, such interactions appear to be highly prevalent,[63,82,98,132,133] being involved in the natural biological activity of many vectors as well as having a potential role in import. A primary example of the biological significance of the vector-polyanion interaction is observed with FGF-1, whose interaction with cell-surface heparin sulfate has been shown to be required for activity of the FGF protein-receptor complex.[77] VP22 has been shown to bind cellular RNA, apparently mediating its translocation during HSV-1 infection.[134] The function of the full-length Tat protein appears to be modified by heparan sulfate.[98,135,136] The interaction of cationic vectors with polyanions may, however, be more general than the specific interactions normally observed in biological systems. For example, the Penetratin peptide has been observed to interact nonspecifically with a number of anionic lipids,[137] the VP22 protein interacts preferentially with both anionic liposomes,[65] and negatively charged fatty acid mimics as well as heparin and polynucleotides,[63] and FGF-1 has been shown to bind to a large number of quite different polyanions.[81]

The wide variety of vector-polyanion interactions described above suggests a potential role for polyanions in nonclassical transport. This could be similar to the role that proteoglycans have been suggested to play in the cellular entry of cationic gene delivery.[138] The idea that cell surface polyanions act as the initial site of interaction for nonclassical vectors with cell membranes appears to be widely accepted, based on a number of critical studies. The internalization of both Penetratin[49,105] and the Tat protein[99] appears to be dependent on the cell-surface polyanions heparan sulfate and polysialic acid, respectively. Similar studies with other arginine-rich cell-penetrating peptides show that an absence of cell-surface proteoglycans reduces transport.[101] Polyanions other than proteoglycans have also been implicated in transport. Some Tat PTD studies suggest that the negative charge at the membrane surface produced by negatively charged phospholipid headgroups may also play a role.[100,101] This is further supported by a Penetratin study suggesting that charge neutralization through the interaction of the vector with negatively charged membrane lipids is the initial and, in fact, an essential component in the internalization pathway.[139] It is interesting to note that many viruses may also use negatively charged moieties as their initial contact point with the cell surface.[138]

These data lead to the further question of additional interactions between the vectors and other polyanions throughout the cell, a question that has not yet been critically addressed. A variety of polyanions, as well as dense areas of negative charge, can be found throughout the cell, including small polyanions such as nucleotides as well as cytoskeleton components, RNA, and DNA, all of which could serve as potential sites of interaction. Whether these intracellular polyanions are also steps in the transport pathway, perhaps serving as transport avenues, as has already been postulated for microtubule and microfilament networks in vesicular transport, remains to be seen.

13.4.1.3. The Role of Nonnative States in Protein Vectors In the previous sections (13.4.1.1 and 13.4.1.2), the focus has been primarily on the highly charged nature of nonclassical vectors and the interaction of these vectors with the polar domain of the lipid bilayer. An additional and potentially more significant barrier to transport lies within the apolar domain of the bilayer. Although this barrier appears to be impassable by large hydrophilic molecules, it has been postulated that, by adopting a partially unfolded conformation, protein molecules can interact directly with the lipid bilayer, facilitating protein-membrane translocation.[10,140] These partially unfolded conformations, often labeled "molten globule" states, involve a loss of tertiary structure with retention of secondary structure, exposing interior hydrophobic surfaces of the protein for interaction with bilayer interior.[141]

Although definitive evidence that these partially unfolded conformations are a key to transport is not yet available, their potential involvement in nonclassical protein transport is becoming increasingly apparent. This is primarily due to evidence that rigid tertiary and/or secondary structure is consistently absent in protein and peptide vehicles and, in some cases, has been found to hinder transport. For example, both VP22 and FGF-1 adopt partially disrupted, MG-like conformations under physiological solution conditions.[63,142] These intermediately folded conformations also appear to dramatically enhance the interactions of VP22,[63,65] and FGF-1,[142] as well as other proteins, with model lipid systems. Wiedlocha and colleagues have shown that FGF-1/diphtheria toxin fusion proteins are transport competent only when in a less ordered conformation.[37] Initial studies of transport by the full-length Tat protein suggest that at least partial unfolding of the cargo protein is required for transport. Additionally, as mentioned previously, the peptide vector Penetratin does not require specific secondary structure for internalization;[106,107] all of these studies support the hypothesis that disordered structure is necessary for bilayer translocation. Further support of this hypothesis is found in studies of the translocation of diphtheria toxin (DT) and *Pseudomonas* exotoxin, both of which possess transmembrane (T) domains that are independently capable of membrane translocation.[143,144] Once again, MG-like states play an important role;

formation of such membrane interactive states appears to be the initial step in the active part of the translocation process.[144,145]

Although the exact mechanism of MG-assisted transport has not been defined, a simple model can be postulated. This involves an initial attraction between the cationic vector and the negatively charged lipid surface, followed by insertion of the vector directly into the membrane through association of the exposed hydrophobic surfaces of the vector with the apolar bilayer interior. Whether the membrane-associating form of the vector and/or cargo must preexist or can be induced by conditions at the cell surface, such as the lower pH present at the membrane interfacial microenvironment[146] or by other components of the lipid bilayer itself (e.g., membrane proteins), is unclear; either situation, however, has the potential to satisfy this model. Actual transport through the membrane may use simple diffusion as a driving force. The idea of diffusion of large molecules through such a restricted space is supported by recent work showing a "jumping" of flexible macromolecules within a highly restricted network while undergoing Brownian motion.[147] This jumping could be facilitated by attraction between the cationic vector and other electrostatic surfaces on the other side (i.e., the cellular interior) of the membrane. A similar hypothesis has been proposed for the DT transmembrane (T) domain-assisted transport of MG-like proteins. In this scheme, it is suggested that translocation is achieved through a series of nonspecific binding events between the cargo protein and the T domain.[148] The T domain then acts as a stabilizing platform for the partially unfolded cargo, perhaps similarly to cellular molecular chaperones.

13.4.2. Proposed Mechanism of Transport

13.4.2.1. Inverted Hexagonal Phase Induction One of the first models of nonclassical transport, proposed by Prochiantz and colleagues, involves the induction of an inverted hexagonal phase within the lipid bilayer.[49,112,149] As shown in Figure 13.2, the vector-cargo complex is proposed to initially bind to the cell surface

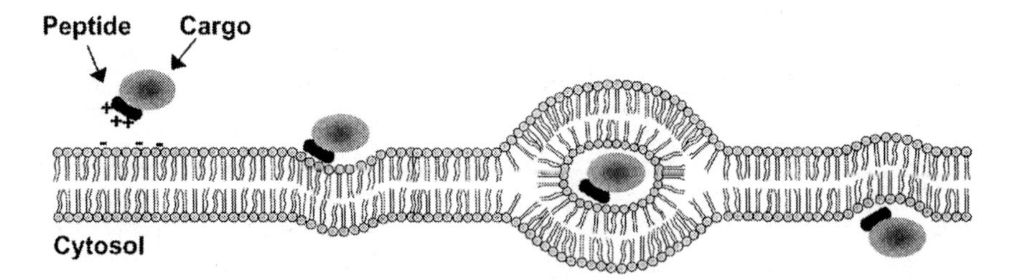

Figure 13.2. Inverted hexagonal phase model of nonclassical transport. The vector-cargo complex initially binds to the cell surface via electrostatic interactions. This induces an inverted hexagonal phase within the lipid bilayer, permitting the complex to become encapsulated within an inverted micelle and subsequently shuttled to the opposite side of the membrane. (*Potential use of non-classical pathways for the transport of macromolecular drugs.* Kueltzo, L.A. and Middaugh, C.R. 2001. *Exp Opin Invest Drugs* 9(9): 2039–2050. Used with permission.)

through electrostatic attraction. This binding induces an inverted hexagonal phase in the immediately adjacent lipid bilayer, causing the complex to be trapped in an inverted micelle.[149,150] The resulting vesicle then diffuses to the opposite side of the bilayer, where its contents are released into the surrounding fluid (the cytosol). This is similar to internalization methods suggested for cationic lipid/DNA gene delivery complexes.[151,152] Initial studies with the Penetratin peptide, the vector on which this model is based, do in fact show the induction of inverted hexagonal phases in lipid solutions, as observed by nuclear magnetic resonance.[150] Unfortunately, similar results have not been observed with other vector systems; thus, this model has yet to receive support as a general model of nonclassical transport.

13.4.2.2. A Unified Hypothesis This mechanism, previously proposed by the authors of this chapter,[153] combines the predominant common qualities of the nonclassical vectors. It is reasonable to assume that the cationic nonclassical vectors can bind to a variety of polyanionic species in a somewhat nonspecific manner during their approach to and passage through the cell membrane and interior. Although this binding may be nonspecific, there would certainly be some differences in binding affinity, suggesting the potential existence of a network of polyanions throughout the cell that could guide the transport of polycationic vectors. The existence of such a network as a guiding entity, either through a gradient of the aforementioned relative binding affinities or perhaps by providing somewhat uniform reactions surfaces, remains to be functionally established. The polyanionic moieties of the cell surface (which could be considered the outer edge of the polyanion network), however, serve as the initial site of attraction for the cationic vector (Figure 13.3).

Following initial binding to the cell surface, the vector is able to interact directly with the lipid bilayer through an MG-like or other disordered state by direct interaction between protein/peptide apolar residues with lipid acyl side chains. This state may be normally adopted by the vector under physiological solution conditions, or may be induced by the microenvironment found at the cell surface. The vector must then pass through the supposedly impenetrable apolar lipid bilayer. This barrier may not, however, be as impervious as was previously thought.[154–156] In addition to the diffusion studies mentioned in Section 13.4.1.3, a study of the diphtheria toxin T domain shows that charged groups within the membrane-spanning domain do not appear to provide the barrier to movement across the membrane that might be expected.[157]

These studies support the idea that the bilayer is actually somewhat permeable to macromolecules under certain conditions. The exact mechanism of this permeability is not known, but the arginine specificity of nonclassical vectors (see Section 13.4.1.1) suggests a reasonable pathway. Although these side chains are not expected to interact with the apolar interior of the bilayer, they may interact with the numerous transmembrane proteins that span biological membranes. The helices of membrane-spanning proteins are often found to be studded with aromatic residues,[158–160] providing an ideal environment for the formation of cation-π interactions between arginine (and potentially lysine) residues of a vector and the

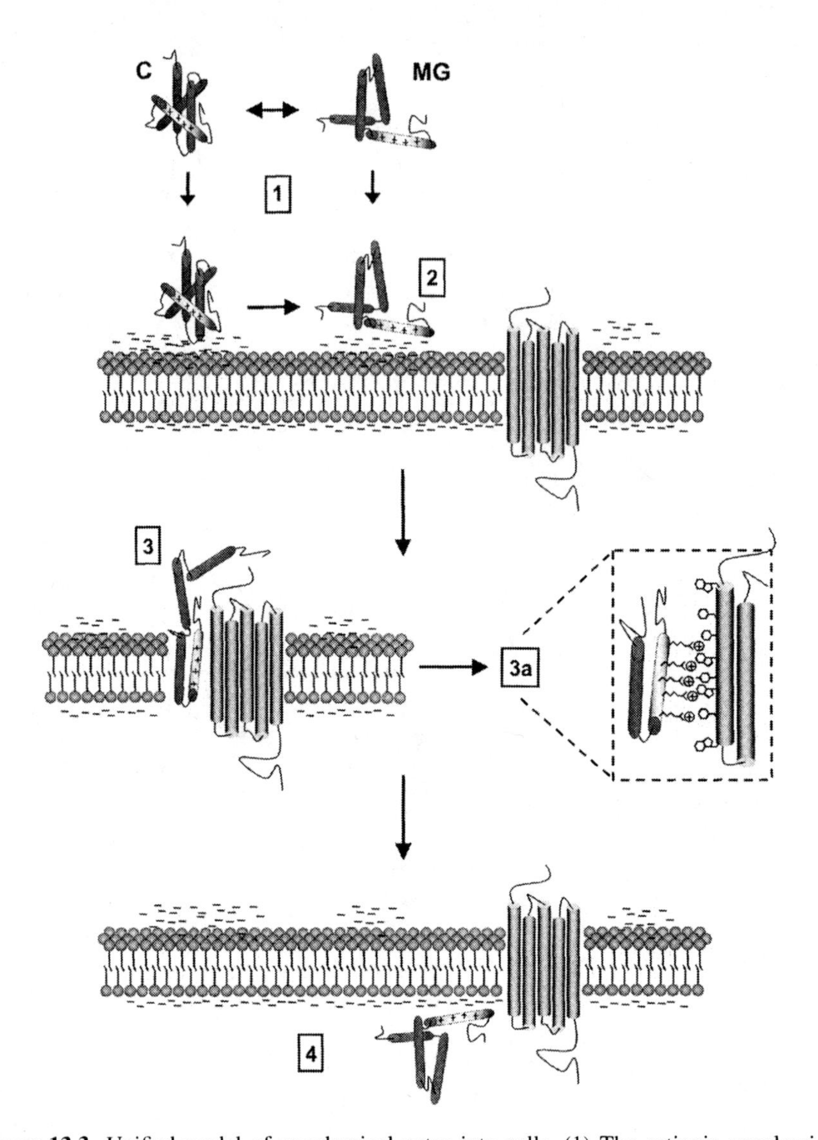

Figure 13.3. Unified model of nonclassical entry into cells. (1) The cationic nonclassical vector, in a compact (**C**) or molten globule-like conformation (**MG**) binds nonspecifically to the negatively charged cell membrane through electrostatic interactions. (2) The acidic microenvironment at the cell surface induces a conversion to an expanded, MG-like conformation for compact vectors. (3) The vector directly penetrates the membrane at the transmembrane protein/lipid bilayer interface, stabilized by interactions between exposed hydrophobic surfaces of the vector and the interior of the bilayer in addition to (3a) cation-π interactions with transmembrane proteins mediated by arginine residues of vectors and aromatic side chains of membrane protein bilayer-spanning helices. (4) The vector passes completely through the bilayer by diffusion, facilitated by an electrostatic attraction for the negatively charged interior surface of the bilayer or other internal polyanions. (*Non-classical transport proteins and peptides: an alternative to classical macromolecule delivery systems*, Kueltzo, L.A. and Middaugh, C.R. *J Pharm Sci.* © 2003, Wiley-Liss, Inc., a Wiley company, and the American Pharmacists Association.)

aforementioned aromatic residues. Such interactions are now thought to play an important role in protein structure as well as in membrane systems.[161] Therefore, one could postulate that the interface between bilayer-spanning protein helices and the lipid bilayer itself could serve as a potential pathway for macromolecular transport (Figure 13.3, Part 3). The flexibility of the lipid bilayer could easily accommodate passage of even very large complexes at this interface, especially if they exist in an extended structure. An additional advantage of extended structure is the potential interaction of the apolar lipid interior with the exposed hydrophobic surfaces of the translocating complex, further reducing the energy barrier to transport. Although this mechanism is currently highly speculative, significant support for this hypothesis is found in the work of Xiang and Anderson, who have shown an increase in the penetration of the lipid bilayer by polar molecules as the transmembrane protein content of the bilayer is increased.[156] Additional supporting evidence comes from studies showing significant insertion of basic peptides into a number of different membrane systems.[162]

The partial neutralization of cationic charges within the lipid bilayer through cation-π interactions need not be the only mechanism. Call and colleagues have recently shown that during the assembly process of transmembrane helices, the helices are stabilized through a multicharge interaction, with two interacting acidic residues on one helix interacting with a single basic residue on a second helix.[163] Additional studies support such interactions, indicating that interactions between acidic residues are indeed favorable in a membrane environment.[164,165] This type of multicharge interaction could also play a part in the neutralization of cationic residues of translocating complexes at the transmembrane protein/lipid bilayer interface.

Following entry into and passage through the bilayer as facilitated by one of the above mechanisms, or perhaps by another not yet recognized process, the complex can then exit the bilayer by a combination of diffusion and electrostatic attraction to the highly negatively charged interior membrane surface. At this point, the complex may continue on its path into the cell via the aforementioned polyanion network or by other as yet unknown pathways. Although this pathway may not be applicable to all examples of nonclassical transport, specifically the transport of large particles, which often possess quite different structural characteristics than vector-protein complexes, it does combine the common features of currently studied vectors. We are presently attempting to test this model using model membrane-spanning helices in liposome-based systems with entrapped polyanions.

13.4.2.3. Secretion The majority of studies involving nonclassical delivery focus exclusively on the import process. The initial work in this field, however, involved proteins that were found to possess not only nonclassical import activity but also leaderless secretory activity (secretion in the absence of a secretion signal peptide or use of the Golgi apparatus). The mechanism of this type of secretion is no clearer than that of import; in fact, it is uncertain at this time whether these mechanisms are even related, especially considering the asymmetry of the cell membrane. It seems reasonable to assume for experimental testing purposes, however, that the common vector properties mentioned above may also be involved in such secretion pathways.

Numerous hypotheses regarding the mechanism of leaderless secretion have been postulated since its initial description. One mechanism suggests secretion via the ABC (ATP-binding cassette) export complex.[166,167] The ABC complex, however, appears to be exclusive to yeast and bacteria, with no evidence that it plays a part in mammalian protein secretion. The secretion of FGF-2, one of the more studied systems, purportedly involves the membrane attack complex of the complement system.[168] A lack of complement factors in the many *in vitro* cell culture systems in which secretion has been demonstrated suggests, however, that they may not be involved in a general secretion pathway. The FGF-1 export pathway also appears to involve a protein complex.[71,75] This complex includes the protein synaptotagmin, which contains a polycationic region, providing a potential link to the nonclassical transport proteins. Rubartelli and colleagues, one of the first groups to study leaderless secretion, have proposed the involvement of endolysosomes in the secretion pathway of interleukin 1β,[10,169,170] followed by translocation of a partially disordered form of the protein complexed with chaperone proteins.[10] All of these potential mechanisms are protein specific; the existence of a general mechanism of leaderless secretion, as well as any potential connection to nonclassical import, remains unclear.

13.4.3. Additional Participants in Uptake and Secretion

The cellular environment is both complex and crowded; as such, it is reasonable to assume that a vector may come into contact with a much wider variety of molecules than those mentioned in the previous sections. These molecules may prove to be necessary elements in the transport pathway; this is a critical consideration in studies that utilize simple lipid models in which such components are absent. For example, in addition to the polyanions mentioned previously, molecular chaperones may play an important role, stabilizing partially unfolded vector and vector-cargo complexes, during and after translocation. In fact, molecular chaperones have already been proposed to be involved in the transport of Tat[97] and VP22,[63] as well as FGF-1,[72] and are well established to play a role in the transport of proteins into mitochondria.[171] Other factors, such as membrane properties (curvature, lipid composition, and spatial distribution) and differences in cell-surface proteoglycans among cell types may also strongly affect transport events. Finally, there may be a number of unknown factors that could be involved, all of which must ultimately be taken into consideration when evaluating the nature of transport.

13.4.4. Potential Artifacts and Other Concerns

Despite the numerous *in vivo* delivery successes demonstrated with nonclassical vehicles that provide strong evidence that the pathway does indeed exist, a few studies suggest that nonclassical transport is actually an artifact of the cell fixation processes employed in many of the *in vitro* studies.[172,173] Some evidence argues that transport may be occurring through traditional endocytotic pathways, rather than via the energy-independent pathway proposed here. For example, a study of the Engrailed vector suggests possible association with caveolae.[110] Inhibition of the caveolae pathway appears to inhibit Tat internalization, the latter study involving Tat-linked lambda phage.[53] The internalization of arginine-based peptides, however, is apparently not affected by caveolae inhibition.[101] It is important to note that internalization by a nonclassical pathway does not preclude parallel internalization by conventional endocytotic pathways. In fact, such endocytotic internalization may well be expected, considering that many delivery studies have been conducted under physiological conditions (where endocytosis is often quite active) following establishment of energy-independent transport during the initial investigation of the vectors. Should the two pathways indeed operate simultaneously, their relative contributions need to be determined to evaluate overall vector efficiency. These factors highlight some of the difficulties in elucidating the nonclassical pathway, and they must be carefully considered in a critical analysis of the data being generated in this field.

In addition to the unresolved mechanistic questions, a number of other issues concerning cationic nonclassical delivery vehicles must also be addressed. A primary issue is cytotoxicity, a common problem with polycationic therapeutics and excipients. Although initial studies suggest that cytotoxicity will not be significant,[40,118] presumably due to the relatively small cationic regions employed, should the polyanion network hypothesis hold true, a number of deleterious effects could occur due to the variety of potential nonspecific interactions. The biological activity of the vectors must also be examined, especially if endogenous protein vectors (e.g., the FGFs) are employed. Many cationic peptides also possess significant antimicrobial activity and are currently under development as human therapeutics. The relationship between their biological activity and potential transport-like properties remains to be better described. Finally, as with any therapeutic, the question of immunogenicity needs to be addressed, since the safety and efficacy issues associated with significant immunogenicity will severely impair the use of these vectors in the clinic, especially with repetitive use.

13.5. CONCLUSIONS

As the number of macromolecular therapeutics in clinical trials increases, the need for efficient delivery vehicles becomes more pressing. The cationic peptides and proteins described here provide an interesting alternative to conventional delivery vehicles. Although the exact mechanism of nonclassical transport has not yet been defined, the *in vitro* and *in vivo* delivery successes of these vehicles are most promising, especially considering the wide array of cargoes and protocols with which they can be employed. Although a number of questions remain to be addressed, research in this field is proceeding at a rapid pace, increasing the possibility that one or more of these vector-cargo complexes could reach the clinic in the not too distant future.

REFERENCES

1. Tan, K. B. *Cytobios.* **1977**, *20*, 143–149.
2. Triguero, D.; Buciak, J. B.; Yang, J.; Pardridge, W. M. *Proc. Natl. Acad. Sci. USA* **1989**, *86*, 4761–4765.
3. Blau, S.; Jubeh, T. T.; Haupt, S. M.; Rubinstein, A. *Crit. Rev. Ther. Drug Carrier Syst.* **2000**, *17*, 425–465.
4. Ehrlich, M.; Sarafyan, L. P.; Myers, D. J. *Biochim. Biophys. Acta* **1976**, *454*, 397–409.
5. Zelphati, O.; Wang, Y.; Kitada, S.; Reed, J. C.; Felgner, P. L.; Corbeil, J. *J. Biol. Chem.* **2001**, *276*, 35103–35110.
6. Lasic, D. D. *Liposomes in Gene Delivery.* CRC Press: Boca Raton, FL, **1997**.
7. Kwoh, D. Y.; Coffin, C. C.; Lollo, C. P.; Jovenal, J.; Banaszczyk, M. G.; Mullen, P.; Phillips, A.; Amini, A.; Fabrycki, J.; Bartholomew, R. M.; Brostoff, S. W.; Carlo, D. J. *Biochim. Biophys. Acta* **1999**, *1444*, 171–190.
8. Godbey, W. T.; Wu, K. K.; Mikos, A. G. *J. Control Release* **1999**, *60*, 149–160.
9. Patri, A. K.; Majoros, I. J.; Baker, J. R. *Curr. Opin. Chem. Biol.* **2002**, *6*, 466–471.
10. Rubartelli, A.; Sitia, R. In *Unusual Secretory Pathways: From Bacteria to Man.*; Kuchler, K., Ed. R.G. Landes Bioscience: Austin, TX, **1997**, pp. 87–114.
11. Kueltzo, L. A.; Middaugh, C. R. *Expert Opin. Invest. Drugs* **2000**, *9*, 2039–2050.
12. Rubartelli, A.; Cozzolino, F.; Talio, M.; Sitia, R. *EMBO J.* **1990**, *9*, 1503–1510.
13. Abraham, J. A.; Mergia, A.; Whang, J. L.; Tumolo, A.; Friedman, J.; Hjerrild, K. A.; Gospodarowicz, D.; Fiddes, J. C. *Science* **1986**, *233*, 545–548.
14. Reiness, C. G.; Seppa, M. J.; Dion, D. M.; Sweeney, S.; Foster, D. N.; Nishi, R. *Mol. Cell. Neurosci.* **2001**, *17*, 931–944.
15. Fawell, S.; Seery, J.; Daikh, Y.; Moore, C.; Chen, L. L.; Pepinsky, B.; Barsoum, J. *Proc. Natl. Acad. Sci. USA* **1994**, *91*, 664–668.
16. Elliott, G.; O'Hare, P. *Cell* **1997**, *88*, 223–233.
17. Perez, F.; Lledo, P. M.; Karagogeos, D.; Vincent, J. D.; Prochiantz, A.; Ayala, J. *Mol. Endocrinol.* **1994**, *8*, 1278–1287.
18. Allinquant, B.; Hantraye, P.; Mailleux, P.; Moya, K.; Bouillot, C.; Prochiantz, A. *J. Cell Biol.* **1995**, *128*, 919–927.
19. Lindgren, M.; Hallbrink, M.; Prochiantz, A.; Langel, U. *Trends Pharmacol. Sci.* **2000**, *21*, 99–103.
20. Futaki, S.; Suzuki, T.; Ohashi, W.; Yagami, T.; Tanaka, S.; Ueda, K.; Sugiura, Y. *J. Biol. Chem.* **2001**, *276*, 5836–5840.
21. Alberts, B.; Bray, D.; Lewis, J.; Raff, M.; Roberts, K.; Watson, J. D. In *Molecular Biology of the Cell*, 3rd ed. Garland Publishing: New York, **1994**, pp. 551–99.
22. Gorlich, D.; Mattaj, I. W. *Science* **1996**, *271*, 1513–1518.
23. Alberts, B.; Bray, D.; Lewis, J.; Raff, M.; Roberts, K.; Watson, J. D. In *Molecular Biology of the Cell*, 3rd ed. Garland Publishing: New York, **1994**, pp. 604–54.
24. Elliott, G.; Mouzakitis, G.; O'Hare, P. *J. Virol.* **1995**, *69*, 7932–7941.
25. Elliott, G.; O'Reilly, D.; O'Hare, P. *Virology* **1996**, *226*, 140–145.
26. Blaho, J. A.; Mitchell, C.; Roizman, B. *J. Biol. Chem.* **1994**, *269*, 17401–17410.
27. Knopf, K. W.; Kaerner, H. C. *J. Gen. Virol.* **1980**, *46*, 405–414.

28. Elliott, G.; O'Hare, P. *J. Virol.* **1998**, *72*, 6448–6455.
29. van Leeuwen, H.; Elliott, G.; O'Hare, P. *J. Virol.* **2002**, *76*, 3471–3481.
30. Derer, W.; Easwaran, H. P.; Knopf, C. W.; Leonhardt, H.; Cardoso, M. C. *J. Mol. Med.* **1999**, *77*, 609–613.
31. Aints, A.; Dilber, M. S.; Smith, C. I. E. *J. Gene Med.* **1999**, *1*, 275–279.
32. Phelan, A.; Elliott, G.; O'Hare, P. *Nat. Biotechnol.* **1998**, *16*, 440–443.
33. Wybranietz, W. A.; Prinz, F.; Speigel, M.; Schenk, A.; Bitzer, M.; Gregor, M.; Lauer, U. M. *J. Gene Med.* **1999**, *1*, 265–274.
34. Caron, N. J.; Torrente, Y.; Camirand, G.; Bujold, M.; Chapdelaine, P.; Leriche, K.; Bresolin, N.; Tremblay, J. P. *Mol. Ther.* **2001**, *3*, 310–318.
35. Schwarze, S. R.; Ho, A.; Vocero-Akbani, A.; Dowdy, S. F. *Science* **1999**, *285*, 1569–1572.
36. Gawiak, S.; Pastan, I.; Siegall, C. *Bioconjug. Chem.* **1993**, *4*, 483–489.
37. Wiedlocha, A.; Madshus, I. H.; Mach, H.; Middaugh, C. R.; Olsnes, S. *EMBO J.* **1992**, *11*, 4835–4842.
38. Dikov, M. M.; Reich, M. B.; Dworkin, L.; Thomas, J. W.; Miller, G. G. *J. Biol. Chem.* **1998**, *273*, 15811–15817.
39. Kilk, K.; Magzoub, M.; Pooga, M.; Eriksson, L. E.; Langel, U.; Graslund, A. *Bioconjug. Chem.* **2001**, *12*, 911–916.
40. Elmquist, A.; Lindgren, M.; Bartfai, T.; Langel, U. *Exp. Cell. Res.* **2001**, *269*, 237–244.
41. Park, J.; Ryu, J.; Kim, K. A.; Lee, H. J.; Bahn, J. H.; Han, K.; Choi, E. Y.; Lee, K. S.; Kwon, H. Y.; Choi, S. Y. *J. Gen. Virol.* **2002**, *83*, 1173–1181.
42. Li, H.; Yao, Z.; Degenhardt, B.; Teper, G.; Papadopoulos, V. *Proc. Natl. Acad. Sci. USA* **2001**, *98*, 1267–1272.
43. Bonfanti, M.; Taverna, S.; Salmona, M.; D'Incalci, M.; Broggini, M. *Cancer Res.* **1997**, *57*, 1442–1446.
44. Giorello, L.; Clerico, L.; Pescarolo, M. P.; Vikhanskaya, F.; Salmona, M.; Colella, G.; Bruno, S.; Mancuso, T.; Bagnasco, L.; Russo, P.; Parodi, S. *Cancer Res.* **1998**, *58*, 3654–3659.
45. Braun, K.; Peschke, P.; Pipkorn, R.; Lampel, S.; Wachsmuth, M.; Waldeck, W.; Friedrich, E.; Debus, J. *J. Mol. Biol.* **2002**, *318*, 237–243.
46. Chen, L.; Wright, L. R.; Chen, C. H.; Oliver, S. F.; Wender, P. A.; Mochly-Rosen, D. *Chem. Biol.* **2001**, *8*, 1123–1129.
47. Normand, N.; van Leeuwen, H.; O'Hare, P. *J. Biol. Chem.* **2001**, *276*, 15042–15050.
48. Astriab-Fisher, A.; Sergueev, D.; Fisher, M.; Shaw, B. R.; Juliano, R. L. *Pharm. Res.* **2002**, *19*, 744–754.
49. Derossi, D.; Chassaing, G.; Prochiantz, A. *Trends Cell. Biol.* **1998**, *8*, 84–87.
50. Kaushik, N.; Basu, A.; Palumbo, P.; Myers, R. L.; Pandey, V. N. *J. Virol.* **2002**, *76*, 3881–3891.
51. Pooga, M.; Soomets, U.; Hallbrink, M.; Valkna, A.; Saar, K.; Rezaei, K.; Kahl, U.; Hao, J. X.; Xu, X. J.; Wiesenfeld-Hallin, Z.; Hokfelt, T.; Bartfai, T.; Langel, U. *Nat. Biotechnol.* **1998**, *16*, 857–861.
52. Cheng, W. F.; Hung, C. F.; Hsu, K. F.; Chai, C. Y.; He, L.; Polo, J. M.; Slater, L. A.; Ling, M.; Wu, T. C. *Hum. Gene Ther.* **2002**, *13*, 553–568.

53. Eguchi, A.; Akuta, T.; Okuyama, H.; Senda, T.; Yokoi, H.; Inokuchi, H.; Fujita, S.; Hayakawa, T.; Takeda, K.; Hasegawa, M.; Nakanishi, M. *J. Biol. Chem.* **2001**, *276*, 26204–26210.
54. Rudolph, C.; Plank, C.; Lausier, J.; Schillinger, U.; Muller, R. H.; Rosenecker, J. *J. Biol. Chem.* **2003**, *278*, 11411–11418.
55. Dilber, M. S.; Phelan, A.; Aints, A.; Mohamed, A. J.; Elliott, G.; Edvard Smith, C. I.; O'Hare, P. *Gene Ther.* **1999**, *6*, 12–21.
56. Wybranietz, W. A.; Gross, C. D.; Phelan, A.; O'Hare, P.; Spiegel, M.; Graepler, F.; Bitzer, M.; Stahler, P.; Gregor, M.; Lauer, U. M. *Gene Ther.* **2001**, *8*, 1654–1664.
57. Lai, Z.; Brady, R. O. *J. Neurosci. Res.* **2002**, *67*, 363–371.
58. Lai, Z.; Han, I.; Zirzow, G.; Brady, R. O.; Reiser, J. *Proc. Natl. Acad. Sci. USA* **2000**, *97*, 11297–11302.
59. Wills, K. N.; Atencio, I. A.; Avanzini, J. B.; Neuteboom, S.; Phelan, A.; Philopena, J.; Sutjipto, S.; Vaillancourt, M. T.; Wen, S. F.; Ralston, R. O.; Johnson, D. E. *J. Virol.* **2001**, *75*, 8733–8741.
60. Zender, L.; Kuhnel, F.; Kock, R.; Manns, M.; Kubicka, S. *Cancer Gene Ther.* **2002**, *9*, 489–496.
61. Torchilin, V. P.; Rammohan, R.; Weissig, V.; Levchenko, T. S. *Proc. Natl. Acad. Sci. USA* **2001**, *98*, 8786–8791.
62. Lewin, M.; Carlesso, N.; Tung, C. H.; Tang, X. W.; Cory, D.; Scadden, D. T.; Weissleder, R. *Nat. Biotechnol.* **2000**, *18*, 410–414.
63. Kueltzo, L. A.; Normand, N.; O'Hare, P.; Middaugh, C. R. *J. Biol. Chem.* **2000**, *275*, 33213–33221.
64. Aints, A.; Guven, H.; Gahrton, G.; Smith, C. I.; Dilber, M. S. *Gene Ther.* **2001**, *8*, 1051–1056.
65. Kueltzo, L. A. In *An Investigation into the Role of Molten Globule-Like States in Non-Classical Protein Transport.* Doctoral Thesis, University of Kansas, Lawrence, **2002**.
66. Ornitz, D. M.; Itoh, N. *Genome Biol.* **2001**, *2*, reviews 3005.1–3005.12.
67. Middaugh, C. R.; Volkin, D.; Thomas, K. *Curr. Opin. Invest. Drugs* **1993**, *2*, 991–1005.
68. Volkin, D. B.; Middaugh, C. R. In *Formulation, Characterization, and Stability of Protein Drugs*; Pearlman, R., Wang, Y. J., Eds. Plenum Press: New York, **1996**, pp. 181–217.
69. Miyakawa, K.; Hatsuzawa, K.; Kurokawa, T.; Asada, M.; Kuroiwa, T.; Imamura, T. *J. Biol. Chem.* **1999**, *274*, 29352–29357.
70. Werner, S. *Cytokine Growth Factor Rev.* **1998**, *9*, 153–165.
71. Prudovsky, I.; Bagala, C.; Tarantini, F.; Mandinova, A.; Soldi, R.; Bellum, S.; Maciag, T. *J. Cell Biol.* **2002**, *158*, 201–208.
72. Edwards, K.; Kueltzo, L.; Fisher, M.; Middaugh, C. *Arch. Biochem. Biophys.* **2001**, *393*, 14–21.
73. Jackson, A.; Friedman, S.; Zhan, X.; Engleka, K. A.; Forough, R.; Maciag, T. *Proc. Natl. Acad. Sci. USA* **1992**, *89*, 10691–10695.
74. Piotrowicz, R. S.; Martin, J. L.; Dillman, W., H.; Levin, E. G. *J. Biol. Chem.* **1997**, *272*, 7042–7047.
75. Tarantini, F.; LaVallee, T.; Jackson, A.; Gamble, S.; Carreira, C. M.; Garfinkel, S.; Burgess, W. H.; Maciag, T. *J. Biol. Chem.* **1998**, *273*, 22209–22216.

76. Landriscina, M.; Soldi, R.; Bagala, C.; Micucci, I.; Bellum, S.; Tarantini, F.; Prodovsky, I.; Maciag, T. *J. Biol. Chem.* **2001**, *276*, 22544–22552.

77. Stauber, D. J.; DiGabriele, A. D.; Hendrickson, W. A. *Proc. Natl. Acad. Sci. USA* **1999**, *97*, 49–54.

78. Citores, L.; Wesche, J.; Kolpakova, E.; Olsnes, S. *Mol. Biol. Cell* **1999**, *10*, 3835–3848.

79. Siegall, C. B.; Epstein, S.; Speir, E.; Hla, T.; Forough, R.; Maciag, T.; Fitzgerald, D. J.; Pastan, I. *Faseb J.* **1991**, *5*, 2843–2849.

80. Siegall, C. B.; Gawlak, S. L.; Chace, D. F.; Merwin, J. R.; Pastan, I. *Bioconjug. Chem.* **1994**, *5*, 77–83.

81. Volkin, D. B.; Tsai, P. K.; Dabora, J. M.; Gress, J. O.; Burke, C. J.; Linhardt, R. J.; Middaugh, C. R. *Arch. Biochem. Biophys.* **1993**, *300*, 30–41.

82. Copeland, R. A.; Ji, H.; Halfpenny, A. J.; Williams, R. W.; Thompson, K. C.; Herber, W. K.; Thomas, K. A.; Bruner, M. W.; Ryan, J. A.; Marquis-Omer, D.; Sanyal, G.; Sitrin, R. D.; Yamazaki, S.; Middaugh, C. R. *Arch. Biochem. Biophys.* **1991**, *289*, 53–61.

83. Plotnikov, A. N.; Schlessinger, J.; Hubbard, S. R.; Mohammadi, M. *Cell* **1999**, *98*, 641–650.

84. Vemuri, S.; Beylin, I.; Sluzky, V.; Stratton, P.; Eberlein, G.; Wang, Y. J. *J. Pharm. Pharmacol.* **1994**, *46*, 481–486.

85. Mach, H.; Middaugh, C. R. *Biochemistry* **1995**, *34*, 9913–9920.

86. Lin, Y.-Z.; Yao, S.; Hawinger, J. *J. Biol. Chem.* **1996**, *271*, 5305–5308.

87. Cao, Y.; Ekstrom, M.; Pettersson, R. *J. Cell Sci.* **1993**, *104*, 77–87.

88. Zhan, X.; Hu, X.; Friedman, S.; Maciag, T. *Biochem. Biophys. Res. Commun.* **1992**, *188*, 982–991.

89. Klein, S.; Carroll, J. A.; Chen, Y.; Henry, M. F.; Henry, P. A.; Ortonowski, I. E.; Pintucci, G.; Beavis, R. C.; Burgess, W. H.; Rifkin, D. B. *J. Biol. Chem.* **2000**, *275*, 3150–3157.

90. Jeang, K.-T.; Xiao, H.; Rich, E. A. *J. Biol. Chem.* **1999**, *274*, 28837–28840.

91. Frankel, A. D.; Pabo, C. O. *Cell* **1988**, *55*, 1189–1193.

92. Green, M.; Loewenstein, P. M. *Cell* **1988**, *55*, 1179–1188.

93. Ensoli, B.; Buonaguro, L.; Barillari, G.; Fiorelli, V.; Gendelman, R.; Morgan, R. A.; Wingfield, P.; Gallo, R. C. *J. Virol.* **1993**, *67*, 277–287.

94. Mann, D. A.; Frankel, A. D. *EMBO J.* **1991**, *10*, 1733–1739.

95. Vives, E.; Brodin, P.; Lebleu, B. *J. Biol. Chem.* **1997**, *272*, 16010–16017.

96. Ezhevsky, S. A.; Nagahara, H.; Vocero-Akbani, A. M.; Gius, D. R.; Wei, M. C.; Dowdy, S. F. *Proc. Natl. Acad. Sci. USA* **1997**, *94*, 10699–10704.

97. Bayley, H. *Nat. Biotechnol.* **1999**, *17*, 1066–1067.

98. Rusnati, M.; Tulipano, G.; Spillmann, D.; Tanghetti, E.; Oreste, P.; Zoppetti, G.; Giacca, M.; Presta, M. *J. Biol. Chem.* **1999**, *274*, 28198–28205.

99. Tyagi, M.; Rusnati, M.; Presta, M.; Giacca, M. *J. Biol. Chem.* **2000**, *276*, 3254–3261.

100. Silhol, M.; Tyagi, M.; Giacca, M.; Lebleu, B.; Vives, E. *Eur. J. Biochem.* **2002**, *269*, 494–501.

101. Suzuki, T.; Futaki, S.; Niwa, M.; Tanaka, S.; Ueda, K.; Sugiura, Y. *J. Biol. Chem.* **2002**, *277*, 2437–2443.

102. Bonifaci, N.; Sitia, R.; Rubartelli, A. *Aids* **1995**, *9*, 995–1000.

103. Schwarze, S. R.; Hruska, K. A.; Dowdy, S. F. *Trends Cell Biol.* **2000**, *10*, 290–295.

104. Ho, A.; Schwarze, S. R.; Mermelstein, S. J.; Waksman, G.; Dowdy, S. F. *Cancer Res.* **2001**, *61*, 474–477.

105. Perez, F.; Joliot, A.; Bloch-Gallego, E.; Zahraoui, A.; Triller, A.; Prochiantz, A. *J. Cell Sci.* **1992**, *102*, 717–722.

106. Drin, G.; Demene, H.; Temsamani, J.; Brasseur, R. *Biochemistry* **2001**, *40*, 1824–1834.

107. Derossi, D.; Calvet, S.; Trembleau, A.; Brunissen, A.; Chassaing, G.; Prochiantz, A. *J. Biol. Chem.* **1996**, *271*, 18188–18193.

108. Derossi, D.; Joliot, A. H.; Chassaing, G.; Prochaintz, A. *J. Biol. Chem.* **1994**, *269*, 10444–10450.

109. Chatelin, L.; Volovitch, M.; Joliot, A. H.; Perez, F.; Prochiantz, A. *Mech. Dev.* **1996**, *55*, 111–117.

110. Joliot, A.; Trembleau, A.; Raposo, G.; Calvet, S.; Volovitch, M.; Prochiantz, A. *Development* **1997**, *124*, 1865–1875.

111. Joliot, A.; Maizel, A.; Rosenberg, D.; Trembleau, A.; Dupas, S.; Volovitch, M.; Prochiantz, A. *Curr. Biol.* **1998**, *8*, 856–863.

112. Prochiantz, A. *Curr. Opin. Neurobiol.* **1996**, *6*, 629–634.

113. Wadia, J. S.; Dowdy, S. F. *Curr. Opin. Biotechnol.* **2002**, *13*, 52–56.

114. Hawiger, J. *Curr. Opin. Immunol.* **1997**, *9*, 189–194.

115. Hawiger, J. *Curr. Opin. Chem. Biol.* **1999**, *3*, 89–94.

116. Schwartz, J. J.; Zhang, S. *Curr. Opin. Mol. Ther.* **2000**, *2*, 162–167.

117. Dunican, D. J.; Doherty, P. *Biopolymers* **2001**, *60*, 45–60.

118. Morris, M. C.; Depollier, J.; Mery, J.; Heitz, F.; Divita, G. *Nat. Biotechnol.* **2001**, *19*, 1173–1176.

119. Langedijk, J. P. *J. Biol. Chem.* **2002**, *277*, 5308–5314.

120. Oess, S.; Hildt, E. *Gene Ther.* **2000**, *7*, 750–758.

121. Pooga, M.; Hallbrink, M.; Zorko, M.; Langel, U. *Faseb J.* **1998**, *12*, 67–77.

122. Oehlke, J.; Scheller, A.; Wiesner, B.; Krause, E.; Beyermann, M.; Klauschenz, E.; Melzig, M.; Bienert, M. *Biochim. Biophys. Acta.* **1998**, *1414*, 127–139.

123. Han, K.; Jeon, M. J.; Kim, S. H.; Ki, D.; Bahn, J. H.; Lee, K. S.; Park, J.; Choi, S. Y. *Mol. Cells* **2001**, *12*, 267–271.

124. Wiethoff, C. M.; Middaugh, C. R. *J. Pharm. Sci.* **2003**, *92*, 203–217.

125. Lee, H. J.; Pardridge, W. M. *Bioconjug. Chem.* **2001**, *12*, 995–999.

126. Eom, K. D.; Miao, Z.; Yang, J. L.; Tam, J. P. *J. Am. Chem. Soc.* **2003**, *125*, 73–82.

127. Mitchell, D. J.; Kim, D. T.; Steinman, L.; Fathman, C. G.; Rothbard, J. B. *J. Pept. Res.* **2000**, *56*, 318–325.

128. Boudon, S.; Wipff, G.; Maigret, B. *J. Phys. Chem.* **1990**, *94*, 6056–6061.

129. Burley, S. K.; Petsko, G. A. *Adv. Protein Chem.* **1988**, *39*, 125–189.

130. Karlin, S.; Zuker, M.; Brocchieri, L. *J. Mol. Biol.* **1994**, *239*, 227–248.

131. Brocchieri, L.; Karlin, S. *Proc. Natl. Acad. Sci. USA* **1994**, *91*, 9297–9301.

132. Dabora, J. M.; Sanyal, G.; Middaugh, C. R. *J. Biol. Chem.* **1991**, *266*, 23637–23640.

133. Richardson, T. P.; Trinkaus-Randall, V.; Nugent, M. A. *J. Biol. Chem.* **1999**, *274*, 13534–13540.

134. Sciortino, M. T.; Taddeo, B.; Poon, A. P.; Mastino, A.; Roizman, B. *Proc. Natl. Acad. Sci. USA* **2002**, *99*, 8318–8323.
135. Rusnati, M.; Coltrini, D.; Oreste, P.; Zoppetti, G.; Albini, A.; Noonan, D.; di Fagagna, F.; Giacca, M.; Presta, M. *J. Biol. Chem.* **1997**, *272*, 11313–11320.
136. Rusnati, M.; Tulipano, G.; Urbinati, C.; Tanghetti, E.; Giuliani, R.; Giacca, M.; Ciomei, M.; Corallini, A.; Presta, M. *J. Biol. Chem.* **1998**, *273*, 16027–16037.
137. Persson, D.; Thoren, P. E.; Herner, M.; Lincoln, P.; Norden, B. *Biochemistry* **2003**, *42*, 421–429.
138. Wiethoff, C. M.; Smith, J. G.; Koe, G. S.; Middaugh, C. R. *J. Biol. Chem.* **2001**, *276*, 32806–32813.
139. Dom, G.; Shaw-Jackson, C.; Matis, C.; Bouffioux, O.; Picard, J. J.; Prochiantz, A.; Mingeot-Leclercq, M. P.; Brasseur, R.; Rezsohazy, R. *Nucleic Acids Res.* **2003**, *31*, 556–561.
140. Bychkova, V. E.; Pain, R. H.; Ptitsyn, O. B. *FEBS Lett.* **1988**, *238*, 231–234.
141. Ptitsyn, O. B. *Adv. Protein Chem.* **1995**, *47*, 83–229.
142. Mach, H.; Ryan, J. A.; Burke, C. J.; Volkin, D. B.; Middaugh, C. R. *Biochemistry* **1993**, *32*, 7703–7711.
143. Oh, K. J.; Senzel, L.; Collier, R. J.; Finkelstein, A. *Proc. Natl. Acad. Sci. USA* **1999**, *96*, 8467–8470.
144. Sanyal, G.; Marquis-Omer, D.; Gress, J. O.; Middaugh, C. R. *Biochemistry* **1993**, *32*, 3488–3497.
145. Chenal, A.; Savarin, P.; Nizard, P.; Guillain, F.; Gillet, D.; Forge, V. *J. Biol. Chem.* **2002**, *277*, 43425–43432.
146. van der Goot, F. G.; Gonzalez-Manas, J. M.; Lakey, J. H.; Pattus, F. *Nature* **1991**, *354*, 408–410.
147. Nykypanchuk, D.; Strey, H. H.; Hoagland, D. A. *Science* **2002**, *297*, 987–990.
148. Ren, J.; Kachel, K.; Kim, H.; Malenbaum, S. E.; Collier, R. J.; London, E. *Science* **1999**, *284*, 955–957.
149. Prochiantz, A. *Ann. NY Acad. Sci.* **1999**, *886*, 172–179.
150. Berlose, J. P.; Convert, O.; Derossi, D.; Brunissen, A.; Chassaing, G. *Eur. J. Biochem.* **1996**, *242*, 372–386.
151. Koltover, I.; Salditt, T.; Radler, J. O.; Safinya, C. R. *Science* **1998**, *281*, 78–81.
152. Safinya, C. R. *Curr. Opin. Struct. Biol.* **2001**, *11*, 440–448.
153. Kueltzo, L. A.; Middaugh, C. R. *J. Pharm. Sci.* **2003**, *92*, 1754–72.
154. Pollack, G. H. In *Cells, Gels and the Engines of Life: A New, Unifying Approach to Cell Function.*; Ebner and Sons: Seattle, 2001, pp. 11–37.
155. Subczynski, W. K.; Wisniewska, A. *Acta Biochim. Pol.* **2000**, *47*, 613–625.
156. Xiang, T.; Anderson, B. D. *J. Membr. Biol.* **2000**, *173*, 187–201.
157. Ren, J.; Sharpe, J. C.; Collier, R. J.; London, E. *Biochemistry* **1999**, *38*, 976–984.
158. Yuen, C. T.; Davidson, A. R.; Deber, C. M. *Biochemistry* **2000**, *39*, 16155–16162.
159. Tsang, S.; Saier, M. H., Jr. *J. Comput. Biol.* **1996**, *3*, 185–190.
160. Espanol, M. J.; Saier, M. H., Jr. *Mol. Membr. Biol.* **1995**, *12*, 193–200.
161. Ma, J. C.; Dougherty, D. A. *Chem. Rev.* **1997**, *97*, 1303–1324.

162. Kim, C. S.; Kweon, D. H.; Shin, Y. K. *Biochemistry* **2002**, *41*, 10928–10933.

163. Call, M. E.; Pyrdol, J.; Wiedmann, M.; Wucherpfennig, K. W. *Cell* **2002**, *111*, 967–979.

164. Gratkowski, H.; Lear, J. D.; DeGrado, W. F. *Proc. Natl. Acad. Sci. USA* **2001**, *98*, 880–885.

165. Zhou, F. X.; Merianos, H. J.; Brunger, A. T.; Engelman, D. M. *Proc. Natl. Acad. Sci. USA* **2001**, *98*, 2250–2255.

166. Kuchler, K. *Trends Cell Biol.* **1993**, *3*, 421–426.

167. Kuchler, K.; Egner, R. In *Unusual Secretory Pathways: From Bacteria to Man*; Kuchler, K., Ed. R.G. Landes Bioscience: Austin, TX, **1997**, pp. 49–86.

168. Acosta, J. A.; Benzaquen, L. R.; Goldstein, D. J.; Tosteson, M. T.; Halperin, J. A. *Mol. Med.* **1996**, *2*, 755–765.

169. Terlecky, S. R.; Dice, J. F. *J. Biol. Chem.* **1993**, *268*, 23490–23495.

170. Andrei, C.; Dazzi, C.; Lotti, L.; Torrisi, M. R.; Chimini, G.; Rubartelli, A. *Mol. Biol. Cell* **1999**, *10*, 1463–1475.

171. Voos, W. *Mol. Cell* **2003**, *11*, 1–3.

172. Fang, B.; Xu, B.; Koch, P.; Roth, J. A. *Gene Ther.* **1998**, *5*, 1420–1424.

173. Richard, J. P.; Melikov, K.; Vives, E.; Ramos, C.; Verbeure, B.; Gait, M. J.; Chernomordik, L. V.; Lebleu, B. *J. Biol. Chem.* **2003**, *278*, 585–590.

14

GENE THERAPY AND GENE DELIVERY

NAOKI KOBAYASHI, MAKIYA NISHIKAWA, AND YOSHINOBU TAKAKURA

Department of Biopharmaceutics and Drug Metabolism, Graduate School of Pharmaceutical Sciences, Kyoto University, Sakyo-ku, Kyoto 606-8501, Japan

14.1. INTRODUCTION

Progress in molecular biology and biotechnology allowed the development of gene therapy in the closing years of the twentieth century. In addition, the completion of

Drug Delivery: Principles and Applications Edited by Binghe Wang, Teruna Siahaan, and Richard Soltero
ISBN 0-471-47489-4

the worldwide project of sequencing the human genome strongly supports this new mode of therapy by supplying information on the genetic basis of many diseases. With the increasing availability of valuable genetic information, gene therapy will be one of the most important techniques in the medical field in the next few decades.

An ultimate aim of gene therapy would be the correction of disease-causing genetic abnormalities. Point mutations or deletions in DNA sequences are the cause of a number of diseases that can theoretically be corrected. Various approaches have been proposed to correct point mutations in genomes, including chimeric RNA-DNA oligonucleotide.[1,2] Separately, a large deletion of DNA sequences could be made less significant through adjustment of the reading frame by a well-designed oligonucleotide. However, these approaches are too low in frequency to be used even in animal models at present. Because this ultimate mode of gene therapy, gene correction, appears to be difficult to achieve successfully in clinical situations, most of the current gene therapy approaches are based on the concept of adding functional genes, in which wild-type or therapeutic genes are introduced

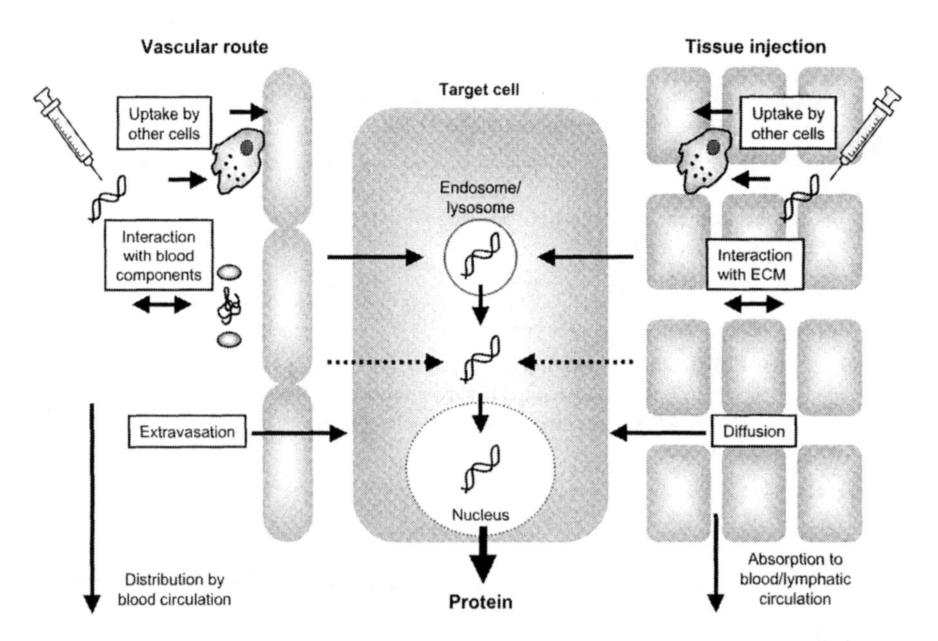

Figure 14.1. Fate of plasmid DNA *in vivo* after intravascular (left) or tissue (right) injection. Upon administration, plasmid DNA can be taken up by various cells, including mononuclear phagocytes such as macrophages. It also interacts with plasma proteins and extracellular matrix (ECM) components. In some cases, extravasation (intravascular route) or diffusion (tissue injection) is required for plasmid DNA to reach the target cell. Cellular uptake of DNA occurrs via an endocytotic route (solid lines) as well as a nonendocytotic route (dashed lines), depending on the vector and delivery method used for gene transfer. When endocytosis occurs, a means of endosomal escape is needed. Only plasmid DNA entering the nucleus has a chance to produce therapeutic protein.

as pharmaceuticals. Due to the difficulties involving genes reaching their targets *in vivo*, delivery is the major challenge in the development of *in vivo* gene therapy protocols. In addition to gene replacement therapy, success of *in vivo* gene transfer will give us an opportunity to carry out DNA vaccination, in which the antigenic protein is delivered in a form of genetic information that ensures the optimal immune response.[3]

A gene is transcribed to mRNA in the nucleus, then translated into a peptide at the ribosome; the peptide then goes through folding, glycosylation, and/or multimeric formation, resulting in the production of a functional protein, that is to say, a process of gene expression. Except for the cases in which genes encoded in DNA do not need to be transcribed before performing their therapeutic functions, gene therapy obviously requires the transgene expression of encoded genes. Thus, the administered gene has no biological significance without reaching its target: the nucleus. On its way to the nucleus from the injection site, DNA encoding a gene will pass through several biological barriers that often preclude successful gene delivery (Figure 14.1).[4] To overcome these barriers, much effort has been made to develop effective vectors for DNA, although most of these are not suitable for clinical use and need to be improved.

In this chapter, the delivery issues of DNA, especially plasmid DNA, are discussed with reference to gene therapy, focusing on existing biological barriers to be overcome and the characteristics of delivery methods and vectors to achieve this. Current progress and the status of nonviral gene delivery approaches are described.

14.2. INTERACTION OF DNA WITH CELLS

Upon administration to the body, plasmid DNA or its complex with any vector, interacts with cells, proteins, and extracellular matrix. The summation of these interactions will determine the *in vivo* fate of plasmid DNA and its complex, which, in turn, determines the location and extent of gene transfer. From a pharmaceutical and biopharmaceutical point of view, plasmid DNA is a huge macromolecule with a strong negative charge, and it is susceptible to attack by nucleases.

14.2.1. Negative Charge-Mediated Uptake

After entering the blood circulation, plasmid DNA encounters various cells, including blood cells, endothelial cells, and macrophages. The tissue distribution after intravenous injection of single- or double-stranded DNA, oligonucleotide, DNA/anti-DNA immune complex, mononucleosome, or chromatin has been examined. Although these studies have shown that the liver is the main organ responsible for the rapid clearance of these forms of DNA from the circulation,[5–9] the uptake mechanism and the cell type contributing to this hepatic uptake remain to be elucidated. We have studied the tissue distribution of plasmid DNA radiolabeled with ^{32}P following intravenous injection into mice.[10,11] After intravenous injection, naked plasmid DNA was rapidly eliminated from the blood circulation and taken up predominantly by the liver nonparenchymal cells, such as liver sinusoidal endothelial cells and Kupffer cells. Being administered in its free form, plasmid DNA was also rapidly degraded by nucleases in serum as well as in tissues. The hepatic uptake, most of which is mediated by the liver nonparenchymal cells, was as high as the plasma flow rate to the organ. The hepatic uptake of plasmid DNA was dependent on the concentration and decreased upon an increase of the dose. In addition, the hepatic uptake was inhibited by calf thymus DNA, polyinosinic acid, dextran sulfate, and heparin, but not by polycytidylic acid and chondroitin sulfate.[10–12] These findings indicate that plasmid DNA is taken up by scavenger receptors, which recognize polyanions in a charge- and/or structure-dependent manner.[13] However, we excluded the possibility of class A scavenger receptors (SRA) being involved in the uptake, based on tissue distribution and uptake experiments using cultured macrophages from SRA-knockout mice.[14] Another study supported the conclusion that SRA are not involved in the uptake of plasmid DNA.[15]

When injected into skeletal muscle or solid tumor tissue, plasmid DNA undergoes degradation as well as absorption into the blood circulation.[16] A fraction of the plasmid DNA is absorbed from the lymphatic system after injection into these tissues. Plasmid DNA and its degraded products exhibit distribution profiles similar to that obtained with intravenous plasmid DNA.

The detailed characteristics of the uptake of plasmid DNA were examined using cultured cells. Various cells have the ability to take up plasmid DNA in a concentration- and temperature-dependent manner. One example involves the brain microvessel endothelial cells,[17] which constitute the blood-brain barrier. However, their contribution to the tissue distribution of plasmid DNA seems to be minor. The main cells responsible for the distribution of plasmid DNA are immune cells such as macrophages. In fact, Kupffer cells, liver macrophages, make a large contribution to the clearance of plasmid DNA administered into the blood circulation. Plasmid DNA is efficiently taken up by cultured mouse peritoneal macrophages.[18] The profile of uptake inhibition is similar to that observed in brain microvessel endothelial cells. Another population of immune cells is dendritic cells (DCs); these are very important as far as both innate and acquired immunity are concerned. The DC cell line (DC2.4 cells) exhibits extensive uptake and degradation of plasmid DNA.[19]

14.2.2. Activation of Immune Cells

The unmethylated CpG sequence, i.e., the CpG motif, is a danger signal for our immune system.[20] A potent CpG motif consists of a central unmethylated CpG dinucleotide flanked by two 5′ purines and two 3′ pyrimidines. Compared with the DNA of eukaryotic cells (frequency of ~1:64), bacterial genomic DNA contains a higher frequency of the dinucleotide sequence CpG (1:16). Prokaryotic DNA is relatively unmethylated compared with the eukaryotic form, in which approximately 80% of the cytosines are methylated, a modification known to eliminate immunostimulation. These differences allow the mammalian immune system to recognize and respond to foreign DNA of bacterial origin. In addition to the immunostimulatory CpG motifs, neutralizing CpG sequences that can neutralize the immune activating properties of the stimulatory motifs have been reported.[21] Because plasmid DNA is derived from bacteria, it contains a number of sequences that can be regarded as immunostimulatory CpG motifs.

14.3. GENE DELIVERY APPROACHES

Successful application of gene therapy to patients requires the development of a gene delivery approach that enables *in vivo* transgene expression that is high enough to produce a therapeutic benefit. Because plasmid DNA is a huge macromolecule with a strong negative charge, it has little access to the nucleus when administered by routine methods. To facilitate the binding to cells, cationic vectors are often used, and the optimization of the structure and function is a major challenge as far as nonviral vector development is concerned. Another strategy is the use of naked plasmid DNA. In this case, the methodologies used to deliver plasmid DNA within cells are critical for gene transfer.

14.3.1. Fundamentals of Plasmid DNA Delivery

The large size of plasmid DNA greatly limits its distribution after *in vivo* administration. The distribution processes have some limitations as far as size is concerned, such as the passage through capillaries (about 5 μm in diameter) and fenestrae (50–300 nm) in the endothelial layers. Tissue distribution of intravascularly administered plasmid DNA is highly restricted by the endothelial wall, which is composed of vascular endothelial cells and basement membrane. Discontinuous endothelium, which is present only in the liver, spleen, bone marrow, and some solid tumors, allows plasmid DNA to come into contact with tissue cells.

Because the cell surface is negatively charged, positively charged compounds interact more intensely with cells than do negatively charged ones. Cationic complexes of plasmid DNA with any cationic vector exhibit very different pharmacokinetic profiles from that of naked plasmid DNA. They avidly bind to endothelial cells and normally accumulate in the lung after intravenous injection due to the first-pass effect.[22,23]

When plasmid DNA or its complex is injected into the extracellular space of a tissue, it distributes in an area that is very close to the injection site. Plasmid DNA within the extracellular space will be delivered into the surrounding cells, probably due to (1) increased pressure from the injection and/or (2) mechanical damage to the cell membrane. Very limited distribution of transfected cells has been reported after tissue injection of both naked and complexed plasmid DNA.[24–26]

14.3.2. Vector Complex

For intravenous administration, formation of a cationic complex greatly increases the interaction of plasmid DNA with the lung endothelial cells.[22,27] Transgene expression is also high in these cells. This might be the result of complicated events occurring *in vivo* after intravenous injection of lipoplex (see below). Serum proteins[28] and blood cells[23] have been reported to affect the tissue distribution of intravascularly administered plasmid DNA complex. When a plasmid DNA complex can avoid first-pass filtration in the lung, only a relatively small DNA complex can pass through the blood vessels and interact directly with the parenchymal cells of each tissue.

Although these nonviral vectors are distinct from their viral counterparts, their advantage of having a high affinity for various types of cells has been used to increase the transfection efficiency by viral vectors.[29,30]

14.3.2.1. Lipoplex So far, an enormous number of cationic lipid/liposome systems have been developed to improve the transfection efficiency of plasmid DNA. Some of the cationic lipids used have been summarized in a recent publication.[31] Cationic liposomes associate with plasmid DNA via an electrostatic interaction, which results in the formation of a complex called "lipoplex."[32] The driving force for lipoplex to introduce genes into cells is its electrostatic binding to negatively charged cellular membranes followed by endocytotic uptake. Some studies have shown that co-lipids, so-called helper lipids, in cationic liposomes are important determinants of transfection efficiency. Intravenous injection of lipoplex can lead to significant *in vivo* transfection activity in the lung when cholesterol is used as a helper lipid,[22,33] although such lipoplex formulations produce less transfection *in vitro* than lipoplex containing dioleoylphosphatidylethanolamine.

To compensate for the lack of specificity of electrostatic interaction of lipoplex, ligands can be introduced into cationic liposomes; sugars are the ligands that have been most extensively investigated so far.[34–37] On the other hand, local injection of lipoplex has also been investigated to ensure site-specific gene transfer. A local injection of lipoplex has been shown to be an effective approach to achieve transgene expression in the lung, brain, tumor, and skin.[38–41] Intratracheal administration of lipoplex is another possible way to introduce genes to the cells in the respiratory tract.[42]

Because the size of lipoplex is a key factor in determining the tissue distribution as well as the cellular uptake, it would be a challenge to reduce the size of lipoplex to increase transfection efficiency. Recently, Dauty et al.[43] succeeded in formulating plasmid DNA into stable nanometric particles with a diameter of less than 40 nm by synthesizing a dimerizable cationic detergent.

A major drawback of lipoplex is the production of inflammatory cytokines. We have shown that Kupffer cells in the liver are the major source of inflammatory cytokines after intravenous injection of lipoplex.[44] A functional depletion of Kupffer cells resulted in less cytokine production after lipoplex injection into mice. Macrophages and DCs can be considered the source of cytokines, because they produce large amounts of cytokines following incubation with lipoplex composed of plasmid DNA.[45] To suppress the immune response, the depletion of CpG sequences from plasmid DNA is a promising option. Yew et al.[46] eliminated 270 of 526 CpG dinucleotides in a reporter plasmid DNA, either by eliminating nonessential regions within the plasmid backbone or by site-directed mutagenesis. A CpG-reduced plasmid DNA was then found to be significantly less immunostimulatory than the original one. The administration of dexamethasone is another approach to reduce the production of such cytokines.[47] A sequential injection of plasmid DNA and cationic liposomes is another way of achieving a level of transgene expression in the lung similar to that following the injection of lipoplex with the production of far fewer inflammatory cytokines.[48,49]

14.3.2.2. Polyplex Cationic polymers are another class of nonviral vectors that can be used to increase gene delivery and transfer to target cells *in vivo*. Various types of polymers have been examined with respect to their ability to protect from nuclease degradation, to deliver target cells, and to increase the transfection efficiency of plasmid DNA. Plasmid DNA-cationic polymer complex, or polyplex, is believed to be taken up by cells via an endocytotic pathway, so its transfection efficiency depends on the release of plasmid DNA into the cytoplasm after cellular uptake. Therefore, polyplex-mediated transfection can be enhanced by the application of polymers possessing buffering ability[50] or fusogenic peptides creating pores on membranes.[51]

Polyethyleneimine (PEI) is one of the most extensively examined polymers because it has a relatively high transfection efficiency. PEI is believed to enter the cell via an endocytotic route and to possess a buffering capacity and the ability to swell when protonated.[52,53] Therefore, at low pH values, it is believed that PEI prevents acidification of the endosome and induces a large inflow of ions and water, subsequently leading to rupture of the endosomal or lysosomal membrane so that the PEI-plasmid DNA complex is delivered to the cytoplasmic space.[54,55] PEI is also reported to undergo nuclear localization while retaining an ordered structure once endocytosed.[53] However, the transfection efficiency of PEI depends on its molecular weight and structure.[56–58]

In another approach, the endosome-lysosome pathway of the polyplex may be circumvented by the use of endosome-disrupting peptides. We have demonstrated that a peptide that mimics the amino terminal of influenza virus hemagglutinin subunit HA-2 can greatly increase transgene expression in the liver after intravenous injection of hepatocyte-targeted polyplex.[59] About a 70-fold increase in transgene expression in the liver was obtained by the covalent binding of the peptide to a galactosylated poly(L-ornithine).

Targeted gene transfer has been widely investigated using polyplex possessing targeting ligands. These include asialoglycoproteins,[60] carbohydrates,[61–63] transferrin,[64,65] antibodies,[66,67] and lung surfactant proteins.[68] The attached ligand can increase the affinity for the target cell through a receptor-ligand interaction. When designed properly, about 80% of a polyplex can be successfully delivered to the target organ after systemic administration.[63]

However, the targeted delivery of polyplex is hampered by interaction with various compounds in the body fluids such as serum proteins.[69] Ogris et al.[70] have shown that, when incubated with plasma, the transferrin-PEI/plasmid DNA complex undergoes aggregation, which leads to reduced delivery to the target. The

PEGylation also appears to be a useful method of prolonging the blood circulation of polyplex after systemic administration, and PEGylated polyplex has resulted in gene transfer to a tumor without significant toxicity after intravenous injection into tumor-bearing mice.[71]

A series of peptides that can be recognized and transported to the nucleus are called "nuclear localization signal" (NLS) peptides. A well-known example is the basic peptide derived from the simian virus 40 large tumor antigen (PKKKRKV), which mediates binding of the karyophilic protein to importin.[72] Because of its nature, its application to gene delivery has been extensively investigated.[73–75] In most cases, transgene expression by plasmid DNA complex was increased by the application of NLS.

14.3.3. Naked Plasmid DNA

Although naked plasmid DNA is considered to have several problems for *in vivo* use as a way of obtaining sufficient gene transfer, the problems involving the use of DNA complex and the development of novel delivery methods allow naked plasmid DNA to be used as an efficient vector for gene therapy. Once it enters the cytoplasm or nucleus, naked plasmid DNA seems to produce much better transgene expression than does any complex. Furthermore, naked plasmid DNA injection using the hydrodynamics-based procedure (see below) can result in efficient transgene expression while producing very few nonspecific inflammatory cytokines,[76] a property that would be favorable for many gene therapy protocols. This is quite unlike the case in which the plasmid DNA complex is used.[44]

14.3.3.1. Direct Injection into Organs Because of its physicochemical properties, naked plasmid DNA has not generally been considered to possess the ability to enter cells without any vector. However, in 1990, Wolff et al.[77] demonstrated that a needle injection of naked plasmid DNA into mouse skeletal muscle resulted in significant transgene expression in the tissue. This was the first report showing that no special delivery system is required for *in vivo* gene transfer using plasmid DNA. Since then, this method of gene transfer has been applied to other sites of the body, such as the heart muscle,[78] liver,[24] brain,[39] skin,[41] urological organs,[79] thyroid,[80] and tumors.[40,81] At least in skeletal muscle, transgene expression in the organs can persist for several months,[82] although the transfected DNA is not thought to undergo chromosomal integration.

The fact that cells with transgene expression are found only in close proximity to the track of the needle injection indicates that dispersion of the injected plasmid DNA within the organ is an important issue and that mechanical force caused by the needle injection is critical for the effective intracellular delivery of naked plasmid DNA. Direct needle injection of plasmid DNA to the target organ is apparently simple and easy. Furthermore, muscle and skin are very attractive tissues as injection sites because surgery is not required for the administration of the DNA to these sites.

Some polymers, which do not form condensed complexes with plasmid DNA, have been reported to exhibit an enhancing effect on the transgene expression of naked plasmid DNA in skeletal muscle. Polyvinyl pyrrolidone and polyvinyl alcohol have been used to increase the extent and level of transgene expression following intramuscular injection of plasmid DNA.[83] In addition, Hartikka et al.[84] reported that a change of injection vehicle from saline or phosphate-buffered saline to 150 mM sodium phosphate buffer could enhance transgene expression by naked plasmid DNA in muscle due to inhibition of DNA degradation.

14.3.3.2. Intravascular Injection/Hydrodynamics-Based Procedure In 1999, Liu et al.[85] and Zhang et al.[86] developed a very powerful technique for *in vivo* gene transfer. In this method of administration, naked plasmid DNA is injected into the vasculature in a large volume of vehicle at high velocity. Liu et al. called this method of gene transfer a "hydrodynamics-based procedure." When injected by this technique, plasmid DNA does not require any vector to protect it from degradation. Although the mechanism of this technique is not fully understood, plasmid DNA can reach the nucleus by this approach. We have examined the tissue distribution characteristics of plasmid DNA administered either under normal conditions or by the hydrodynamics-based procedure.[11] Plasmid DNA injected by the hydrodynamics-based procedure was mainly recovered in the liver, which is similar to the profile obtained after the normal injection of plasmid DNA.[10] However, the hepatic uptake was not inhibited by prior administration of polyanions, including poly I, dextran sulfate, and heparin, suggesting a nonspecific uptake process. This hypothesis was supported by the finding that significant hepatic uptake of proteins such as bovine serum albumin and immunoglobulin G was observed after the hydrodynamics-based procedure.

The large-volume injection of naked plasmid DNA has also been applied to tissue-selective gene transfer, where a large volume of naked plasmid DNA solution was injected into the vasculature that led directly to the target organ with transient occlusion of the outflow. Budker et al.[87] demonstrated successful transgene expression in the liver by injecting plasmid DNA into the portal vein in a large volume of hypertonic solution (20% mannitol) with temporary occlusion of the hepatic vein to increase osmotic and hydrostatic pressure in the organ. The pressure generated at the local site may widen the sinusoid fenestrae and enhance DNA extravasation, followed by DNA uptake by hepatocytes. A similar approach was applied to the skeletal muscle of the rat hindlimb, where naked plasmid DNA was injected into the femoral artery in a large volume of saline over a period of 10 seconds while all blood vessels leading to and out of the hindlimb were occluded.[88] This intravascular injection increased transgene expression in muscle up to 40-fold compared with intramuscular injection. In a similar manner, kidney-targeted gene transfer was achieved by a retrograde injection of plasmid DNA into the renal vein of rats.[89] Transgene expression was detected only in the interstitial fibroblasts near the peritubular capillaries of the injected kidney.

14.3.3.3. Physical Enhancement of Delivery/Expression In addition to the large-volume-induced delivery of plasmid DNA into the target cells, other approaches to achieve intracellular delivery have been widely explored. The retention time of plasmid DNA in the tissue can be a factor determining gene transfer. Transient stoppage of blood flow through the diaphragm of mice greatly increased transgene expression in this muscular tissue following intravenous injection under normal conditions.[90] Significant expression of dystrophin (15–20% of the diaphragm muscle) after peripheral intravenous injection of plasmid DNA encoding the mouse dystrophin gene was detected in dystrophin-deficient *mdx* mice. Recently, significant transgene expression in the liver was achieved by simple mechanical massage after intravenous injection of naked plasmid DNA into mice.[91]

Electroporation has been used to increase the transport of charged molecules, such as plasmid DNA, across biological membranes.[92,93] After initial permeation, the pores close and plasmid DNA is trapped within the cell. Therefore, electroporation following a tissue injection of plasmid DNA increases the chance of DNA uptake by cells adjacent to the injection site. *In vivo* electroporation generally increases transgene expression up to 1000-fold compared with injection of naked plasmid DNA without electroporation in tissues such as skin,[94] liver,[95] melanoma,[96] and muscle.[97]

Shooting gold particles coated with plasmid DNA into target tissues or cells can deliver the DNA into the cytoplasm or even the nucleus. Skin, liver, and muscle have been successfully transfected after surgical exposure of the tissue. The efficiency of transfection varies among tissues, from 10–20% for skin epidermal cells to 1–5% for muscle cells.[98] In many cases, genes for antigens or cytokines have been introduced by this procedure for vaccination and immunotherapy, respectively.[99,100]

The application of ultrasound has also been investigated in an attempt to improve *in vivo* transgene expression, and this facilitated nonendocytotic uptake of plasmid DNA into cells.[101] Recently, the combined use of microbubbles was found to be effective in increasing ultrasound-mediated gene transfer. So far, ultrasound-mediated gene transfer has been carried out in vascular cells[102,103] and heart muscle.[104]

14.4. CONCLUSION

Since the discoveries of target genes that are responsible for monogenetic disorders such as the cystic fibrosis transmembrane conductance regulator gene and the dystrophin gene, the concept of gene therapy has been accepted as a very attractive and promising method of treatment for these diseases. In addition, the recent completion of the Human Genome Project has linked common illnesses such as heart disease, diabetes, and asthma with certain genetic factors, so these too can be treated by gene therapy. However, gene therapy, especially *in vivo*, has not so far been as effective as was initially hoped. Solution of the problems associated with gene delivery is a prerequisite for successful *in vivo* gene therapy. We believe that the latest results of basic research on gene delivery will offer a way to develop effective gene therapy protocols in the near future.

REFERENCES

1. Cole-Strauss, A.; Yoon, K.; Xiang, Y.; Byrne, B. C.; Rice, M. C.; Gryn, J.; Holloman, W. K.; Kmiec, E. B. *Science* **1996**, *273*, 1386–1389.
2. Igoucheva, O.; Yoon, K. *Hum. Gene Ther.* **2000**, *11*, 2307–2312.
3. Gurunathan, S.; Klinman, D. M.; Seder, R. A. *Annu. Rev. Immunol.* **2000**, *18*, 927–974.
4. Nishikawa, M.; Huang, L. *Hum. Gene Ther.* **2001**, *12*, 861–870.
5. Emlen, W.; Mannik, M. *J. Exp. Med.* **1978**, *147*, 684–699.
6. Emlen, W.; Mannik, M. *Clin. Exp. Immunol.* **1984**, *56*, 185–192.
7. Emlen, W.; Burdick, G. *J. Immunol.* **1988**, *140*, 1816–1822.
8. Gauthier, V. J.; Tyler, L. N.; Mannik, M. *J. Immunol.* **1996**, *156*, 1151–1156.
9. Du Clos, T. W.; Volzer, M. A.; Hahn, F. F.; Xiao, R.; Mold, C.; Searles, R. P. *Clin. Exp. Immunol.* **1999**, *117*, 403–411.
10. Kawabata, K.; Takakura, Y.; Hashida, M. *Pharm. Res.* **1995**, *12*, 825–830.
11. Kobayashi, N.; Kuramoto, T.; Yamaoka, K.; Hashida, M.; Takakura, Y. *J. Pharmacol. Exp. Ther.* **2001**, *297*, 853–860.
12. Yoshida, M.; Mahato, R. I.; Kawabata, K.; Takakura, Y.; Hashida, M. *Pharm. Res.* **1996**, *13*, 599–603.
13. Terpstra, V.; van Amersfoort, E. S.; van Velzen, A. G.; Kuiper, J.; van Berkel, T. J. *Arterioscler Thromb. Vasc. Biol.* **2000**, *20*, 1860–1872.
14. Takakura, Y.; Takagi, T.; Hashiguchi, M.; Nishikawa, M.; Yamashita, F.; Doi, T.; Imanishi, T.; Suzuki, H.; Kodama, T.; Hashida, M. *Pharm. Res.* **1999**, *16*, 503–508.
15. Zhu, F. G.; Reich, C. F.; Pisetsky, D. S. *Immunology* **2001**, *103*, 226–234.
16. Kawase, A.; Nomura, T.; Yasuda, K.; Kobayashi, N.; Hashida, M.; Takakura, Y. *J. Pharm. Sci.* **2003**, *92*, 1295–1304.
17. Nakamura, M.; Davila-Zavala, P.; Tokuda, H.; Takakura, Y.; Hashida M. *Biochem. Biophys. Res. Commun.* **1998**, *245*, 235–239.
18. Takagi, T.; Hashiguchi, M.; Mahato, R. I.; Tokuda, H.; Takakura, Y.; Hashida, M. *Biochem. Biophys. Res. Commun.* **1998**, *245*, 729–733.
19. Yoshinaga, T.; Yasuda, K.; Ogawa, Y.; Takakura, Y. *Biochem. Biophys. Res. Commun.* **2002**, *299*, 389–394.
20. Krieg, A. M. *Annu. Rev. Immunol.* **2002**, *20*, 709–760.
21. Krieg, A. M.; Wu, T.; Weeratna, R.; Efler, S. M.; Love-Homan, L.; Yang, L.; Yi, A. K.; Short, D.; Davis, H. L. *Proc. Natl. Acad. Sci. USA* **1998**, *95*, 12631–12636.
22. Liu, Y.; Mounkes, L. C.; Liggitt, H. D.; Brown, C. S.; Solodin, I.; Heath, T. D.; Debs, R. J. *Nat. Biotechnol.* **1997**, *15*, 167–173.
23. Sakurai, F.; Nishioka, T.; Saito, H.; Baba, T.; Okuda, A.; Matsumoto, O.; Taga, T.; Yamashita, F.; Takakura, Y.; Hashida, M. *Gene Ther.* **2001**, *8*, 677–686.
24. Hickman, M. A.; Malone, R. W.; Lehmann-Bruinsma, K.; Sih, T. R.; Knoell, D.; Szoka, F. C.; Walzem, R.; Carlson, D. M.; Powell, J. S. *Hum. Gene Ther.* **1994**, *5*, 1477–1483.
25. Nomura, T.; Nakajima, S.; Kawabata, K.; Yamashita, F.; Takakura, Y.; Hashida, M. *Cancer Res.* **1997**, *57*, 2681–2686.
26. O'Hara, A. J.; Howell, J. M.; Taplin, R. H.; Fletcher, S.; Lloyd, F.; Kakulas, B.; Lochmuller, H.; Karpati, G. *Muscle Nerve* **2001**, *24*, 488–495.

27. Mahato, R. I.; Kawabata, K.; Nomura, T.; Takakura, Y.; Hashida, M. *J. Pharm. Sci.* **1995**, *84*, 1267–1271.
28. Li, S.; Tseng, W. C.; Stolz, D. B.; Wu, S. P.; Watkins, S. C.; Huang, L. *Gene Ther.* **1999**, *6*, 585–594.
29. Bouri, K.; Feero, W. G.; Myerburg, M. M.; Wickham, T. J.; Kovesdi, I. Hoffman, E. P.; Clemens, P. R. *Hum. Gene Ther.* **1999**, *10*, 1633–1640.
30. Porter, C. D. *J. Gene Med.* **2002**, *4*, 622–633.
31. Audouy, S. A.; de Leij, L. F.; Hoekstra, D.; Molema, G. *Pharm. Res.* **2002**, *19*, 1599–1605.
32. Felgner, P. L.; Barenholz, Y.; Behr, J. P.; Cheng, S. H.; Cullis, P.; Huang, L.; Jessee, J. A.; Seymour, L.; Szoka, F.; Thierry, A. R.; Wagner, E.; Wu, G. *Hum. Gene Ther.* **1997**, *8*, 511–512.
33. Templeton, N. S.; Lasic, D. D.; Frederik, P. M.; Strey, H. H.; Roberts, D. D.; Pavlakis, G. N. *Nat. Biotechnol.* **1997**, *15*, 647–652.
34. Remy, J. S.; Kichler, A.; Mordvinov, V.; Schuber, F.; Behr, J. P. *Proc. Natl. Acad. Sci. USA* **1995**, *92*, 1744–1748.
35. Hara, T.; Aramaki, Y.; Takada, S.; Koike, K.; Tsuchiya, S. *Gene Ther.* **1995**, *2*, 784–788.
36. Kawakami, S.; Fumoto, S.; Nishikawa, M.; Yamashita, F.; Hashida, M. *Pharm. Res.* **2000**, *17*, 306–313.
37. Kawakami, S.; Sato, A.; Nishikawa, M.; Yamashita, F.; Hashida, M. *Gene Ther.* **2000**, *7*, 292–299.
38. Brigham, K. L.; Meyrick, B.; Christman, B.; Magnuson, M.; King, G.; Berry, L. C., Jr. *Am. J. Med. Sci.* **1989**, *298*, 278–281.
39. Ono, T.; Fujino, Y.; Tsuchiya, T.; Tsuda, M. *Neurosci. Lett.* **1990**, *117*, 259–263.
40. Plautz, G. E.; Yang, Z. Y.; Wu, B. Y.; Gao, X.; Huang, L.; Nabel, G. J. *Proc. Natl. Acad. Sci. USA* **1993**, *90*, 4645–4649.
41. Raz, E.; Carson, D. A.; Parker, S. E.; Parr, T. B.; Abai, A. M.; Aichinger, G.; Gromkowski, S. H.; Singh, M.; Lew, D.; Yankauckas, M. A.; Baird, S. M.; Rhodes, G. H. *Proc. Natl. Acad. Sci. USA* **1994**, *91*, 9519–9523.
42. McCluskie, M. J.; Chu, Y.; Xia, J. L.; Jessee, J.; Gebyehu, G.; Davis, H. L. *Antisense Nucleic Acid Drug Dev.* **1998**, *8*, 401–414.
43. Dauty, E.; Remy, J. S.; Blessing, T.; Behr, J. P. *J. Am. Chem. Soc.* **2001**, *123*, 9227–9234.
44. Sakurai, F.; Terada, T.; Yasuda, K.; Yamashita, F.; Takakura, Y.; Hashida, M. *Gene Ther.* **2002**, *9*, 1120–1126.
45. Yasuda, K.; Ogawa, Y.; Kishimoto, M.; Takagi, T.; Hashida, M.; Takakura, Y. *Biochem. Biophys. Res. Commun.* **2002**, *293*, 344–348.
46. Yew, N. S.; Zhao, H.; Wu, I. H.; Song, A.; Tousignant, J. D.; Przybylska, M.; Cheng, S. H. *Mol. Ther.* **2000**, *1*, 255–262.
47. Tan, Y.; Li, S.; Pitt, B. R.; Huang, L. *Hum. Gene Ther.* **1999**, *10*, 2153–2161.
48. Song, Y. K.; Liu, F.; Liu, D. *Gene Ther.* **1998**, *5*, 1531–1537.
49. Tan, Y.; Liu, F.; Li, Z.; Li, S.; Huang, L. *Mol. Ther.* **2001**, *3*, 673–682.
50. Boussif, O.; Lezoualch, F.; Zanta, M. A.; Mergny, M. D.; Scherman, D.; Demeneix, B.; Behr, J. P. *Proc. Natl. Acad. Sci. USA* **1995**, *92*, 7297–7301.

51. Wagner, E.; Plank, C.; Zatloukal, K.; Cotten, M.; Birnstiel, M. L. *Proc. Natl. Acad. Sci. USA* **1992**, *89*, 7934–7938.
52. Boletta, A.; Benigni, A.; Lutz, J.; Remuzzi, G.; Soria, M. R.; Monaco, L. *Hum. Gene Ther.* **1997**, *8*, 1243–1251.
53. Godbey, W. T.; Wu, K. K.; Mikos, A. G. *Proc. Natl. Acad. Sci. USA* **1999**, *96*, 5177–5181.
54. Wagner, E. *J. Control. Release* **1998**, *53*, 155–158.
55. Klemm, A. R.; Young, D.; Lloyd, J. B. *Biochem. Pharmacol.* **1998**, *56*, 41–46.
56. Ouatas, T.; Le Mevel, S.; Demeneix, B. A.; de Luze, A. *Int. J. Dev. Biol.* **1998**, *42*, 1159–1164.
57. Fischer, D.; Bieber, T.; Li, Y.; Elsasser, H. P.; Kissel, T. *Pharm. Res.* **1999**, *16*, 1273–1279.
58. Morimoto, K.; Nishikawa, M.; Kawakami, S.; Nakano, T.; Hattori, Y.; Fumoto, S.; Yamashita, F.; Hashida, M. *Mol. Ther.* **2003**, *7*, 254–261.
59. Nishikawa, M.; Yamauchi, M.; Morimoto, K.; Ishida, E.; Takakura, Y.; Hashida, M. *Gene Ther.* **2000**, *7*, 548–555.
60. Wu, G. Y.; Wu, C. H. *J. Biol. Chem.* **1988**, *263*, 14621–14624.
61. Midoux, P.; Mendes, C.; Legrand, A.; Raimond, J.; Mayer, R.; Monsigny, M.; Roche, A. C. *Nucleic Acids Res.* **1993**, *21*, 871–878.
62. Perales, J. C.; Ferkol, T.; Beegen, H.; Ratnoff, O. D.; Hanson, R. W. *Proc. Natl. Acad. Sci. USA* **1994**, *91*, 4086–4090.
63. Nishikawa, M.; Takemura, S.; Takakura, Y.; Hashida, M. *J. Pharmacol. Exp. Ther.* **1998**, *287*, 408–415.
64. Wagner, E.; Zenke, M.; Cotten, M.; Beug, H.; Birnstiel, M. L. *Proc. Natl. Acad. Sci. USA* **1990**, *87*, 3410–3414.
65. Kircheis, R.; Schuller, S.; Brunner, S.; Ogris, M.; Heider, K. H.; Zauner, W.; Wagner, E. *J. Gene Med.* **1999**, *1*, 111–120.
66. Ferkol, T.; Kaetzel, C. S.; Davis, P. B. *J. Clin. Invest.* **1993**, *92*, 2394–2400.
67. Li, S.; Tan, Y.; Viroonchatapan, E.; Pitt, B. R.; Huang, L. *Am. J. Physiol.* **2000**, *278*, L504–L511.
68. Ross, G. F.; Morris, R. E.; Ciraolo, G.; Huelsman, K.; Bruno, M.; Whitsett, J. A.; Baatz, J. E.; Korfhagen, T. R. *Hum. Gene Ther.* **1995**, *6*, 31–40.
69. Opanasopit, P.; Nishikawa, M.; Hashida, M. *Crit. Rev. Ther. Drug Carrier Syst.* **2002**, *19*, 191–233.
70. Ogris, M.; Brunner, S.; Schuller, S.; Kircheis, R.; Wagner, E. *Gene Ther.* **1999**, *6*, 595–605.
71. Kircheis, R.; Blessing, T.; Brunner, S.; Wightman, L.; Wagner, E. *J. Control. Release* **2001**, *72*, 165–170.
72. Adam, S. A.; Gerace, L. *Cell* **1991**, *66*, 837–847.
73. Kaneda, Y.; Iwai, K.; Uchida, T. *Science* **1989**, *243*, 375–378.
74. Fritz, J. D.; Herweijer, H.; Zhang, G.; Wolff, J. A. *Hum. Gene Ther.* **1996**, *7*, 1395–1404.
75. Ludtke, J. J.; Zhang, G.; Sebestyen, M. G.; Wolff, J. A. *J. Cell Sci.* **1999**, *112*, 2033–2041.
76. Kobayashi, N.; Kuramoto, T.; Chen, S.; Watanabe, Y.; Takakura, Y. *Mol. Ther.* **2002**, *6*, 737–744.

77. Wolff, J. A.; Malone, R. W.; Williams, P.; Chong, W.; Acsadi, G.; Jani, A.; Felgner, P. L. *Science* **1990**, *247*, 1465–1468.
78. Lin, H.; Parmacek, M. S.; Morle, G.; Bolling, S.; Leiden, J. M. *Circulation* **1990**, *82*, 2217–2221.
79. Yoo, J. J.; Soker, S.; Lin, L. F.; Mehegan, K.; Guthrie, P. D.; Atala, A. *J. Urol.* **1999**, *162*, 1115–1118.
80. Sikes, M. L.; O'Malley, B. W., Jr.; Finegold, M. J.; Ledley, F. D. *Hum. Gene Ther.* **1994**, *5*, 837–844.
81. Nomura, T.; Yasuda, K.; Yamada, T.; Okamoto, S.; Mahato, R. I.; Watanabe, Y.; Takakura, Y.; Hashida, M. *Gene Ther.* **1999**, *6*, 121–129.
82. Acsadi, G.; Dickson, G.; Love, D. R.; Jani, A.; Walsh, F. S.; Gurusinghe, A.; Wolff, J. A.; Davies, K. E. *Nature* **1991**, *352*, 815–818.
83. Alila, H.; Coleman, M.; Nitta, H.; French, M.; Anwer, K.; Liu, Q.; Meyer, T.; Wang, J.; Mumper, R.; Oubari, D.; Long, S.; Nordstrom, J.; Rolland, A. *Hum. Gene Ther.* **1997**, *8*, 1785–1795.
84. Hartikka, J.; Bozoukova, V.; Jones, D.; Mahajan, R.; Wloch, M. K.; Sawdey, M.; Buchner, C.; Sukhu, L.; Barnhart, K. M.; Abai, A. M.; Meek, J.; Shen, N.; Manthorpe, M. *Gene Ther.* **2000**, *7*, 1171–1182.
85. Liu, F.; Song, Y.; Liu, D. *Gene Ther.* **1999**, *6*, 1258–1266.
86. Zhang, G.; Budker, V.; Wolff, J. A. *Hum. Gene Ther.* **1999**, *10*, 1735–173.
87. Budker, V.; Zhang, G.; Knechtle, S.; Wolff, J. A. *Gene Ther.* **1996**, *3*, 593–598.
88. Budker, V.; Zhang, G.; Danko, I.; Williams, P.; Wolff, J. *Gene Ther.* **1998**, *5*, 272–276.
89. Maruyama, H.; Higuchi, N.; Nishikawa, Y.; Hirahara, H.; Iino, N.; Kameda, S.; Kawachi, H.; Yaoita, E.; Gejyo, F.; Miyazaki, J. *Hum. Gene Ther.* **2002**, *13*, 455–468.
90. Liu, F.; Nishikawa, M.; Clemens, P. R.; Huang, L. *Mol. Ther.* **2001**, *4*, 45–51.
91. Liu, F.; Huang, L. *Hepatology* **2002**, *35*, 1314–1319.
92. Mir, L. M.; Banoun, H.; Paoletti, C. *Exp. Cell Res.* **1988**, *175*, 15–25.
93. Somiari, S.; Glasspool-Malone, J.; Drabick, J. J.; Gilbert, R. A.; Heller, R.; Jaroszeski, M. J.; Malone, R. W. *Mol. Ther.* **2000**, *2*, 178–187.
94. Titomirov, A. V.; Sukharev, S.; Kistanova, E. *Biochim. Biophys. Acta* **1991**, *1088*, 131–134.
95. Heller, R.; Jaroszeski, M.; Atkin, A.; Moradpour, D.; Gilbert, R.; Wands, J.; Nicolau, C. *FEBS Lett.* **1996**, *389*, 225–228.
96. Rols, M. P.; Delteil, C.; Golzio, M.; Dumond, P.; Cros, S.; Teissie, J. *Nat. Biotechnol.* **1998**, *16*, 168–171.
97. Aihara, H.; Miyazaki, J. *Nat. Biotechnol.* **1998**, *16*, 867–870.
98. Yang, N. S.; Burkholder, J.; Roberts, B.; Martinell, B.; McCabe, D. *Proc. Natl. Acad. Sci. USA* **1990**, *87*, 9568–9572.
99. Lin, M. T.; Pulkkinen, L.; Uitto, J.; Yoon, K. *Int. J. Dermatol.* **2000**, *39*, 161–170.
100. Yoshida, S.; Kashiwamura, S. I.; Hosoya, Y.; Luo, E.; Matsuoka, H.; Ishii, A.; Fujimura, A.; Kobayashi, E. *Biochem. Biophys. Res. Commun.* **2000**, *271*, 107–115.
101. Huber, P. E.; Pfisterer, P. *Gene Ther.* **2000**, *7*, 1516–1525.

102. Taniyama, Y.; Tachibana, K.; Hiraoka, K.; Namba, T.; Yamasaki, K.; Hashiya, N.; Aoki, M.; Ogihara, T.; Yasufumi, K.; Morishita, R. *Circulation* **2002**, *105*, 1233–1239.

103. Teupe, C.; Richter, S.; Fisslthaler, B.; Randriamboavonjy, V.; Ihling, C.; Fleming, I.; Busse, R.; Zeiher, A. M.; Dimmeler, S. *Circulation* **2002**, *105*, 1104–1109.

104. Shohet, R. V.; Chen, S.; Zhou, Y. T.; Wang, Z.; Meidell, R. S.; Unger, R. H.; Grayburn, P. A. *Circulation* **2000**, 101, 2554–2556.

15

PARENTERAL FORMULATION FOR PEPTIDES, PROTEINS, AND MONOCLONAL ANTIBODIES DRUGS: A COMMERCIAL DEVELOPMENT OVERVIEW

JOHN A. BONTEMPO

Biopharmaceutical Product Development Consultant, 18 Benjamin Street, Somerset, NJ 08873

Drug Delivery: Principles and Applications Edited by Binghe Wang, Teruna Siahaan, and Richard Soltero
ISBN 0-471-47489-4

15.1. DEVELOPMENT OF BIOPHARMACEUTICAL PARENTERAL DOSAGE FORMS

15.1.1. Introduction

The objectives of this chapter are to give aspiring formulation scientists a brief yet useful and meaningful overview of conventional parenteral technology as it is applied today and the development of new delivery systems of biopharmaceutical drugs to yield successful, sterile parenteral products for human and veterinary use.

Some of the key factors in considering specific delivery systems are safety, stability, and efficacy. The parenteral administration of proteins and peptides today offers assured levels of bioavailability and the ability of the product to reach the marketplace first. It is safe to assume that over 95% of the protein therapeutics approved by the Food and Drug Administration (FDA) today are injectable products since parenteral administration avoids physical and enzymatic degradation.

However, intensive studies are in progress today in the biopharmaceutical industry for the application of alternate delivery systems, and future advances in medical therapies will depend on new deliveries. These new deliveries may focus on modified parenteral system(s), as well as cavitational, respiratory, gastrointestinal, nasal, dermal, and other areas. In each case, membrane and gastrointestinal barriers must be extensively studied in order to increase bioavailability.

The success of the development of biopharmaceutical therapeutic agents requires close interactions of interdisciplinary sciences encompassing molecular biology, fermentation, process development, protein chemistry, analytical biochemistry, pharmacology, toxicology, preformulations, formulations, clinical development, quality assurance, scale-up, bulk manufacturing, packaging, aseptic manufacturing, regulatory affairs, marketing, and others.[1]

Considerations for the most applicable route of parenteral administration are of the utmost importance, as we will see later. Most of these routes are:

- Intramuscular
- Intravenous
- Subcutaneous
- Epidural
- Intrathecal

Why choose a parenteral system? This system ensures that the drug will reach specific target areas of the body via blood and lymphatic systems. It allows the researcher to have control over pharmacological parameters,[2] serum levels, tissue concentrations, elimination of the drug from the body, and other factors.

15.1.1.1. Key Requirements to Consider Before Preformulations and Formulations of Biopharmaceuticals Begin Applied formulation scientists today face formidable challenges in their quest to formulate stable recombinant protein therapeutics. Proteins possess unique characteristics. We are dealing with very

TABLE 15.1 Physicochemical Factors to Be Considered for Protein Drug Formulations

Structure of the protein drug	Agents affecting stability
Isoelectric point	pH
	Temperature
Molecular weight	Light
	Oxygen
Amino acid composition	Metal ions
	Freeze-thaw
Disulfide bonds	Mechanical stress
Spectral properties	
Agents affecting solubility	Polymorphism
Detergents	Stereoisomers
Salts	Filtration media compatibility
Metal ions	Shear
pH	Surface denaturation

complex, high molecular weight, highly purified, heat-unstable molecules with a high tendency to aggregate. As this happens, chemical and physical changes occur, leading to a great deal of instability.

Protein instability mechanisms have been reviewed by several investigators.[3–13] Chemical reactions such as oxidation, deamidation, proteolysis, racemization, isomerization, disulfide exchange, photolysis, and others will give rise to chemical instability. It is critical that when this happens, the denaturation mechanisms must be identified in order to select appropriate stabilizing excipients. These chemical excipients may be in the form of amino acids, surfactants, polyhydric alcohols, antioxidants, phospholipids, chelating agents, and others.

Before preformulations begin, it is critical for the formulation scientist to do extensive research on the physicochemical properties of the active drug substance(s) protein, peptide, or monoclonal antibody.

15.1.1.2. Preformulation Physicochemical Factors All the factors listed in Table 15.1 should be reviewed.

15.1.1.3. Preformulations From an industrial point of view, preformulation studies are designed to cover a wide range of properties in a short time and to learn as much as possible, but not in great depth. The studies should focus on the identification of potential problems early enough to evaluate potential alternatives to stabilize future formulations that could lead to a product.

Table 15.2 summarizes the bioanalytical methods employed to evaluate some initial preformulation breakdown products.

15.1.1.4. Initial Variables to Be Tested Perhaps five to eight initial preformulation combinations should be considered. Variables to be tested with varying protein

TABLE 15.2 Bioanalytical Methods Used to Evaluate Initial Preformulation Development

Method	Function
Bioassay	Measure of activity throughout the shelf life of a formulation
Immunoassay	Purity assessment and measures concentration of a particular molecular species
pH	Chemical stability
Sodium dodecyl sulfate–polyacrylamide gel electrophoresis (SDS-PAGE reduced and nonreduced)	Separation by molecular weight, characterization of proteins and purity
Reverse phase high-performance liquid chromatography	Estimation of purity, identity, and stability of proteins Separation and analysis of protein digests
Iso electric focusing (IEF)	Determines the isoelectric point of the protein and detects modifications of the protein
–high-performance liquid chromatography (SE-HPLC)	Method of separating molecules according to their molecular size and purity determination
N-terminal Sequencing	Elucidation of the C terminus, identity
Ultraviolet	Detection of individual components, concentration, and aggregation
Circular dichroism in the UV region	Detects secondary and tertiary conformation and quantitates various structures

concentrations are the effects of buffer species, ionic strength, pH range, temperature, initial shear, surface denaturation, and agitation.

15.1.1.5. Experimental Conditions for Initial Preformulation Studies

- Protein drugs have been tested at varying ranges of activity. Their respective concentrations are at the nanogram to microgram to milligram level and vary from protein to protein.
- **pH Range**
  Ranges of pH from 3, 5, 7 and 9 should be selected. Specific pH units will be determined during the final formulation studies. pH changes may have varying impact on the solubility and stability of the formulation. pH control in pharmaceutical dosage forms is critical.[13] The proper pH selection is one of the key factors in developing a stable product.

- **Buffers**
  Buffers should be selected from the United States Pharmacopeia (USP) physiological buffer list based upon their optimal pH range. Some of these buffers and their pH ranges are:

Phosphates	6.2–8.2
Succinate	3.2–6.6
Citrate	2.0–6.2
Tris	7.1–9.1

  Buffer concentrations[14] most frequently used should be in the range of 10, 15, or 20 mM. As buffer concentrations increase, so does the pain upon injection.
- **Chelating Agents**
  During fermentation, purification of the bulk active protein residual metal ions could be present as it contacts stainless steel, iron or copper surfaces. The use of a chelating agent is a requirement, and recommended dosages may range from 0.01 to 0.05% to bind or chelate the metal ions present in the solution.
- **Antioxidants**
  Since oxidation is one of the major factors in protein degradation, the use of specific antioxidants may be required. Ascorbic acid, monothio-glycerol, and alpha tocopherols have been used for this purpose. A recommended antioxidant dose[14] would range from 0.05 to 0.1%.
- **Preservatives**
  If a multidose formulation is required, an antimicrobial agent, also called a "preservative," must be added to the formulation for regulatory compliance. The preservative must meet USP requirements and the "Antimicrobial Effectiveness Test" as stated in the USP. The most often used preservatives for proteins or recombinant products are phenol at 0.3 to 0.5%; chlorobutanol, also at 0.3 to 0.5%, and benzyl alcohol at 1.0 to 3.0%.
- **Glass Vial Selection**
  The vial selected should be of type I glass, as stated in the USP. It is highly recommended that all experimental work begin with this type of glass, and that early in the process the interactions of the proteins with the glass surfaces should be determined. Glass is *not* an inert material. Glass surfaces must be taken into consideration to study the adsorptive properties of the respective protein. Adsorption of proteins will be treated later in this chapter.
- **Rubber Stopper Selection**
  Equally important in screening initial preformulations is the selection of a rubber stopper compatible with protein solutions. The variety of composition of rubber stoppers in parenteral formulations of biopharmaceuticals requires studies on compatibility with proteins, involving chemical extractants from the rubber composition into the protein solution over periods of stability at varying temperatures. We must also consider particle shedding from the stoppers into the protein solution, adsorption, absorption, permeation through the

stoppers, and flexibility of stopper properties for machinability. More information on rubber composition is presented later in this chapter.

- **Membrane Filter Selections**
  Since all protein formulations are aseptically filled for final sterilization of the product, selection of the membrane filter and its media composition is very important. The chemical nature of the filter membrane and the pH of the protein adsorption[16] are perhaps two of the most important parameters to study.
  Several issues require investigations. Membrane filters shed particles or fibers released during filtration. This is an area where potential extractables may occur. The potential toxicity of the filters and the product's compatibility with the membrane must be determined. Of all the filters tested (unpublished data), polyvinylidene difluoride, polycarbonate, and polysulfone were found to be most compatible with several proteins, with minimal amounts of protein binding and deactivation.
- **Degradation Mechanisms**
  There are physical and chemical degradations to be observed and recorded during the period of initial stability at varying temperature levels. The formulators must select, from among many bioanalytical methods, those that can differentiate physical and chemical degradation both quantitatively and qualitatively. In Table 15.2 the preliminary methods are listed. Both physical and chemical degradations are found in the literature cited.
  From the summary data of the preformulation studies, scientists should obtain potential leads based on stability conditions designed in order to consider them for further development. The next stage of development, which we call "formulation," should take into account all the parameters that may achieve one or more stable formulations for marketability with acceptable industrial stability.
- **Formulation**
  The design of experimental formulations should be based on supportive results obtained from preformulations. These key results must be taken into consideration:

  1. Initial compatibility testing of the active drug substance with excipients tested.
  2. Favorable stability factors such as temperature, light, and packaging components.
  3. Initial degradation products of the preformulations.
  4. Performance of stability methods indicating perhaps "stability-indicating" testing at this time.

- **Formulation of the Bulk Active Drug**
  Select a bulk active drug which may have as much characterization as possible. Know the levels of homogeneity and, whenever possible, lot-to-lot reproducibility. If any impurities are detected, there should be qualitative and quantitative identifications.

- **Characterization of the Bulk Active Drug**
  Characterization of the bulk active drug is of paramount importance to a successful formulation project. Physicochemical properties should be elucidated in order to select the most compatible excipients with the active drug and with the other formulation excipients. Table 15.3 summarizes the physicochemical properties and stabilizers most often used in protein, peptide, and monoclonal antibody formulations.[17–27]
- **Product Formulation**
  Bringing a potentially stable preformulation to an experimental formulation, and then to a successful marketable product, is the result of all the scientific

TABLE 15.3 Stabilizers Used in Protein Formulations

Stabilizer	Action/Uses
Proteins	
Human serum albumin (HSA)	Prevents surface adsorption Conformational stabilizer Complexing agent Cryoprotectant
Amino Acids	
Glycine	Stabilizer
Alanine	Solubilizer
Arginine	Buffer
Leucine	Inhibits aggregation
Glutamic acid	Thermostabilizer
Aspartic acid	Isomerism inhibitor
Surfactants	
Polysorbate 20 and 80	Retard aggregation
Poloxamer 407	Prevents denaturation, stabilize cloudiness
Fatty Acids	
Phosphotidyl choline	Stabilizer
Ethanolamine	
Acethyltryptophanate	
Polymers	
Polyethylene glycol (PEG)	Stabilizer
Polyvinylpyrrolidone (PVP) 10, 24, 40	Prevent aggregation
Polyhydric Alcohol	
Sorbitol	Prevents denaturation aggregation
Mannitol	Cryoprotectant
Glycerin	May act as an antioxidant
Sucrose	
Glucose	Strengthens conformational
Propylene glycol	Prevents aggregation
Ethylene glycol	
Lactose	
Trehalose	

(*Continued*)

TABLE 15.3 (*Continued*)

Stabilizer	Action/Uses
Antioxidants	
Ascorbic acid	Retards oxidation
Cysteine HCI	
Thioglycerol	
Thioglycolic acid	
Thiosorbitol	
Gluthathione	
Reducing Agents	
Several thiols	Inhibit disulfide bond formation Prevent aggregation
Chelating Agents	
EDTA salts	Inhibit oxidation by removing metal ions
Glutamic acid	
Aspartic acid	
Metal Ions	
Ca^{2+}, Ni^{2+}, Mg^{2+}, MN^{2+}	Stabilize protein conformation

disciplines working together. Perhaps one of the most demanding and exact technologies that a formulation group must use is analytical technology. The development of quantitative methods of testing any active protein or peptide is of paramount importance in the successful evaluation of chemical and physical stress.

These quantitative methods must eventually detect potential chemical degradants, contaminants, and impurities induced by oxidation, deamidation, proteolysis, and disulfide exchange. Physical instability such as aggregation, denaturation, adsorption, and precipitation must also be detected and quantitated.

The analytical methods used must be reproducibly validatable to ensure regulatory compliance for the FDA and confidence in the quality of a potentially successful, marketable product. Some of the most often used analytical methods are the chromatographic techniques such as size exclusion, ion exchange, affinity chromatography, and reverse-phase chromatography.

Other equally important methods are the electrophoretic techniques, such as polyacrylamide gel electrophoresis, isoelectric focusing, Western blots, combined electrophoresis, and isoelectric focusing (two-dimensional electrophoresis). Bioactivity methods are other key methods used in biotechnology product development. These include *in vivo* whole animal bioassay, cell culture bioassay, immunoassay, and biochemical assay. Many references and several textbooks are available in many industrial and academic libraries to provide additional and up-to-date information.

- **Package Selection**
  Another critical component is the package. Glass vials, and more recently plastic vials, are the containers with which the product comes in contact. As stated previously, glass is not inert, and there are several key points to consider in choosing the correct one. Some of these factors are type I glass, USP specifications, size requirements, protection from oxygen and moisture, compatibility, and adsorption as well as considerations.[28,29] All possible forms of degradation that could occur, such as adsorption, aggregation, structure, composition of silicate glasses, depyrogenation, and others must be dealt with to ensure industrial stability for the product.
- **Elastomeric Closures**
  The formulator should review and select appropriate rubber stoppers for the potential product. The screening and final selection should be based on these considerations: the stopper should be essentially nonreactive, physically and chemically, and with a complete barrier to vapor/gas permeation; it should have compressibility and resealability, be resistant to coring and fragmentation, and maintain the seal interface and packaging integrity. There are also other desirable properties[30–33] to take into consideration.
- **Membrane Filtration**
  Membrane filtration application to biopharmaceutical product development is extremely important since sterile protein-peptide products can only be prepared via sterile filtration and gamma radiation; steam cannot be used under pressure. There are several excellent works in the field of sterile membrane filtration.[34–36] The filter media most often tested for protein formulations with minimum adsorption and maximum compatibility are mixed esters of cellulose acetate, cellulose nitrate, polysulfone, and nylon 66. Membrane filters must be tested for compatibility with the active drug substance and selected for formulations if they have the lowest adsorption and maximum compatibility with the product.
- **Glass Vials, Elastomeric Closures, and Filtration Membrane Extractables**
  All pharmaceuticals and material for medical items are carefully screened and tested for extractables, as required by regulatory agencies. Suppliers are also responsible for developing extractable procedures and conducting toxicity studies to be shared with users in the industry and government.

  Some pertinent degradants from glass are silica lamination, especially in phosphate buffer after 6 or more months of stability. Citrate and EDTA can induce complexing agents as well as high levels of sodium, aluminum, barium, and iron.

  Elastomeric closures can leach out accelerators such as mercaptobenzothiazole and tetramethyl thiuran disulfide and activators such as zinc oxide. Lubricants will excrete stearic acid as inert components, and antioxidants will excrete hindered phenol.

  Commercial membrane filters must be tested for regulatory compliance. Manufacturers supply extractables data to the FDA in their Drug Master File

(DMF). Usually these extractables[37] are so small that quantitatively they fall below the level of detection of the test method(s). Another significant review[38] has focused on "Extractables/Leachables Substances from Plastic Materials used as Pharmaceutical Product Container/Devices." This is a very comprehensive and useful work.

- **Multidose Formulation**
  If a pharmaceutical company manufactures a multidose formulation, the timeline for the development of this dose is significantly greater than for a unit dose formulation. A multidose will require the incorporation of an antimicrobial agent (preservative) in the protein formulation.

  Why does a multidose formulation require more time for development?

  1. Several antimicrobial agents must be screened for compatibility with the proteins and the excipients of the formulation. Table 15.4 summarizes the various characteristics of preservatives.
  2. Antimicrobial agents are known to be protein inhibitors; therefore, it is important to titrate the amount of the agent to act as a preservative. This preservative must be able to kill or inhibit microbial agents that could contaminate the product and adulterate it. The other issue is that the concentration of the preservative or antimicrobial agent may inhibit the protein drug. A significant amount of laboratory time must be devoted to this project—much more than a unit dose will require.
  3. The multidose formulation will be tested to determine its efficacy according to the Antimicrobial Effectiveness Test required by the USP.[39] If the results support the USP requirements, international requirements must then be met. International markets require different preservatives, different concentrations, and different excipients of the formulation. In addition, the period required for the inhibition and/or killing of the challenge microorganisms may be different. International regulatory requirements for compliance should be well researched and understood by the scientific and the management staff.

- **Stability Program**
  Formal stability studies must be performed to determine key parameters to support an acceptable and successful marketed product. The following are some of the parameters required:

  1. Robustness of the formulation.
  2. Shipping tests designed, performed, and validated at different temperatures, and under shaking and freeze-thaw conditions.
  3. Compatibility of the physical and chemical variables.
  4. Recommended product specifications.
  5. Characterization of degradation products.
  6. Real-time stability data supporting expiration dating.

TABLE 15.4 Characteristics of Preservatives

	Benzyl Alcohol	Chlorobutanol	Metacresol	Parabens (Hydroxybenzoates: Methyl, Propyl)	Phenol
Antimicrobial activity	Bacteria, weak against fungi	Bacteria, fungi	Bacteria, fungi	Primarily fungi and gram-positive bacteria Poor vs. pseudomonads	Bacteria, fungi
Use concentrations	1.0–3.0%	Up to 0.5%	0.3–0.5%	Methylparaben 0.18% Propylparaben 0.02%	0.3–0.5%
Solubility	1:25 in water	Soluble in water (1:125), more soluble in hot water Soluble in ethanol	1:50 in water	Methylparaben (in water) 1:400 Propylparaben (in water) 1:2000 Alcohol 1:2:5	1:15 in water
Optimum pH	4–7	Up to 4	2–8	3–8	Wide range (2–8)
Stability	Slowly oxides to benzaldehyde	Decomposed by alkalis	Activity decrease at high pH	Essentially good	Activity decreases at high pH
Compatibility/inactivation	Inactivated by nonionic surfactants (Tween 80)	Incompatible with some nonionic surfactants (10% Tween 80) Decomposes at 65°C	May be inactivated by iron and certain nonionic surfactants	Serum reduces activity; also nonionic surfactants	May be inactivated by iron, albumin, and oxidizing agents May be incompatible with some nonionic surfactants
Comments	Bacteriostatic Used for parenteral and local anesthetic action	Wide range of compatibility Local anesthetic action Widely used	The meta isomer is most effective and least toxic; the ortho isomer is the weakest Mode of action is apparently related to solubility in fatty portions of organisms Combine with and denature proteins	Binds to PEG Slightly soluble Stable and nonirritating Proposed to block essential enzyme system of organism	Mode of action is physical damage of the cell wall and enzyme inactivation by free hydroxyl group

TABLE 15.5 Formulation Development Stability Program

Preformulation	
Time:	0, 1W, 2W, 1M, 2M, 3M
Temp:	°C: 2–8, 25, 37, 45
Experimental Formulation	
Time:	0, 1M, 3M, 6M, 9M, 12M, 18M, 24M
Temp:	°C: 2–8, 25, 37, 45
Primary Formulation	
Time:	0, 1M, 3M, 6M, 12M, 18M, 24M, 36M, 48M, 60M
Temp:	°C: 2–8, 25, 37, 45
Market Formulation	
Time:	0, 1M, 3M, 6M, 12M, 24M, 36M, 48M, 60M
Temp:	°C: 2–8, 25, 37, 45
	Lots from this formulation can be qualified as conformity lots
Relative Humidity (RH) in Percent	
At 25°C and 30°C, use 60% RH; at 40°C, use 75% RH	

W: week; M: month.

TABLE 15.6 Proposed ICH Storage Conditions

Temperature	Time
25°C/60% RH	0, 3, 6, 9, 12, 18, 24, 36 months
30°C/60% RH	0, 3, 6, 9, 12, 24, 36 months
40°C/75% RH	0, 1, 3, 6 months

Source: Federal Register, September 22, 1994.

TABLE 15.7 ICH Stability Testing

Long-term testing storage conditions	(25°C/60% RH)
12 months required at time of submission	
Accelerated testing storage conditions	(40°C/75% RH)
6 months required at time of submission	
If significant changes occurred during accelerated testing, conduct at intermediate conditions	(30°C/60% RH)
6 months of the 12 months required	
Photostability	
UVA/Fluorescent Cool White or Artificial Daylight ID 65 monitored by (1) lux meter, (2) radiometer	

Table 15.5 summarizes appropriate stability programs. Table 15.6 presents the proposed International Conferences on Harmonization (ICH) storage conditions, and Table 15.7 describes the ICH stability testing program.

Table 15.8 summarizes the most frequently accepted specifications for purified bulk drug concentrate. Table 15.9 'summarizes the most frequently accepted specifications for a finished dosage form.

- **Formulation Development Scale-up Considerations**
  At this stage of formulation development, if the data demonstrate acceptable stability for a potentially marketable product, the scale-up process should begin. There are several considerations to be addressed:

1. Design of the scale-up.
2. Reproducibility from lot to lot.
3. Selection of the final container, such as vials, syringes, or ampules.
4. Selection of the key excipients.
5. International acceptance of the excipients.
6. Last, but not least, cost considerations.

TABLE 15.8 Specifications for Purified Bulk Drug Concentrate

Test Methods
1. *Physical Evaluation*
Appearance
pH
2. *Identity*
Bioassay
Peptide mapping
Amino acid analysis
3. *Protein Potency*
Nitrogen content
HPLC
4. *Biological Potency*
Specific activity
5. *Purity*
HPLC
SDS-PAGE
(CZE)
IEF
DNA
Endotoxins
Sterility
Other specifications can be included, depending on the specific requirements of the individual protein.

TABLE 15.9 Specifications for a Protein/Peptide Drug in Finished Dosage Form

Test Methods
1. *Physical Evaluation*
Appearance
pH
Volume/container
Moisture (lyophilized product)
Total protein
Particulates (for both liquid and lyophilized formulations)
2. *Potency Tests*
In vitro assays
Radioimmunoassays
Enzyme immunoassays
Chromatographic methods
Bioassays (animal model or cell-line derived)
Protein content
3. *Identity*
Peptide mapping
NH_2 terminal analysis
Western blot
Isoelectric focusing
SDS-PAGE
Coomassie stain (reduced and unreduced)
Biological activity
4. *Purity*
SDS-PAGE
Coomassie stain
HPLC-RP
HPLC-SEC
HPLC-gel filtration
DNA contamination
Other specifications can be included, depending on the specific requirement of the protein
5. *Microbiological Tests*
Sterility
Pyrogens
Mycoplasma
6. *Safety*
7. *Degradation Assays*
SDS-PAGE
Enzyme-linked immunosorbent assay (ELISA)
HPLC
Electrophoresis

Formulation and manufacturing, at this stage of development, need to focus on the physical and chemical problems associated with biopharmaceuticals. Physical adsorption of the product on surfaces such as glass, metal, plastic, and any prefilters could mean significant loss of the product, inconsistent concentrations per unit container, poor yield, and, ultimately, rejection of a lot.

15.2. NOVEL PARENTERAL FORMULATIONS AND ALTERNATE DELIVERY SYSTEMS FOR PROTEINS, PEPTIDES, AND MONOCLONAL ANTIBODIES DRUGS

15.2.1. Introduction

Alternate delivery systems are the focus of many laboratories throughout the world. The major routes selected are oral, pulmonary, intranasal, transdermal, vaginal, and others. Successful drug delivery discoveries are the result of interdisciplinary efforts of biochemists, chemists, engineers, physicists, pharmaceutical scientists, and clinical investigators.

Why is the biopharmaceutical community making such monumental efforts to discover novel delivery technologies? Alternate delivery systems may have highly desirable attributes and improved utilities[40] for the clinical administration of novel medicinal and chemotherapeutic agents. These attributed included the following:

- Targeting delivery to specific tissues
- Improved safety and efficacy
- Decreases in the frequency of dosing and possibly in the amount of drug needed
- Reduction of toxicity
- Reduction and/or elimination of pain-related administration

15.2.2. Selection of Novel Delivery Systems

How does a pharmaceutical company select one or more of these novel delivery systems? In almost all cases, a pharmaceutical company's objectives are based upon[41]

- Clinical application
- Physicochemical properties of the drug
- Bioavailability
- Efficacy
- Simple formulation
- Patients acceptability
- Extending patent life
- Maximizing future product value
- Last, but not least, profitability

The literature has cited several key delivery systems. Some of these are:[41–45]

- PEGylation—Increased potency of a protein, improved stability, compatibility, and higher bulk drug purity.
- Liposomes—Lipid vesicle–containing drug(s) designed to target specific tissues. They can be charged with water-insoluble drugs (hydrophobic membrane) and water-soluble (hydrophilic) drugs.
- Microparticles—Large molecules can be made transformed very small particles and delivered via a biodegradable matrix.
- Microspheres—Very small particles placed in a polymer matrix or a microsphere encapsulation.
- Nanoparticles—With spray drying, protein powders can be controlled at a uniform size requirement for injection.
- Micelles—Polymeric chains that can carry poorly soluble drugs to the target area.
- Pumps—Devices filled with controlled release to site-specific organs.
- Osmotic implants—Controlled drug concentration release at physiological temperature for extended periods of time.
- Inhalation—Widely used for lung delivery treatment studies of aerodynamics of varying particle size.
- Transdermal—Small lipophilic drugs designed to cross the skin barrier with proper solvents and penetration enhancers.
- Needleless injections—Liquid or powdered drugs propelled with pressure through the skin.

Medicaments derived from these delivery systems are targeting diseases like cancer, infections, cardiovascular problems, inflammations, transplants, and several other conditions.

15.2.3. Obstacles to Be Overcome by This Technology

Although this technology has already made significant contributions in novel delivery to treat diabetes, tuberculosis, some forms of cancer, growth deficiency, pain control, lung fibrosis, systemic inflammation, angina, hypertension, and many other conditions, biopharmaceutical companies must overcome significant obstacles[46] to meet higher clinical expectations and achieve greater financial results. This novel technology needs to address the following key product development issues:

1. Establishing the physicochemical properties of the active drug substance.
2. Preformulation of the active drug substance in various physiological buffers.
3. Determining initial compatibility of the active drug substance with preformulation excipients.

4. Selecting compatible solvent(s) for the active drug substance and excipients in order to establish optimal solubility.
5. Developing initial analytical and quantitative assay methods that are specific and reproducible.
6. Designing an appropriate initial stability profile at various temperatures.
7. Evaluating and identifying initial degradation products.
8. From preformulation data, by selecting the most promising preformulation results, a formulator can design some formulation candidates with minimal toxicity.
9. Packaging selection and compatibility of the active drug substance with formulation excipients.
10. Determining particle size control of the active drug substance, rate control, and study movement of the drug particles.
11. Determining appropriate methods of sterilization, i.e., sterile filtration and gamma radiation treatment.
12. Establishing a reproducible process in place, scaling up, and developing the respective specifications.
13. Ensuring that the product is manufacturable.
14. Ensuring that the product meets the specifications of the regulatory agencies.

Specific product development references are listed in the References section for several delivery systems cited in this section. Each delivery system has its own unique procedure, from research and development to product scale-up and, finally, manufacture. Review of some of these references can be very beneficial, as they describe various methods, materials, and experimental procedures selected and why.

15.3. SUMMARY

Key phases of successful product development place emphasis on close, collaborative, and productive interactions of the interdisciplinary sciences within a biopharmaceutical group.

Physicochemical properties of proteins, peptides, and monoclonal antibodies must be identified in order to develop preformulation and formulation studies with rational designs. Specific stabilizers play a major role in stabilizing the product under specific experimental conditions.

Analytical methods that determine the potential stability or degradation product of a formulation must be developed, validated, and qualified as "stability indicators."

The ultimate goals of the pharmaceutical scientist working on discovery, process, purification, and production are to deliver to the health field protein drugs which are safe, effective, pure, stable, elegant, suitable for production, cost effective, and marketable.

REFERENCES

1. Bontempo, J. A., Ed. *Development of Biopharmaceutical Parenteral Dosage Forms.* Marcel Dekker: New York, **1997**, pp. 1–9.
2. Avis, K. E.; Lachman, L.; Lieberman, H. A., Eds. *Pharmaceutical Dosage Forms.* Marcel Dekker: New York, **1997**, pp. 1–9.
3. Hageman, M. J. *Drug Dev. Industr. Pharm.* **1988**, *14*, 2047.
4. Wang, Y. C.; Hansen, M. A. *J. Parenteral Sci. Tech.* **1988**, 42–53.
5. Manning, M. C.; Patel, K.; Borchardt, R. T. *Pharm. Res.* **1989**, *6*, 903.
6. Geigert, J. J. *Parenteral Tech.* **1989**, *43*, 220.
7. Privalov, P. L.; Gill, S. I. *Adv. Protein Chem. 39*, 191.
8. Nozhaev, V. V.; Berezin, I. V.; Martinek, K. *CRC Crit. Rev. Biochem.* **1988**, *23*, 235.
9. Pace, C. N.; Shirley, B. A.; Thomson, J. H., Eds. *Protein Structure: A Practical Approach.* IRL Press: Oxford, **1989**, p. 311.
10. Timasheff, N. S.; Arakawa, T.; Creighton, T. E., Eds., *Protein Structure: A Practical Approach.* IRL Press: Oxford. **1989**, p. 34.
11. Arakawa, T., et al. *Pharm. Res.* **1991**, *8*, 225.
12. Schein, C. H. *Biotechnology*, **1990**, *8*, 308.
13. Glynn, L. G. *J. Parenteral Drug Assoc.* **1980**, *34*, 139.
14. Powell, E. M.; Nguyen, T.; Baloian, L. *J. Parenteral Drug Assoc.* **1998**, 238–311.
15. Hensen, F. A., et al. *J. Colloid Interface Sci.* **1970**, *32*, 162.
16. Hansen, A. M.; Rowan, S. K. In *Stability of Proteins Pharmaceuticals, Part B.*; Achern, J. T., Manning, M., Eds. Plenum Press: New York, **1992**, p. .
17. Stewart, W. E., et al. *Biochem. Biophys. Acta* **1974**, 359.
18. Quinn, R.; Andrade, D. J. *J. Pharm. Sci* **1983**, 72.
19. Fukushima, T., et al. *Eur Pat 37078,* **1981**.
20. Mikaelson, M. E., et al. *Chem. Rev.* **1982**, 82.
21. Busby, T. E.; Ingham K.C. *Biochim. Biophy. Acta.* **1984**, 779.
22. Anderson, L.; Hahn-Hagerdol, F. *Biochim. Biophys. Acta* **1987**, 912.
23. Arakawa, T.; Timasheff, N. S. *Biochemistry* **1982**, 21.
24. Geigert, J. *J. Parenteral Sci. Tech.* **1989**, 43.
25. Privalov, P. L.; Gill, S. J. *Advances in Protein Chemistry.* Academic Press: New York, **1988**, p. 191.
26. Yang, Y. C. J.; Hanson, M. A. *J. Parenteral Sci. Tech* **1988**, 42.
27. Akers, M. J. *J. Parenteral Sci. Tech.* **1982**, 36.
28. Gatlin, A. L. *IBC Conference on Formulation Strategies for Biopharmaceuticals—Ensuring Success to Market*, **2001**, 1–17.
29. Swift, W. R. *Conference on Formulation Strategies for Biopharmaceuticals—Ensuring Success to Market.* **2001**, 1–19.
30. Hopkins, G. J. *J. Pharm. Sci.* **1965**, *54*, 138.
31. Parenteral Drug Association. *Tech. Methods Bull.* **1981**, 2.
32. Avis, E. K. In *Parenteral Medications*, 2nd ed., Vol. 1. Marcel Dekker: New York, **1992**.

33. Wang, W. J.; Chien, Y. W. *Sterile Pharmaceutical Packaging: Compatibility and Stability.* Parenteral Drug Association, **1984**, Tech. Rep. 5.
34. Meltzer, T. H. *Filtration in the Pharmaceutical Industry.* Marcel Dekker: New York, **1987**.
35. Nema, S.; Avis, K. *J. Parenteral Sci. Technol.* **1993**, *47*, 16–21.
36. Olson, W. P. *Separation Technology Pharmaceutical and Biotechnology Applications.* Interpharm: Prairie View, IL, **1996**.
37. Stone, T.; Goel, V.; Leszczak. *Pharm. Technol.* **1994**, *18*, 116–130.
38. Jenke, D. *J. Pharm. Sci. Tech.* **2002**, *56*, 332–371.
39. Scott, V. W.; Porter, D. *J. Pharm. Sci. Tech.* **2002**, *56*, 300–311.
40. Langer, R. *Nature, Suppl. 1*, **1998**, *392*, 5–10.
41. Davis, S. S. *Am. Pharm. Rev.* **2002**, *5*, 29–36.
42. Baker, D. E. *Rev. Gastrointest. Disord.* **2001**, *1*, 87–99.
43. Stevenson, C. L.; Tan, M. M. *J. Pept. Res.* **2000**, *55*, 129–139.
44. Russo, P. L., Harrington, G. A.; Spelman, D. W. *J. Infect. Control* **1999**, *27*, 431–434.
45. Horspool, K. *IBC 2nd Int. Conf.* **2002**, 1–17.
46. Kannan, V.; Kandarapu, R.; Garg, S. *Pharm. Tech.* **2003**, *27*, 74–88.

16

PULMONARY DRUG DELIVERY: PHARMACEUTICAL CHEMISTRY AND AEROSOL TECHNOLOGY

ANTHONY J. HICKEY

School of Pharmacy, University of North Carolina, Kerr Hall, Chapel Hill, NC 27599

Drug Delivery: Principles and Applications Edited by Binghe Wang, Teruna Siahaan, and Richard Soltero
ISBN 0-471-47489-4

16.1. INTRODUCTION

Aerosol delivery of drugs to the lungs may be viewed sequentially as involving physical chemistry, aerosol physics, physiological/anatomical, and pharmacological strategies.[1] The multidisciplinary nature of the approach required to develop effective therapies is daunting but reflects a high standard of achievement in both science and technology. At least three generations of scientists, engineers, and clinicians have expended their energies in pursuit of the most effective, safe, and elegant solution to the problems of pulmonary drug delivery. However, in terms of sheer numbers of researchers and quantities of resources, there has been nothing to match the intense activity of the past decade. This is an appropriate time to ponder the achievements that define the field and the potential for future developments.

At the end of the 1980s, the aerosol products that were available reflected approximately 30 years of research and development.[2] During this period, it was clear that the propellant-driven metered dose inhaler (pMDI) had revolutionized asthma therapy, largely due to the invention of the metering valve and actuator combination.[3] In contrast, dry powder inhalers (DPIs) were very primitive and somewhat ineffective systems that deservedly did not compete with MDIs in terms of physician and patient acceptance.[4] Consequently, they were not a great commercial success. Nebulizers had been available for a number of years and had already found their role in acute care.[5,6] Some questions were being asked about the susceptibility of nebulizer performance to operating conditions against a backdrop of few, if any, manufacturer specifications and little government regulation.

A growing movement for change was developing throughout the late 1980s. In the 1990s, the scene was set for dramatic developments in the nature of aerosol products driven by restrictions on the use of chlorofluorocarbon (CFC) propellants and by an upsurge in novel therapeutic agents which were intended for lung delivery.[7] Alternative, non-ozone-depleting, propellant systems appeared to replace CFC products, extending the application of MDI technology.[8] A host of new DPIs was under development based on a greater understanding of the requirements for adequate aerosol dispersion.[4] It is worth noting that very few of these products have appeared commercially. Finally, a range of more efficient nebulizers and hand-held aqueous aerosol delivery systems were developed, bringing a standard of performance to the systems that had previously been absent.[7,9]

In the following sections, the performance of recently developed devices will be contrasted with that of earlier systems in order to emphasize the progress that has been made. The Conclusion will focus on reasonable expectations for the future and indicate areas in which further fundamental observations may facilitate the design and production of new aerosol delivery systems.

16.2. AEROSOL TECHNOLOGY

The devices that are employed to deliver drugs to the lungs may be divided into three categories: pMDIs, DPIs, and nebulizers.[10] Each of these systems delivers aerosols by a different principle, and the chemistry associated with the product varies significantly among them.

16.2.1. Particle Preparation

Before drug particles can be incorporated into aerosol products, they must be prepared in size ranges and with structures suitable for delivery to the lungs. A variety of methods have been developed for the preparation of particles.[11,12] The most common method of particle size reduction is one in which bulk drug product, most of which is prepared by conventional crystallization/precipitation followed by drying techniques, is air jet milled at high pressure. Attrition of particles occurs, leading to the micronization of the product. In recent years, methods have been used which combine conversion of the solution to a solid with size reduction or crystal engineering. The first of these methods, spray drying, involves forcing the drug solution through a nozzle at high pressure into a drying airstream from which small particles can be recovered that are suitable for inclusion in aerosol products. These particles occasionally have unique properties that will be discussed later. Supercritical fluid manufacture is a particle construction or crystal engineering method. In its simplest form, this involves dispersing drug in a supercritical fluid (usually carbon dioxide), and by controlling the conditions of temperature, pressure, volume, or the presence of an antisolvent, the drug may be crystallized to form morphologically well-defined particles.

16.2.2. pMDIs

Drugs delivered from pMDIs are initially prepared as suspensions or solutions in a selected propellant.[13,14] Often other components are added to aid in suspension of particles or drug dispersion into solution. These additives may be cosolvents, such as ethanol, or surfactants, such as oleic acid.

The original pMDIs employed CFC propellants 11, 114, and 12 [e.g., beclomethasone dipropionate (BDP) products Vanceril and Beclovent.][15] In recent years, these propellants have been replaced with hydrofluoroalkane (HFA) propellants 134a or 227 (e.g., BDP product Q_{var}) due to concern about atmospheric ozone depletion associated with CFCs.

Containers are filled on a large scale with drug formulation by a variety of techniques, based on high pressure or low temperature to control the state of the propellant during filling.[16] Valves are crimped on the opening to the container either before (pressure filling) or after (cold filling) propellant filling occurs.

The principle of aerosol delivery from pMDIs is based on the following sequence of events.[14,17] A small volume of a homogeneous dispersion of the drug, in solution or suspension, in a high vapor pressure propellant or a propellant blend from a reservoir, is isolated. The small-volume container (the metering chamber) is opened through an actuator nozzle. The metering chamber filling and opening to the atmosphere are achieved by means of a metering valve. Once opened to the atmosphere, the high vapor pressure contents of the metering valve immediately begin to equilibrate with atmospheric pressure. This has the effect of propelling the contents rapidly through the nozzle, which causes shear and droplet formation. Throughout this process the propellant is evaporating propelling, shearing, and ultimately reducing the size of the droplets produced. The components of an pMDI are shown in Figure 16.1A.

16.2.3. DPIs

DPIs have been through a number of evolutionary changes over the past 40 years.[4] All approved inhalers have been passive, in the sense that they employ the patient's inspiratory flow as the means of dispersion and entrainment of the aerosol into the lungs. The majority of powder products are blends of respirable drug particles and large lactose carrier particles. Early designs employed a unit dose gelatin capsule metering system (Rotahaler, Spinhaler). This has to some extent been superseded by multiple unit dose blister discs (Diskhaler) and rolls (Diskus) or reservoir powder devices (Turbuhaler, Clickhaler). The general principle of powder delivery is shown in Figure 16.1B. A powder bed is exposed to a shearing air supply (usually the inspiratory airflow) that entrains particles. A blend will employ the fluidizing effects of large lactose particles to help disperse the respirable particles associated with their surfaces. The small drug particles will be carried to the lungs of the patient, while the large carrier particles will be deposited in the mouthpiece of the inhaler or the oropharynx of the patient.

A variety of mechanisms for assisting with the dispersion of the powder have been adopted, including impellors (Spiros), compressed air assist (Nektar), vibration (Oriel, Microdose), and impact hammers (3M, DelSys).

16.2.4. Nebulizers

Nebilizers are among the oldest devices used for delivery of therapeutic agents.[18] They employ energy from compressed gas or piezoelectric ceramics to generate droplets of water containing drug. The principle of air jet dispersion is shown in Figure 16.1C. Drug solution (or occasionally suspension) is drawn from a reservoir

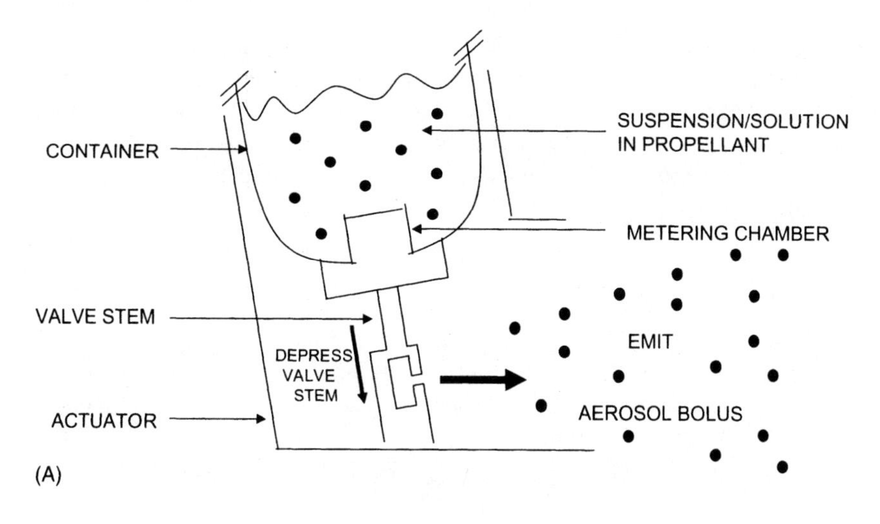

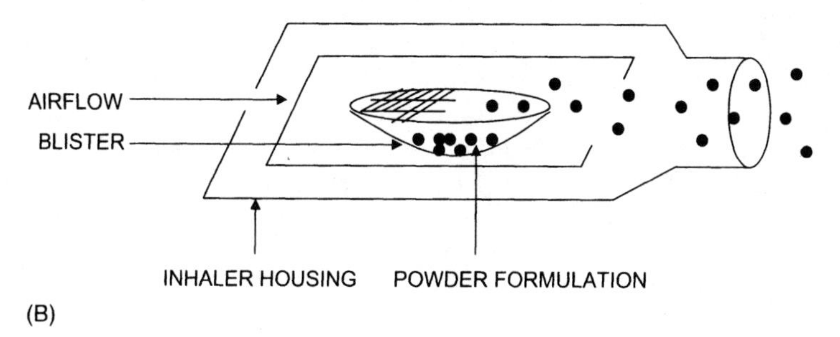

(B)

BAFFLE

AIRFLOW

SOLUTION
RESERVOIR

COMPRESSED AIR

(C)

Figure 16.1. (A) Schematic diagram of a propellant-driven metered dose inhaler. (B) Schematic diagram of a dry powder inhaler. (C) Schematic diagram of a air-jet nebulizer.

through a capillary tube by the Venturi (Bernoulli) effect. In principle, a low-pressure region is created at the exit from the capillary tube when compressed gas is passed at high velocity over the tube, drawing liquid into the air, where droplets are formed. Large droplets are projected onto a baffle, where they are collected, and small ones pass around the baffle and are delivered to the patient's lungs on their inspiratory flow.

16.3. DISEASE THERAPY

The lungs have been a route of drug delivery for millennia. Modern medical applications of aerosol delivery can be traced to the development of the pMDI in the middle of the twentieth century.[19]

Initially drugs for asthma therapy were the prominent therapeutic category of interest. With increased understanding of pulmonary biology and the pathogenesis of disease, agents such as proteins and peptides, to achieve local and systemic therapeutic effect,[20] and antimicrobials for infectious disease therapy[21] have been studied. There are a number of classical texts that outline the pharmacology of the lungs, including Goodman and Gilman,[22,23] and the medicinal chemistry of drugs.[24] These general references have been drawn on for the following discussion. Thorough descriptions of agents delivered to the lungs may also be found in the literature.[25,26]

16.3.1. Asthma

The first drugs developed for asthma therapy were used to bronchodilate patients by acting on the sympathetic or parasympathetic receptors. These agents fall broadly into two categories: β-adrenergic agonists and anticholinergics. Subsequently, other agents were added that acted on other manifestations of the disease. Notably, steroids acted on the underlying inflammation. Figure 16.2 depicts the action of some of the drugs commonly used to treat asthma by therapeutic category.[27]

16.3.1.1. β-Adrenergic Agonists (BAAs) The receptors for BAAs are distributed throughout the lung but occur in increasing numbers as one approaches the periphery. Consequently, aerosol BAAs must be delivered to the periphery of the lungs to produce their pharmacodynamic effect.

A number of BAAs were evaluated approximately 50 years ago for their ability to induce bronchodilation. The molecules are analogs of epinephrine, a very-short-acting bronchodilator. Epinephrine is currently marketed in an over-the-counter product, Primatine Mist (Whitehall-Robins, Richmond, VA). The safety of this product is related to its very short duration of action in the lung following delivery. However, early examples of nonspecific agents such as isoproterenol were longer-acting in the lungs and at other sites were capable of inducing serious, in some cases life-threatening, side effects. In the 1960s, more specific β_2-adrenergic agonists were developed, the most notable of which was albuterol (salbutamol, 3M, GSK,

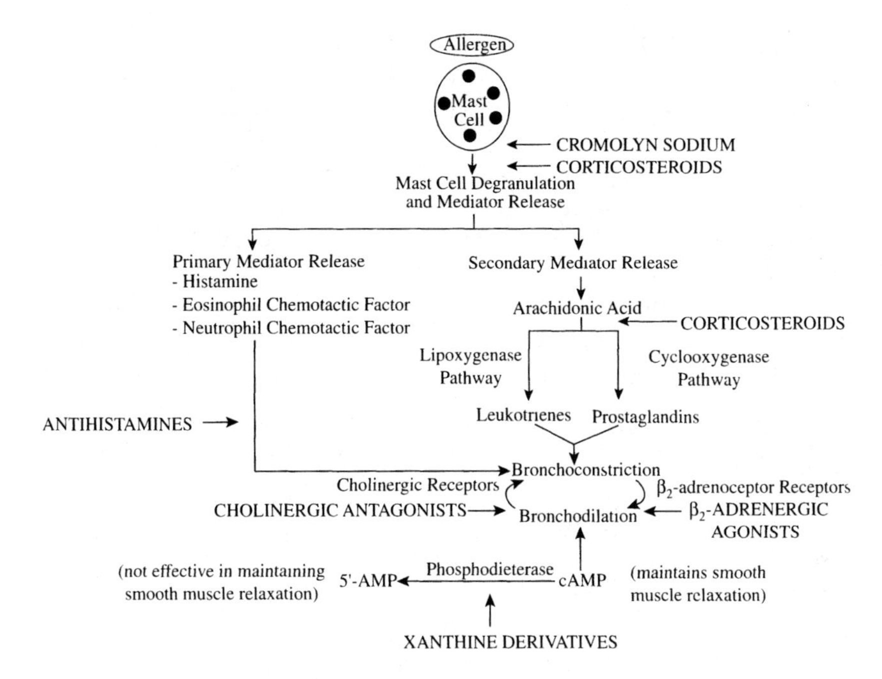

Figure 16.2. Mechanism of action of drugs for asthma therapy.

Schering Plough), which targeted airway receptors and exhibited reduced systemic side effects. A series of short-acting agents following the same general structure (shown in Figure 16.3A) were developed, including terbutaline (AstraZeneca) and fenoterol (Boehringer Ingelheim).

As the chemistry of these drugs came to be understood, they were modified to achieve several things, including local metabolism to an active agent, increased residence time at the receptor (extended duration of action), and selection for receptor binding to reduce toxicity. The first prodrugs were bitolterol and bambuterol, which are both metabolized to active adrenergic agonists. Two long-acting agents have been developed based on increased residence time at the receptor, salmeterol (GSK) and formoterol (BI) (shown in Figure 16.3B). Recently, it was noted that selected isomeric forms of BAAs may exhibit lower toxicity and enhanced efficacy. The most prominent example of this is levalbuterol (Xopenex, Sepracor).

16.3.1.2. Anticholinergics Anticholinergic agents act on the central airways, the location of the majority of receptors. A small number of anticholinergic agents have been developed in the past 30 years. Ipratropium (BI) was the first of these agents to be approved, followed by oxitropium (see Figure 16.4). Asthma was the target disease state for these agents. The success of tiotropium in the treatment of chronic obstructive pulmonary disease, for which there had previously been no aerosol therapy, has created the opportunity for a new generation of anticholinergics.

16.3.1.3. Corticosteroids Corticosteroids have long been known to be therapeutically beneficial in the treatment of asthma. They have the significant advantage of treating the underlying inflammation of the lungs that causes the disease. However, the systemic use of corticosteroids is associated with substantial systemic side effects. Consequently, orally ingested steroids are given infrequently.

Steroids, which are active following oral ingestion, such as prednisone, prednisolone, and dexamethasone, cause significant systemic immunosuppression, with consequent risks to the health of the patient. Since the inflammation associated with asthma is localized in the lungs, local topical delivery with aerosols has the

General structure:

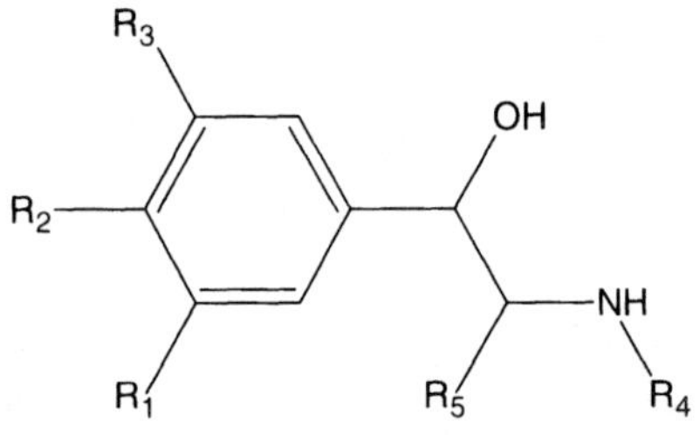

Albuterol:
R_1 – H
R_2 – OH
R_3 – CH_2OH
R_4 – $C(CH_3)_3$
R_5 – H

Terbutaline:
R_1 – OH
R_2 – H
R_3 – OH
R_4 – $C(CH_3)_3$
R_5 – H

Fenoterol:
R_1 – OH
R_2 – H
R_3 – OH 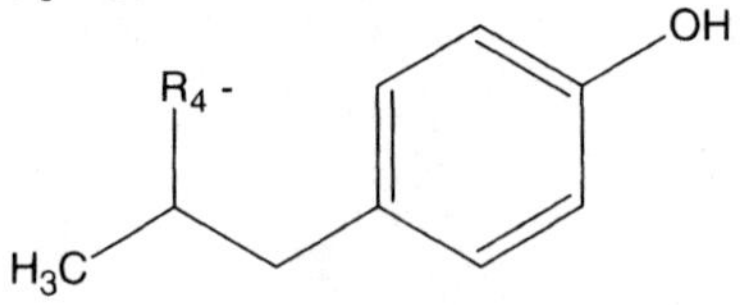
R_4 -
R_5 – H

(A)

Figure 16.3. (A) Short-acting β-agonists. (B) Long acting β-agonists.

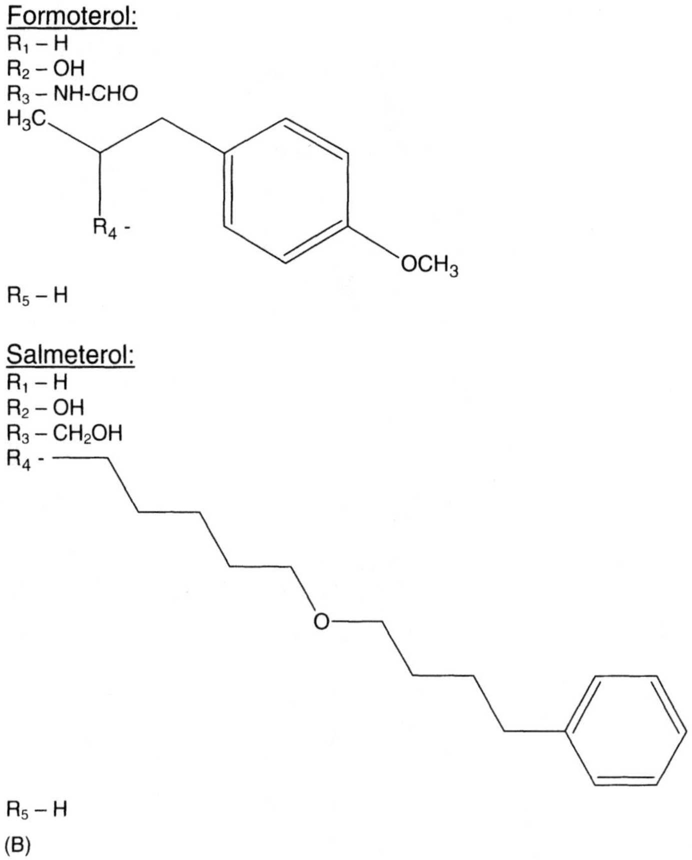

Formoterol:
R_1 – H
R_2 – OH
R_3 – NH-CHO
R_4 - H_3C … OCH_3
R_5 – H

Salmeterol:
R_1 – H
R_2 – OH
R_3 – CH_2OH
R_4 - … O …
R_5 – H

(B)

Figure 16.3. (*Continued*)

advantage of requiring low doses sufficient to achieve a local therapeutic effect but resulting in small circulating concentrations of drug capable of causing systemic side effects. The first drug in this class was beclomethasone dipropionate (GSK, 3M, Schering Plough). This was followed by triamcinolone acetonide (Aventis). While these agents exhibited some preferential local effects, the introduction of budesonide (AstraZeneca), flunisolide, and fluticasone (GSK) produced maximum local effects and minimum systemic effects based on the local binding of the steroids in the lungs. Some examples of steroid structures are shown in Figure 16.5.

16.3.1.4. Cromones Cromones (cromolyn sodium and nedocromil sodium, Aventis; see Figure 16.6) are a unique class of compounds that are known to cause mast cell stabilization, thereby preventing the histamine release involved in local hypersensitivity of the lungs. In addition, these agents are implicated in preventing the release of other inflammatory mediators and the sensitivity of myelinated nerves

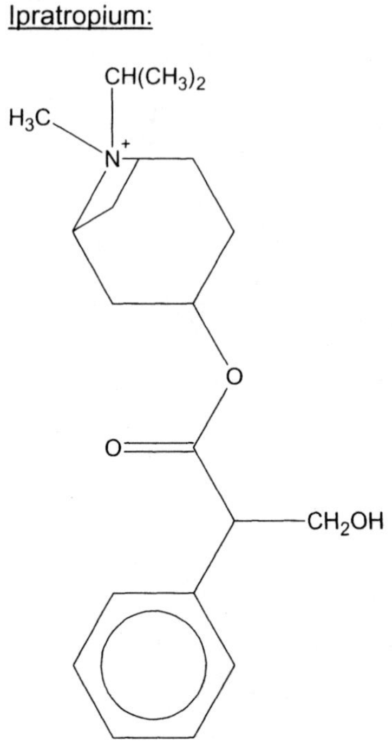

Ipratropium:

Figure 16.4. Anticholinergic.

in the airways. The major action of these compounds is still obscure, but they have found a particular application in exercise-induced asthma.

16.3.2. Emphysema

A number of agents have been employed for the treatment of emphysema, associated with the action of elastase in the lungs. The most prominent example of a drug for the treatment of emphysema is α1-antitrypsin. Others include peptidyl carbamates and Eglin C. These drugs reduce the free elastase in the lungs, thereby preventing the structural remodeling of the lungs, which leads to poor gaseous exchange and severe disability associated with the disease. Some individuals exhibit a genetic predisposition for emphysema; in others, it is the result of a history of smoking.

16.3.3. Cystic Fibrosis

Cystic fibrosis is characterized by an imbalance in airway chloride ion concentrations due to the absence of cystic fibrosis transmembrane regulator (CFTR) receptor. Three approaches have been taken to the use of aerosols to treat this

General structure:

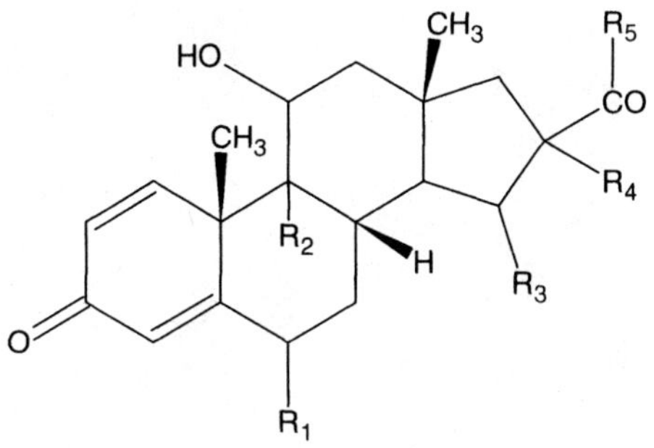

Beclomethasone dipropionate:

R_1 – H
R_2 – Cl
R_3 – CH_3
R_4 – $COOCC_2H_5$
R_5 – $CH_2COOC_2H_5$

Triamcinolone acetonide:

R_1 – H
R_2 – F 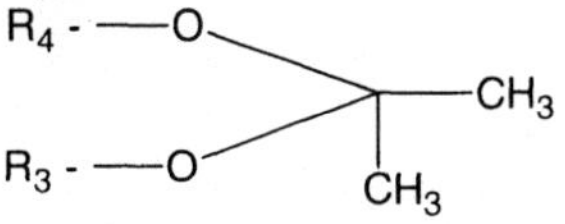
R_4 - —O
R_3 - —O (—C(—CH_3)(CH_3))
R_5 – CH_2OH

Budesonide:

R_1 – H
R_2 – H 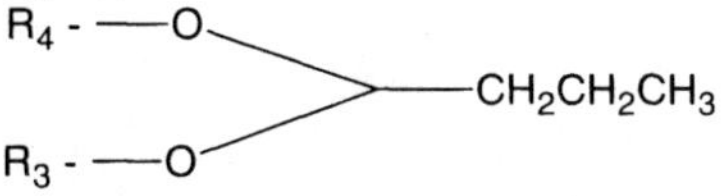
R_4 - —O
R_3 - —O (—CH—$CH_2CH_2CH_3$)
R_5 – CH_2OH

Fluticasone propionate:

R_1 – F
R_2 – F
R_3 – CH_3
R_4 – $COOCC_2H_5$
R_5 – SCH_2F

Figure 16.5. Steroids.

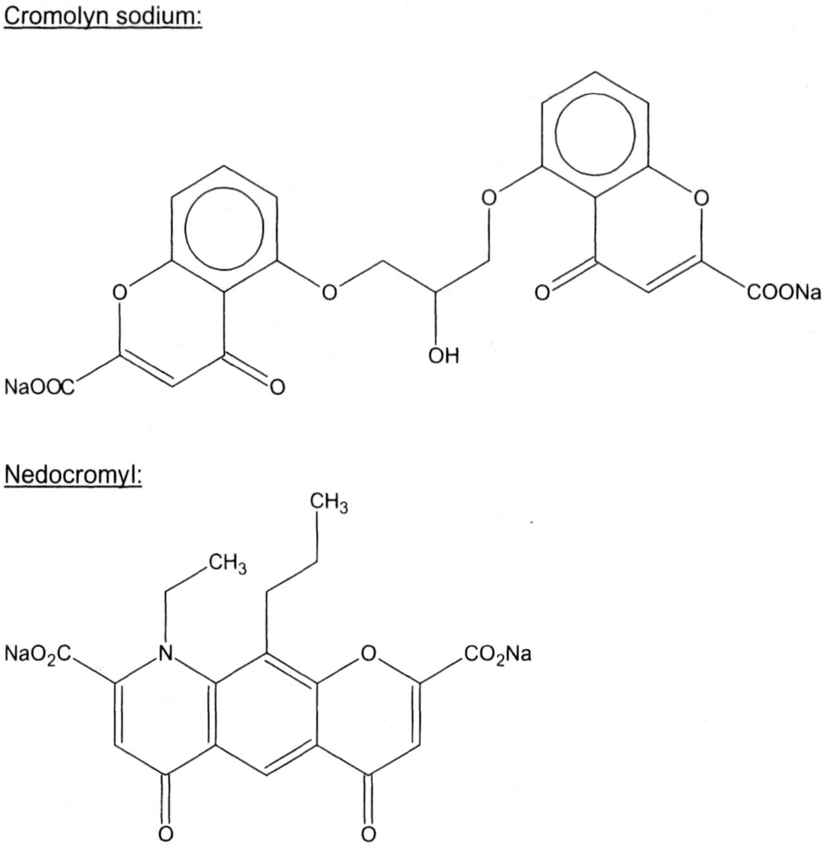

Cromolyn sodium:

Nedocromyl:

Figure 16.6. Cromones.

disease.[28] The first involves the delivery of recombinant human DNase to cleave the tangle of leukocyte DNA that results from cell infiltration into the lungs, reducing the viscosity of the mucus layer in the lungs and facilitating expectoration. In the second approach, an acrosol is employed to deliver an antibacterial agent, tobramycin (Figure 16.7), which acts against *Pseudomonas aeruginosa*, which grows on mucus plaques in the lungs of patients. Finally, nucleic acid is employed to correct the genetic imbalance in cystic fibrosis expression and thereby correct chloride ion transport.[29]

16.3.4. Other Locally Acting Agents

Amikacin and amphotericin B have both been prepared in liposomal formulations for different reasons. Amikacin can be targeted to macrophages for the treatment of intracellular microorganisms such as *Mycobacterium avium* complex (MAC).[30,31]

Pentamidine:

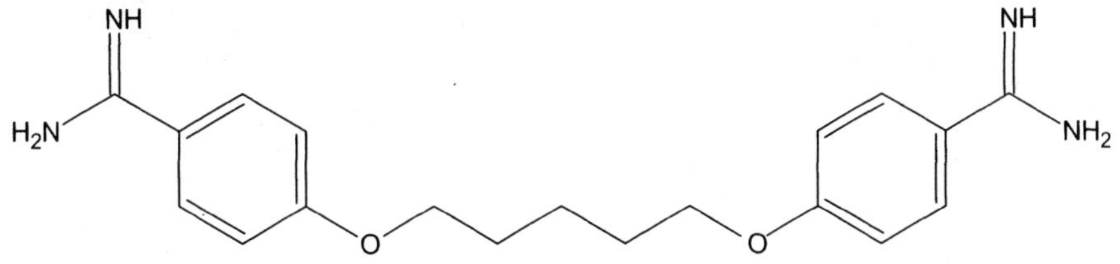

Tobramycin:

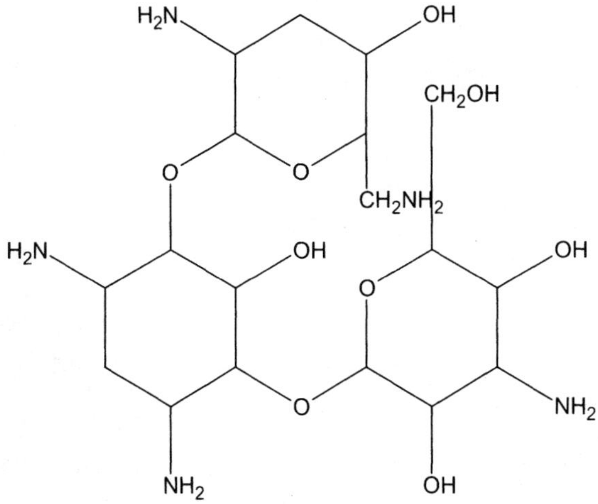

Figure 16.7. Antimicrobials.

Amphotericin B exhibits increased solubility in liposomes capable of delivering a dose to the lungs for the treatment of aspergillosis.[32]

Pentamidine (Figure 16.7) and its analogs have been delivered to the lungs for the treatment of *Pneumocystis carinii* pneumonia, a secondary infection associated with acquired immune deficiency syndrome.[33]

16.3.5. Systemically Acting Agents

A variety of systemically acting agents have been evaluated for delivery via the lungs. Among these are insulin, leuprolide acetate, calcitonin, parathyroid hormone, and growth hormone.[34] Each of these agents targets a different disease state. Insulin is employed in the control of diabetes. Leuprolide acetate is employed for the treatment of prostate cancer and endometriosis. Calcitonin and parathyroid hormone are used to prevent osteoporosis. Growth hormone, as its name implies, controls deficiencies in the normal growth of children.

16.4. FORMULATION VARIABLES

The above agents are delivered in a variety of forms. Their compatibility with various solvents (propellants, alcohol, water), liquids (glycerol, polyethylene glycol, oleic acid, sorbitan trioleate, lecithin), and solid (lactose) phase excipients is key to the chemical and physical stability of the products. The *Handbook of Pharmaceutical Excipients* lists most of these materials.[35]

16.4.1. Excipients

16.4.1.1. Propellant-Based Systems Propellants are relatively inert materials in which most drugs exhibit limited solubility. Consequently, there are two favored approaches to incorporating drug into propellants. The first is to use surfactants to disperse respirable particles in a suspension. Three excipients were approved for this purpose in CFC propellants oleic acid (Figure 16.8), sorbitan trioleate, and lecithin (Figure 16.9). HFA systems are much more limited in their use of excipients, and oleic acid is the only excipient currently employed in these systems. The second approach is to use the co-solvent ethanol to bring the drug into solution and thereby achieve a molecularly homogeneous distribution in the propellant.

Suspension formulations are prepared to achieve a controlled flocculation (DLVO theory, ref), which will allow ease of redispersion upon shaking. A number of physicochemical phenomena may occur to disrupt the stability of the product. Moisture ingress to the container (and association with the drug particles) will lead to hydrolysis for molecules that are susceptible to this mechanism of degradation. Because of the inert hydrophobic nature of propellant, this can be controlled to some extent. The presence of moisture will also give rise to interactions between particles, which may result in irreversible aggregation. Related forms of aggregation may result

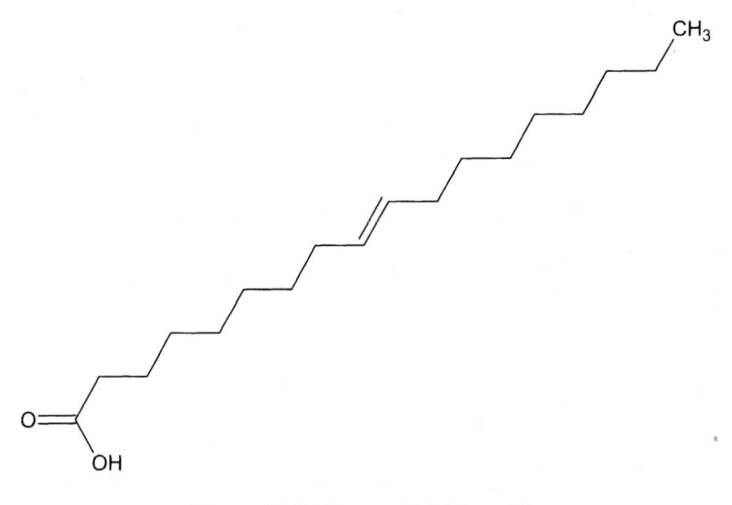

Figure 16.8. Fatty acid (oleic acid).

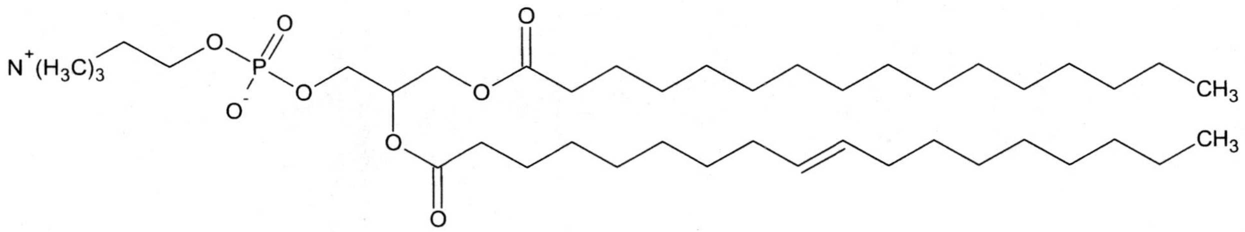

Figure 16.9. Lecithin.

from electrostatic effects. The presence of small quantities of moisture and a low but defined solubility of drug in propellant may also result in particle growth by Ostwald ripening. Aggregation and particle growth will change the dose delivery and proportion of particles in the respirable range, which in turn will impact on lung deposition and therapeutic effect.

Solution formulations generally require a semipolar cosolvent to achieve dissolution, and these cosolvents are more susceptible to moisture uptake. Since the drug is molecularly dispersed, side groups are exposed which may be susceptible to rapid degradation.

In both solution and suspension formulations, the propellant can potentially leach extractables (nitrosamines, elastomers) from the gaskets in the containers, which may result in instability.

16.4.1.2. Dry Powder Systems Dry powder systems are occasionally prepared from pure drug substance. More frequently, blends with lactose (Figure 16.10) are prepared. The lactose blends consist of respirable drug particles and large (~50–150 μm) excipient particles. The excipient is included as a diluent to aid in dispensing the drug and as a fluidizing agent to assist dispersion. Tertiary blends have been prepared in which small lactose particles (<50 μm) have been used to aid flow and dispersion of drug from the inhaler. Other sugars have been evaluated as exipients (e.g., mannitol, trehalose), but none has yet been approved by the Food and Drug Administration for marketed products.

Some spray-dried particles have been prepared using other generally regarded as safe (GRAS) substances such as lecithin, human serum albumin, polylysine, and polyarginine, but these have also not yet been included in approved products for delivery to the lungs.

16.4.1.3. Nebulizer Solutions Solutions intended for delivery from nebulizers are subject to the rules guiding general solution chemistry. Susceptibility to hydrolysis mitigates against solution formulation. Light and heat may accelerate degradation. The presence of complexing agents will influence solubility. Using concentrations near the solubility limit of the drug may result in precipitation (e.g. pentamidine).

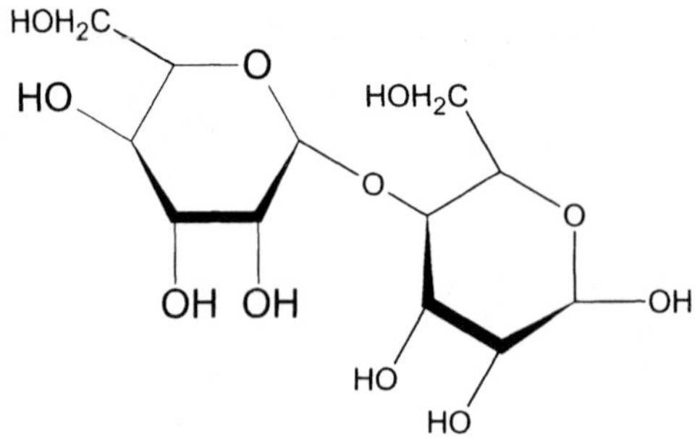

HOH$_2$C
O
HO
HOH$_2$C
O
O
OH OH
OH
HO
HO

Figure 16.10. Lactose.

Nebulizer solutions are now prepared as sterile products due to the degradation of drugs and the potential for nosocomiosis from contaminating microorganisms.[36] Historically, preservatives, such as benzalkonium chloride, were employed, but this approach is no longer favored due to adverse events associated with these agents.

16.4.2. Interactions

The combination of drugs and excipients often results in unforeseen interactions, which are detrimental to the product and potentially to the health of the patient. These interactions may result in physical or chemical changes which impact the stability, efficacy, and toxicity of the product. It is impossible to include a comprehensive review of all of the circumstances in which such interactions could occur. However, examples are given below for each of the major delivery systems.

16.4.2.1. pMDIs The incorporation of surfactants in pMDIs contributes to suspension stability and lubricates the valve to facilitate the delivery of consistent doses throughout the life span of the product. It has been noted in a model system that side chain interactions may occur which will ultimately influence the physicochemical properties of the product.[37] In this context, disodium fluorescein, a suitable model hygroscopic material for certain drugs, was shown to interact with selected fatty acids in nonaqueous solution. This interaction was concentration dependent and potentially initiated several interfacial phenomena ranging from adsorption to precipitation.[38] It was noted that a specific interaction occurred between the phenolic sodium of the disodium fluorescein and the carboxylate group of the fatty acid, which, in extreme cases, could result in the production of the salt form of the fatty acid and a monoanion of fluorescein. This indirect observation was superseded by studies of a variety of drugs and their interaction with the approved excipients oleic acid and sorbitan trioleate.[39] In these studies examining albuterol (salbutamol) and two salt forms of isoproterenol (isoprenaline), it was postulated that a proton exchange occurred between the adsorbed surfactant and the drug, which resulted in a charge effect capable of contributing to the stability of the suspension. A reduction in susceptibility to hygroscopic growth, both on storage and in transit through the airways, was an incidental effect of the adsorption/association of the surfactant with the surface of model drug particles.[40,41]

Another clear effect of the presence of surfactant in a nonaqueous suspension of particles is the potential for changes in crystal habit[42] or Ostwald ripening.[43]

16.4.2.2. Dry Powder Systems Dry powder formulations are susceptible to a number of potential interactions. Since there is currently only one approved excipient, the drugs have to be compatible with lactose.[4] In addition, dry powders are prone to moisture sorption, which can give rise to chemical degradation by hydrolysis or physical instability due to capillary forces.[44] As other excipients, such as lecithin, are explored as excipients in dry powder products, a hydrophobic effect may become the source of aggregation and the stability of the excipient itself may become a source of concern.

16.4.2.3. Nebulizer Solutions The potential interactions that can occur in nebulizer solutions are subject to simple solution chemistry. A few excipients are employed in nebulizer products, notably sodium chloride to achieve isotonicity with body fluids. There are some rare examples in which interactions with sodium chloride and other solutes influence nebulizer performance. In studies of the delivery of amiloride hydrochloride and trisodium uridine triphosphate (UTP), both used in the treatment of cystic fibrosis, an interaction occurred in which a precipitate of the two drugs was formed in the ratio 3 amiloride: 1 UTP.[45] The presence of sodium chloride suppressed the solubility of amiloride further by the common ion effect.

Suspension nebulizer formulations exhibit a complex behavior in which the particle or aggregate size in suspension may influence the size of the droplet delivered from an air jet nebulizer, and as the size of the particles approaches and exceeds the droplet size produced, size selectivity in dispersion of the particles may occur.[46] Ultrasonic nebulizers are also susceptible to particle size in dispersion.[47]

16.4.3. Stability

The potential for drugs to interact with solvent, excipient, packaging materials, atmospheric moisture, oxygen, light, and heat contributes to the overall stability of the final product. These factors have been recognized by international regulatory bodies. Notable among these is the Food and Drug Administration. Clear guidance has been given for the Chemistry, Manufacturing, and control section of any submission to new drug approval.[48,49]

The performance of pMDIs and DPIs is scrutinized in terms of efficiency and reproducibility of dose delivery, particle size, and distribution under a range of storage conditions, with respect to temperature and humidity, for extended periods of time (up to 2 years).[50] Since nebulizer products do not bring the device in contact with the drug until the point of use, a slightly different approach is taken to their approval. Recommended devices and conditions of operation for the delivery of a particular drug must now be stated. The solution formulation is then viewed as a sterile parenteral product and requires concomitant testing.

16.5. FUTURE DEVELOPMENTS

In the relatively near future, the focus for aerosol delivery of drugs may shift from the physics and chemistry associated with the product's development to the biology of identifying new drug candidates and targets. The driving force for such a shift is multifaceted. As device technology is refined and the limits to efficiency and reproducibility of drug delivery are approached, the ability to improve disease therapy will refocus on the desired pharmacological effect. In this context, the desire to locate receptor, enzyme, and transport targets will become stronger, and the notion of combining therapies to achieve a particular goal will become more widespread. This opens up a significant opportunity for those involved in quantitative structure-activity relationship research and bioinformatics to search existing chemical libraries and reexamine the use of previously overlooked potential drugs. It is predicted that this approach will lead to improved disease management and new generations of drugs with desired efficacy and reduced toxicity.

16.6. CONCLUSION

The importance of generating drug particles in the 1–5 μm size range is central to the efficient delivery of drugs. A variety of devices and mechanisms have been developed over the past half century to achieve this goal.

A steady increase in new aerosol products has occurred in the past 10 years. The majority of these products have come from large pharmaceutical companies with a history of achievement in this field. It can only be hoped that the other products currently in development will soon be commercially available.

As it becomes increasingly probable that delivery of drugs as aerosols can be achieved readily, the focus can shift to the nature of the therapeutic agent and its physical and chemical stability in the required dosage forms. New chemical entities can be considered for delivery to the lungs to facilitate the control of pulmonary diseases or diseases that may be treated by pulmonary drug delivery.

The chemical composition and structure of drug and excipient particles play a significant role in the success of therapeutic agents in a number of diseases for which the lungs are a target or portal. Consideration of these issues is essential to promoting effective disease management. There are some specific sources of drug instability, degradation, and physical properties that detract from product performance. If these issues are considered, the likelihood of developing a commercially viable new therapy that will meet stringent regulatory requirements is very good.

REFERENCES

1. Hickey, A. J. *Inhalation Aerosols: Physical and Biological Basis for Therapy*, Vol. 94. Marcel Dekker: New York, **1996**.
2. Ganderton, D.; Jones, T. C. *Drug Delivery to the Respiratory Tract*. John Wiley & Sons: New York, **1988**.
3. Thiel, C. In *Respiratory Drug Delivery V*; Dalby, R. N., Byron, P. R., Farr, S. J., Eds., Interpharm Press: Phoenix, AZ, **1996**, pp. 115–123.
4. Dunbar, C. A.; Hickey, A. J.; Holzner, P. *KONA* **1998**, *16*, 7–44.
5. Matthys, H.; Köhler, D. *Respiration* **1985**, *48*, 269–276.
6. Clay, M. M.; Pavia, D.; Newman, S. P.; Lennard-Jones, T. L.; Clarke, S. W. *Lancet* **1983**, *2*, 592–594.
7. Dunbar, C.; Hickey, A. J. *Pharm. Tech.* **1997**, *21*, 116–125.
8. Davies, R.; Leach, C.; Lipworth, B.; Shaw, R. *Hosp. Med.* **1999**, *60*, 263–270.

9. Crowder, T. M.; Louey, M. D.; Sethuraman, V. V.; Smyth, H. D. C.; Hickey, A. J. *Pharm. Tech.* **2001**, *25*, 99–113.
10. Hickey, A. J. In *Pharmaceutical Inhalation Aerosol Technology*, Hickey, A. J., Ed. Marcel Dekker: New York, **2003**, pp. 385–421.
11. Hickey, A. J.; Ganderton, D. *Pharmaceutical Process Engineering*, Vol. 112. Marcel Dekker: **2001**.
12. Sacchetti, M.; Oort, M. M. V. In *Inhalation Aerosols: Physical and Biological Basis for Therapy*, Vol. 94; Hickey, A. J., Ed. Marcel Dekker: New York, **1996**, pp. 337–384.
13. Johnson, K.; In *Inhalation Aerosols: Physical and Biological Basis for Therapy*, Vol. 94; Hickey, A. J., Ed. Marcel Dekker: New York, **1996**, pp. 385–415.
14. Hickey, A. J.; Evans, R. M. In *Inhalation Aerosols: Physical and Biological Basis for Therapy*, Vol. 94, Hickey, A. J., Ed. Marcel Dekker: New York, **1996**, pp. 417–439.
15. Purewal, T. S.; Grant, D. J. *Metered Dose Inhaler Technology*. CRC Press: Boca Raton, FL, **1997**.
16. Sirand, C.; Varlet, J.-P.; Hickey, A. J. In *Pharmaceutical Inhalation Aerosol Technology*; Hickey, A. J., Ed. Marcel Dekker: New York, **2003**, pp. 311–343.
17. Dunbar, C. A.; Watkins, A. P.; Miller, J. F. *J. Aerosol Med.* **1997**, *10*, 351–368.
18. Niven, R. W. In *Inhalation Aerosols: Physical and Biological Basis for Therapy*, Vol. 94; Hickey, A. J., Ed. Marcel Dekker: New York, **1996**, pp. 273–312.
19. Dalby, R.; Suman, J. *Adv. Drug Deliv. Rev.* **2003**, *55*, 779–791.
20. Adjei, A.; Garren, J. *Pharm. Res.* **1990**, *7*, 565–569.
21. Williams, D. W. In *Pharmaceutical Inhalation Aerosol Technology*; Hickey, A. J., Ed. Marcel Dekker: New York, **2003**, pp. 473–488.
22. Goodman, L. S.; Gilman, A. G.; Limbird, L. E.; Hardman, J. G.; *Gilman, A. G. Goodman & Gilman's The Pharmacological Basis of Therapeutics*. McGraw-Hill: New York, **2001**.
23. Katzung, B. G. *Basic and Clinical Pharmacology (Lange Series)*. Lange Medical Books/ McGraw-Hill: New York, **2000**.
24. Gringauz, A. *Introduction to Medicinal Chemistry: How Drugs Act and Why*, Wiley-VCH: New York, **1997**.
25. Crooks, P. A.; Al-Ghananeem, A. M. In *Pharmaceutical Inhalation Aerosol Technology*; Hickey, A. J.; Ed. Marcel Dekker: New York, **2003**, pp. 89–154.
26. Smith, S. J.; Bernstein, J. A. In *Inhalation Aerosols: Physical and Biological Basis for Therapy*, Vol. 94; Hickey, A. J., Ed. Marcel Dekker: New York, **1996**, pp. 233–269.
27. Olivieri, D.; Barnes, P. J.; Hurd, S. S.; Folco, G. C. *Asthma Treatment: A Multidisciplinary Approach*, Vol. 229. Plenum Press: New York, **1992**.
28. Garcia-Contreras, L.; Hickey, A. J. *Adv. Drug Deliv. Rev.* **2002**, *54*, 1491–1504.
29. Hanes, J.; Dawson, M.; Har-el, Y.-e.; Suh, J.; Fiegel, J. In *Pharmaceutical Inhalation Aerosol Technology*; Hickey, A. J., Ed. Marcel Dekker: New York, **2003**, pp. 489–539.
30. Ausborn, M.; Wichert, B. V.; Carvajal, M. T.; Niven, R. W.; Soucy, D. M.; Gonzalez-Rothi, R. J.; Gao, X. Y.; Schreier, H. *Proc. Program Int. Symp. Controlled Release Bioact. Mater.* **1991**, *18*, 371–372.
31. Wichert, B. V.; Gonzalez-Rothi, R. J.; Straub, L. E.; Wichert, B. M.; Schreier, H. *Int. J. Pharm.* **1992**, *78*, 227–235.
32. BitMansour, A.; Brown, J. M. Y. *J. Infect. Dis.* **2002**, *186*, 134–137.

33. Hickey, A. J.; Montgomery, A. B. In *Pharmaceutical Inhalation Aerosol Technology*; Hickey, A. J., Ed. Marcel Dekker: New York, **2003**, pp. 459–472.

34. Byron, P. R.; Patton, J. S. *J. Aerosol Med.* **1994**, *7*, 49–75.

35. Wade, A.; Weller, P. J. *Handbook of Pharmaceutical Excipients*, American Pharmaceutical Association: Washington, **1994**.

36. Simberkoff, M. S.; Santos, M. R. *Curr. Opin. Pulmon. Med.* **1996**, *2*, 228–235.

37. Hickey, A. J.; Jackson G. V.; Fildes, F. J. T. *J. Pharm. Sci.* **1988**, *77*, 804–809.

38. Cooney, D. J.; Hickey, A. J. *J. Pharm. Sci.* **2003**, *92*, 2341–2344.

39. Farr, S. J.; McKenzie, L.; Clarke, J. G. In *Respiratory Drug Delivery IV*; Byron, P. R., Dalby, R. N., Farr, S. J., Eds. Interpharm Press: Buffalo Grove, IL, **1994**, pp. 221–230.

40. Hickey, A. J.; Byron, P. R. *Drug Des. Deliv.* **1987**, *2*, 35–39.

41. Hickey, A. J.; Gonda, I.; Irwin, W. J.; Fildes, F. J. T. *J. Pharm. Sci.* **1990**, *79*, 1009–1014.

42. Fults, K. A.; Miller, I. F.; Hickey, A. J. *Pharm. Dev. Tech.* **1997**, *2*, 67–79.

43. Phillips, E. M. *J. Biopharm. Sci.* **1992**, *3*, 11–18.

44. Concessio, N. M.; Hickey, A. J. *Pharm. Tech.* **1994**, *18*, 88–98.

45. Pettis, R. J.; Knowles, M. R.; Olivier, K. N.; Hickey, A. J. *J. Pharm. Sci.* **2003**, .

46. Hickey, A. J.; Kuchel, K.; Masinde, L. E. In *Repiratory Drug Delivery IV*; Byron, P. R., Dalby, R. N., Farr, S. J., Eds. Interpharm Press: Buffalo Grove, IL, **1994**, pp. 259–263.

47. Dalby, R. N.; Hickey, A. J.; Tiano, S. L. In *Inhalation Aerosols*, Vol. 94; Hickey, A. J., Ed. Marcel Dekker: New York, **1996**, pp. 441–473.

48. USDHHS. *Draft Guidance for Industry—Nasal Spray and Inhalation Solution, Suspension and Spray Drug Products Chemistry, Manufacturing, and Controls.* Food and Drug Administration, Center for Drug Evaluation and Research: **May 26, 1999**.

49. USP24. *The United States Pharmacopoeia and National Formulary*: **2000**, pp. 1895–1912.

50. Jones, L. D.; McGlynn, P.; Bovet, L.; Hickey, A. J. *Pharm. Tech.* **2000**, *24*, 40–54.

17

ANTIBODY-DIRECTED DRUG DELIVERY

HERVÉ LE CALVEZ, JOHN MOUNTZOURIS, AND KOSI GRAMATIKOFF

Abgent Inc., 6310 Nancy Ridge Dr., Ste 106. San Diego, CA 92121

FANG FANG

NexBio, Inc., 6330 Nancy Ridge Dr., Ste 105, San Diego, CA 92121

Drug Delivery: Principles and Applications Edited by Binghe Wang, Teruna Siahaan, and Richard Soltero
ISBN 0-471-47489-4

17.1. INTRODUCTION

The approval of 12 recombinant antibodies (Table 17.1) for therapeutic applications by the Food and Drug Administration (FDA) since 1996 has fully established antibodies in the realm of human medicine. Indeed, antibodies currently comprise 30% of clinical trials being conducted by biopharmaceutical companies, with more than 30 drug candidates in late-stage development. The market for antibody-based drugs is estimated to reach $50 billion by 2020. Such unbridled optimism represents a remarkable change of fate for antibody-based therapeutics from the earlier failures in the 1970s and 1980s. It also reflects the increasing importance of antibodies in the postgenome era; use of antibodies is probably the quickest way to identify, characterize, and validate hundreds or thousands of potential therapeutic targets revealed by the sequence of the human genome.

The success of antibodies as therapeutic molecules has been made possible by some significant technological progress in antibody engineering and manufacture. In particular, methods of humanizing murine monoclonal antibodies (mAbs) have successfully reduced the immunogenicity of mAbs. In addition, completely human mAbs can now be generated from transgenic mice. The phage display antibody library technology offers another way to isolate high-affinity antibodies, circumventing the traditional time- and labor-intensive hybridoma method. Finally, technologies that mimic molecular evolution *in vitro* have been successfully applied to improve the potency, stability, and manufacturing yield of therapeutic antibody candidates. All these areas have been reviewed extensively and therefore will not be addressed in this chapter.

A vast majority of the therapeutic antibodies approved by the FDA are naked antibody molecules that exert most of their therapeutic effects by merely binding to selected cellular targets. However, recent approval of Zevalin (IDEC Pharmaceuticals) for cancer radioimmunotherapy marked a new direction in using antibodies

TABLE 17.1 Approved Antibody Drugs

Antibodies	Indication
Bexxar	Relapsed non-Hodgkin's lymphoma
Campath	B-cell chronic lymphocytic leukemia
Herceptin	$HER2^+$ breast cancer
Mylotarg	Relapsed acute myelogenous leukemia
Orthoclone OKT3	Transplant rejection
Remicade	Arthritis, Crohn's disease
ReoPro	Blood clots
Rituxan	Non-Hodgkin's lymphoma
Simulect	Transplant rejection
Synagis	Respiratory syncytial virus infection
Zenapax	Transplant rejection
Zevalin	Non-Hodgkin's lymphoma

as drug delivery molecular devices. In this regard, active research and development is being pursued on customized antibodies conjugated to toxins, radioisotopes, small drugs, enzymes, and genes that can be deployed to selectively destroy harmful cells in the body. This exciting new technology is finding initial applications in oncology, where current chemotherapy drugs cause high toxicity due to lack of specific targeting to tumor tissues and cells. Recent progress in this rapidly moving area is the main focus of this chapter. As for future trends and challenges in the field, some emerging technologies that may still be years away from the clinic will also be covered in the last part of this review.

17.2. ANTIBODY PHARMACOKINETICS

As described later in this chapter, few antibody-based reagents have succeeded in living up to their anticipated role as highly specific targeting agents for cancer therapy. This has likely been the result of suboptimal delivery of antibody to tumors due to a number of factors, including the physiology of the tumors and the large size of immunoglobulin G (IgG) molecules. In many experiments using radiolabeled antibodies, their pharmacokinetic behavior and homing to the target tissue have been evaluated. In general, intact antibodies administrated intravenously bind nonspecifically to the liver via Fc interactions and are then cleared following degradation and/or excretion of immune complexes through the kidneys. It has been estimated that only 0.001 to 0.01% of an injected dose of antitumor mAb accumulates specifically in each gram of tumor.[1,2]

IgGs are large molecules (150 kDa) which diffuse slowly into tumors and are slowly cleared from the circulation, resulting in low tumor:normal organ ratios.[3,4] When effector functions carried by the Fc fragment are not required, smaller antigen-binding constructs can be used. Smaller single-chain Fv antibody fragments (scFv, 25 kDa) penetrate tumors better than IgG, are cleared more rapidly from the circulation, and provide greater targeting specificity.[1,5,6] For instance, radiolabeled anti-tumor c-erbB2 scFv penetrates deeply into human tumor xenografts in mice and is cleared rapidly from circulation and normal tissue, resulting in highly specific tumor retention by as early as 4 hours after administration.[7] scFv are typically constructed from the heavy (V_H) and light (V_L) chain variable region genes of murine IgG and thus are still potentially immunogenic. In addition, scFv are monovalent and dissociate from tumor antigen faster than bivalent IgG molecules, which exhibit a higher apparent affinity due to avidity.[8] Loss of avidity, combined with rapid clearance from blood, results in significantly lower quantitative retention of scFv in tumor.[9] Significant tumor retention beyond 24 hours requires a dissociation rate constant (k_{off}) of less than 10^{-4} s^{-1} ($t_{1/2} = 1.8$ hours). Since antibodies typically have rapid ($>10^5$ $M^{-1}s^{-1}$) association rate constants (k_{on}), this requires a $K_d < 10^{-9}$ M, which is rarely achievable by murine immunization.[10] Antibody engineering and variable domain repertoire phage display techniques have overcome some of the full IgG molecule limitations.

17.3. ANTIBODY-DIRECTED DRUG DELIVERY: THERAPEUTIC IMMUNOCONJUGATES

Therapeutic immunoconjugates consist of a specifically tumor-targeting antibody covalently linked or chelated to a toxic effector molecule. They can be categorized into three groups defined by the nature of the effector molecule:

1. Radioimmunoconjugate: antibody molecules coupled to a radioisotope;
2. Immunotoxin conjugate: antibody molecules coupled to a protein toxin;
3. Drug immunoconjugate: antibody molecules coupled to a small drug or prodrug (tumor-activated prodrug).

Radioimmunoconjugates can be prepared by direct coupling of a radionuclide to an antibody through a bifunctional chelating agent [namely, polyaminocarboxylic acids, e.g., ethylenediaminetetraacetic acid (EDTA), diethylenetriaminepentaacetic acid (DTPA), or 1,4,7,10-tetrazazcyclododecane-N, N', N'', N'''-tetraacetic acid (DOTA)].[11] The chelating agent can also be coupled to a polymer that is then linked to an antibody. Use of a polymer allows binding of a large number of radionuclides per antibody molecule and also provides possibilities for additional modification of antibody and conjugate properties.[11] One strict requirement these bifunctional chelating agents must meet is that the metal-chelate complex must remain intact *in vivo* to avoid release of the radionuclide at areas other than the targeted sites.

Preparation of immunotoxin conjugates or drug immunoconjugates involves the coupling of drug to an antibody. A number of approaches of drug attachment to antibody have been reviewed recently.[12] A drug can be attached to an antibody directly or through a spacer (linker). A spacer may be used merely to make the chemistry of the coupling possible, but may have the secondary function of allowing a specific type of release mechanism. The sites of an antibody for drug attachment can be the ϵ-amino group of lysine residues, sugar residues, and intrachain disulfide bonds. The impacts of drug attachment to these sites on the structure, solubility, and immunoreactivity of the antibody are reviewed in ref. 12 and will not be discussed here. A drug can be coupled to an antibody through a peptide bond, ester bond, aldehyde/Schiff base linkage, sulfhydryl linkage, acetal and ketal, or hydrazone linkage.[12]

17.3.1. Radioimmunoconjugates

The beta-emitting radioisotopes yttrium-90 (^{90}Y) and iodine-131 (^{131}I) exhibit superior cell killing activity toward rapidly growing tumor cells, whereas the much slower-growing normal cells are less sensitive to radiation. The ideal characteristics of a radionuclide used for therapeutic applications include radiation emissions of a type and a energy level such that the path length (χ_{90}) in tissue results in optimal local energy deposition within tumors and minimal dose to distant organs.[13] ^{131}I was the earliest radionuclide investigated for therapeutic radioimmunoconjugates. The development of methods for attaching metal chelates to proteins has made possible the investigation of radioimmunotherapy treatments utilizing other potentially more effective radionuclides, such as ^{90}Y. As a pure high-energy beta emitter, ^{90}Y offers distinct advantages over ^{131}I. The higher energy of ^{90}Y (maximum = 2.3 MeV, mean = 0.94 MeV) when compared to ^{131}I (maximum = 0.61 MeV, mean = 0.18 MeV), and the longer path length ($^{90}Y\chi_{90}$ = 5 mm, $^{131}I\chi_{90}$ = 2 mm) allow delivery of a cytotoxic radiation dose to tumor cells more distant from the antibody-bound cell.[13] These characteristics may be especially advantageous in the treatment of bulky or poorly vascularized tumors. The shorter half-life of ^{90}Y (64 hours) compared to ^{131}I (193 hours) approximates the biological half-life of the radiolabeled antibody (47 hours), which may minimize radiotoxicity to nontarget organs.[13] Moreover, the property of emitting penetrating gamma rays (0.36 MeV) by ^{131}I requires measures to be taken such as shielding, counseling, occupancy rates or proximity to others, and possibly hospitalization when an increased whole body dose of ^{131}I is used.[13] Conversely, the pure beta-emitting ^{90}Y can be given on an outpatient basis with few radiation precautions.[13] Nevertheless, both ^{131}I and ^{90}Y have been employed in radioimmunotherapy (Table 17.2).

Radioimmunoconjugates have the advantages of allowing a radionuclide to be delivered and concentrated to tumor sites to spare the normal cells and tissues, especially those with a high growth rate such as bone marrow. Antibody molecules confer superior binding specificity and are therefore a logical solution for specific delivery of radioisotopes to tumor cells. A number of antibodies against various tumor-specific antigens have been coupled with radioisotopes and moved into animal and human testing in the past two decades. Although proving drug candidate efficacy in humans has not been easy, the field is thriving, with many clinical trials underway. The most remarkable success of radioimmunoconjugates has been in the treatment of hematological cancers. For example, Zevalin, a ^{90}Y-anti-CD20 antibody (ibritumomab tiuxetan; IDEC Pharmaceuticals) for the treatment of non-Hodgkin's lymphoma, is the first radioimmunotherapeutic mAb to gain FDA approval, and Bexxar, a ^{131}I-anti-CD20 antibody (tositumomab, Corixa Corp. and GlaxoSmithKline) has recently been approved (Table 17.2).[14–16] These drugs are delivered via intravenous injection to patients in the clinic.

Radioimmunoguided surgery (RIGS) is a very interesting new application of radioimmunoconjugates. Specifically, radiolabeled antibodies are given intravenously before surgery to precisely mark the location of tumor. During surgery, a hand-held gamma-detecting probe is used to locate tumor in the operative field, and the tumor is treated by either surgical removal or local implantation of radiation devices. For this application, scFv offer potential advantages over larger antibody molecules due to the rapid blood clearance and good tumor penetration of scFv antibody fragments. A Phase I clinical trial with colorectal cancer patients is reported on RIGS with an scFv (MFE-23-his) to carcinoembryonic antigen (CEA).[17] Iodinated-MFE-23-his showed good tumor localization, with an 84% accuracy compared to histological findings. The short interval between injection and operation, the lack of significant toxicity, and the relatively simple production in bacteria make this antibody suitable for RIGS.

TABLE 17.2 Radioimmunoconjugates in Clinical Development

Radioimmuno-conjugate	Specificity/ Drug	Cancer	Company	Development Status	Reference
Zevalin, ibritumomab tiuxetan	^{90}Y-anti-CD20	Non-Hodgkin's lymphoma	IDEC Pharmaceuticals	FDA approved (February 2002)	14
Bexxar, tositumomab	^{131}I-anti-CD20	Non-Hodgkin's lymphoma	Corixa Corp, and GlaxoSmithKline	FDA approved (June 2003)	15
MFE-23-his scFv	^{131}I-anti-CEA	Colorectal		Phase I	17
Theragyn pemtumomab	^{90}Y-anti-Muc1	Gastric, ovary	Antisoma and Hoffman-LaRoche	Phase II/III	18
81C6 Ab	^{131}I-anti-tenascin	Glioma	NCI	Phase II	19

Besides intravenous injection, some radioimmunoconjugate-based drug candidates have also been directly injected into the tumor sites in clinical trials and have reached Phase II and phase III development. TheraGyn (^{90}Y -anti-Muc1) is administered to colon cancer patients by intraperitoneal injection.[18] A ^{131}I-anti-tenascin radioimmunoconjugate is injected into a cranial resection cavity of glioma patients.[19] Although benefits for the patient were observed in these trials, the absorbed dose of radiation was still below the estimated dose needed to eradicate solid tumors and will need to be increased in future evaluations.

17.3.2. Immunotoxin Conjugates

Some well-characterized bacterial toxins kill cells effectively upon direct contact. Researchers have been attempting to harness this cell-killing activity and direct it to tumor cells. Again, due to the specific binding capability of antibody molecules, antibodies have been selected as the molecular device to direct toxin to the site of tumors. Antibodies against some tumor-specific antigens have been conjugated to several bacterial toxins, including *Pseudomonas* exotoxin (PE), staphylococcal enterotoxin A, and ricin toxin A. Because of the better tissue penetration and shorter systemic half-life of antibody scFv fragments compared with whole antibody molecules, most recombinant immunotoxins are composed of an scFv fragment fused to toxin. Earlier trials using murine scFv-fused toxin resulted in human anti-mouse antibodies (HAMA) and a human antitoxin response (HATA). Encouraging results from improved conjugates have since been reported (Table 17.3).

A group led by Ira Pastan reported using PE conjugate to treat leukemia and lymphoma. PE has three domains: a cell binding domain, a translocation domain, and an ADP-ribosylation domain that inactivates elongation factor 2 and leads to cell death. In this case, the PE ribosylation domain is fused to an anti-Tac [anti-CD25 or interleukin-2 (IL2) receptor] Fv fragment. In a Phase I trial, 35 patients with CD25-positive malignancies received intravenous injections of immunotoxin. The trial resulted in one complete response in a patient with hairy cell leukemia (HCL) and seven partial responses in patients with HCL (three), cutaneous lymphocytic leukemia (CTCL) (one), Hodgkin's disease (one), and adult T-cell leukemia (one). All patients with HCL and CTCL responded. Nine patients tolerated the treatment at the maximum dose and only six patients developed neutralizing HAMAs.[20,21]

Antibody conjugates with staphylococcal enterotoxin A were tested in earlier studies and were shown to carry unacceptable toxicity due to the extraordinary potency of staphylococcal enterotoxin A as a superantigen, which nonspecifically stimulates the host immune system, causing systemic inflammation and an autoimmune response.[22,23] Another serious side effect of immunotoxin therapeutics is the potentially fatal vascular leak syndrome (VLS) due to damage to vascular endothelium by the toxins. Efforts have been made to engineer toxins to eliminate toxicity to vascular endothelial cells without compromising their tumor cell–killing potency. Preliminary success has been reported with engineered ricin toxin A. In this case, mutated ricin toxin A linked to an antibody that targets lymphoma induced

TABLE 17.3 Immunotoxin Conjugates in Clinical Development

Immunotoxin Conjugate	Specificity/Drug	Cancer	Company	Development Status	Reference
BL22 (RFB4 (dsFv)-PE38)	α-CD22 PE38 (ds scFv fusion to PE38)	Hairy cell leukemia	NCI	Phase I	27
SGN10 (BR96 scFv-PE40/BR96-SCIT)	α-Lewis^{Y}-SCIT (SCA)	Breast, colon, lung, prostate	Seattle Genetics/ Aventis	Phase I completed	28
LMB-9 (B3 (ds scFv)-PE38)	α-Lewis^{Y}-PE38 (ds scFv fusion to PE38	Colorectum, pancreas, esophagus, stomach, breast, NSCLC, GIC, bladder, ovarian	NCI/IVAX	Phase I	29
LMB-2 (α-Tac(Fv)-PE38)	α-CD25-PE38 (ds scFv fusion to PE38)	Hematopoietic malignancies	NCI	Phase I completed	20,21
SS1 (dsFv)-PE38	α-mesothelin-PE38 (ds scFv fusion to PE38)	Mesothelioma, ovatian, squamous cell NSCLC	NeoPharm	Phase I	30

ds scFv: disulfide-stabilized, single-chain variable domain fragment; GIC: gastrointestinal cancer; NSCLC: non-small-cell lung carcinoma; PE38 and PE40: 38 kDa and 40 kDa truncated pE polypeptide; SCIT: single-chain immunotoxin PE40.

significantly less VLS than unmodified ricin toxin A in a mouse model.[24] A new generation of immunotoxins with modified ricin toxin A should have much lower vascular toxicity, thereby making a higher dosage more tolerable to the patients and significantly improving their clinical outcome.[25]

Besides bacterial toxins, antibodies have also been conjugated with certain cytokine molecules in an attempt to promote a tumor-selective local immune response. For instance, a fusion protein containing the human scFv fragment ML3.9 (anti-HER2/neu) and the human complement fragment C5a was created and tested. The molecule retains anti-HER2/neu binding capacity and promotes *in vitro* neutrophile migration and degranulation.[26] Despite some encouraging clinical responses to immunotoxins, HATA and HAMA have been commonly observed in clinical trials, especially with solid tumor targeting, in which higher circulating doses of drugs are needed. Therefore, there is considerable work to be done before immunotoxins move to the marketplace as a new class of therapeutics.

17.3.3. Drug Immunoconjugates

Antibody-drug conjugates direct small drugs to the tumor through the targeting agent. Their purpose is to minimize the drug dosage and to avoid killing healthy cells. However, their clear superiority to nontargeted versions of the same drugs has not been demonstrated *in vivo*. Besides, conjugation of the cytotoxic agent to an antibody often renders it noncytotoxic to cells. Antibody-drug-based therapy has evolved toward tumor-activated prodrug therapy (TAP), in which the drug is activated upon internalization and cleavage of the complex in the cell. Using linkers such as a disulfide bond releases the active drug into an intracellular compartment under acidic conditions during internalization of the drug immunoconjugate.[31] The small, highly cytotoxic drugs currently used are inhibitors of tubulin polymerization, such as maytansinoids, dolastatins, auristatin, and crytophycin; DNA-alkylating agents such as CC-1065 analogs and duocarmycin; or enediyene antibiotics such as calicheamin and esperamicin.[32] Drugs like antifolates, vinca alkaloids, and anthracyclines (doxorubicin) have shown lack of potency in clinical trials.[33] The FDA-approved gemtuzumab ozogamicin (Mylotarg) consists of calicheamicin conjugated to a humanized anti-CD33 antibody.[34] In one study, 30% of patients with CD33-positive acutemyelogenous leukemia (AML) who were treated with Mylotarg achieved remission. The treatment is restricted to patients with a specific clinical profile (age 60 or older and not suitable for cytotoxic chemotherapy). Severe myelosuppression and hepatotoxicity are observed in some patients.[35]

Conjugates currently in clinical trials comprise humanized antibodies linked via a disulfide linker to the highly potent maytansine derivative called DM1. The conjugate is nontoxic in the blood system and becomes active when it reaches the tumor site upon antigen binding. After the immunoconjugates are internalized, the chemical moiety of the conjugate is cleaved from the antibody and then binds to intracellular targets to exert its cytotoxic effects. The TAPs have a clear advantage over conventional drugs in the dosage given to patients in producing a tumor growth inhibitory effect.[32,36–38] Clinical trials are in different phases right now

TABLE 17.4 Drug Immunoconjugates in Clinical Development

Drug Immunoconjugate	Specificity/Drug (Antibody)	Cancer	Company	Development Status	Reference
Mylotarg, gemtuzumab ozogamicin	α-CD33-calicheamycin (humanized by CDR grafting)	AML	Wyeth-Ayrst/ Celltech group	FDA approval (5/18/00)	35
Cantuzumab mertansine (huC242-DM1/SB-408075)	a-CanAg-DM1 TAP (humanized by resurfacing)	Colorectal, pancreatic	ImmunoGen	Phase I completed	39–41
BB-10901/huN901-DM1	α-CD56-DM1 TAP (humanized by CDR grafting)	SCLC	British Biotech/ ImmunoGen	In Phase I (UK); Phase I/II (US)	42
MLN2704 (MLN591-DM1)	α-PSMA-DM1 TAP (deimmunized)	Prostate	Millenium[a]	Initiated Phase I (11/21/02)	
Bivatuzumab mertansine	α-CD44v6-DM1 TAP (humanized)	Unspecified	Boehringer Ingelheim[a]	Initiated clinical trials (10/15/02)	
Trastuzumab-DM1/ Herceptin-DM1	α-Her2/neu-DM1 TAP (humanized by CDR grafting and framework changes)	Breast	Genentech[a]	Preclinical development	38
My9–6-DM1	α-CD33-DM1 TAP (humanized by resurfacing)	AML	ImmunoGen	Preclinical development	43
SGN-15 (BMS-182248/ BR96-doxorubicin)	α-Lewis^{Y}-doxorubicin (chimeric)	Breast,colon, prostate, lung	Seattle Genetics/ Aventis	Three Phase II clinical trials; breast trial completed; clinical development will not be pursued	44,45
SGN-25 (BR96-auristatin E)	α-Lewis^{Y}-auristatin E (chimeric)	Breast,colon, prostate, lung	Seattle Genetics	Preclinical development	46
SGN-35	α-CD30-auristatin E	Hematological malignancies, lymphomas	Seattle Genetics	Preclinical development	31

[a]ImmunoGen Inc. technology, AML: acute myelogenous leukemia; DM1: N2′-deacetyl-N2′-(3-mercapto-1-oxopropyl)-maytansine; SCA: single-chain antibody; SCLC: small cell lung carcinoma; TAP: tumor-activated prodrug.

(Table 17.4). TAPs seem to be effective in limiting disease progression and do not trigger anti-human antibodies. A cantuzumab mertansine is in Phase I testing for the treatment of patients with Can-Ag-positive malignancies.[39–41] An anti-CD56 DM1 is in Phase I/II testing for treatment of lung carcinoma and neuroendocrine-derived tumors.[42]

17.3.4. Antibody-Directed Enzyme Prodrug Therapy (ADEPT)

Enzyme-activating prodrug therapy is a two-step approach. In the first step, a drug-activating enzyme is targeted and expressed in tumors. In the second step, a nontoxic prodrug, which is a substrate of the exogenous enzyme expressed in tumors, is administered systemically.[47–49] The net gain is that a systemically administered prodrug can be converted to a high local concentration of an active anticancer drug in tumors. To be clinically successful, both enzymes and prodrugs should meet certain requirements for this strategy. The enzymes should be either of nonhuman origin or a human protein that is absent or expressed only at low concentrations in normal tissues. The protein must achieve sufficient expression in tumors and have high catalytic activity. The prodrug should be a good substrate for the expressed enzyme in tumors but not activated by endogenous enzyme in nontumor tissues. It must be able to cross the tumor cell membrane for intracellular activation, and the cytotoxicity differential between the prodrug and its corresponding active drug should be as high as possible. It is preferable that the activated drug be highly diffusible or actively taken up by adjacent cancer cells to produce a "bystander" killing effect. In addition, the half-life of the active drug should be long enough to induce a bystander effect but short enough to avoid leakage of the drug out into the systemic circulation. Currently, delivery methods for an enzyme/prodrug strategy can be divided into two major classes: (1) delivery of genes that encode prodrug-activating enzymes into tumor tissues [gene-directed enzyme prodrug therapy (GDEPT), enzyme prodrug therapy (VDEPT), etc.] and (2) delivery of active enzymes to tumor tissues (ADEPT). Details on clinical applications of ADEPT are described in Chapter 11.

17.4. FUTURE CHALLENGES AND OPPORTUNITIES: ANTIBODY-DIRECTED INTRACELLULAR DRUG DELIVERY

An antibody is a useful tool to deliver a drug or radioisotope to the vicinity of a specific cell type; antigenic cellular receptors on the cell surface condense antibody at the exterior face of the cell. However, due to the relative impermeability of the cell membrane to large proteins, delivering a macromolecule to the putative cytoplasmic or nuclear target within the cell remains a substantial technical challenge. Several approaches used to solve this problem are described. Although it probably will be many years before drug candidates will be identified and tested in humans, interest in their potential is fueling current research.

17.4.1. Protein Transduction Domains

The critical observation that some proteins can enter cells in the absence of endocytosis led to the identification of key basic peptide sequences about 10–16 residues in length from these proteins that can traverse the plasma membrane and thereby smuggle the full protein across the membrane boundary.[50] Well-characterized protein transduction domains (PTDs) include those identified from the *Drosophilae* homeotic transcription factor Antennapedia (Antp), the herpes simplex virus (HSV) protein VP22, and the human immunodeficiency virus 1 (HIV)-1 transcriptional activator Tat.[51–54] Similarities in the tertiary structure of these PTDs are not apparent, although the propensity for multiple charged lysine and arginine residues has been noted. Cellular ingress of the PTDs is highly specific, nondestructive to the cell, independent of protein size, and takes place on the order of minutes. These domains, which can be reversibly or even noncovalently attached to proteins, are able to promote intracellular delivery of the protein at temperatures ranging from 4°C to 37°C and function in the presence of cellular transport inhibitors. A number of synthetic PTDs have been designed to optimize the length and efficiency of protein delivery *in vivo*. Pep-1 is a 21-residue-long carrier consisting of three domains: a hydrophobic, tryptophan-rich motif that targets the cell membrane and forms hydrophobic interactions with proteins; a hydrophilic, lysine-rich domain derived from the simian virus 40 large T antigen nuclear localization sequence, which improves intracellular delivery and solubility of the peptide vector; and a spacer sequence.[55] Proteins up to 500 kDa, including antibodies and protein-DNA complexes, have been successfully transduced into cultured cell lines by fusing with Pep-1.[55] The ability of Pep-1 to deliver proteins *in vivo* is yet to be demonstrated and is a focus of current research.

17.4.2. Liposomal Carriers

Lipid-based carriers have been established to transport DNA into cells, but the efficiency of protein delivery based on conventional liposomal formulations is below 5%. Zelphati et al., however, have developed a new lipid formulation that interacts rapidly and noncovalently with protein, creating a protective vehicle for delivery.[56] The protein encapsulated in the formulation binds to the negatively charged membrane, is internalized in endosomal vesicles by endocytosis, and is then released inside the cell. The system displays no significant toxicity under optimal conditions.

In spite of some interesting preliminary results, transduction of antibodies into cells via either PTDs or liposomes remains at an early experimental stage. However, the potential application will be very attractive when the technical problems can be surmounted.

17.4.3. Antibody-Mediated Translocation

In 1978, Alarcon-Segovia et al. reported that antibodies to nuclear ribonucleoprotein (RNP) penetrated live human cells.[57] Internalization of certain antibodies has been shown to be mediated by antibody interaction with antigens expressed on the cell surface. Following endocytosis, changes in cell functions have been observed for antibodies to nuclear RNP and DNA.[58] Other groups have confirmed these results and have demonstrated that anti-RNP antibodies from human systemic lupus erythematosus (SLE) patients, IgG anti-DNA antibodies, and anti-neural antibodies (anti-Hu) from patients with neuropathy can penetrate lymphocytes, epithelial cells, and hepatocytes.[59,60] A key characteristic of antibodies possessing this ability is polyreactivity against self antigens such as double-stranded and single-stranded DNA and various proteins (actin, myoglobin, myosin, tubulin, histones).[61] Although various germ-line V_H gene families are used, an overrepresentation of tyrosine, lysine, and arginine is found in the CDR3 of anti-DNA mAb with the capacity to enter cells. Haptens (biotin, fluorescein, oligonucleotides) and macromolecules (peroxidase, IgG) covalently coupled to the mAb or their F(ab′)2 and Fab fragments were translocated through the cytoplasm and into the cell nucleus. Further work on minimal peptides derived from antibody-CDRs has shown the ability to penetrate cells and may provide a vehicle for more controlled delivery.[61]

17.4.4. Identification of New Target Molecules to Assist Intracellular Antibody Delivery

One potential approach to improving antibody-directed intracellular drug delivery is to find novel cell surface antigens that have the property of rapidly internalizing any bound molecule, besides being highly specific for the targeted tissue or cell types. Identification of appropriate targeting molecules has so far been performed largely by individually screening receptor ligands or antibodies.

Recently, it has proven possible to directly select peptides and antibody fragments binding cell-surface receptors from filamentous phage libraries.[62–66] This has led to a marked increase in the number of potential targeting molecules. The ability of bacteriophage to undergo receptor-mediated endocytosis indicates that phage libraries can be selected not only for cell binding but also for internalization into mammalian cells.[63,67,68] If the single-stranded phage genome can be transcribed and translated, then it should prove possible to screen or select for phage that bind receptors in a manner that leads to endocytosis and delivery of the phage genome into the correct trafficking pathway, leading to expression. It has been shown that phage can enter mammalian cells after chemical alteration of the cell membrane, leading to reporter gene expression. More recently, Larocca et al. demonstrated in 1998 that indirect bacteriophage-mediated gene delivery could occur by targeting biotinylated phage via streptavidin and biotinylated fibroblast growth factor (FGF) to mammalian cells expressing FGF receptor.[69]

Poul and Marks showed that filamentous phage displaying the anti-ErbB2 scFv F5 as genetic fusion with the phage minor coat protein pIII can directly infect mammalian cells expressing ErbB2, leading to expression of a reporter gene contained in the phage genome.[70] This offers a new way to discover targeting molecules for intracellular drug delivery or gene therapy by directly screening phage antibodies to identify those capable of undergoing endocytosis and delivering a gene or drug into a cellular compartment. This should significantly facilitate the identification of appropriate targets and targeting of molecules for gene therapy and other applications when delivery in the cytosol is required.

17.5. CONCLUSION

Antibody-directed drug delivery has firmly established its remarkable value in the marketplace in recent years. In addition to several drugs that are already approved by the FDA, many more drug candidates are in the development pipeline. Although initial clinical indications for this class of therapeutic molecules are limited to oncology, much broader applications of antibody drug delivery are anticipated in the future, as the improved specificity and potency offered by this approach continues to be highly desirable in many therapeutic areas.

REFERENCES

1. Yokota, T.; Milenic, D.; Whitlow, M.; Schlom, J. *Cancer Res.* **1992**, *52*, 3402–3408.
2. Jain, R. K. *Sci. Am.* **1994**, *271*, 42–49.
3. Clauss, M. A.; Jain, R. K. *Cancer Res.* **1990**, *50*, 3487–3492.
4. Sharkey, R. M.; Gold, D. V.; Aninipot, R.; Vagg, R.; Ballance, C.; Newman, E. S.; Ostella, F.; Hansen, H. J.; Goldenberg, D. M. *Cancer Res.* **1990**, *50*, 828s–834s.
5. Colcher, D.; Minelli, F. M.; Roselli, M.; Muraro, R.; Simpson-Milenic, D.; Schlom, J. *Cancer Res.* **1988**, *48*, 4597–4603.
6. Milenic, D. E.; Yokota, T.; Filpula, D. R.; Finkelman, M. A. J.; Dodd, S. W.; Wood, J. F.; Whitlow, M.; Snow, P.; Schlom, J. *Cancer Res.* **1991**, *51*, 6363–6371.
7. Adams, G. P.; McCartney, J. E.; Tai, M-S.; Opperman, H.; Huston, J. S.; Stafford, W. F.; Bookman, M. A.; Fand, I.; Houston, L. L.; Weiner, L. M. *Cancer Res.* **1993**, *53*, 4026–4034.
8. Crothers, D. M.; Metzger, H. *Immunochemistry* **1972**, *9*, 341–357.
9. Adams, G. P.; DeNardo, S. J.; Amin, A.; Kroger, L. A.; DeNardo, G. L.; Hellstrom, I.; Hellstrom, K. E. *Antibody Immunoconj. Radiopharm.* **1992**, *5*, 81–95.
10. Foote, J.; Eisen, H. N. *Proc. Natl. Acad. Sci. USA* **1995**, *92*, 1254–1256.
11. Torchilin, V. P.; Klibanov, A. L. *Crit. Rev. Ther. Drug Carrier Syst.* **1991**, *7*, 275–308.
12. Garnett, M. C. *Adv. Drug Deliv. Rev.* **2001**, *53*,171–216.
13. Wiseman, G. A.; White, C. A.; Sparks, R. B.; Erwin, W. D.; Podoloff, D. A.; Lamonica, D.; Bartlett, N. L.; Parker, J. A.; Dunn, W. L.; Spies, S. M.; Belanger, R.; Witzig, T. E.; Leigh; B.R. *Crit. Rev. Oncol. Hematol.* **2001**, *39*, 181–194.
14. Ansell, S. M.; Ristow, K. M.; Habermann, T. M.; Wiseman, G. A.; Witzig, T. E. *J. Clin. Oncol.* **2002**, *20*, 3885–3890.
15. Kaminski, M. S.; Zelenetz, A. D.; Press, O. W.; Saleh, M.; Leonard, J.; Fehrenbacher, L.; Lister, T. A.; Stagg, R. J.; Tidmarsh, G. F.; Kroll, S.; Wahl, R. L.; Knox, S. J.; Vose, J. M. *J. Clin. Oncol.* **2001**, *19*, 3918–3928.

16. Cheson, B. D. *Blood.* **2003**, *101*, 391–398.
17. Mayer, A.; Tsiompanou, E.; O'Malley, D.; Boxer, G. M.; Bhatia, J.; Flynn, A. A.; Chester, K. A.; Davidson, B. R.; Lewis, A. A.; Winslet, M. C.; Dhillon, A. P.; Hilson, A. J.; Begent, R. H. *Clin. Cancer. Res.* **2000**, *6*, 1711–1719.
18. Hird, V.; Maraveyas, A.; Snook, D.; Dhokia, B.; Soutter, W. P.; Meares, C.; Stewart, J. S.; Mason, P.; Lambert, H. E.; Epenetos, A. A. *Br. J. Cancer* **1993**, *68*, 403–406.
19. Reardon, D. A.; Akabani, G.; Coleman, R. E.; Friedman, A. H.; Friedman, H. S.; Herndon, J. E. II; Cokgor, I.; McLendon, R. E.; Pegram, C. N.; Provenzale, J. M.; Quinn, J. A.; Rich, J. N.; Regalado, L. V.; Sampson, J. H.; Shafman, T. D.; Wilkstrand, C. J.; Wong, T. Z.; Zhao, X. G.; Zalutsky, M. R.; Bigner, D. D. *J. Clin. Oncol.* **2002**, *20*, 1389–1397.
20. Kreitman, R. J.; Wilson, W. H.; Robbins, D.; Marguiles, I.; Stetler-Stevenson, M.; Waldmann, T. A.; Pastan, I. *Blood.* **1999**, *94*, 3340–3348.
21. Kreitman, R. J.; Wilson, W. H.; Bergeron, K.; Raggio, M.; Stetler-Stevenson, M.; FitzGerald, D. J.; Pastan, I. *N. Engl. J. Med.* **2001**, *345*, 241–247.
22. Giantono, B. J.; Alpaugh, R. K.; Schultz, J.; McAleer, C.; Newton, D. W.; Shanon, B.; Guedez, Y.; Koth, M.; Vitek, L.; Persson, R.; Gunnarsson, P. O.; Kalland, T.; Dohlsten, M.; Persson, B.; Weiner, L. M. *J. Clin. Oncol.* **1997**, *15*, 1994–2007.
23. Alpaugh, R. K.; Schultz, J; McAleer, C.; Giantonio, B. J.; Persson, R.; Burnite, M.; Nielsen, S. E.; Vitek, L.; Persson, B.; Weiner, L. M. *Clin. Cancer Res.* **1998**, *4*, 1903–1914.
24. Smallshaw, J. E.; Ghetie, V.; Rizo, J.; Fulmer, J. R.; Trahan, L. L.; Ghetie, M. A.; Vietta, E. S. *Nat. Biotechnol.* **2003**, *21*, 387–391.
25. Kreitman, R. J. *Nat. Biotech.* **2003**, *21*, 372–374.
26. Alpaugh, R. K.; Simmons, H.; Adams, G.; Weiner, L. M. *IBC Antibody Eng. Conf.* **1999**, Dec. 6–9, 617.
27. Kreitman, R. J.; Wilson, W. H.; White, J. D.; Stetler-Stevenson, M.; Jaffe, E. S.; Giardina, S.; Waldmann, T. A.; Pastan, I. *J. Clin. Oncol.* **2000**, *18*, 1622–1636.
28. Posey, J.A; Khazaeli, M. B.; Bookman, M. A.; Nowrouzi, A.; Grizzle, W. E.; Thornton, J.; Carey, D. E.; Lorenz, J. M.; Sing, A. P.; Siegall, C. B., LoBuglio, A. F.; Saleh, M. N. *Clin. Cancer Res.* **2002**, *8*, 3092–3099.
29. Kuan, C. T.; Pai, L. H.; Pastan, I. *Clin. Cancer Res.* **1995**, *1*, 1589–1594.
30. Chang, K.; Pai, L. H.; Batra, J. K.; Pastan, I.; Willingham, M. C. *Cancer Res.* **1992**, *52*, 181–186.
31. Francisco, J. A.; Cerveny, C. G.; Meyer, D. L.; Siegall, C. B.; Senter, P. D.; Wahl, A. F. *Proc. Annu. Meet. AACR*, **2003**.
32. Blattler, W. A.; Chari, R. V. J. In *Anticancer Agents—Frontiers in Cancer Chemotherapy*; Ojima, I., Vite, G. D., and Altmann, K.-H., Eds. American Chemical Society: Washington, DC, **2001**, pp. 317–338.
33. Chari, R. V. *Adv. Drug. Deliv. Rev.* **1998**, *31*, 89–104.
34. Sievers, E. L.; Linenburger, M. *Curr. Opin. Oncol.* **2001**, *13*, 522–527.
35. Giles, F.; Estey, E.; O'Brien, S. *Cancer.* **2003**, *98*, 2095–2104.
36. Liu, C.; Tadayoni, B. M.; Bourret, L. A.; Mattocks, K. M.; Derr, S. M.; Widdison, W. C.; Kedersha, N. L.; Ariniello, P. D.; Goldmacher, V. S.; Lambert, J. M.; Blattler, W. A.; Chari, R. V. *Proc. Natl. Acad. Sci. USA*, **1996**, *93*, 8618–8623.

37. Ross, S.; Spencer, S. D.; Holcomb, I; Tan, C.; Hongo, J.; Devaus, B.; Rangell, L; Keller, G. A.; Schow, P.; Steeves, R. M.; Lutz, R. J.; Frantz, G.; Hillan, K; Peale, F.; Tobin, P.; Eberhard, D.; Rubin, M. A.; Lasky, L. A.; Koeppen, H. *Cancer Res.* **2002**, *62*, 2546–2553.
38. Schwall, R. H.; Dugger, D.; Erickson, S. L.; Yee, S.; Philips, G.; Chari, R.; Sliwkowski, M. X. *AARC-NCI-EORTC Int. Conf.*. Miami Beach, FL. **2001**, p. 132.
39. Helft, P. R.; Schilsky, R. L.; Kindler, H. L.; Bertucci, D.; Dewitte, M. H.; Martino, H. K.; Friedman, C. J.; Erickson, J. C.; Lambert, J. M.; Ratain, M. J. *AACR-NCI-EORTC Int. Conf.* Miami Beach, FL, **2001**, p. 134, Abstract #657.
40. Rowinsky, E. K.; Ochoa, L.; Patnaik, A.; de Bono, J. S.; Hammond, L. A.; Takimoto, C.; Garrison, M. A.; Schwartz, G.; Smith, L.; Hao, D.; et al. American Society of Clinical Oncology: Orlando, FL, **2002**, p. 30a, Abstract #118.
41. Tolcher, A. W.; Ochoa, L.; Hammond, L. A.; Patnaik, A.; Edwards, T.; Takimoto, C.; Smith, L.; de Bono, J.; Schwartz, G.; Mays, T.; Jonak, Z. L.; Johnson, R.; DeWitte, M.; Martino, H.; Audette, C.; Maes, K.; Chari, R. V.; Lambert, J. M.; Rowinsky, E. K. *J. Clin. Oncol.* **2003**, *21*, 211–222.
42. Fosella, F. V.; Tolcher, A.; Elliott, M.; Lambert, J. M.; Lu, R.; Zinner, R.; Lu, C.; Oh, Y.; Forouzesh, B.; McCreary, H.; et al. American Society of Clinical Oncology: Orlando, FL, **2002**, p. 309a, Abstract #1232.
43. Lutz, R. J., Xie, H.; Dionne, C.; Steeves, R. M.; Goldmacher, V. S.; Leece, B.; Bartle, L.; Chari, R. American Association for Cancer Research: San Francisco, **2002**, p. 257, Abstract #1279.
44. Ajani, J. A.; Kelsen, D. P.; Haller, D.; Hargraves, K.; Healey, D. *Cancer J.* **2000**, *6*, 78–81.
45. Tolcher, A. W.; Sugarman, S.; Gelmon, K. A.; Cohen, R.; Saleh, M.; Isaacs, C.; Young, L.; Healy, D.; Onetto, N.; Slichenmeyer, W. *J. Clin Oncol.* **1999**, *17*, 478–484.
46. Senter, P. H.;Doronina, S. P., Cerveny, C.; Chace, D.; Francisco, J.; Klussman, K.; Mendelsohn, B.; Meyer, D.; Siegall, C. B.; Thompson, J.; et al. American Association for Cancer Research: San Francisco, **2002**, p. 414, Abstract #2062.
47. Weyel, D.; Sedlacek, H. H.; Muller, R.,; Brusselbach, S. *Gene Ther.* **2000**, *7*, 224–231.
48. Springer, C. J.; Niculescu-Duvaz, I. *J. Clin. Invest.* **2000**, *105*, 1161–1167.
49. Hamstra, D. A.; Rehemtulla, A. *Hum. Gene Ther.* **1999**, *10*, 235–248.
50. Schwarze, S. R.; Dowdy, S. F. *Trends Pharmacol. Sci.* **2000**, *21*, 45–48.
51. Han, K.; Jeon, M. J.; Kim, K. A.; Park, J.; Choi, S. Y. *Mol. Cells.* **2000**, *10*, 728–732.
52. Torchilin, V. P. *Proc. Natl. Acad. Sci. USA* **2001**, *98*, 8786–8791.
53. Ho, A.; Schwarze, S. R.; Mermelstein, S. J.; Waksman, G.; Dowdy, S. F. *Cancer Res.* **2001**, *61*, 474–477.
54. Bonetta, L. *The Scientist*, **2002**, *16*, 38–40.
55. Morris, M. C.; Depollier, J.; Mery, J.; Heitz, F.; Divita, G. *Nat. Biotech.* **2001**, *19*, 1173–1176.
56. Zelphati, O.; Wang, Y.; Kitada, S.; Reed, J. C.; Felgner, P. L.; Corbeil, J. *J. Biol. Chem.* **2001**, *276*, 35103–35110.
57. Alarcon-Segovia, D.; Ruiz-Arguelles, A.; Fishbein, E. *Nature* **1978**, *271*, 67–69.
58. Alarcon-Segovia, D.; Ruiz-Arguelles, A.; Llorente, L. *Immunol. Today* **1996**, *17*, 163–164.
59. Ma, J.; Chapman, G. V.; Chen, S. L.; Melick, G.; Penny, R.; Breit, S. N. *Clin. Exp. Immunol.* **1991**, *84*, 83–91.

60. Hormingo, A.; Leiberman, F. *J. Neuroimmunol.* **1994**, *55*, 205–212.

61. Ternynck, T.; Avrameas, A.; Ragimbeau, J.; Buttin, G.; Avrameas, S. *J. Autoimmun.* **1998**, *11*, 511–521.

62. Andersen, P. S.; Stryhn, A.; Hansen, B. E.; Fugger, L.; Engberg, J.; Buus, S. *Proc. Natl Acad. Sci. USA* **1996**, *93*, 1820–1824.

63. Barry, M. A.; Dower, W. J.; Johnston, S. A. *Nat. Med.* **1996**, *2*, 299–305.

64. Cai, X.; Garen, A. *Proc. Natl Acad. Sci USA* **1995**, *92*, 6537–6541.

65. de Kruif, J.; Terstappen, L.; Boel, E.; Logtenberg, T. *Proc. Natl Acad. Sci. USA* **1995**, *92*, 3938–3942.

66. Marks, J. D.; Ouwehand, W. H.; Bye, J. M.; Finnern, R.; Gorick, B. D.; Voak, D.; Thorpe, S.; Hughes-Jones, N. C.; Winter, G. *Bio/Tech.* **1993**, *11*, 1145–1149.

67. Hart, S. L.; Knight, A. M.; Harbottle, R. P.; Mistry, A.; Hunger, H. D.; Cutler, D. F.; Williamson, R.; Coutelle, C. *J. Biol. Chem.* **1994**, *269*, 12468–12474.

68. Becerill, B.; Poul, M. A.; Marks, J. D. *Biochem. Biophys. Res. Commun.* **1999**, *255*, 386–393.

69. Larocca, D.; Witte, A.; Johnson, W.; Pierce, F. G.; Baird, A. *Gene Ther.* **1998**, *9*, 2393–2399.

70. Poul, M. A.; Marks, J. D. *J. Mol. Biol.* **1999**, *288*, 203–211.

18

EFFLUX TRANSPORTERS IN DRUG EXCRETION

SHUZHONG ZHANG AND MARILYN E. MORRIS

Department of Pharmaceutical Sciences, School of Pharmacy and Pharmaceutical Sciences, University at Buffalo, State University of New York, Amherst, NY 14260

18.1. INTRODUCTION

Therapeutic agents or other xenobiotic compounds exert their pharmacological or toxicological actions only when sufficient concentrations of these compounds are present at the site of action, where they can bind to the targeted receptors or enzymes. Therefore, the ability of drug molecules to cross biological membranes represents an important determinant of their absorption, distribution, elimination, and, ultimately, their therapeutic or toxic effects. It is clear that the complex biological membrane system is not just pure lipid bilayers, but lipid bilayers embedded with numerous proteins, including transporters. Thus, for a large number of drug

Drug Delivery: Principles and Applications Edited by Binghe Wang, Teruna Siahaan, and Richard Soltero
ISBN 0-471-47489-4

TABLE 18.1 Characteristics of Efflux Transporters

Member	HUGO Symbol	Alternative Name	Tissue Localization	Subcellular Level	Associated Disease
MDR1*	ABCB1	PGY1, P-gp	Liver, gut, kidney, adrenal gland, blood-brain barrier, placenta	Apical	Drug resistance
MDR3*	ABCB4	PGY3, MDR2/MDR3	Liver, canalicular membrane	Apical	PFIC3
MRP1*	ABCC1	MRP, GS-X	Ubiquitous	Basolateral	Drug resistance
MRP2*	ABCC2	cMOAT	Liver, intestine, kidney	Apical	Dubin-Johnson syndrome
MRP3*	ABCC3	cMOAT2, MLP2, MOAT-D	Liver, intestine, kidney, adrenal gland	Basolateral	?
BCRP*	ABCG2	ABCP, MXR	Placenta, liver, intestine, apical membrane	Apical	Drug resistance
BSEP†	ABCB11	SPGP	Liver canalicular membrane	Apical	PFIC2

Source: Data from Litman et al.[3] * or Trauner and Boyer.[274] †

molecules, their ability to pass through biological membranes is not solely determined by their physicochemical parameters such as lipophilicity, but also governed by the transporter activities. Among these transporters, a group of so-called efflux transporters (Table 18.1), including P-glycoprotein, multidrug resistance–associated proteins (MRPs), and breast cancer–resistant protein (BCRP) are of particular interest in that they actively remove a wide range of structurally and functionally distinct molecules from the cells against a concentration gradient. Their transport activities toward a number of clinically important anticancer agents, such as doxorubicin, paclitaxel, and vinblastine, prevent the intracellular accumulation of these cytotoxic agents and lead to inefficient cell killing, a phenomenon known as "multidrug resistance" (MDR), which remains the primary obstacle to successful cancer chemotherapy.[1–4] In addition, studies characterizing the molecular and functional properties and physiological functions of these transporters have revealed that these efflux transporters, apart from mediating MDR, play an essential role in governing the absorption and the intestinal, hepatobiliary, and renal excretion of a variety of endogenous and exogenous compounds.[5–9] The localization of these efflux transporters on the luminal side of the blood-brain barrier, blood-testis barrier, and placenta suggests their central role in regulating the entry of potentially harmful compounds into these pharmacological sanctuaries. It is widely accepted that at least some of these transporters constitute an essential component for the barrier functions between the blood and various tissues and determine the passage of drug molecules or other compounds into these tissues.[5–9] Furthermore, considering the impact of these transporters on drug disposition, their wide substrate spectrum, and their potential saturability, adverse drug interactions due to competitive inhibition or induction of these efflux transporters by coadministered drugs, ingested food, or environmental compounds could be expected, and this has been proven in a number of *in vivo* animal or clinical studies.[10–14] On the other hand, these transport interactions may also result in beneficial interactions and improve the therapeutic efficacy of a particular drug of interest. For example, the low bioavailability of some anticancer agents could be improved by inhibiting intestinal P-glycoprotein or other efflux transporters.[15–17] Lastly, it has been shown that the expression of these efflux transporters varies substantially among individuals, and this variability could be due to age and gender differences, genetic polymorphism, or prior exposure to drugs, food, and environmental compounds.[13,18–22] The impact of this variability in the expression of these transporters on drug pharmacokinetics remains the topic of extensive investigation, and the results of these studies will have significant impact on future therapy. To appreciate the importance of the efflux transporters in drug therapy, an understanding of the molecular and functional characteristics of these transporters and their tissue distribution, as well as an appreciation of their impact on drug disposition, is essential. This is the focus of the present overview.

18.2. P-GLYCOPROTEIN

P-glycoprotein is a membrane efflux transporter protein discovered by Juliano and Ling in 1976 in the plasma membrane of Chinese hamster ovary cells selected for resistance to colchicine.[23] These cells also displayed cross-resistance to a wide range of structurally and functionally unrelated drugs, a phenomenon nowadays known as multidrug resistance (MDR). The consistent observation of this membrane protein in several MDR cell lines selected with different drugs[23–26] and the positive correlation found between the level of P-glycoprotein expression and drug resistance in a variety of MDR cell lines[27,28] strongly suggested that P-glycoprotein may play a key role in mediating MDR. This was subsequently confirmed by studies,[29,30] in which transfection of P-glycoprotein cDNA was shown to confer the MDR phenotype upon otherwise drug-sensitive cells. The mechanism by which P-glycoprotein mediates MDR is believed to be that P-glycoprotein functions as an ATP-dependent efflux pump, actively extruding a wide range of cytotoxic agents, such as anthracyclines, vinca alkaloids, epipodophyllotoxins, and taxol, from inside the cell to the extracellular space, resulting in inadequate intracellular accumulation of these agents for efficient cell killing.[1,31–34] It is well established that P-glycoprotein overexpression is one of the major mechanisms responsible for the development of MDR.[2,35] The clinical relevance of this MDR mechanism was substantiated by the findings that P-glycoprotein was often detected in numerous resistant human tumors and that the expression of this protein represents a poor prognosis factor.[36–44]

The genes encoding P-glycoprotein have been cloned and belong to a small family of closely related genes designated *mdr*. The family consists of two members (MDR1 and MDR3) in humans and three members (*mdr1a, mdr1b, and mdr2*) in rodents.[45–48] Despite the high homology between different members of the family, only human MDR1 and its mouse homolog *mdr1a* and *mdr1b* protein can confer MDR and drug transport capabilities, while human MDR3 and its mouse homolog, *mdr2* protein, apparently cannot.[29,30,47,49–54] The latter was shown to be more concentrated in the liver canalicular membranes and functions as a phosphatidylcholine translocase or flippase.[55–58] Human P-glycoprotein has 1280 amino acids, and the polypeptide component of the protein has a molecular weight of 120 to 140 kDa.[45] The apparent molecular weight of P-glycoprotein, however, could vary between 130 and 190 kDa, depending on the level of glycosylation. The molecular structure of the protein was predicted to consist of two homologous halves, each consisting of six transmembrane domains, and a hydrophilic nucleotide binding domain with Walker A, Walker B and ABC signature sequences, characteristic of ABC proteins (Figure 18.1). The nucleotide binding sites are located intracellularly and exhibit ATPase activity, which hydrolyzes ATP and provides the energy for the pumping function of the protein.[59,60]

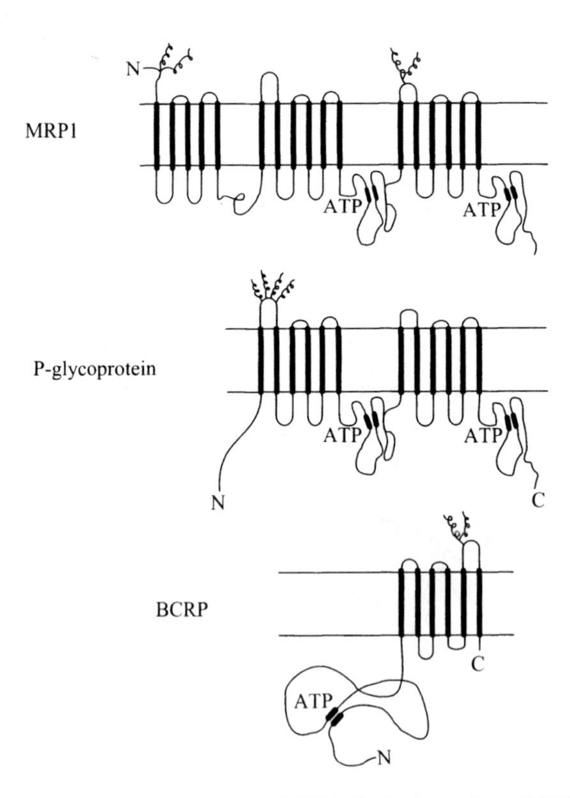

Figure 18.1. Topology of efflux transporters MRP1, P-glycoprotein and BCRP (Reproduced with permission from ref. 3).

TABLE 18.2 Common Substrates and Inhibitors for the Efflux Transporters

Transporter	Substrates	Inhibitors
MDR1	Anthracyclines, vinca alkaloids, epipodophyllotoxins, paclitaxel, topotecan,* mitoxantrone,* HIV protease inhibitors, digoxin, Rhodamine123, methotrexate	Verapamil, diltiazem, trifuoperazine, quinidine, reserpine, cyclosporin A, valinomycin, terfenidine, PSC833, VX710*, PAK-104P, GF120918, LY35979*, XR9576*
MDR3	Phosphatidylcholine, digoxin (?), paclitaxel (?), vinblastine (?)	verapamil$^+$, cyclosporin A$^+$, and PSC833$^+$
MRP1	Aflatoxin B1, doxorubicin, etoposide, vincristine, methotrexate, and various lipophilic GSH, glucuronide, and sulfate conjugates	MK571, cyclosporin A, VX710*, PA-104P*
MRP2	GSH conjugates, glucuronides, sulfate conjugates, methotrexate, temocaprilat, CPT11 carboxylate, SN38 carboxylate, cisplatin, pravastatin	MK571, cyclosporin A
MRP3	GSH conjugates, glucuronides, sulfate conjugates, methotrexate, monoanionic bile acids (taurocholate, glycocholate), vincristine, etoposide	MK571 (?)
BCRP	Anthracyclines, epipoxophyllotoxins, camptothecins or their active metabolites, mitoxantrone, bisantrene, methotrexate, flavopiridol, zidovudine, lamivudine	FTC, GF120918, Ko-134
BSEP	Bile salts	

Source: Data from Litman et al.[3] * or Taipalensuu et al.[263]†

One of the distinctive features of P-glycoprotein from conventional drug transporters is its broad spectrum of substrate specificity (Table 18.2). These substrates include anticancer agents (e.g., anthracyclines, vinca alkaloids, epipodophylotoxins, and taxol),[2] cardiac drugs (e.g., digoxin, quinidine),[61,62] human immunodeficiency virus (HIV) protease inhibitors (e.g., saquinavir, indinavir, ritonavir),[63] immunosupressants (e.g., cyclosporine),[64] antibiotics (e.g., actinomycin D),[65] steroids (e.g., cortisol, aldosterone, dexamethasone),[66,67] and cytokines [e.g., interleukin-2 (IL-2), IL-4, interferon-4 (IFN-4)].[68] The list of P-glycoprotein substrates could be expanded to include many more compounds. The only common characteristics of these substrates are that most of these compounds are hydrophobic, positively charged or neutral compounds with planar structure[2,69]; however, negatively charged compounds, such as methotrexate and phenytoin, can also serve as substrates under certain circumstances.[70–72] How P-glycoprotein recognizes such a wide range of structurally unrelated chemical entities still remains an enigma, but could be partly due to the multiple drug binding sites present in the transmembrane domains of the protein.[73–76] The proposed mechanism by which P-glycoprotein performs its transport function is the "hydrophobic vacuum cleaner" model or the "flippase" model.[2,77,78] In the hydrophobic vacuum cleaner model, P-glycoprotein binds directly to its substrates within the plasma membrane and pump them out of the cells.[2] In the flippase model, the binding of substrates takes place in the inner leaflet of the plasma membrane bilayer and the substrates are flipped by P-glycoprotein to the outer leaflet, from which they diffuse into the extracellular space.[77,78] In either case, the substrates are removed directly from the cell membrane by P-glycoprotein before their entry into the cytoplasmic solution. The high local concentrations of the hydrophobic compounds in the lipid membrane may facilitate the transport by P-glycoprotein even in the absence of high-affinity binding, and this may also help to explain such a diverse substrate spectrum.[79]

A wide range of P-glycoprotein inhibitors, which are as chemically diverse as the substrates, has also been identified. These inhibitors include calcium channel blockers (e.g., verapamil, diltiazem),[80] calmodulin antagonists (e.g., trifluoperazine, fluphenazine),[81,82] steroidal compounds (e.g., progesterone, tamoxifen),[83,84] immunosuppressive agents (e.g., cyclosporin A, FK506),[85,86] antibiotics (e.g., cefoperazone, erythromycin),[87,88] and nonionic detergents (e.g., Triton-X100, Nonidet P-40).[89] Interestingly, a number of pharmaceutical excipients such as cremophor EL, Tween 80, and polyethylene glycols were also shown to inhibit P-glycoprotein.[90,91] More recently, the list of these inhibitors has been extended to include many dietary compounds in a variety of natural products, such as flavonoids,[92–95] curcumin,[96] and piperine.[97] Many of these inhibitors have undergone clinical testing for their ability to restore tumor responsiveness to chemotherapeutic agents by blocking P-glycoprotein; however, the toxicities associated with the high concentrations of these inhibitors required for significant P-glycoprotein inhibition have prevented their clinical use. The newly developed second and third generations of P-glycoprotein inhibitors such as PSC833,[98] GF120918,[99] LY335979,[100] and XR9576[101] have very high potency and low toxicity, and clinical trials using these agents as chemosensitizers have produced some promising results.[102–105]

The expression of P-glycoprotein is not limited to MDR tumor cells. High levels of expression also have been detected in a number of normal tissues, such as the liver, kidney, gastrointestinal tract, and the blood-brain and blood-testis barriers, as well as the adrenal glands.[106–109] At the subcellular level, P-glycoprotein has been shown to be predominantly located on the apical surface of the epithelial (or endothelial) cells with a specific barrier function, such as the endothelial cells of the blood capillaries in the brain, the canalicular membranes of the hepatocytes, the brush border membranes of renal proximal tubules, and the luminal membrane of the enterocytes in the colon and jejunum.[106,108,109] The polarized expression of this protein in the excretory organs (liver, kidney, and intestine) and blood-tissue barriers, together with its ability to transport a wide diversity of chemicals, indicates that the protein may play an important role in protecting the body or certain tissues (such as brain and testis) from the insult of ingested toxins and toxic metabolites by actively excreting these toxic agents into bile, urine, and intestine or by restricting their entry into the brain and other pharmacological sanctuaries. P-glycoprotein was also found in placental trophoblasts from the first trimester of pregnancy to full term, indicating that it may be also involved in the protection of the developing fetus.[2]

The role of P-glycoprotein in manipulating excretion and distribution of xenobiotics was initially supported by a number of *in vivo* animal or clinical studies using a combination of P-glycoprotein substrate drugs and inhibitors, in which reduced elimination and increased tissue accumulation of the substrate drugs by the coadministered inhibitors were often observed.[110–112] However, due to the possible interactions between these inhibitors and other drug transporters or drug-metabolizing enzymes and the fact that the inhibitors used in these early studies were relatively nonspecific, other interpretations could not be excluded. The most convincing evidence is from a series of elegant studies conducted by Schinkel et al.[62,113,114] using knockout mice. Both *mdr1a* (-/-) and *mdr1a/1b* (-/-) knockout mice have been created by disruption of *mdr1a* or both *mdr1a* and *mdr1b* genes. These knockout mice appeared to be viable, healthy, and fertile, with normal histological, hematological, and immunological parameters, indicating that mdr1-type P-glycoprotein may not be essential for basic physiology.[113,114] However, the mice lacking mdr1-type P-glycoprotein did show hypersensitivity to xenobiotic toxins. For example, the *mdr1a* (-/-) mice were 50–100-fold more sensitive to ivermectin, an acaricide and anthelmintic drug, compared to the wild-type mice, and this increased toxicity could be explained by the 90-fold increase in the brain accumulation of ivermectin in the knockout mice, since the toxicity of ivermectin results from its interaction with a neurotransmitter system in the central nervous system (CNS).[113] Another interesting example is related to the antidiarrheal drug loperamide, a P-glycoprotein substrate. Although loperamide is a typical opioid drug, in humans and animals this drug demonstrates peripheral opiate-like effects only on the gastrointestinal tract, with little effect in the CNS due to its inability to pass through the blood-brain barrier. After oral administration of loperamide, the *mdr1a* (-/-) mice demonstrated markedly increased CNS opiate-like effects compared with the wild-type mice, consistent with a dramatic increase in the brain accumulation of this drug in the knockout mice (13-fold; $p < .001$).[115] Interestingly, CNS effects of loperamide in humans were also observed when it was co-administered with quinidine, a competitive inhibitor of P-glycoprotein.[10] An increased brain accumulation of many other P-glycoprotein substrate drugs such as vinblastine, cyclosporine, and digoxin have also been observed in the *mdr1a* (-/-) or *mdr1a/1b* (-/-) mice.[62,113–115] Taken together, these data clearly indicate that mdr1-type P-glycoprotein plays a very important role in regulating the entry of xenobiotics or endogenous compounds into the brain. In addition to the marked alterations in the brain accumulation of these P-glycoprotein substrates in the knockout mice, the blood concentrations and the accumulation of these substrates in other tissues such as the liver, heart, and intestine were also shown to be significantly elevated, albeit to a lesser extent, indicating a diminished elimination of these compounds in the knockout mice.[62,113–117] The high level of P-glycoprotein found in the excretory organs in the body and the diminished elimination of P-glycoprotein substrates observed in P-glycoprotein-deficient mice point to an important role of the protein in the elimination of xenobiotics by these excretory routes.

Xenobiotics can be eliminated from the body by fecal excretion if they are poorly absorbed after oral administration or following secretion into the intestinal lumen. The polarized expression of P-glycoprotein on the apical membrane of the enterocytes lining the intestinal wall[106] suggests that this efflux transporter is involved in the active secretion of P-glycoprotein substrates into the intestinal lumen and thus facilitates their fecal excretion. In addition, the P-glycoprotein-mediated active efflux of its substrates from the intestinal epithelial cells back to the lumen limits the absorption/bioavailability of orally dosed drugs or other compounds that are P-glycoprotein substrates. Significant P-glycoprotein-mediated effects on intestinal secretion and absorption/bioavailability have been observed in a number of studies. In mice, mdr1a P-glycoprotein is the major isoform expressed in the intestine and brain.[48,113] The area under the plasma concentration-time curve (AUC) of paclitaxel, a known P-glycoprotein substrate, has been shown to be two- and sixfold higher in *mdr1a* (-/-) knockout mice than in the wild-type mice after IV and oral adiministration, respectively. The cumulative intestinal secretion of paclitaxel (0–96 hours) was dramatically decreased from 40% in the wild-type animal to <3% in the knockouts after IV dosing and the bioavailability of paclitaxel increased from 11% in the wild-type mice to 35% in the knockouts after oral dosing (10 mg/kg).[118] Similar results have been obtained for a number of other P-glycoprotein substrates, such as digoxin, grepafloxacin, vinblastine, and HIV protease inhibitors.[119–123] For example, the direct intestinal secretion of ^{3}H-digoxin was only 2% of the dose in *mdr1a* (-/-) mice, in contrast to 16% in the wild-type animals.[119] Collectively, the results obtained from these knockout animal studies provide convincing evidence of the important contribution of P-glycoprotein to intestinal secretion and absorption of substrate compounds. The clinical relevance of these observations in the animal studies has been demonstrated in several human studies. For example, the intestinal secretion of talinolol, a β_1-adrenergic receptor blocker, was shown to be against a concentration gradient (5.5 (lumen): 1 (blood)), after its IV administration, indicating the involvement of an active process. In addition, the secretion rate of talinolol in the presence of a simultaneous intraluminal perfusion of R-verapamil, a known P-glycoprotein inhibitor, dropped to 29–59% of the values obtained in the absence of R-verapamil.[124] Similar results have also been obtained for digoxin.[125] Furthermore, the intestinal secretion of talinolol was increased significantly in human subjects treated with rifampin, and the increased secretion can be attributed to the 4.2-fold increase in the intestinal P-glycoprotein expression induced by treatment with rifampin.[13] The oral bioavailability of P-glycoprotein substrates in humans was also shown to be at least partly limited by intestinal P-glycoprotein,[20,126–131] and coadministration of P-glycoprotein inhibitors or competitive substrates could increase the bioavailability of these substrates.[11,12]

Biliary excretion represents another important route for the elimination of drugs and other xenobiotics. Following the uptake of xenobiotics into the hepatocytes, compounds may undergo metabolic modification, or the parent compound, as well as the formed metabolites, may be excreted into bile through the canalicular membrane or effluxed back across the sinusoidal membrane into blood. The relatively small surface area of the canalicular membrane (10–15% of the hepatocyte surface area), in contrast to the sinusoidal membrane (at least 70%), and the small intracanalicular fluid volume suggest that carrier-mediated transport may contribute significantly to the biliary excretion of both endogenous and exogenous compounds.[7,132] Indeed, many active transporters have been identified in the canalicular membrane to mediate this process,[132–134] including P-glycoprotein and MRP2.[106,132] The contribution of P-glycoprotein to biliary secretion has been demonstrated by several investigations. For example, the biliary excretion of unchanged doxorubicin decreased from 13.3% of the dose in wild-type mice to only 2.4% in *mdr1a* (-/-) knockout mice after a 5 mg/kg IV dose.[121] Similar results have been obtained for a number of amphiphilic model substrates, which exhibited markedly reduced biliary excretion in both *mdr1a* (-/-) and *mdr1a/1b* (-/-) knockout mice compared to normal mice.[117,123] Studies using P-glycoprotein inhibitors also provided results consistent with the important contribution of P-glycoprotein to biliary excretion. In an isolated perfused rat liver study, erythromycin significantly decreased the biliary excretion of fexofenadine, a P-glycoprotein substrate.[135] Cyclosporin A and its analog, PSC833, have been reported to decrease the biliary excretion of both colchicine and doxorubicin[136,137] *in vivo*. Similar results were observed for doxorubicin and grepafloxacin when the competitive substrates erythromycin (for both doxorubicin and grepafloxacin) and cyclosporin (for grepafloxacin) were administered simultaneously.[138,139] In addition, the biliary excretion of P-glycoprotein substrates was shown to depend on the expression level of this protein, and a significant increase in the biliary excretion of vinblastine was observed in rats with increased levels of P-glycoprotein, which was induced by 2-acetylaminofluorene and phenothiazine, respectively, in two independent studies.[140,141] These data suggest that P-glycoprotein plays an important role in biliary excretion. However, other studies have failed to find significant effects on P-glycoprotein-mediated biliary excretion in knockout mice. For example, while the intestinal secretion and bioavailability of paclitaxel were markedly altered in *mdr1a* (-/-) knockout mice, the biliary excretion of this model substrate in the knockout mice was not significantly different from that in the wild-type animals.[118] Even in the *mdr1a/1b* (-/-) double knockouts, the biliary excretion of both digoxin and vinblastine was not substantially changed.[114] One possible explanation of these conflicting results is the presence of alternative transport processes responsible for the secretion of these substrates into bile. P-glycoprotein may act in concert with other transporters in excreting certain substrates into bile, and the loss of P-glycoprotein function could be compensated for by other transport processes under certain circumstances. Indeed, it has been shown that *mdr1b* expression in the liver and kidney was consistently increased in *mdr1a* (-/-) knockout mice compared to the wild-type animals, indicating that the loss of *mdr1a* function could be compensated for by mdr1b protein for their common substrates.[113] Other canalicular membrane transporters may also exhibit overlapping substrate specificity for certain P-glycoprotein substrates.

Renal clearance represents an important route for the elimination of a large number of xenobiotic compounds. This dynamic process includes glomerular filtration, renal tubular secretion, and tubular reabsorption. Renal secretion usually takes place against a concentration gradient and thus is mainly an active process involving a variety of transporter mechanisms.[142] In addition to the two major carrier systems responsible for the renal handling of organic cations and organic anions, several ATP-dependent transporters, including P-glycoprotein and MDR-associated proteins, have been detected in the kidney.[142] The transport function and the localization of P-glycoprotein on the apical membrane of the proximal tubule cells[106] suggest the involvement of this protein in the renal secretion of its substrates into urine. The observation that a classic P-glycoprotein inhibitor, cyclosporin, decreased colchicine renal clearance after IV administration from 6.23 ± 0.46 to 3.58 ± 0.31 ml/(min·kg) (mean ± SD; $p < .05$) without affecting glomerular filtration and the secretion of the organic cation ranitidine or the organic anion *p*-aminohippurate, provided the first *in vivo* demonstration for this functional role of P-glycoprotein.[143] Subsequently, a significant reduction of the renal secretion of digoxin (in rats), vinblastine, and vincristine (in dogs) by cyclosporin A was also observed by using the isolated perfused rat kidney or the single-pass multiple indicator dilution method.[144,145] In humans, the renal clearance of digoxin was decreased 20% by the concomitant use of itraconazole ($p < .01$). Since digoxin is mainly excreted unchanged into urine, this reduction is most likely mediated by the inhibition of P-glycoprotein.[146] Similarly, the renal clearance of quinidine was also decreased by 50% ($p < .001$) by itraconazole in a double-blind, randomized, two-phase crossover study, and inhibition of P-glycoprotein is thought to be the most likely underlying mechanism.[147] Taken together, these studies demonstrated that P-glycoprotein significantly contributes to the renal excretion of its substrates.

18.3. MULTIDRUG RESISTANCE–ASSOCIATED PROTEINS (MRPs)

The family of MRPs is another group of ABC transporters, so far consisting of nine members. Among these members, MRP1, MRP2, and MRP3 have been characterized in some detail in terms of their capability of conferring MDR and their possible physiological functions[148] and so will be the focus of this discussion. The founding member of this family, MRP1, was cloned in 1992 from the resistant human small cell lung cancer cell line,[149] which does not overexpress P-glycoprotein.[150–153] Subsequent transfection studies demonstrated that overexpression of this 190 kDa membrane protein can confer MDR on a number of natural product anticancer agents such as the anthracyclines, vinca alkaloids, and epipodophyllotoxins by causing the active efflux of these cytotoxic agents from cells and thus lowering their intracellular concentrations.[154–157] Later, MRP2 (cMOAT) and other members were also identified and characterized to varying extents.[158–167] Among these MRPs, MRP3 is the member most closely related to MRP1, with 58% amino acid identity, followed by MRP2 (49%).[168] These three MRPs have similar topology, containing a typical ABC core structure of two segments, each consisting of six transmembrane domains and an ATP binding domain, similar to P-glycoprotein, and an extra N-terminal segment of five transmembrane domains linked to the core structure through an intracellular loop[148] (Figure 18.1). Similar to MRP1, both MRP2 and MRP3 have also been shown to be able to confer MDR on several anticancer drugs.[169–172] The clinical relevance for MRP1-mediated MDR has been a topic of extensive investigation, and there is some evidence suggesting that overexpression of MRP1 might represent a poor prognostic factor.[173–181] The clinical relevance of MRP2- and MRP3-mediated MDR is currently unknown.

In contrast to P-glycoprotein, which mainly transports large, hydrophobic cationic compounds, MRP1 mainly transports amphiphilic anions, preferentially lipophilic compounds conjugated with glutathione (e.g., leukotrene C4, DNP-SG), glucuronate (e.g., bilirubin, 17β-estradiol), or sulfate[5] (Table 18.2). Some unconjugated amphiphilic anions such as methotrexate and Fluo-3, a penta-anionic fluorescent dye, can also serve as substrates, and they are transported in unchanged form.[182,183] In addition to the anionic compounds, MRP1 accepts amphiphilic cations or neutral compounds, such as anthracyclines, etoposide, and vinca alkaloids, as its substrates. But paclitaxel, a good P-glycoprotein substrate, appears not to be transported.[154–157] These cationic or neutral substrates are thought to be transported intact but need reduced glutathione (GSH) as a cotransporting factor.[184–187] As such, depletion of intracellular GSH by buthionine sulphoximine (BSO), an inhibitor of GSH synthesis, can increase the intracellular accumulation of these substrates in MRP1- overexpressing cells.[188,189] Both MRP2 and MRP3 share a similar substrate spectrum with MRP1. They also transport conjugates of lipophilic substances with GSH, glucuronate, and sulfate such as glutathione S-conjugate leukotriene C4, glucuronosyl bilirubin, and the anticancer agents methotrexate, vincristine, and etoposide.[5] In transporting cationic substrates, MRP2 and MRP3 seem to function by the same mechanism as MRP1 and need GSH as a cosubstrate.[187] However, the substrate specificity of these three MRP isoforms is not identical, and for their common substrates, the transporting efficiency of these isoforms varies substantially[5]; there are substrates that can be recognized by one isoform but not by others. For example, cisplatin has been shown to be a substrate for MRP2 but not for MRP1[154,190,191]; the conjugated monoanionic bile acids glycocholate and taurocholate are substrates for MRP3 but not for MRP1 and MRP2.[192–194] In contrast to P-glycoprotein, for which many inhibitors have been identified, there are only a few compounds known to inhibit MRP to a significant degree. Well-known potent P-glycoprotein inhibitors such as GF120918 and LY335979 have little effect on MRP, while verapamil, cyclosporin A, and PSC833 have been shown to be, at best, moderate MRP inhibitors.[5,195,196] The best-known MRP inhibitor so far appears to be MK571, a leukotriene D4 receptor antagonist. MK571 inhibits both MRP1 and MRP2, but a mild stimulatory effect on MRP3-mediated transport of 17β-estradiol glucuronide has been reported.[159,197,198]

To understand the physiological function of these energy-dependent MRP efflux transporters, their normal tissue distribution has been extensively investigated. While MRP1 appears to be distributed in a wide range of tissues throughout the body, MRP2 and MRP3 have been detected mainly in the gut, liver, and kidney.[6,148] At the subcellular level, MRP1 is predominantly located in the cell plasma membrane, and in polarized epithelial cells such as hepatocytes, enterocytes, and endothelial cells, where its distribution is confined to the basolateral membranes.[6,148,199] The active transport function of MRP1 toward a number of exogenous and endogenous toxic substrates, and its ubiquitous tissue distribution indicates that MRP1 may represent a detoxifying mechanism, protecting some tissues or organs from exposure to toxic substances.[168] Recent studies using *mrp1* (-/-) knockout mice have provided convincing evidence for this important function. It has been shown that mice with disrupted Mrp1 (*Mrp1* (-/-)) are viable, fertile, and have no physiological or histological abnormalities, indicating that Mrp1 may not be essential for normal mouse physiology. However, these *mrp1* (-/-) mice did show a two-fold higher sensitivity to a cytotoxic agent, etoposide, with increased bone marrow toxicity.[200,201] A similar observation has also been made by Johnson et al.,[202] who demonstrated that a therapeutic dose of vincristine, which normally does not express bone marrow toxicity and gastrointestinal damage, caused extensive damage to these tissues in both *mrp1 (-/-)* and *mdr1a/1b* (-/-) knockout mice, indicating that Mrp1, mdr1-type P-glycoprotein, and probably other related efflux transporters work in concert as a detoxifying mechanism to protect tissue from damage induced by toxic agents. In addition, the polarized localization of MRP1 in the basolateral membrane of the choroid plexus epithelium[203] suggests that it may significantly contribute to blood-CSF (cerebrospinal fluid) barrier function, preventing the entry of ampiphilic anions or anticancer drug substrates into CSF. This has also been convincingly demonstrated in a knockout mouse study conducted by Wijnholds et al.,[204] in which the investigators found that after an IV dose of etoposide, the CSF concentration was about 10-fold higher in *mdr1a/mdr1b/mrp1* (-/-/-) triple knockout mice than in *mdr1a/mdr1b* (-/-) double knockout mice, indicating the important contribution of Mrp1 to the blood-CSF barrier function in mice. Taken together, there is strong evidence indicating that MRP1 plays an important role in protecting tissues from damage induced by both exogenous and endogenous toxic substances, and contributes significantly to maintaining the blood-CSF barrier function.

Unlike MRP1, the distribution of MRP2 and MRP3 is restricted to certain tissues such as liver, intestine, and kidney.[6,148] Similar to P-glycoprotein, MRP2 is exclusively localized to the apical membrane of the polarized cells such as hepatocytes, intestinal epithelial cells, and renal proximal tubule cells,[5,159,205] suggesting that it may also play a similar role in the secretion of xenobiotics and endobiotics by these excretory routes. The loss of MRP2 in humans is associated with Dubin-Johnson syndrome, a benign hereditary disorder characterized by mild conjugated hyperbilirubinemia and pigment disposition in the liver due to impairment in MRP2-mediated transport function.[206–209] Two naturally occurring mutants, GY/TR$^-$ and EHBR rats from the Wistar and Sprague-Dawley rat colonies, respectively, also lack Mrp2 expression and are considered animal models for the human Dubin-Johnson syndrome.[158,160,210,211] Many functional characterization and substrate identification studies for MRP2 have been performed using these Mrp2-

deficient rats. It has been shown that the AUC (0–6 hours) of ^{14}C-temocapril was dramatically increased and the biliary clearance, as measured by total radioactivity, was markedly decreased (0.25 ml/min/kg vs. 5.00 ml/min/kg) in EHBR rats compared with the control Sprague Dawley rats after IV administration. Since the active metabolite temocaprilat accounted for >95% of the total radioactivity, these data indicate that Mrp2 plays a central role in the biliary excretion of the metabolites of this drug.[212] The biliary excretion of grepafloxacin was also markedly decreased for both the parent compound (0.52 vs. 1.79 ml/min/kg) and glucuronide metabolites (0.09 vs. 15.53 ml/min/kg) in EHBR rats compared with the Sprague Dawley rats.[213] Recently, Chen et al.[214] reported that the biliary excretion of methotrexate and probenecid was decreased 39- and 37-fold, respectively, in EHBR rats compared to control rats. Similar results were also observed for several other drugs or metabolites such as cefodizime, acetaminophen glucuronide, acetaminophen glutathione conjugate and acetaminophen mercapturate, pravastatin, and indomethacin glucuronide.[215–219] Interestingly, it was shown that the biliary excretion of CPT11, the active metabolite SN-38, and its glucuronide conjugate can be substantially decreased by probenecid, an MRP2 inhibitor, with concomitant elevation of plasma concentrations of these compounds in normal rats, resulting in decreased gastrointestinal toxicity.[220] Collectively, these data strongly suggest the essential role of MRP2 in the biliary excretion of xenobiotics or their metabolites that are MRP2 substrates. The polarized localization of MRP2 in intestinal epithelial cells also suggests its potential contribution to intestinal secretion and to limiting the intestinal absorption of its substrates, leading to decreased bioavailability. This hypothesis has been supported by the results of a number of studies. For example, after IV administration of CDNB (1-chloro-2,4-dinitrobenzene), the intestinal secretion of DNP-SG (2,4-dinitrophenyl-*S*-glutathione) was negligible in EHBR rats, whereas a small amount of secretion was observed in Sprague Dawley rats, indicating the involvement of MRP2 in the active secretion of DNP-SG into intestinal lumen. This was also confirmed by Ussing chamber studies, in which the serosal-to-mucosal flux of DNP-SG was shown to be 1.5-fold higher than the mucosal-to-serosal flux in Sprague Dawley rats, and no difference in the flux in both directions was observed in EHBR rats.[221] The decreased intestinal secretion of grepafloxacin in EHBR rats was observed in a study by Naruhashi et al.,[222] which was also confirmed by the twofold higher flux in the serosal-to-mucosal direction compared with that in the mucosal-to-serosal direction in the Sprague Dawley rats and no differences in the EHBR rats. By analogy with P-glycoprotein, the impact of MRP2-mediated efflux of its substrates from the enterocytes into the lumen can be illustrated by the twofold higher absorption of PhIP (*p*-2-amino-1-methyl-6-phenylimidazo-[4,5-b]pyridine), a food-derived carcinogen, in Mrp2-deficient rats compared with the normal rats, and the increased bioavailability of PhIP in normal rats treated with BSO, an inhibitor of GSH synthesis.[223,224] All these studies provide convincing evidence for the important contribution of MRP2 to the intestinal secretion and absorption of drugs. Whether MRP2 is also present in the brain capillary endothelial cells may still need further investigation, but current evidence suggests that it most likely is.[225] Mouse Mrp2 was detected on the luminal surface of the brain capillary endothelium[226] and was shown to actively transport sulforhodamine 101 and fluorescein methotrexate into the luminal compartment of isolated brain capillary.[226] This transport process can be inhibited by leukotriene C_4, 1-chloro 2,4-dinitrobenzene (a precursor of DNP-SG), and vanadate (an ATPase inhibitor), but not by P-glycoprotein inhibitors such as PSC833 and verapamil. Therefore, the evidence suggests that human MRP2 may also contribute to the blood-brain barrier function in a manner similar to that of P-glycoprotein. MRP3 has a tissue distribution similar to that of MRP2 but is located on the basolateral surface of the polarized cells.[171,199,227]

The expression of MRP3 in the basolateral membrane of intestinal epithelial cells, hepatocytes, and renal proximal tubule cells suggests that it tends to remove the substrates from the cytosol into blood. The impact of this process on drug disposition still remains to be clarified, especially considering its limited expression in excretory organs under normal physiological conditions. Interestingly, it has been shown that MRP3 is significantly up-regulated in the liver of MRP2-deficient rats and in patients with Dubin-Johnson syndrome or in patients with primary biliary cirrhosis,[227,228] indicating that MRP3 may serve as a compensatory mechanism to remove the conjugates from the hepatocytes through the sinusoidal membrane under the condition in which MRP2-mediated biliary excretion is impaired.[5]

18.4. BREAST CANCER RESISTANCE PROTEIN (BCRP)

BCRP is a new member of the ABC transporter superfamily initially cloned from a doxorubicin-resistant breast cancer cell line (MCF-7/AdrVp) selected with a combination of adriamycin and verapamil.[229] Two other groups also independently identified this transporter from human placenta[230] and human colon carcinoma cells (S1-M1-80),[231] and named the protein ABCP (ABC transporter in placenta) and MXR (mitoxantrone resistance-associated protein), respectively. Molecular characterization revealed that BCRP consists of 655 amino acids with a molecular weight of 72.1 kDa. In contrast to P-glycoprotein and MRP1 or MRP2, which contain a typical core structure of 12 transmembrane domains and 2 ATP binding sites, BCRP only has 6 transmembrane domains and 1 ATP binding site (Figure 18.1), and therefore appears to be a half ABC transporter.[230] BCRP is the second member of the ABCG subfamily, containing members such as *drosophila white, brown*, and *scarlet* genes, and thus, "ABCG2" was recommended by the Human Genome Nomenclature Committee (HUGO) to refer to this newly identified transporter.[3] As a half transporter, BCRP most likely forms a homodimer to transport its substrates out of the cells, utilizing the energy derived from ATP hydrolysis.[3,229,232–234] The murine homolog of BCRP, Bcrp1, has also been cloned and shown to be highly similar (81%) to BCRP, with a virtually superimposable hydrophobicity profile.[235] In addition, another gene closely related to *Bcrp1* has been identified in mice and named *Bcrp2*, which shares 54% identity with *Bcrp1*.[236] Whether *Bcrp1* forms a heterodimer with *Bcrp2* to perform its transport function remains unknown; however, the different expression patterns of these two genes indicate that *Bcrp2* is not a necessary component for the transport function of *Bcrp1*. Distinct from other half transporters such as TAP1 and TAP2 (the transporters associated with antigen presentation), which are localized in the intracellular membranes,[237] both human BCRP and murine *Bcrp1* were shown to be present mainly in the plasma membrane.[16,238,239] Similar to P-glycoprotein and MRP1, both BCRP and *Bcrp1* can be overexpressed *in vitro* upon drug selection or by transfection of cDNAs encoding these proteins, and they confer multidrug resistance by the energy-dependent efflux of its substrates out of cells.[229,232,233,235,240,241] Significant and variable expressions of BCRP have been detected in human tumors such as acute leukemia and breast cancer; however, the contribution of this efflux transporter to clinical MDR needs to be further investigated.[242–248]

There is considerable overlap in the substrate specificity among P-glycoprotein, MRP1 or MRP2, and BCRP, although the binding affinity of a particular substrate to these transporters may vary substantially.[3] The BCRP/*Bcrp1* substrates identified so far include a number of anticancer agents such as anthracyclines (e.g., doxorubicin, daunorubicin, epirubicin), epipoxophyllotoxins (e.g., etoposide, teniposide), camptothecins or their active metabolites (e.g., topotecan, SN-38, 9-aminocamptothecin, CPT11), mitoxantrone, bisantrene, methotrexate, flavopiridol, and HIV-1 nucleoside reverse transcriptase inhibitors (e.g., zidovudine, lamivudine); vincristine, paclitaxel, and cisplatin appear not to be substrates.[8,229,235,249–253] The amino acid at the 482 position seems to be critical in defining substrate specificity because mutated forms of BCRP with arginine at the 482 position changed to threonine or glycine have shown different substrate preferences.[251] Whether these mutated forms of BCRP also occur *in vivo*, especially in normal physiological situations, is currently unknown. However, a similar phenomenon observed in mouse cell lines selected with doxorubicin indicates that the 482 position appears to be a hot mutation spot; thus, similar mutations might also occur in human tumors upon drug treatment.[254] For investigating the pharmacological and physiological functions of BCRP and for MDR reversal, substantial efforts have been made to search for and develop potent BCRP/*Bcrp1* inhibitors. Fumitremorgin C (FTC) derived from *Aspergillus fumigatus* cultures appears to be the first identified potent and specific inhibitor for BCRP/*Bcrp1*[255,256]; however, its *in vivo* application is limited by its neurotoxicity. The typical P-glycoprotein inhibitors, GF120918 and reserpine, were also shown to be potent BCRP inhibitors,[235,257] but many of the other P-glycoprotein inhibitors such as LY335979, cyclosporin A, PSC833, and verapamil have little effect.[8,258] So far, the most potent specific BCRP inhibitor appears to be Ko134, an analog of FTC.[259] The compound has been used *in vivo* and has demonstrated little or low toxicity in mice at high oral or intraperitoneal doses and could potentially be used *in vivo* for BCRP inhibition.[259]

Interestingly, the distribution of BCRP in normal tissues is similar to that of P-glycoprotein. High levels of BCRP expression were detected in the human placenta syncytiotrophoblast plasma membrane, facing the maternal bloodstream, in the canalicular membrane of the liver hepatocytes, the apical membrane of the epithelium in the small and large intestines, in the ducts and lobules of the breast, and in the luminal surface of brain capillaries.[260,261] In addition, significant amounts of

BCRP were also found in venous and capillary, but not arterial, endothelial cells in almost all tissues investigated.[261] By analogy with P-glycoprotein, it is reasonable to speculate that one, if not the major, physiological function of BCRP is to protect the body or certain tissues from the exposure to toxic endogenous or exogenous compounds. The localization of BCRP in the placenta, brain, and testis may regulate the entry of its substrates into the developing fetus, brain, and other pharmacological sanctuaries, and therefore represents an important component of the blood-placenta, blood-brain, and blood-testis barriers. The expression of BCRP in the luminal side of the intestinal epithelial cells and canalicular membrane of the hepatocytes suggests that the protein may play a significant role in intestinal secretion or back efflux to the intestinal lumen and in biliary excretion, thus limiting the entry of xenobiotic toxins into the systemic circulation or facilitating their elimination. A recent study conducted by Jonker et al.,[262] using *Bcrp1* (-/-) knockout mice, strongly supports this speculation. In the study, the authors demonstrated that the oral bioavailability of topotecan increased about sixfold in *Bcrp1* (-/-) mice compared with wild-type mice and the accumulation of topotecan in Bcrp1 (-/-) fetuses was elevated twofold compared with the accumulation in wild-type fetuses (following normalization by the maternal plasma concentration), indicating that *Bcrp1* plays a critical role in protecting the fetus from exposure to harmful substances. In addition, the authors elegantly demonstrated that without functional *Bcrp1*, mice become at least 100-fold more sensitive to pheophorbide a, a dietary chlorophyll breakdown product and *Bcrp1* substrate, resulting in phototoxicity. The hypersensitivity could be explained by the markedly elevated plasma concentrations of pheophorbide a in these knockout mice, and therefore illustrates the importance of this transporter in providing protection against natural toxins. Studies from the same group also demonstrated that coadministration of topotecan with GF120918, a *Bcrp1* inhibitor, increased the AUC of topotecan more than sixfold due to the increased uptake from the intestine and the decreased biliary excretion in the *mdr1a* (-/-) mice (i.e., in the absence of P-glycoprotein) (Figure 18.2).[16] Similar results were obtained when topotecan was coadministered with Ko134.[259] In humans, it has been shown that the apparent oral bioavailability of topotecan was dramatically increased, from 40.0% to 97.1%, following the coadministration of GF120918. This change most likely resulted from the inhibition of BCRP; it is known that topotecan is only a weak substrate of P-glycoprotein.[15] Taken together, there is convincing evidence that BCRP plays an important role in governing the body disposition of xenobiotics. It also should be noted that the expression of BCRP and MRP2 in the human intestine is even higher than that of P-glycoprotein[263]; therefore, it is possible that the contribution of BCRP to the intestinal secretion and oral absorption of xenobiotics can be comparable with, if not greater than, that of P-glycoprotein.

18.5. OTHER EFFLUX TRANSPORTERS (MDR3, BSEP)

MDR3 is the other human P-glycoprotein isoform, with a molecular structure virtually identical to that of the human MDR1 and mouse *mdr1b* genes.[46] MDR3 is

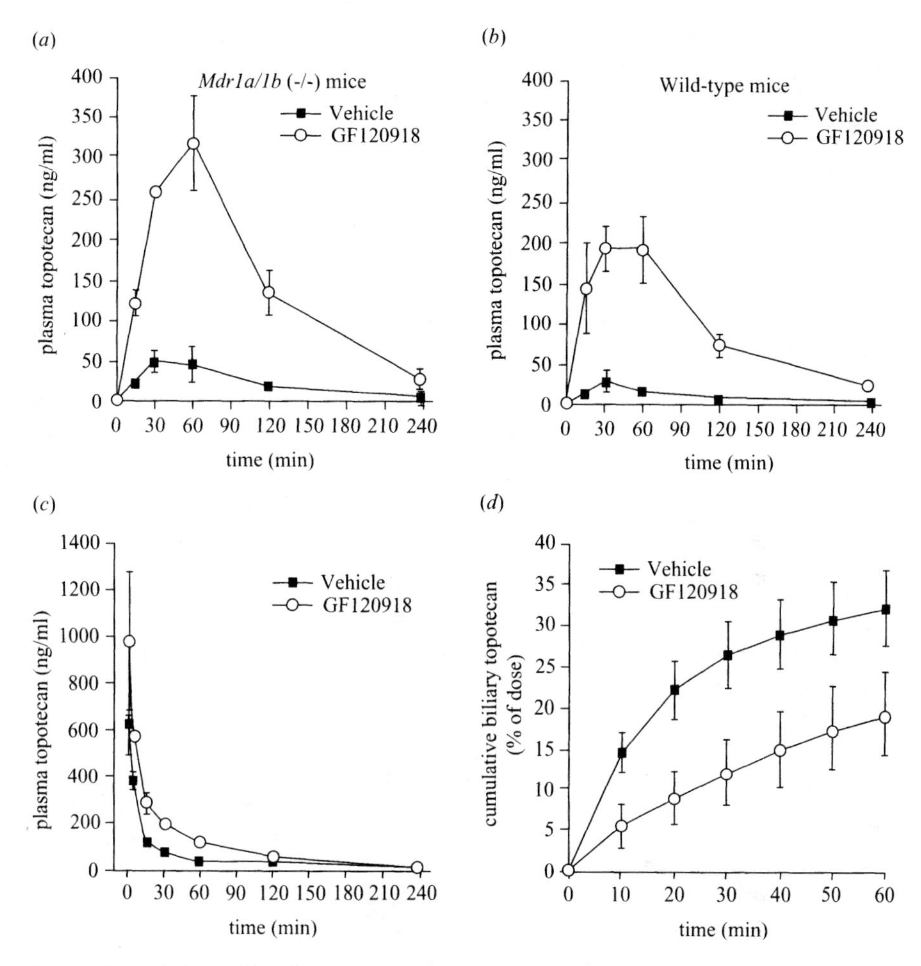

Figure 18.2. Effects of GF120918 on the plasma concentration and biliary excretion of topotecan in mice. *Mdr1a/1b* (-/-) (*a*) or wild-type (*b*) mice were given an oral dose of GF120918 (50 mg/kg) or vehicle 15 minutes before an oral dose of topotecan (1 mg/kg). (*c*) *Mdr1a/1b* (-/-) mice were given an i.v. dose of topotecan in combination of an oral GF120918 or vehicle. (*d*) Cumulative biliary excretion of topotecan in *mdr1a/1b* (-/-) mice treated in the same way as (*c*). Results are the means ± SD (n 3). (Reproduced with permission from ref. 16).

mainly present in the canalicular membrane of liver hepatocytes and functions as an ATP-dependent phosphotidylcholine translocator.[55–58] Initially, it was thought that MDR3 protein and its mouse homolog *mdr2* cannot transport drugs and confer multidrug resistance.[53,54] But more recently, it has been shown that MDR3 is also capable of transporting several cytotoxic drugs, such as digoxin, paclitaxel, and vinblastine, but with low efficiency.[264] A defect in MDR3 is believed to be associated with an autosomal recessive hereditary disorder, progressive familial intrahepatic cholestasis type 3 (PFIC3).[265,266] BSEP (SPGP, ABCB11) is another homolog of MDR P-glycoproteins, initially identified from the pig and named "Sister of P-glycoprotein" (spgp).[267] Subsequently, the rat *Bsep* gene was also cloned and shown to be an ATP-dependent bile salt exporter with a K_m value of about 5 μM for transporting taurocholate.[268] *Bsep* is almost exclusively present in the liver and localized to the canalicular microvilli and subcanalicular vesicles of the hepatocytes. It functions as a major bile salt export pump in mammalian livers.[267] The functional characterization of human BSEP has also been carried out recently and is reported to have a K_m value similar to that of rat Bsep for taurocholate.[269,270] A defect in BSEP in humans was associated with type 2 PFIC (PFIC2).[271–273] At this time, it is generally believed that both MDR3 and BSEP may not play a significant role in terms of drug disposition.

18.6. CONCLUSIONS

The molecular and functional characterization of efflux transporters, over the past 15 years has facilitated our understanding of how these transporters control the passage of a diverse range of substrates through biological membranes. The characterization of their tissue localization and function suggests a significant impact of these transporters on the absorption, elimination, and distribution of xenobiotic compounds as a body defense mechanism against the exposure of both endogenous and exogenous toxins. The generation of knockout mice lacking specific transporter(s) and the identification of specific inhibitors have greatly enhanced our ability to understand the physiological and pharmacological functions of these transporters. It has been clearly demonstrated by the studies presented here, as well as others, that these efflux transporters play an essential role in intestinal absorption, biliary excretion, and renal secretion and contribute to the barrier functions between the blood and various tissues such as brain, testis, and placenta. Considering the important impact of these efflux transporters on drug disposition, identification of substrates and inhibitors from commonly prescribed drugs or food-derived compounds, and characterization of their kinetic parameters will help to predict potential drug interactions mediated by these transport mechanisms. However, the full appreciation of the impact of these transporters on drug disposition will depend on our understanding of the mechanism(s) by which these transporters recognize such a wide range of structurally distinct substances, the mechanism(s) by which these transporters are regulated, and the influence of multiple coexisting transporters, as well as the interplay of these transporters with drug-metabolizing enzymes. These aspects are all largely unknown and remain to be investigated.

ACKNOWLEDGMENT

We acknowledge research support through the Susan G. Komen Breast Cancer Foundation, U. S. Army Breast Cancer Research Program Contract DAMD17-00-1-0376 and Pfizer Global Research and Development.

REFERENCES

1. Gottesman, M. M.; Pastan, I. *J. Biol. Chem.* **1988**, *263*, 12163–12166.
2. Gottesman, M. M.; Pastan, I. *Annu. Rev. Biochem.* **1993**, *62*, 385–427.
3. Litman, T.; Druley, T. E.; Stein, W. D.; Bates, S. E. *Cell. Mol. Life Sci.* **2001**, *58*, 931–959.
4. Hipfner, D. R.; Deeley, R. G.; Cole, S. P. *Biochim. Biophys. Acta* **1999**, *1461*, 359–376.
5. Konig, J.; Nies, A. T.; Cui, Y.; Leier, I.; Keppler, D. *Biochim. Biophys. Acta* **1999**, *1461*, 377–394.
6. Borst, P.; Evers, R.; Kool, M.; Wijnholds, J. *J. Natl. Cancer Inst.* **2000**, *92*, 1295–1302.
7. Ayrton, A.; Morgan, P. *Xenobiotica* **2001**, *31*, 469–497.
8. Allen, J. D.; Schinkel, A. H. *Mol. Cancer Ther.* **2002**, *1*, 427–434.
9. Schinkel, A. H. *Int. J. Clin. Pharmacol. Ther.* **1998**, *36*, 9–13.
10. Sadeque, A. J.; Wandel, C.; He, H.; Shah, S.; Wood, A. J. *Clin. Pharmacol. Ther.* **2000**, *68*, 231–237.
11. Schwarz, U. I.; Gramatte, T.; Krappweis, J.; Oertel, R.; Kirch, W. *Int. J. Clin. Pharmacol. Ther.* **2000**, *38*, 161–167.
12. Westphal, K.; Weinbrenner, A.; Giessmann, T.; Stuhr, M.; Franke, G.; Zschiesche, M.; Oertel, R.; Terhaag, B.; Kroemer, H. K.; Siegmund, W. *Clin. Pharmacol. Ther.* **2000**, *68*, 6–12.
13. Westphal, K.; Weinbrenner, A.; Zschiesche, M.; Franke, G.; Knoke, M.; Oertel, R.; Fritz, P.; von Richter, O.; Warzok, R.; Hachenberg, T.; Kauffmann, H. M.; Schrenk, D.; Terhaag, B.; Kroemer, H. K.; Siegmund, W. *Clin. Pharmacol. Ther.* **2000**, *68*, 345–355.
14. Johne, A.; Brockmoller, J.; Bauer, S.; Maurer, A.; Langheinrich, M.; Roots, I. *Clin. Pharmacol. Ther.* **1999**, *66*, 338–345.
15. Kruijtzer, C. M.; Beijnen, J. H.; Rosing, H.; ten Bokkel Huinink, W. W.; Schot, M.; Jewell, R. C.; Paul, E. M.; Schellens, J. H. *J. Clin. Oncol.* **2002**, *20*, 2943–2950.
16. Jonker, J. W.; Smit, J. W.; Brinkhuis, R. F.; Maliepaard, M.; Beijnen, J. H.; Schellens, J. H.; Schinkel, A. H. *J. Natl. Cancer. Inst.* **2000**, *92*, 1651–1656.
17. Kimura, Y.; Aoki, J.; Kohno, M.; Ooka, H.; Tsuruo, T.; Nakanishi, O. *Cancer Chemother. Pharmacol.* **2002**, *49*, 322–328.
18. Morris, M. E.; Lee, H.; Predko, L. M. *Pharmacol. Rev.* **2003**, *55*, 229–240.
19. Lown, K. S.; Mayo, R. R.; Leichtman, A. B.; Hsiao, H. L.; Turgeon, D. K.; Schmiedlin-Ren, P.; Brown, M. B.; Guo, W.; Rossi, S. J.; Benet, L. Z.; Watkins, P. B. *Clin. Pharmacol. Ther.* **1997**, *62*, 248–260.
20. Hoffmeyer, S.; Burk, O.; von Richter, O.; Arnold, H. P.; Brockmoller, J.; Johne, A.; Cascorbi, I.; Gerloff, T.; Roots, I.; Eichelbaum, M.; Brinkmann, U. *Proc. Natl. Acad. Sci. USA* **2000**, *97*, 3473–3478.
21. Brumme, Z. L.; Dong, W. W.; Chan, K. J.; Hogg, R. S.; Montaner, J. S.; O'Shaughnessy, M. V.; Harrigan, P. R. *Aids* **2003**, *17*, 201–208.
22. Hamman, M. A.; Bruce, M. A.; Haehner-Daniels, B. D.; Hall, S. D. *Clin. Pharmacol. Ther.* **2001**, *69*, 114–121.
23. Juliano, R. L.; Ling, V. *Biochim. Biophys. Acta* **1976**, *455*, 152–162.
24. Beck, W. T.; Mueller, T. J.; Tanzer, L. R. *Cancer Res.* **1979**, *39*, 2070–2076.
25. Kartner, N.; Shales, M.; Riordan, J. R.; Ling, V. *Cancer Res.* **1983**, *43*, 4413–4419.

26. Giavazzi, R.; Kartner, N.; Hart, I. R. *Cancer Chemother. Pharmacol.* **1984**, *13*, 145–147.
27. Kartner, N.; Evernden-Porelle, D.; Bradley, G.; Ling, V. *Nature* **1985**, *316*, 820–823.
28. Kartner, N.; Riordan, J. R.; Ling, V. *Science* **1983**, *221*, 1285–1288.
29. Ueda, K.; Cardarelli, C.; Gottesman, M. M.; Pastan, I. *Proc. Natl. Acad. Sci. USA* **1987**, *84*, 3004–3008.
30. Gros, P.; Ben Neriah, Y. B.; Croop, J. M.; Housman, D. E. *Nature* **1986**, *323*, 728–731.
31. Dano, K. *Biochim. Biophys. Acta* **1973**, *323*, 466–483.
32. Skovsgaard, T. *Cancer Res.* **1978**, *38*, 1785–1791.
33. Horio, M.; Gottesman, M. M.; Pastan, I. *Proc. Natl. Acad. Sci. USA* **1988**, *85*, 3580–3584.
34. Shapiro, A. B.; Ling, V. *J. Biol. Chem.* **1995**, *270*, 16167–16175.
35. Goldstein, L. J.; Pastan, I.; Gottesman, M. M. *Crit. Rev. Oncol. Hematol.* **1992**, *12*, 243–253.
36. Dalton, W. S. *Curr. Opin. Oncol.* **1994**, *6*, 595–600.
37. Gregorcyk, S.; Kang, Y.; Brandt, D.; Kolm, P.; Singer, G.; Perry, R. R. *Ann. Surg. Oncol.* **1996**, *3*, 8–14.
38. Koh, E. H.; Chung, H. C.; Lee, K. B.; Lim, H. Y.; Kim, J. H.; Roh, J. K.; Min, J. S.; Lee, K. S.; Kim, B. S. *Yonsei Med. J.* **1992**, *33*, 137–142.
39. List, A. F. *Leukemia* **1996**, *10*, 937–942.
40. Nooter, K.; Sonneveld, P. *Leuk. Res.* **1994**, *18*, 233–243.
41. Chan, H. S.; Haddad, G.; Thorner, P. S.; DeBoer, G.; Lin, Y. P.; Ondrusek, N.; Yeger, H.; Ling, V. *N. Engl. J. Med.* **1991**, *325*, 1608–1614.
42. Zochbauer, S.; Gsur, A.; Brunner, R.; Kyrle, P. A.; Lechner, K.; Pirker, R. *Leukemia* **1994**, *8*, 974–977.
43. van der Zee, A. G.; Hollema, H.; Suurmeijer, A. J.; Krans, M.; Sluiter, W. J.; Willemse, P. H.; Aalders, J. G.; de Vries, E. G. *J. Clin. Oncol.* **1995**, *13*, 70–78.
44. Marie, J. P.; Zittoun, R.; Sikic, B. I. *Blood* **1991**, *78*, 586–592.
45. Chen, C. J.; Chin, J. E.; Ueda, K.; Clark, D. P.; Pastan, I.; Gottesman, M. M.; Roninson, I. B. *Cell* **1986**, *47*, 381–389.
46. Lincke, C. R.; Smit, J. J.; van der Velde-Koerts, T.; Borst, P. *J. Biol. Chem.* **1991**, *266*, 5303–5310.
47. Gros, P.; Raymond, M.; Bell, J.; Housman, D. *Mol. Cell. Biol.* **1988**, *8*, 2770–2778.
48. Croop, J. M.; Raymond, M.; Haber, D.; Devault, A.; Arceci, R. J.; Gros, P.; Housman, D. E. *Mol. Cell. Biol.* **1989**, *9*, 1346–1350.
49. Pastan, I.; Gottesman, M. M.; Ueda, K.; Lovelace, E.; Rutherford, A. V.; Willingham, M. C. *Proc. Natl. Acad. Sci. USA* **1988**, *85*, 4486–4490.
50. Hsu, S. I.; Lothstein, L.; Horwitz, S. B. *J. Biol. Chem.* **1989**, *264*, 12053–12062.
51. Devault, A.; Gros, P. *Mol. Cell Biol.* **1990**, *10*, 1652–1663.
52. van der Bliek, A. M.; Kooiman, P. M.; Schneider, C.; Borst, P. *Gene* **1988**, *71*, 401–411.
53. Schinkel, A. H.; Roelofs, E. M.; Borst, P. *Cancer Res.* **1991**, *51*, 2628–2635.
54. Buschman, E.; Gros, P. *Mol. Cell Biol.* **1991**, *11*, 595–603.
55. Smith, A. J.; Timmermans-Hereijgers, J. L.; Roelofsen, B.; Wirtz, K. W.; van Blitterswijk, W. J.; Smit, J. J.; Schinkel, A. H.; Borst, P. *FEBS Lett.* **1994**, *354*, 263–266.

56. Smit, J. J.; Schinkel, A. H.; Oude Elferink, R. P.; Groen, A. K.; Wagenaar, E.; van Deemter, L.; Mol, C. A.; Ottenhoff, R.; van der Lugt, N. M.; van Roon, M. A.; et al. *Cell* **1993**, *75*, 451–462.
57. Smit, J. J.; Schinkel, A. H.; Mol, C. A.; Majoor, D.; Mooi, W. J.; Jongsma, A. P.; Lincke, C. R.; Borst, P. *Lab. Invest.* **1994**, *71*, 638–649.
58. Ruetz, S.; Gros, P. *Cell* **1994**, *77*, 1071–1081.
59. Germann, U. A. *Eur. J. Cancer* **1996**, *32A*, 927–944.
60. Ambudkar, S. V.; Lelong, I. H.; Zhang, J.; Cardarelli, C. O.; Gottesman, M. M.; Pastan, I. *Proc. Natl. Acad. Sci. USA* **1992**, *89*, 8472–8476.
61. Tsuruo, T.; Iida, H.; Kitatani, Y.; Yokota, K.; Tsukagoshi, S.; Sakurai, Y. *Cancer Res.* **1984**, *44*, 4303–4307.
62. Schinkel, A. H.; Wagenaar, E.; van Deemter, L.; Mol, C. A.; Borst, P. *J. Clin. Invest.* **1995**, *96*, 1698–1705.
63. Lee, C. G.; Gottesman, M. M.; Cardarelli, C. O.; Ramachandra, M.; Jeang, K. T.; Ambudkar, S. V.; Pastan, I.; Dey, S. *Biochemistry* **1998**, *37*, 3594–3601.
64. Meador, J.; Sweet, P.; Stupecky, M.; Wetzel, M.; Murray, S.; Gupta, S.; Slater, L. *Cancer Res.* **1987**, *47*, 6216–6219.
65. Horio, M.; Chin, K. V.; Currier, S. J.; Goldenberg, S.; Williams, C.; Pastan, I.; Gottesman, M. M.; Handler, J. *J. Biol. Chem.* **1989**, *264*, 14880–14884.
66. van Kalken, C. K.; Broxterman, H. J.; Pinedo, H. M.; Feller, N.; Dekker, H.; Lankelma, J.; Giaccone, G. *Br. J. Cancer* **1993**, *67*, 284–289.
67. Ueda, K.; Okamura, N.; Hirai, M.; Tanigawara, Y.; Saeki, T.; Kioka, N.; Komano, T.; Hori, R. *J. Biol. Chem.* **1992**, *267*, 24248–24252.
68. Drach, J.; Gsur, A.; Hamilton, G.; Zhao, S.; Angerler, J.; Fiegl, M.; Zojer, N.; Raderer, M.; Haberl, I.; Andreeff, M.; Huber, H. *Blood* **1996**, *88*, 1747–1754.
69. Kusuhara, H.; Suzuki, H.; Sugiyama, Y. *J. Pharm. Sci.* **1998**, *87*, 1025–1040.
70. de Graaf, D.; Sharma, R. C.; Mechetner, E. B.; Schimke, R. T.; Roninson, I. B. *Proc. Natl. Acad. Sci. USA* **1996**, *93*, 1238–1242.
71. Norris, M. D.; De Graaf, D.; Haber, M.; Kavallaris, M.; Madafiglio, J.; Gilbert, J.; Kwan, E.; Stewart, B. W.; Mechetner, E. B.; Gudkov, A. V.; Roninson, I. B. *Int. J. Cancer* **1996**, *65*, 613–619.
72. Potschka, H.; Loscher, W. *Epilepsia* **2001**, *42*, 1231–1240.
73. Martin, C.; Berridge, G.; Higgins, C. F.; Mistry, P.; Charlton, P.; Callaghan, R. *Mol. Pharmacol.* **2000**, *58*, 624–632.
74. Dey, S.; Ramachandra, M.; Pastan, I.; Gottesman, M. M.; Ambudkar, S. V. *Proc. Natl. Acad. Sci. USA* **1997**, *94*, 10594–10599.
75. Shapiro, A. B.; Ling, V. *Eur. J. Biochem.* **1997**, *250*, 130–137.
76. Shapiro, A. B.; Fox, K.; Lam, P.; Ling, V. *Eur. J. Biochem.* **1999**, *259*, 841–850.
77. Higgins, C. F.; Gottesman, M. M. *Trends Biochem. Sci.* **1992**, *17*, 18–21.
78. Raviv, Y.; Pollard, H. B.; Bruggemann, E. P.; Pastan, I.; Gottesman, M. M. *J. Biol. Chem.* **1990**, *265*, 3975–3980.
79. Shapiro, A. B.; Ling, V. *Acta Physiol. Scand. Suppl.* **1998**, *643*, 227–234.
80. Cornwell, M. M.; Pastan, I.; Gottesman, M. M. *J. Biol. Chem.* **1987**, *262*, 2166–2170.

81. Akiyama, S.; Cornwell, M. M.; Kuwano, M.; Pastan, I.; Gottesman, M. M. *Mol. Pharmacol.* **1988**, *33*, 144–147.
82. Ford, J. M.; Prozialeck, W. C.; Hait, W. N. *Mol. Pharmacol.* **1989**, *35*, 105–115.
83. Naito, M.; Yusa, K.; Tsuruo, T. *Biochem. Biophys. Res. Commun.* **1989**, *158*, 1066–1071.
84. Mansouri, A.; Henle, K. J.; Nagle, W. A. *SAAS Bull. Biochem. Biotech.* **1992**, *5*, 48–52.
85. Pourtier-Manzanedo, A.; Boesch, D.; Loor, F. *Anticancer Drugs* **1991**, *2*, 279–283.
86. Gaveriaux, C.; Boesch, D.; Boelsterli, J. J.; Bollinger, P.; Eberle, M. K.; Hiestand, P.; Payne, T.; Traber, R.; Wenger, R.; Loor, F. *Br. J. Cancer* **1989**, *60*, 867–871.
87. Gosland, M. P.; Lum, B. L.; Sikic, B. I. *Cancer Res.* **1989**, *49*, 6901–6905.
88. Hofsli, E.; Nissen-Meyer, J. *Int. J. Cancer* **1989**, *44*, 149–154.
89. Zordan-Nudo, T.; Ling, V.; Liu, Z.; Georges, E. *Cancer Res.* **1993**, *53*, 5994–6000.
90. Friche, E.; Jensen, P. B.; Sehested, M.; Demant, E. J.; Nissen, N. N. *Cancer Commun.* **1990**, *2*, 297–303.
91. Hugger, E. D.; Audus, K. L.; Borchardt, R. T. *J. Pharm. Sci.* **2002**, *91*, 1980–1990.
92. Zhang, S.; Morris, M. E. *J. Exp. Pharm. Ther.* **2003**,
93. Ferte, J.; Kuhnel, J. M.; Chapuis, G.; Rolland, Y.; Lewin, G.; Schwaller, M. A. *J. Med. Chem.* **1999**, *42*, 478–489.
94. Conseil, G.; Baubichon-Cortay, H.; Dayan, G.; Jault, J. M.; Barron, D.; Di Pietro, A. *Proc. Natl. Acad. Sci. USA* **1998**, *95*, 9831–9836.
95. de Wet, H.; McIntosh, D. B.; Conseil, G.; Baubichon-Cortay, H.; Krell, T.; Jault, J. M.; Daskiewicz, J. B.; Barron, D.; Di Pietro, A. *Biochemistry* **2001**, *40*, 10382–10391.
96. Romiti, N.; Tongiani, R.; Cervelli, F.; Chieli, E. *Life Sci.* **1998**, *62*, 2349–2358.
97. Bhardwaj, R. K.; Glaeser, H.; Becquemont, L.; Klotz, U.; Gupta, S. K.; Fromm, M. F. *J. Pharmacol. Exp. Ther.* **2002**, *302*, 645–650.
98. Twentyman, P. R. *Biochem. Pharmacol.* **1992**, *43*, 109–117.
99. Hyafil, F.; Vergely, C.; Du Vignaud, P.; Grand-Perret, T. *Cancer Res.* **1993**, *53*, 4595–4602.
100. Dantzig, A. H.; Shepard, R. L.; Cao, J.; Law, K. L.; Ehlhardt, W. J.; Baughman, T. M.; Bumol, T. F.; Starling, J. J. *Cancer Res.* **1996**, *56*, 4171–4179.
101. Roe, M.; Folkes, A.; Ashworth, P.; Brumwell, J.; Chima, L.; Hunjan, S.; Pretswell, I.; Dangerfield, W.; Ryder, H.; Charlton, P. *Bioorg. Med. Chem. Lett.* **1999**, *9*, 595–600.
102. Advani, R.; Saba, H. I.; Tallman, M. S.; Rowe, J. M.; Wiernik, P. H.; Ramek, J.; Dugan, K.; Lum, B.; Villena, J.; Davis, E.; Paietta, E.; Litchman, M.; Sikic, B. I.; Greenberg, P. L. *Blood* **1999**, *93*, 787–795.
103. Advani, R.; Fisher, G. A.; Lum, B. L.; Hausdorff, J.; Halsey, J.; Litchman, M.; Sikic, B. I. *Clin. Cancer Res.* **2001**, *7*, 1221–1229.
104. Chico, I.; Kang, M. H.; Bergan, R.; Abraham, J.; Bakke, S.; Meadows, B.; Rutt, A.; Robey, R.; Choyke, P.; Merino, M.; Goldspiel, B.; Smith, T.; Steinberg, S.; Figg, W. D.; Fojo, T.; Bates, S. *J. Clin. Oncol.* **2001**, *19*, 832–842.
105. Thomas, H.; Coley, H. M. *Cancer Control* **2003**, *10*, 159–165.
106. Thiebaut, F.; Tsuruo, T.; Hamada, H.; Gottesman, M. M.; Pastan, I.; Willingham, M. C. *Proc. Natl. Acad. Sci. USA* **1987**, *84*, 7735–7738.
107. Sugawara, I.; Kataoka, I.; Morishita, Y.; Hamada, H.; Tsuruo, T.; Itoyama, S.; Mori, S. *Cancer Res.* **1988**, *48*, 1926–1929.

108. Cordon-Cardo, C.; O'Brien, J. P.; Casals, D.; Rittman-Grauer, L.; Biedler, J. L.; Melamed, M. R.; Bertino, J. R. *Proc. Natl. Acad. Sci. USA* **1989**, *86*, 695–698.

109. Thiebaut, F.; Tsuruo, T.; Hamada, H.; Gottesman, M. M.; Pastan, I.; Willingham, M. C. *J. Histochem. Cytochem.* **1989**, *37*, 159–164.

110. Fedeli, L.; Colozza, M.; Boschetti, E.; Sabalich, I.; Aristei, C.; Guerciolini, R.; Del Favero, A.; Rossetti, R.; Tonato, M.; Rambotti, P.; et al. *Cancer* **1989**, *64*, 1805–1811.

111. Horton, J. K.; Thimmaiah, K. N.; Houghton, J. A.; Horowitz, M. E.; Houghton, P. J. *Biochem. Pharmacol.* **1989**, *38*, 1727–1736.

112. Lum, B. L.; Kaubisch, S.; Yahanda, A. M.; Adler, K. M.; Jew, L.; Ehsan, M. N.; Brophy, N. A.; Halsey, J.; Gosland, M. P.; Sikic, B. I. *J. Clin. Oncol.* **1992**, *10*, 1635–1642.

113. Schinkel, A. H.; Smit, J. J.; van Tellingen, O.; Beijnen, J. H.; Wagenaar, E.; van Deemter, L.; Mol, C. A.; van der Valk, M. A.; Robanus-Maandag, E. C.; te Riele, H. P.; et al. *Cell* **1994**, *77*, 491–502.

114. Schinkel, A. H.; Mayer, U.; Wagenaar, E.; Mol, C. A.; van Deemter, L.; Smit, J. J.; van der Valk, M. A.; Voordouw, A. C.; Spits, H.; van Tellingen, O.; Zijlmans, J. M.; Fibbe, W. E.; Borst, P. *Proc. Natl. Acad. Sci. USA* **1997**, *94*, 4028–4033.

115. Schinkel, A. H.; Wagenaar, E.; Mol, C. A.; van Deemter, L. *J. Clin. Invest.* **1996**, *97*, 2517–2524.

116. van Asperen, J.; Schinkel, A. H.; Beijnen, J. H.; Nooijen, W. J.; Borst, P.; van Tellingen, O. *J. Natl. Cancer Inst.* **1996**, *88*, 994–999.

117. Smit, J. W.; Schinkel, A. H.; Muller, M.; Weert, B.; Meijer, D. K. *Hepatology* **1998**, *27*, 1056–1063.

118. Sparreboom, A.; van Asperen, J.; Mayer, U.; Schinkel, A. H.; Smit, J. W.; Meijer, D. K.; Borst, P.; Nooijen, W. J.; Beijnen, J. H.; van Tellingen, O. *Proc. Natl. Acad. Sci. USA* **1997**, *94*, 2031–2035.

119. Mayer, U.; Wagenaar, E.; Beijnen, J. H.; Smit, J. W.; Meijer, D. K.; van Asperen, J.; Borst, P.; Schinkel, A. H. *Br. J. Pharmacol.* **1996**, *119*, 1038–1044.

120. Kim, R. B.; Fromm, M. F.; Wandel, C.; Leake, B.; Wood, A. J.; Roden, D. M.; Wilkinson, G. R. *J. Clin. Invest.* **1998**, *101*,

121. van Asperen, J.; van Tellingen, O.; Beijnen, J. H. *Drug Metab. Dispos.* **2000**, *28*, 264–267.

122. Yamaguchi, H.; Yano, I.; Saito, H.; Inui, K. *J. Pharmacol. Exp. Ther.* **2002**, *300*, 1063–1069.

123. Smit, J. W.; Schinkel, A. H.; Weert, B.; Meijer, D. K. *Br. J. Pharmacol.* **1998**, *124*, 416–424.

124. Gramatte, T.; Oertel, R. *Clin. Pharmacol. Ther.* **1999**, *66*, 239–245.

125. Drescher, S.; Glaeser, H.; Murdter, T.; Hitzl, M.; Eichelbaum, M.; Fromm, M. F. *Clin. Pharmacol. Ther.* **2003**, *73*, 223–231.

126. Hebert, M. F. *Adv. Drug Deliv. Rev.* **1997**, *27*, 201–214.

127. Robbins, D. K.; Castles, M. A.; Pack, D. J.; Bhargava, V. O.; Weir, S. J. *Biopharm. Drug Dispos.* **1998**, *19*, 455–463.

128. Greiner, B.; Eichelbaum, M.; Fritz, P.; Kreichgauer, H. P.; von Richter, O.; Zundler, J.; Kroemer, H. K. *J. Clin. Invest.* **1999**, *104*, 147–153.

129. Guns, E. S.; Denyssevych, T.; Dixon, R.; Bally, M. B.; Mayer, L. *Eur. J. Drug Metab. Pharmacokinet.* **2002**, *27*, 119–126.

130. Verstuyft, C.; Strabach, S.; El-Morabet, H.; Kerb, R.; Brinkmann, U.; Dubert, L.; Jaillon, P.; Funck-Brentano, C.; Trugnan, G.; Becquemont, L. *Clin. Pharmacol. Ther.* **2003**, *73*, 51–60.
131. Wetterich, U.; Spahn-Langguth, H.; Mutschler, E.; Terhaag, B.; Rosch, W.; Langguth, P. *Pharm. Res.* **1996**, *13*, 514–522.
132. Keppler, D.; Arias, I. M. *Faseb. J.* **1997**, *11*, 15–18.
133. Hooiveld, G. J.; van Montfoort, J. E.; Meijer, D. K.; Muller, M. *Eur. J. Pharm. Sci.* **2000**, *12*, 13–30.
134. Kim, R. B. *Toxicology* **2002**, *181–182*, 291–297.
135. Milne, R. W.; Larsen, L. A.; Jorgensen, K. L.; Bastlund, J.; Stretch, G. R.; Evans, A. M. *Pharm. Res.* **2000**, *17*, 1511–1515.
136. Speeg, K. V.; Maldonado, A. L. *Cancer Chemother. Pharmacol.* **1994**, *34*, 133–136.
137. Speeg, K. V.; Maldonado, A. L.; Liaci, J.; Muirhead, D. *Hepatology* **1992**, *15*, 899–903.
138. Kiso, S.; Cai, S. H.; Kitaichi, K.; Furui, N.; Takagi, K.; Nabeshima, T.; Hasegawa, T. *Anticancer Res.* **2000**, *20*, 2827–2834.
139. Zhao, Y. L.; Cai, S. H.; Wang, L.; Kitaichi, K.; Tatsumi, Y.; Nadai, M.; Yoshizumi, H.; Takagi, K.; Hasegawa, T. *Clin. Exp. Pharmacol. Physiol.* **2002**, *29*, 167–172.
140. Schrenk, D.; Gant, T. W.; Preisegger, K. H.; Silverman, J. A.; Marino, P. A.; Thorgeirsson, S. S. *Hepatology* **1993**, *17*, 854–860.
141. Watanabe, T.; Suzuki, H.; Sawada, Y.; Naito, M.; Tsuruo, T.; Inaba, M.; Hanano, M.; Sugiyama, Y. *J. Hepatol.* **1995**, *23*, 440–448.
142. Perri, D.; Ito, S.; Rowsell, V.; Shear, N. H. *Can. J. Clin. Pharmacol.* **2003**, *10*, 17–23.
143. Speeg, K. V.; Maldonado, A. L.; Liaci, J.; Muirhead, D. *J. Pharmacol. Exp. Ther.* **1992**, *261*, 50–55.
144. Okamura, N.; Hirai, M.; Tanigawara, Y.; Tanaka, K.; Yasuhara, M.; Ueda, K.; Komano, T.; Hori, R. *J. Pharmacol. Exp. Ther.* **1993**, *266*, 1614–1619.
145. de Lannoy, I. A.; Mandin, R. S.; Silverman, M. *J. Pharmacol. Exp. Ther.* **1994**, *268*, 388–395.
146. Jalava, K. M.; Partanen, J.; Neuvonen, P. J. *Ther. Drug Monit.* **1997**, *19*, 609–613.
147. Kaukonen, K. M.; Olkkola, K. T.; Neuvonen, P. J. *Clin. Pharmacol. Ther.* **1997**, *62*, 510–517.
148. Borst, P.; Evers, R.; Kool, M.; Wijnholds, J. *Biochim. Biophys. Acta* **1999**, *1461*, 347–357.
149. Cole, S. P.; Bhardwaj, G.; Gerlach, J. H.; Mackie, J. E.; Grant, C. E.; Almquist, K. C.; Stewart, A. J.; Kurz, E. U.; Duncan, A. M.; Deeley, R. G. *Science* **1992**, *258*, 1650–1654.
150. Marsh, W.; Sicheri, D.; Center, M. S. *Cancer Res.* **1986**, *46*, 4053–4057.
151. Marsh, W.; Center, M. S. *Cancer Res.* **1987**, *47*, 5080–5086.
152. McGrath, T.; Center, M. S. *Biochem. Biophys. Res. Commun.* **1987**, *145*, 1171–1176.
153. McGrath, T.; Latoud, C.; Arnold, S. T.; Safa, A. R.; Felsted, R. L.; Center, M. S. *Biochem. Pharmacol.* **1989**, *38*, 3611–3619.
154. Cole, S. P.; Sparks, K. E.; Fraser, K.; Loe, D. W.; Grant, C. E.; Wilson, G. M.; Deeley, R. G. *Cancer Res.* **1994**, *54*, 5902–5910.
155. Grant, C. E.; Valdimarsson, G.; Hipfner, D. R.; Almquist, K. C.; Cole, S. P.; Deeley, R. G. *Cancer Res.* **1994**, *54*, 357–361.

156. Zaman, G. J.; Flens, M. J.; van Leusden, M. R.; de Haas, M.; Mulder, H. S.; Lankelma, J.; Pinedo, H. M.; Scheper, R. J.; Baas, F.; Broxterman, H. J.; et al. *Proc. Natl. Acad. Sci. USA* **1994**, *91*, 8822–8826.

157. Kruh, G. D.; Chan, A.; Myers, K.; Gaughan, K.; Miki, T.; Aaronson, S. A. *Cancer Res.* **1994**, *54*, 1649–1652.

158. Ito, K.; Suzuki, H.; Hirohashi, T.; Kume, K.; Shimizu, T.; Sugiyama, Y. *Am. J. Physiol.* **1997**, *272*, G16–22.

159. Buchler, M.; Konig, J.; Brom, M.; Kartenbeck, J.; Spring, H.; Horie, T.; Keppler, D. *J. Biol. Chem.* **1996**, *271*, 15091–15098.

160. Paulusma, C. C.; Bosma, P. J.; Zaman, G. J.; Bakker, C. T.; Otter, M.; Scheffer, G. L.; Scheper, R. J.; Borst, P.; Oude Elferink, R. P. *Science* **1996**, *271*, 1126–1128.

161. Taniguchi, K.; Wada, M.; Kohno, K.; Nakamura, T.; Kawabe, T.; Kawakami, M.; Kagotani, K.; Okumura, K.; Akiyama, S.; Kuwano, M. *Cancer Res.* **1996**, *56*, 4124–4129.

162. Hopper, E.; Belinsky, M. G.; Zeng, H.; Tosolini, A.; Testa, J. R.; Kruh, G. D. *Cancer Lett.* **2001**, *162*, 181–191.

163. Lee, K.; Belinsky, M. G.; Bell, D. W.; Testa, J. R.; Kruh, G. D. *Cancer Res.* **1998**, *58*, 2741–2747.

164. Lee, K.; Klein-Szanto, A. J.; Kruh, G. D. *J. Natl. Cancer Inst.* **2000**, *92*, 1934–1940.

165. Kool, M.; de Haas, M.; Scheffer, G. L.; Scheper, R. J.; van Eijk, M. J.; Juijn, J. A.; Baas, F.; Borst, P. *Cancer Res.* **1997**, *57*, 3537–3547.

166. Kool, M.; van der Linden, M.; de Haas, M.; Baas, F.; Borst, P. *Cancer Res.* **1999**, *59*, 175–182.

167. Chen, Z. S.; Hopper-Borge, E.; Belinsky, M. G.; Shchaveleva, I.; Kotova, E.; Kruh, G. D. *Mol. Pharmacol.* **2003**, *63*, 351–358.

168. Leslie, E. M.; Deeley, R. G.; Cole, S. P. *Toxicology* **2001**, *167*, 3–23.

169. Cui, Y.; Konig, J.; Buchholz, J. K.; Spring, H.; Leier, I.; Keppler, D. *Mol. Pharmacol.* **1999**, *55*, 929–937.

170. Zeng, H.; Bain, L. J.; Belinsky, M. G.; Kruh, G. D. *Cancer Res.* **1999**, *59*, 5964–5967.

171. Kool, M.; van der Linden, M.; de Haas, M.; Scheffer, G. L.; de Vree, J. M.; Smith, A. J.; Jansen, G.; Peters, G. J.; Ponne, N.; Scheper, R. J.; Elferink, R. P.; Baas, F.; Borst, P. *Proc. Natl. Acad. Sci. USA* **1999**, *96*, 6914–6919.

172. Kawahara, M.; Sakata, A.; Miyashita, T.; Tamai, I.; Tsuji, A. *J. Pharm. Sci.* **1999**, *88*, 1281–1287.

173. Tada, Y.; Wada, M.; Migita, T.; Nagayama, J.; Hinoshita, E.; Mochida, Y.; Maehara, Y.; Tsuneyoshi, M.; Kuwano, M.; Naito, S. *Int. J. Cancer* **2002**, *98*, 630–635.

174. Laupeze, B.; Amiot, L.; Drenou, B.; Bernard, M.; Branger, B.; Grosset, J. M.; Lamy, T.; Fauchet, R.; Fardel, O. *Br. J. Haematol.* **2002**, *116*, 834–838.

175. Ito, K.; Fujimori, M.; Nakata, S.; Hama, Y.; Shingu, K.; Kobayashi, S.; Tsuchiya, S.; Kohno, K.; Kuwano, M.; Amano, J. *Oncol. Res.* **1998**, *10*, 99–109.

176. Filipits, M.; Suchomel, R. W.; Dekan, G.; Haider, K.; Valdimarsson, G.; Depisch, D.; Pirker, R. *Clin. Cancer Res.* **1996**, *2*, 1231–1237.

177. Nooter, K.; Brutel de la Riviere, G.; Look, M. P.; van Wingerden, K. E.; Henzen-Logmans, S. C.; Scheper, R. J.; Flens, M. J.; Klijn, J. G.; Stoter, G.; Foekens, J. A. *Br. J. Cancer* **1997**, *76*, 486–493.

178. Ota, E.; Abe, Y.; Oshika, Y.; Ozeki, Y.; Iwasaki, M.; Inoue, H.; Yamazaki, H.; Ueyama, Y.; Takagi, K.; Ogata, T.; et al. *Br. J. Cancer* **1995**, *72*, 550–554.

179. Oshika, Y.; Nakamura, M.; Tokunaga, T.; Fukushima, Y.; Abe, Y.; Ozeki, Y.; Yamazaki, H.; Tamaoki, N.; Ueyama, Y. *Mod. Pathol.* **1998**, *11*, 1059–1063.

180. Sugawara, I.; Yamada, H.; Nakamura, H.; Sumizawa, T.; Akiyama, S.; Masunaga, A.; Itoyama, S. *Int. J. Cancer* **1995**, *64*, 322–325.

181. Young, L. C.; Campling, B. G.; Voskoglou-Nomikos, T.; Cole, S. P.; Deeley, R. G.; Gerlach, J. H. *Clin. Cancer Res.* **1999**, *5*, 673–680.

182. Zeng, H.; Chen, Z. S.; Belinsky, M. G.; Rea, P. A.; Kruh, G. D. *Cancer Res.* **2001**, *61*, 7225–7232.

183. Keppler, D.; Cui, Y.; Konig, J.; Leier, I.; Nies, A. *Adv. Enzyme Regul.* **1999**, *39*, 237–246.

184. Loe, D. W.; Almquist, K. C.; Deeley, R. G.; Cole, S. P. *J. Biol. Chem.* **1996**, *271*, 9675–9682.

185. Loe, D. W.; Deeley, R. G.; Cole, S. P. *Cancer Res.* **1998**, *58*, 5130–5136.

186. Renes, J.; de Vries, E. G.; Nienhuis, E. F.; Jansen, P. L.; Muller, M. *Br. J. Pharmacol.* **1999**, *126*, 681–688.

187. Evers, R.; de Haas, M.; Sparidans, R.; Beijnen, J.; Wielinga, P. R.; Lankelma, J.; Borst, P. *Br. J. Cancer* **2000**, *83*, 375–383.

188. Schneider, E.; Yamazaki, H.; Sinha, B. K.; Cowan, K. H. *Br. J. Cancer* **1995**, *71*, 738–743.

189. Versantvoort, C. H.; Broxterman, H. J.; Bagrij, T.; Scheper, R. J.; Twentyman, P. R. *Br. J. Cancer* **1995**, *72*, 82–89.

190. Kawabe, T.; Chen, Z. S.; Wada, M.; Uchiumi, T.; Ono, M.; Akiyama, S.; Kuwano, M. *FEBS Lett.* **1999**, *456*, 327–331.

191. Itoh, Y.; Tamai, M.; Yokogawa, K.; Nomura, M.; Moritani, S.; Suzuki, H.; Sugiyama, Y.; Miyamoto, K. *Anticancer Res.* **2002**, *22*, 1649–1653.

192. Kruh, G. D.; Zeng, H.; Rea, P. A.; Liu, G.; Chen, Z. S.; Lee, K.; Belinsky, M. G. *J. Bioenerg. Biomembr.* **2001**, *33*, 493–501.

193. Zeng, H.; Liu, G.; Rea, P. A.; Kruh, G. D. *Cancer Res.* **2000**, *60*, 4779–4784.

194. Hirohashi, T.; Suzuki, H.; Takikawa, H.; Sugiyama, Y. *J. Biol. Chem.* **2000**, *275*, 2905–2910.

195. Dantzig, A. H.; Law, K. L.; Cao, J.; Starling, J. J. *Curr. Med. Chem.* **2001**, *8*, 39–50.

196. Germann, U. A.; Ford, P. J.; Shlyakhter, D.; Mason, V. S.; Harding, M. W. *Anticancer Drugs* **1997**, *8*, 141–155.

197. Bodo, A.; Bakos, E.; Szeri, F.; Varadi, A.; Sarkadi, B. *Toxicol. Lett.* **2003**, *140–141*, 133–143.

198. Gekeler, V.; Ise, W.; Sanders, K. H.; Ulrich, W. R.; Beck, J. *Biochem. Biophys. Res. Commun.* **1995**, *208*, 345–352.

199. Evers, R.; Zaman, G. J.; van Deemter, L.; Jansen, H.; Calafat, J.; Oomen, L. C.; Oude Elferink, R. P.; Borst, P.; Schinkel, A. H. *J. Clin. Invest.* **1996**, *97*, 1211–1218.

200. Lorico, A.; Rappa, G.; Finch, R. A.; Yang, D.; Flavell, R. A.; Sartorelli, A. C. *Cancer Res.* **1997**, *57*, 5238–5242.

201. Wijnholds, J.; Evers, R.; van Leusden, M. R.; Mol, C. A.; Zaman, G. J.; Mayer, U.; Beijnen, J. H.; van der Valk, M.; Krimpenfort, P.; Borst, P. *Nat. Med.* **1997**, *3*, 1275–1279.

202. Johnson, D. R.; Finch, R. A.; Lin, Z. P.; Zeiss, C. J.; Sartorelli, A. C. *Cancer Res.* **2001**, *61*, 1469–1476.

203. Rao, V. V.; Dahlheimer, J. L.; Bardgett, M. E.; Snyder, A. Z.; Finch, R. A.; Sartorelli, A. C.; Piwnica-Worms, D. *Proc. Natl. Acad. Sci. USA* **1999**, *96*, 3900–3905.

204. Wijnholds, J.; deLange, E. C.; Scheffer, G. L.; van den Berg, D. J.; Mol, C. A.; van der Valk, M.; Schinkel, A. H.; Scheper, R. J.; Breimer, D. D.; Borst, P. *J. Clin. Invest.* **2000**, *105*, 279–285.

205. Schaub, T. P.; Kartenbeck, J.; Konig, J.; Spring, H.; Dorsam, J.; Staehler, G.; Storkel, S.; Thon, W. F.; Keppler, D. *J. Am. Soc. Nephrol.* **1999**, *10*, 1159–1169.

206. Kartenbeck, J.; Leuschner, U.; Mayer, R.; Keppler, D. *Hepatology* **1996**, *23*, 1061–1066.

207. Paulusma, C. C.; Kool, M.; Bosma, P. J.; Scheffer, G. L.; ter Borg, F.; Scheper, R. J.; Tytgat, G. N.; Borst, P.; Baas, F.; Oude Elferink, R. P. *Hepatology* **1997**, *25*, 1539–1542.

208. Kajihara, S.; Hisatomi, A.; Mizuta, T.; Hara, T.; Ozaki, I.; Wada, I.; Yamamoto, K. *Biochem. Biophys. Res. Commun.* **1998**, *253*, 454–457.

209. Toh, S.; Wada, M.; Uchiumi, T.; Inokuchi, A.; Makino, Y.; Horie, Y.; Adachi, Y.; Sakisaka, S.; Kuwano, M. *Am. J. Hum. Genet.* **1999**, *64*, 739–746.

210. Jansen, P. L.; Peters, W. H.; Lamers, W. H. *Hepatology* **1985**, *5*, 573–579.

211. Kuipers, F.; Enserink, M.; Havinga, R.; van der Steen, A. B.; Hardonk, M. J.; Fevery, J.; Vonk, R. J. *J. Clin. Invest.* **1988**, *81*, 1593–1599.

212. Ishizuka, H.; Konno, K.; Naganuma, H.; Sasahara, K.; Kawahara, Y.; Niinuma, K.; Suzuki, H.; Sugiyama, Y. *J. Pharmacol. Exp. Ther.* **1997**, *280*, 1304–1311.

213. Sasabe, H.; Tsuji, A.; Sugiyama, Y. *J. Pharmacol. Exp. Ther.* **1998**, *284*, 1033–1039.

214. Chen, C.; Scott, D.; Hanson, E.; Franco, J.; Berryman, E.; Volberg, M.; Liu, X. *Pharm. Res.* **2003**, *20*, 31–37.

215. Sathirakul, K.; Suzuki, H.; Yamada, T.; Hanano, M.; Sugiyama, Y. *J. Pharmacol. Exp. Ther.* **1994**, *268*, 65–73.

216. Xiong, H.; Turner, K. C.; Ward, E. S.; Jansen, P. L.; Brouwer, K. L. *J. Pharmacol. Exp. Ther.* **2000**, *295*, 512–518.

217. Yamazaki, M.; Akiyama, S.; Ni'inuma, K.; Nishigaki, R.; Sugiyama, Y. *Drug Metab. Dispos.* **1997**, *25*, 1123–1129.

218. Chen, C.; Hennig, G. E.; Manautou, J. E. *Drug Metab. Dispos.* **2003**, *31*, 798–804.

219. Kouzuki, H.; Suzuki, H.; Sugiyama, Y. *Pharm. Res.* **2000**, *17*, 432–438.

220. Horikawa, M.; Kato, Y.; Sugiyama, Y. *Pharm. Res.* **2002**, *19*, 1345–1353.

221. Gotoh, Y.; Suzuki, H.; Kinoshita, S.; Hirohashi, T.; Kato, Y.; Sugiyama, Y. *J. Pharmacol. Exp. Ther.* **2000**, *292*, 433–439.

222. Naruhashi, K.; Tamai, I.; Inoue, N.; Muraoka, H.; Sai, Y.; Suzuki, N.; Tsuji, A. *Antimicrob. Agents Chemother.* **2002**, *46*, 344–349.

223. Dietrich, C. G.; de Waart, D. R.; Ottenhoff, R.; Schoots, I. G.; Elferink, R. P. *Mol. Pharmacol.* **2001**, *59*, 974–980.

224. Dietrich, C. G.; de Waart, D. R.; Ottenhoff, R.; Bootsma, A. H.; van Gennip, A. H.; Elferink, R. P. *Carcinogenesis* **2001**, *22*, 805–811.

225. Dombrowski, S. M.; Desai, S. Y.; Marroni, M.; Cucullo, L.; Goodrich, K.; Bingaman, W.; Mayberg, M. R.; Bengez, L.; Janigro, D. *Epilepsia* **2001**, *42*, 1501–1506.

226. Miller, D. S.; Nobmann, S. N.; Gutmann, H.; Toeroek, M.; Drewe, J.; Fricker, G. *Mol. Pharmacol.* **2000**, *58*, 1357–1367.

227. Konig, J.; Rost, D.; Cui, Y.; Keppler, D. *Hepatology* **1999**, *29*, 1156–1163.

228. Hirohashi, T.; Suzuki, H.; Sugiyama, Y. *J. Biol. Chem.* **1999**, *274*, 15181–15185.

229. Doyle, L. A.; Yang, W.; Abruzzo, L. V.; Krogmann, T.; Gao, Y.; Rishi, A. K.; Ross, D. D. *Proc. Natl. Acad. Sci. USA* **1998**, *95*, 15665–15670.

230. Allikmets, R.; Schriml, L. M.; Hutchinson, A.; Romano-Spica, V.; Dean, M. *Cancer Res.* **1998**, *58*, 5337–5339.

231. Miyake, K.; Mickley, L.; Litman, T.; Zhan, Z.; Robey, R.; Cristensen, B.; Brangi, M.; Greenberger, L.; Dean, M.; Fojo, T.; Bates, S. E. *Cancer Res.* **1999**, *59*, 8–13.

232. Ozvegy, C.; Litman, T.; Szakacs, G.; Nagy, Z.; Bates, S.; Varadi, A.; Sarkadi, B. *Biochem. Biophys. Res. Commun.* **2001**, *285*, 111–117.

233. Honjo, Y.; Hrycyna, C. A.; Yan, Q. W.; Medina-Perez, W. Y.; Robey, R. W.; van de Laar, A.; Litman, T.; Dean, M.; Bates, S. E. *Cancer Res.* **2001**, *61*, 6635–6639.

234. Kage, K.; Tsukahara, S.; Sugiyama, T.; Asada, S.; Ishikawa, E.; Tsuruo, T.; Sugimoto, Y. *Int. J. Cancer* **2002**, *97*, 626–630.

235. Allen, J. D.; Brinkhuis, R. F.; Wijnholds, J.; Schinkel, A. H. *Cancer Res.* **1999**, *59*, 4237–4241.

236. Mickley, L.; Jain, P.; Miyake, K.; Schriml, L. M.; Rao, K.; Fojo, T.; Bates, S.; Dean, M. *Mamm. Genome* **2001**, *12*, 86–88.

237. Townsend, A.; Trowsdale, J. *Semin. Cell Biol.* **1993**, *4*, 53–61.

238. Rocchi, E.; Khodjakov, A.; Volk, E. L.; Yang, C. H.; Litman, T.; Bates, S. E.; Schneider, E. *Biochem. Biophys. Res. Commun.* **2000**, *271*, 42–46.

239. Scheffer, G. L.; Maliepaard, M.; Pijnenborg, A. C.; van Gastelen, M. A.; de Jong, M. C.; Schroeijers, A. B.; van der Kolk, D. M.; Allen, J. D.; Ross, D. D.; van der Valk, P.; Dalton, W. S.; Schellens, J. H.; Scheper, R. J. *Cancer Res.* **2000**, *60*, 2589–2593.

240. Ross, D. D.; Yang, W.; Abruzzo, L. V.; Dalton, W. S.; Schneider, E.; Lage, H.; Dietel, M.; Greenberger, L.; Cole, S. P.; Doyle, L. A. *J. Natl. Cancer Inst.* **1999**, *91*, 429–433.

241. Maliepaard, M.; van Gastelen, M. A.; de Jong, L. A.; Pluim, D.; van Waardenburg, R. C.; Ruevekamp-Helmers, M. C.; Floot, B. G.; Schellens, J. H. *Cancer Res.* **1999**, *59*, 4559–4563.

242. Ross, D. D.; Karp, J. E.; Chen, T. T.; Doyle, L. A. *Blood* **2000**, *96*, 365–368.

243. Kanzaki, A.; Toi, M.; Nakayama, K.; Bando, H.; Mutoh, M.; Uchida, T.; Fukumoto, M.; Takebayashi, Y. *Jpn. J. Cancer Res.* **2001**, *92*, 452–458.

244. Sargent, J. M.; Williamson, C. J.; Maliepaard, M.; Elgie, A. W.; Scheper, R. J.; Taylor, C. G. *Br. J. Haematol.* **2001**, *115*, 257–262.

245. Faneyte, I. F.; Kristel, P. M.; Maliepaard, M.; Scheffer, G. L.; Scheper, R. J.; Schellens, J. H.; van de Vijver, M. J. *Clin. Cancer Res.* **2002**, *8*, 1068–1074.

246. van der Kolk, D. M.; Vellenga, E.; Scheffer, G. L.; Muller, M.; Bates, S. E.; Scheper, R. J.; de Vries, E. G. *Blood* **2002**, *99*, 3763–3770.

247. Sauerbrey, A.; Sell, W.; Steinbach, D.; Voigt, A.; Zintl, F. *Br. J. Haematol.* **2002**, *118*, 147–150.

248. Steinbach, D.; Sell, W.; Voigt, A.; Hermann, J.; Zintl, F.; Sauerbrey, A. *Leukemia* **2002**, *16*, 1443–1447.

249. Kawabata, S.; Oka, M.; Shiozawa, K.; Tsukamoto, K.; Nakatomi, K.; Soda, H.; Fukuda, M.; Ikegami, Y.; Sugahara, K.; Yamada, Y.; Kamihira, S.; Doyle, L. A.; Ross, D. D.; Kohno, S. *Biochem. Biophys. Res. Commun.* **2001**, *280*, 1216–1223.

250. Schellens, J. H.; Maliepaard, M.; Scheper, R. J.; Scheffer, G. L.; Jonker, J. W.; Smit, J. W.; Beijnen, J. H.; Schinkel, A. H. *Ann. NY Acad. Sci.* **2000**, *922*, 188–194.

251. Robey, R. W.; Medina-Perez, W. Y.; Nishiyama, K.; Lahusen, T.; Miyake, K.; Litman, T.; Senderowicz, A. M.; Ross, D. D.; Bates, S. E. *Clin. Cancer Res.* **2001**, *7*, 145–152.

252. Wang, X.; Furukawa, T.; Nitanda, T.; Okamoto, M.; Sugimoto, Y.; Akiyama, S.; Baba, M. *Mol. Pharmacol.* **2003**, *63*, 65–72.

253. Volk, E. L.; Farley, K. M.; Wu, Y.; Li, F.; Robey, R. W.; Schneider, E. *Cancer Res.* **2002**, *62*, 5035–5040.

254. Allen, J. D.; Jackson, S. C.; Schinkel, A. H. *Cancer Res.* **2002**, *62*, 2294–2299.

255. Rabindran, S. K.; He, H.; Singh, M.; Brown, E.; Collins, K. I.; Annable, T.; Greenberger, L. M. *Cancer Res.* **1998**, *58*, 5850–5858.

256. Rabindran, S. K.; Ross, D. D.; Doyle, L. A.; Yang, W.; Greenberger, L. M. *Cancer Res.* **2000**, *60*, 47–50.

257. de Bruin, M.; Miyake, K.; Litman, T.; Robey, R.; Bates, S. E. *Cancer Lett.* **1999**, *146*, 117–126.

258. Shepard, R. L.; Cao, J.; Starling, J. J.; Dantzig, A. H. *Int. J. Cancer* **2003**, *103*, 121–125.

259. Allen, J. D.; van Loevezijn, A.; Lakhai, J. M.; van der Valk, M.; van Tellingen, O.; Reid, G.; Schellens, J. H.; Koomen, G. J.; Schinkel, A. H. *Mol. Cancer Ther.* **2002**, *1*, 417–425.

260. Cooray, H. C.; Blackmore, C. G.; Maskell, L.; Barrand, M. A. *NeuroReport* **2002**, *13*, 2059–2063.

261. Maliepaard, M.; Scheffer, G. L.; Faneyte, I. F.; van Gastelen, M. A.; Pijnenborg, A. C.; Schinkel, A. H.; van De Vijver, M. J.; Scheper, R. J.; Schellens, J. H. *Cancer Res.* **2001**, *61*, 3458–3464.

262. Jonker, J. W.; Buitelaar, M.; Wagenaar, E.; Van Der Valk, M. A.; Scheffer, G. L.; Scheper, R. J.; Plosch, T.; Kuipers, F.; Elferink, R. P.; Rosing, H.; Beijnen, J. H.; Schinkel, A. H. *Proc. Natl. Acad. Sci. USA* **2002**, *99*, 15649–15654.

263. Taipalensuu, J.; Tornblom, H.; Lindberg, G.; Einarsson, C.; Sjoqvist, F.; Melhus, H.; Garberg, P.; Sjostrom, B.; Lundgren, B.; Artursson, P. *J. Pharmacol. Exp. Ther.* **2001**, *299*, 164–170.

264. Smith, A. J.; van Helvoort, A.; van Meer, G.; Szabo, K.; Welker, E.; Szakacs, G.; Varadi, A.; Sarkadi, B.; Borst, P. *J. Biol. Chem.* **2000**, *275*, 23530–23539.

265. Deleuze, J. F.; Jacquemin, E.; Dubuisson, C.; Cresteil, D.; Dumont, M.; Erlinger, S.; Bernard, O.; Hadchouel, M. *Hepatology* **1996**, *23*, 904–908.

266. de Vree, J. M.; Jacquemin, E.; Sturm, E.; Cresteil, D.; Bosma, P. J.; Aten, J.; Deleuze, J. F.; Desrochers, M.; Burdelski, M.; Bernard, O.; Oude Elferink, R. P.; Hadchouel, M. *Proc. Natl. Acad. Sci. USA* **1998**, *95*, 282–287.

267. Childs, S.; Yeh, R. L.; Georges, E.; Ling, V. *Cancer Res.* **1995**, *55*, 2029–2034.

268. Gerloff, T.; Stieger, B.; Hagenbuch, B.; Madon, J.; Landmann, L.; Roth, J.; Hofmann, A. F.; Meier, P. J. *J. Biol. Chem.* **1998**, *273*, 10046–10050.

269. Boyer, J. L. *Gastroenterology* **2002**, *123*, 1733–1735.

270. Byrne, J. A.; Strautnieks, S. S.; Mieli-Vergani, G.; Higgins, C. F.; Linton, K. J.; Thompson, R. J. *Gastroenterology* **2002**, *123*, 1649–1658.

271. Strautnieks, S. S.; Bull, L. N.; Knisely, A. S.; Kocoshis, S. A.; Dahl, N.; Arnell, H.; Sokal, E.; Dahan, K.; Childs, S.; Ling, V.; Tanner, M. S.; Kagalwalla, A. F.; Nemeth, A.; Pawlowska, J.; Baker, A.; Mieli-Vergani, G.; Freimer, N. B.; Gardiner, R. M.; Thompson, R. J. *Nat. Genet.* **1998**, *20*, 233–238.

272. Strautnieks, S. S.; Kagalwalla, A. F.; Tanner, M. S.; Knisely, A. S.; Bull, L.; Freimer, N.; Kocoshis, S. A.; Gardiner, R. M.; Thompson, R. J. *Am. J. Hum. Genet.* **1997**, *61*, 630–633.

273. Wang, L.; Soroka, C. J.; Boyer, J. L. *J. Clin. Invest.* **2002**, *110*, 965–972.

274. Trauner, M.; Boyer, J. L. *Physiol. Rev.* **2003**, *83*, 633–671.

19

LIPOSOMES AS DRUG DELIVERY VEHICLES

GUIJUN WANG

Department of Chemistry, University of New Orleans, New Orleans, LA 70148

19.1. INTRODUCTION

Liposomes or vesicles are cell-like spherical aggregates formed by self-assembling of amphiphilic molecules such as phospholipids. Liposomes were first discovered

Drug Delivery: Principles and Applications Edited by Binghe Wang, Teruna Siahaan, and Richard Soltero
ISBN 0-471-47489-4

by A.D. Bangham in the early 1960s.[1,2] They were envisioned as ideal drug delivery systems because of their high degree of biocompatibility and their ability to encapsulate a large amount of material inside the vesicle. The lamellar vesicles are important as models for biomembranes, for drug formulation, for gene delivery, and as vehicles for delivering various agents used in cancer therapy. Liposomes are especially effective in delivering drugs to the phagocytes of the immune system.

Liposome drug delivery systems can improve the therapeutic index and bioavailability, increase efficacy, and reduce toxicity. They have been used as carriers for anticancer drugs[3] and antimicrobial agents,[4] the delivery of macromolecules including DNA[5] and proteins,[6] and the delivery of drugs for diabetes and cardiovascular diseases. Since their discovery, many liposomal formulations have been in clinical trials and a few successful liposome-based drugs have been approved by the Food and Drug Administration (FDA).[7–12] These approved formulations include the delivery of the anticancer agents doxorubicin (Doxil and Myocet), daunorubicin (DaunoXome) and the antifungal agent amphotericin B (AmBisome, Amphotect, ABELCET). The liposome formulations of these drugs have achieved significant reduction in the toxicity of the drugs and have maintained or improved the efficacy of the active compounds.

However, the therapeutic potential of liposome drug delivery systems has not been fully reached. The main reason for the limited success is that there are many design and application problems with drug delivery systems. These include stability, time of release, cost of preparation, short shelf life, and poor interaction with certain drugs. It is still difficult to selectively increase the bioavailability of the drug at the target tissue while maintaining stability during circulation. This chapter will focus mainly on the most recent developments in using liposomes as drug delivery vehicles including stabilized liposomes, triggered release, and targeted delivery systems.

19.2. CONVENTIONAL LIPOSOMES

Conventional liposomes (CLs) are lipid vesicles composed of various phospholipids, glycolipids, and other lipids without derivatization to increase the circulation time. Figure 19.1 shows the general structures of some common membrane lipids. Among the most important phospholipids, phosphatidylcholine **1** (PC) usually forms a bilayer structure. Phosphatidyl ethanolamine **2** (PE) tends to form micelles or inverted hexagonal structures because of its small head group. Because of its strong propensity to form bilayer structures or vesicles, PC (lecithin) is the most extensively studied phospholipid, especially as a drug delivery carrier.[13–16] PE has been studied extensively because protonation of the amino group can lead to the transition to a lamellar packing mode, and PE is the only natural amino acid that can be modified easily in the head group region. However, the uses of natural membrane lipids have been limited by difficulties in their fabrication and by their poor stability. Many synthetic lipids have been developed to form self-assembled aggregates.

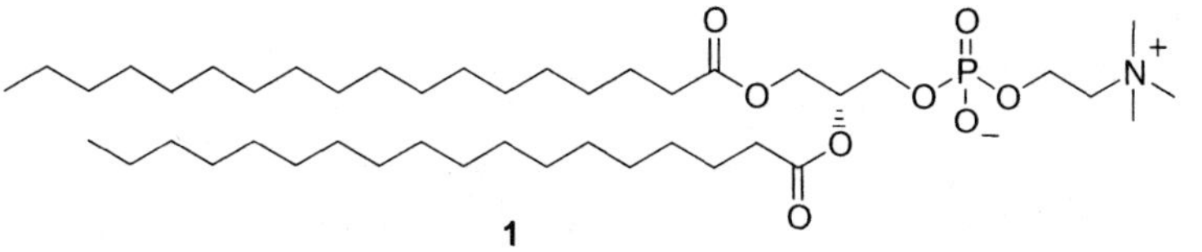

Figure 19.1. The general structures of some phospholipids, distearoyl-phosphatidylcholine (DSPC), **1**, and distearoylphosphatidylethanolamine (DSPE), **2**.

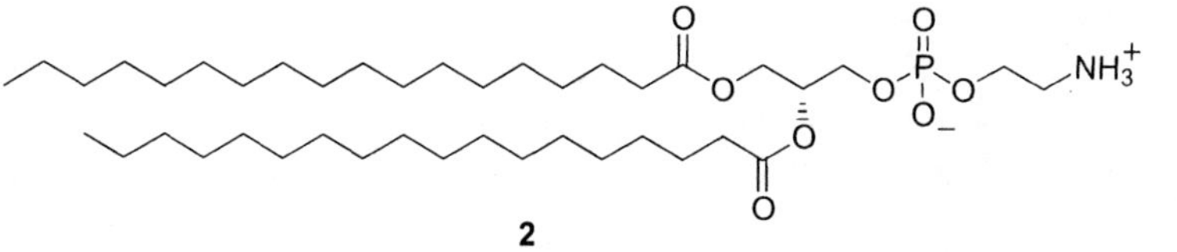

Depending on the nature of the amphiphilic lipids, different types of liposomes, ranging from very small to very large vesicles, unilamellar liposomes or multilamellar liposomes (Figure 19.2*a–c*) can be formed. Because liposomes contain both hydrophobic and hydrophilic layers alternatively, hydrophobic molecules can be contained within the bilayer and water-soluble materials within the aqueous compartments. Highly nonpolar drugs can be trapped within the nonpolar bilayer, whereas more polar molecules can be encapsulated within the aqueous cavity (Figure 19.2*d*). Inside a liposome, a drug is protected from the cellular enzymes that would destroy it. This increases the lifetime of the drug in the body and therefore reduces the dosage required to obtain a desired effect.

Since their discovery by Bangham, a vast amount of technology has been developed using liposomes. Many applications of liposomes as delivery systems have been developed, but only a few of them have reached the commercial stage. These

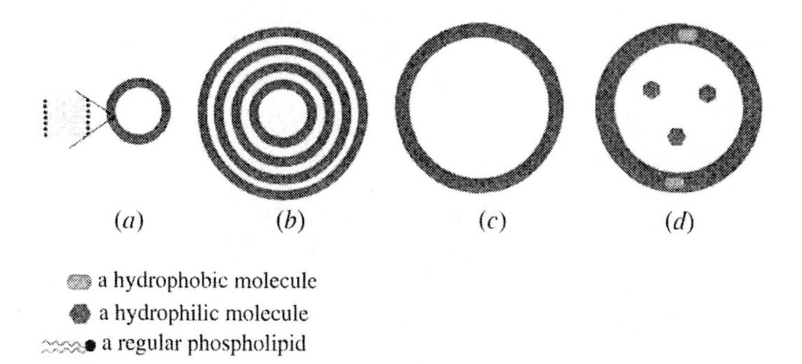

Figure 19.2. (*a*) Small unilamellar liposome (SUV, 20–50 nm). (*b*) Multilamellar vesicles (MLV, up to 10 um), (*c*) Large unilamellar vesicle (LUV, 50 nm–10 um). (*d*) A liposome can encapsulate water-soluble drugs inside the cavity and nonpolar molecules within the bilayers.

include enzyme-targeted delivery of anticancer drugs,[17–19] a bacteriacide,[20] and moisturizers and anti-inflammatory agents[21] for skin. Cationic liposomes have been widely explored for gene therapy.[22–24] The potential of liposomal delivery seems very promising. But there are some problems with conventional liposomes. Although the idea of using them as drug delivery systems has been around since the early 1970s, conventional liposomes have proven to be poor carriers for drugs because they are readily broken by contact with other surfaces (like soap bubbles) and are easily degraded by cellular enzymes. They can also be easily disrupted by changes in pH, temperature, or salt strength.

Conventional liposomes suffer from low stability and rapid clearance from blood. They are rapidly taken up mainly by macrophages in the liver and spleen and to a lesser extent in the reticuloendothelial system (RES) or the mononuclear phagocytic system (MPS). Although liposomes have been explored to treat diseases that affect the phagocytes of the immune system, rapid clearance has hampered their development as drug delivery systems. The major limitations of using conventional liposomes as drug delivery vehicles are rapid clearance from blood, restricted control of encapsulated molecule release, low or nonreproducible drug loading, physical and chemical instability, and large-scale sterile preparation. Many of these problems have been addressed during the past two decades of research. However, liposomal delivery systems still need further development studies, and the precise mechanisms of their actions in the body need to be elucidated. The targeting to specific disease sites is still challenging.

19.3. STABILIZED LIPOSOMES

Because of the stability problems with conventional liposomes, scientists have sought many methods to stabilize them. One important development is the sterically stabilized liposomes (SSLs), which are sometimes called "stealth liposomes."[25–28] Synthetic polymers are used for steric stabilization. Another approach involves cross-linking membrane components covalently or by the polymerization of polymerizable lipids.[29,30] A third approach utilizes unusually stable archaebacterial membrane lipids mimics.[31]

19.3.1. Sterically Stabilized Liposomes

Protecting the liposome surface by polyethylene glycol (PEG) or other polar surface ligands such as carbohydrates has led to the development of stealth liposomes. These long-circulating liposomes have improved pharmacokinetics compared to CLs.[32–34] Surface components on stealth liposomes prevent the liposomes from sticking to each other as well as to blood cells or to vascular walls. They are invisible to the immune system and have shown promising results in cancer therapy. They are also less prone to uptake by the liver. They can therefore remain in circulation longer than conventional liposomes.[35] A small percentage of the PEG-modified PE **3** and some normal PCs can form sterically stabilized liposomes.

Prolonged circulation and reduced MPS uptake has been achieved with PEG molecules in the range of 1000 to 5000 daltons.

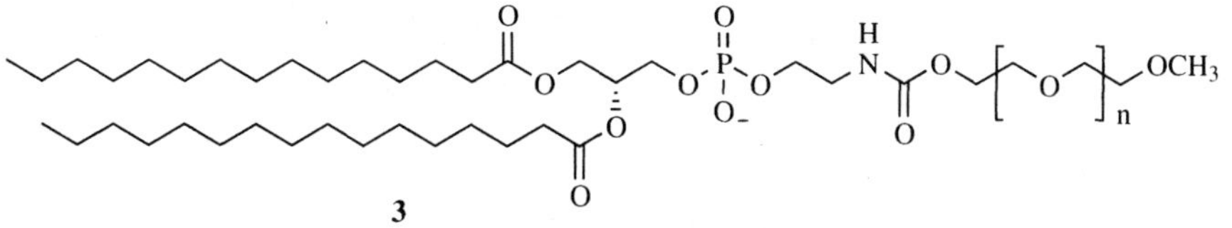

3

SSLs have also been created in which the lipid bilayer contains glycolipids, principally monosialoganglioside GM (1). The resulting liposomes showed decreased uptake into the MPS, increased circulation half-lives, increased stability to contents leakage, and dose-independent pharmacokinetics.[36] Sterically stabilized liposomes can stay in the blood up to 100 times longer than conventional liposomes; thus, they can increase the pharmacological efficacy of encapsulated agents. The mechanism of the SSLs is that the surface-grafted chains of flexible and hydrophilic polymers can form dense "conformational clouds," preventing other macromolecules from interacting with the surface even at low concentrations of the protecting polymer. The incorporation of PEG and other hydrophilic polymers into lipid bilayers gives rise to sterically stabilized liposomes that exhibit reduced blood clearance and concomitant changes in tissue distribution largely because of reduced phagocytic uptake. The polymer forms a surface coating which has been characterized by physical measurements. It appears to function through steric inhibition of the protein binding and cellular interactions which lead to phagocytic uptake.[37,38] These systems have revived the possibility of ligand-dependent targeting to specific cells by incorporating targeting ligands on their surface, because they are much less subject to nonspecific uptake than are the conventional liposomes.[39–41] Site-specific drug delivery can be achieved by using sterically stabilized drug carriers where ligands bearing targeting information are attached on the carrier surface or when a phase transition is induced by an external stimulus.

Some unexpected observations have been made during *in vivo* studies of PEG-modified liposomes.[42] Studies with PEG liposomes in patients have shown that liposome can induce side effects such as flushing and tightness of the chest. The PEG liposomes were cleared relatively rapidly from the bloodstream without the supposed long circulation property.[42] Despite this, the development of SSLs represented a major advance in using liposomes as drug delivery systems. Many ongoing research efforts are devoted to bring SSLs to practical applications.[43,44]

19.3.2. Polymerization-Stabilized Vesicles

The practical use of self-assembled systems in technological applications requires that the supramolecular assembly survive and function under a variety of conditions. Another strategy for obtaining stability is by polymerization of the microstructures following self-assembly. There have been intensive studies in this area.[45,46] Fatty acids with polymerizable groups have been used to prepare polymerizable

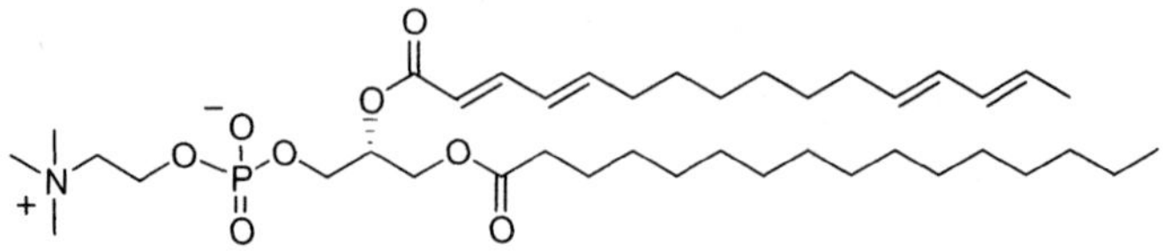

Figure 19.3. Structures of some polymerizable lipids with butadiene (**4**), methacryloyl (**5**), and diacetylene (**6**, **7**) functional groups.

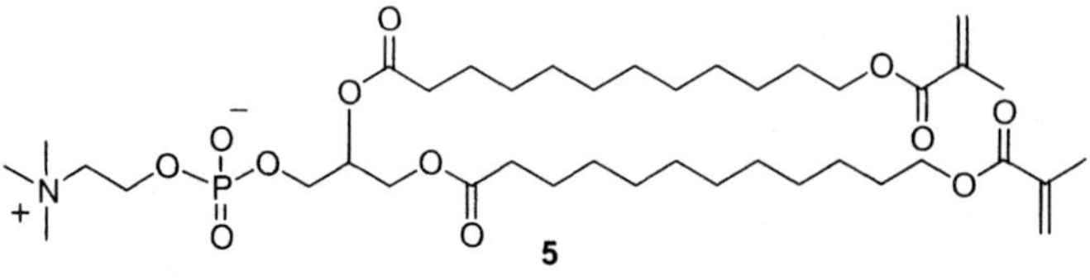

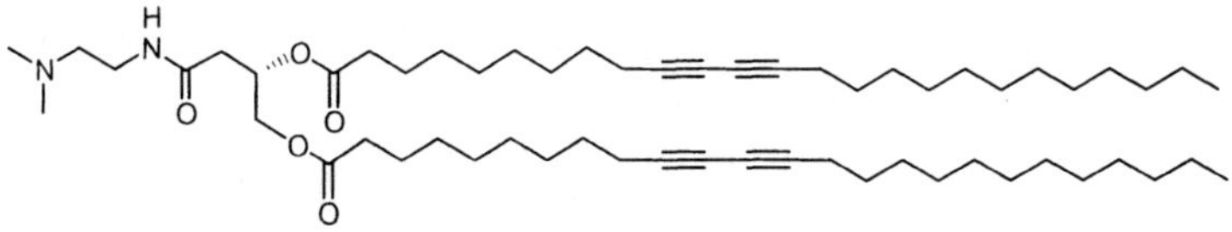

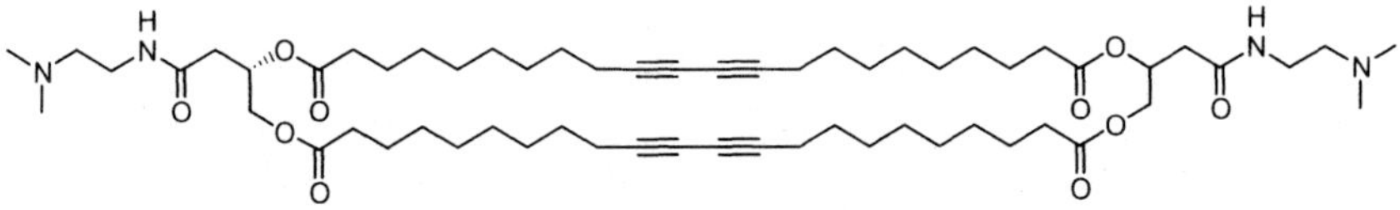

vesicles.[47,49] Phospholipids containing diacetylenes in the fatty acyl chains [50–53] and diacetylenic lipids with other polar head groups[54–56] have also been synthesized and characterized. Other polymerizable groups that have been explored include butadienes,[57] terminal vinyl and methacryloyl functionalities,[58,59] and acrylates.[60,61] Some structures of lipids containing polymerizable groups including methacrylate and diacetylene groups are shown in Figure 19.3. Lipids **4** and **5** contain polymerizable diene and methacryloyl functional groups;[57–59] lipids **6** and **7** contain diacetylene groups that can be cross-linked with ultraviolet (UV) light.[54–56] These materials have been investigated for producing stabilized microstructures for biosensors and encapsulation applications. As noted earlier, one general drawback to the use of phospholipids in the applications discussed above is the general instability of the supramolecular structures they form. These vesicles tend to lyse too easily and leak the materials that are trapped therein. The use of polymeric vesicles and micelles as drug carriers has also been reviewed recently.[62,63] Cross-linking of monomer units after formation of vesicles will lead to stabilized vesicles with improved stability. This approach holds great potential for fabricating drug delivery carriers, but it has not been investigated fully. Further studies to optimize the flexibility and controlled stability need to be carried out before the application in drug or gene delivery systems can be developed.

19.3.3. Archaebacterial Membrane Lipids and Analogs

The third method of obtaining stabilized liposomes is to use the archaebacterial membrane lipids or their analogs as the lipid components of the liposomes. Archaebacteria and some gram-positive bacteria can survive in extreme environments such as high temperatures, in organic solvents, and at extremes of pH (2 to 10).[64–66] These organisms survive by synthesizing membrane lipids containing very long fatty acid chains that go through the entire membrane instead of just from the middle of the membrane to the outside.[67–69] The fatty acids are very long α,ω-dicarboxylic acids (28 to 36 carbons) that are esterified to a glycerol molecule on both ends. The head groups are phosphatidyl glycerol, monoglucosyl diacyl glycerols, or any of the common functionalities found in typical membranes (Figure 19.4). Studies have shown that much of the stability of the membranes of such extremophilic organisms stems from the presence of the transmembrane fatty acyl group.

The archaebacterial lipids have been used to prepare liposomes. They form monolayer lipid membranes[70] rather than the normal bilayer type because there are polar groups on both ends of the α,ω-dicarboxylic acids. Membrane lipids extracted from these bacteria have also been used to form liposomes. Physical studies showed positive results such as extra stability at high temperature, at high or low pH, and in other harsh conditions.[71–74]

Because isolation from bacteria yields material only in very small quantities, it is impractical to use them for applications such as targeting drug delivery systems and in the manufacture of biocompatible surfaces. In recent years, due to the importance of these special types of membrane lipids, many synthetic efforts have been made in this area, including the total synthesis of these archaebacterial lipids.[75–78] Some bolaform amphiphilic models of archaebacterial membrane lipids [79–83] which contain transmembrane alkyl chains and ether or ester linkage to the head groups have also been prepared. The structures and properties of these types of lipids were studied systematically. These bipolar lipids have the advantage of forming lamellar systems that are stable at extreme pH,[84] high temperatures,[85] and high ionic strengths.[86] We have also synthesized a dimeric phosphatidyl ethanolamine joined by a very long chain fatty acid component.[87] This lipid **6** can form uniform, stable liposomes the surface of which can be functionalized by specific receptors.

The potential applications of archaebacterial lipid liposomes (archaeosomes) as novel vaccine and drug delivery systems have been reviewed recently.[88–90] In general, this type of liposomes exhibited higher stability to oxidation, high temperature, alkaline pH, and action of phopholipases. The safety profile study of these type of liposomes had shown that they are well tolerated after intravenous or oral delivery in mice. The stability and safety profile of these liposomes indicated that they may offer a superior alternative to the use of conventional liposomes.

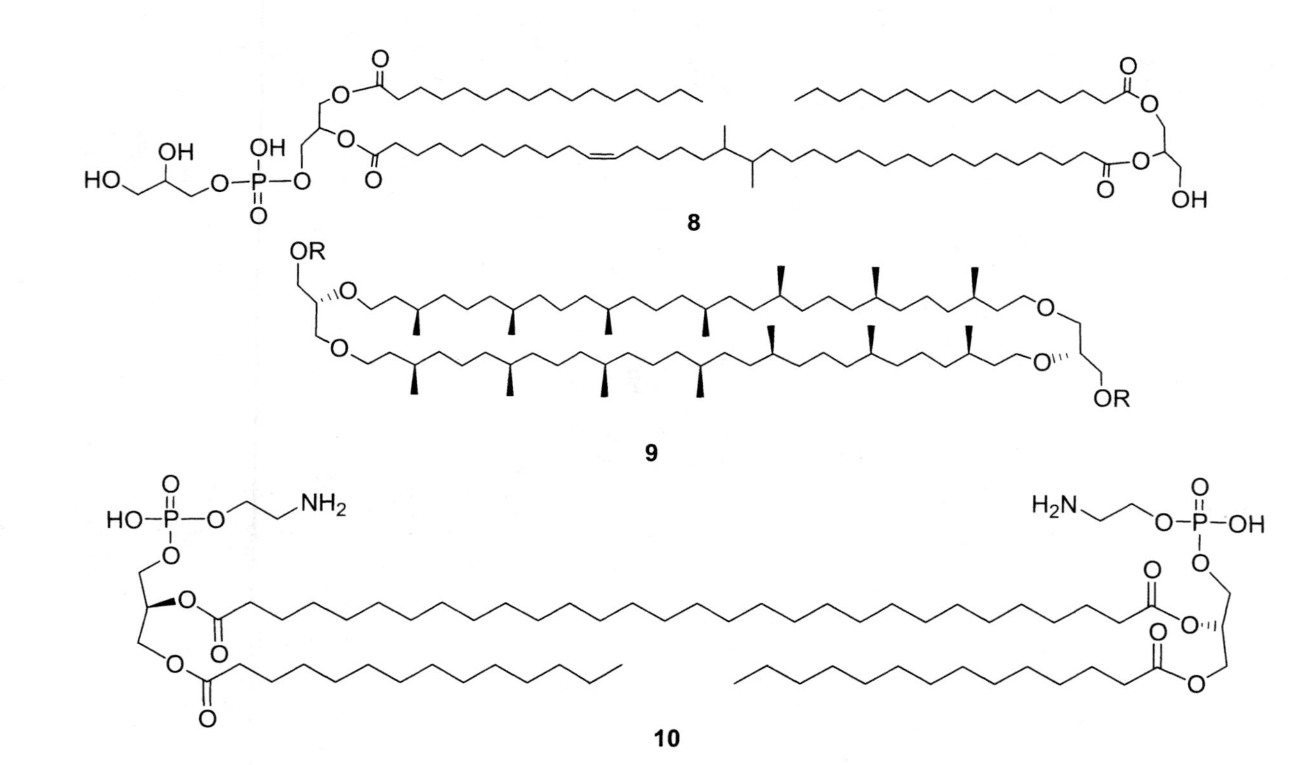

Figure 19.4. Structures of two archaebacterial lipid models (**8**, **9**) and a synthetic extremophile lipid analog with PE-type head groups (**10**).

19.4. CATIONIC LIPOSOMES

Liposomes hold especially great potential for use in gene therapy.[91,92] This is a new approach to the treatment of disease that offers promise for tomorrow's medicine where aberrant genes are replaced with functional ones. Once inside a liposome, the polar DNA molecule can traverse the protective membrane of the target cell. It is also protected from nucleases, enzymes that degrade DNA. Suitable gene delivery carriers or vectors are crucial for successful gene therapy. It is difficult to deliver macromolecules to cells while maintaining their activity. A suitable carrier should be able to encapsulate the gene and protect it from nucleases and other enzymes, deliver it to the cell, release it, and then be degraded safely. Viruses have the capability to deliver their own genes to human cells in a pathogenic manner. They are being exploited as gene therapy vectors. Though they are effective gene delivery vectors, viruses have some problems. These include safety concerns, the high cost of manufacture, difficulty of preparation, and difficulty in maintaining activity. Viral vectors very often evoke an undesirable and potentially lethal immune response. Synthetic vectors such as liposomes offer an alternative to viral carriers.[93–109]

The most popular synthetic liposome carriers are often cationic-lipid-based systems having amino groups in the head group region or quaternary ammonium salts.[93] Under physiological pH, the amino groups are protonated and therefore have positive charges. The first few important lipids synthesized for use in gene transfection are DOTAP (**11**), DOTMA (**12**), and DDAB (**13**) (Figure 19.5). DOTMA was commercialized as Lipofectin as a one-to-one mixture with DOPE and has been widely used to transfect a variety of animal and plant eukaryotic cells. The lipopolyamine DOGS has also been commercialized as Transfectam and has been shown to transfect many animal cells efficiently. The problem associated with cationic liposome delivery systems is that they are rapidly cleared from the blood before they can deliver the genes to the target sites. Also, toxicity and low transfection efficiency are major barriers limiting their clinical applications. Over the past decades, many new cationic lipids have been synthesized and the problems involved in cationic liposome gene delivery have been addressed; the recent research results about the structures of DNA-liposome complexes and strategies used to design different cationic lipids have been reviewed in the literature.[93–108] Several human gene therapy clinical trials using cationic liposomes have been conducted, and more trials will begin in the near future. Their simplicity, efficiency, and relative safety make cationic liposomes attractive vehicles for human gene therapy.

19.5. LIPOSOMES WITH TRIGGERED RELEASE FUNCTIONALITIES

During the past few decades, numerous chemical methodologies have been developed to address drug encapsulation, retention, and stability of liposomes. Strategies or lipid structures that allow optimal release at the target site are just beginning to gain attention. Currently there are no such formulations in clinical trials. However,

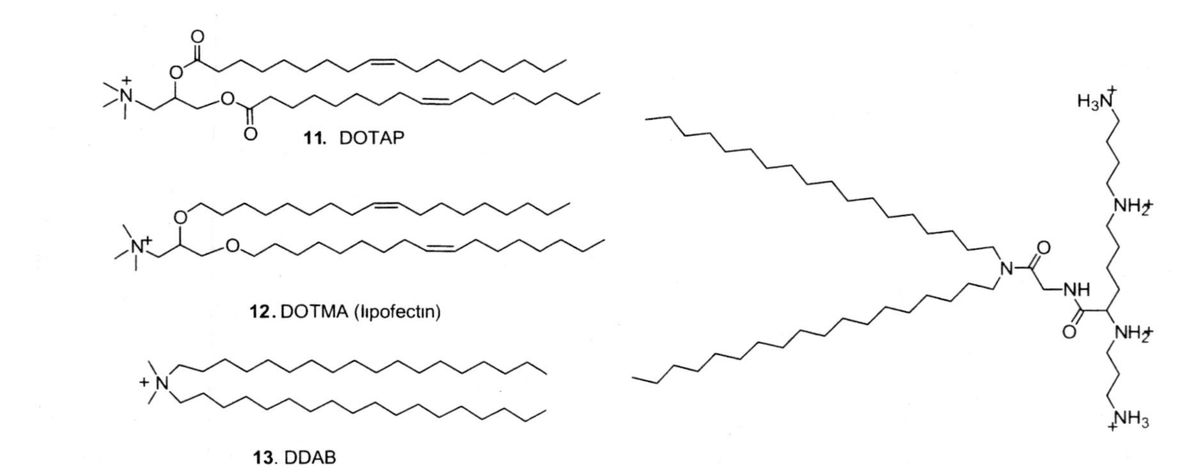

Figure 19.5. The common structures of the most widely studied cationic lipids used in gene delivery. Dioleoyloxy-3-(trimethylammonio)propane (DOTAP) **11**, Dioleoyloxypropyl-trimethylammoinium chloride (DOTMA) **12**, dimethyloctadecylammonium bromide (DDAB) **13**, and dioctadecylamidoglicylspermin (DOGS) **14**.

to fully realize the drug and gene delivery potential of liposomes, a triggering mechanism that allows the drug to be delivered in a controlled fashion is important. Several types of triggerable liposomes have emerged in the literature recently. These include triggering release in response to external or cellular stimuli such as changes of temperature, pH, and light radiation, as well as the presence of enzyme.

19.5.1. Thermosensitive Liposomes

Temperature sensitive liposomes leak more readily above the phase transition temperature of their membrane lipids.[109–111] These liposomes are designed to be stable up to 37°C but will break down as they pass through an area of the body where the temperature is above 40°C. Such sites include the interior of a tumor or an area subjected to external local heating. These liposomes have been prepared with lipids whose membranes undergo a gel-to-liquid crystalline phase transition a few degrees above physiological temperature. Temperature sensitization of liposomes has been attempted using thermosensitive polymers as well. Thermosensitive liposomes have found applications in the treatment of benign prostatic hyperplasia (BPH). The Prolieve Thermodilation system was approved by the FDA in February 2004 for the treatment of BPH. Other ongoing clinical trials exploring the applications of heat-sensitive liposomes are being carried out. These include a Phase I study using Thermodox (doxorubicin encapsulated in a heat-activated liposome).[110] Animal studies have shown that heat-sensitive liposomes can improve chemotherapy delivery to tumors. Thermosensitve liposomes[110,111] represent one of the advanced delivery methods for anticancer agents; however, they require external heat to activate the liposome surrounding, including the tumor area. This is not applicable to the treatment of distant metastases. It is technically difficult to produce thermosensitve liposomes that can retain encapsulated drugs. Even though they constitute a promising triggered delivery method, further studies are necessary to regulate the mechanism and increase the efficacy and reliability of heat-sensitive liposomes.

19.5.2. pH-Sensitive Liposomes

The endosomal compartment has an acidic pH (5.0–5.5), and elevated acidity is also observed in tumor cells.[112] Because of this, there has been great interest in using pH-sensitive liposomes as drug delivery systems.[113–116] pH-sensitive liposomes were initially designed to exploit the acidic environment of tumors to trigger destabilization of the liposome membrane. However, the tumor interstitium (pH 6.5) is often not acidic enough. This makes the engineering of liposomes to be disrupted at this narrow window problematic. Endosomes and lysosomes often have acidic pH conditions lower than 5.0. pH-sensitive liposomes can be recognized and taken up by endocytosis; one competing process is the delivery to lysosomes where its contents can be hydrolyzed and degraded. In order to circumvent this, pH-sensitive liposomes are designed to release their contents prior to reaching the lysosomes. The most extensively studied pH-sensitive liposomes are various PE derivatives in combination with mildly acidic amphiphiles. The PE head group has a primary amino group which can be protonated in low-pH conditions. The rationale is that after protonation of PE at lower pH, the membrane packing will be disrupted, and the contents released. However, though extensively studied, PE derivatives are not likely to be useful *in vivo* due to their serum instability and rapid clearance. Most investigations are still at the laboratory stage; no clinical studies have been done.

Besides the PE derivatives, another type of acid-labile or "caged" lipids containing acid-cleavable groups has been investigated as triggerable drug delivery vehicles. Triggerable lipids that can respond to different pHs might be suitable drug delivery systems. A few systems utilizing acid-labile or caged liposomes have been developed. These are mainly based on derivatives of naturally available lipids such as modified PE and plasmalogens. Synthetic orthoester derivatives have also been developed as acid-cleavable lipids. The anhydride-modified DOPEs such as *N*-maleyl and *N*-citraconyl DOPE are sensitive to pH 5.5–6.5; the liposomes hydrolyze rapidly to give DOPE (Figure 19.6). The one potential problem can be the stability of this type of liposome. Plasmalogens,[114,117] another naturally occurring lipid, have also been investigated. In a plasmalogen molecule, one of the hydrocarbon chains is attached to the head group by an enol ether linkage. This linkage is broken at low pH, causing disruption of the liposome (Figure 19.7).

One drawback of these natural lipid-based systems is that they are difficult to synthesize or purify in large quantity. PE is not available in large quantities, and it is difficult to isolate or synthesize. The synthesis of plasmalogens is also not trivial. In order for the acid-sensitive systems to be practical, the problem of producing material on a large scale at a reasonable cost has to be addressed. To increase the stability of acid-labile or acid-sensitive liposomes, PEG has been used in the design of diortho ester conjugates of PEG and distearoyl glycerol (Figure 19.8).[116] Despite the progress that has been made, further research is necessary for the medical applications of these acid-triggerable liposomes.

19.5.3. Light-Sensitive Liposomes

Light-activated supramolecular systems[118,119] should have potential applications as drug delivery carriers. Light-triggered release of liposomal contents offers an attractive alternative to temperature- or pH-triggered release. There are a few approaches to photoresponsive vesicles based on the disruption of lipid bilayer integrity.[120] These include photopolymerization of lipid tails in mixed-lipid systems that results in phase separations, thus releasing the contents, and azo benzene isomerization,[120] sensitized photo-oxidation cleavage of the lipid tail of plasmalogen and diplasmalogen[121,122] that results in increased membrane permeability and a photocleavable DOPE derivative,[123] NOVC-DOPE. The mechanism of its photo-activated cleavage is shown in Figure 19.9. NOVC-DOPE (**20**) is an example of applying caged compounds in biological studies in the design of light-sensitive lipids.[124–126] These systems are mainly based on natural phospholipid derivatives that are restricted of large scale availability. Compound **22** is a synthetic photosensitive lipid that

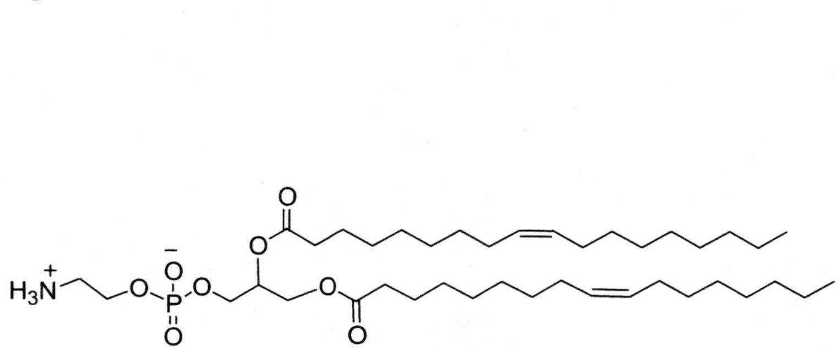

15
DOPE

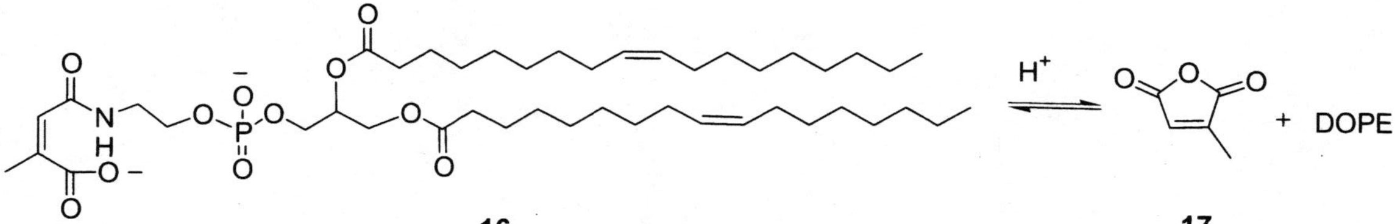

16

H^+

17

+ DOPE

Figure 19.6. A pH-sensitive lipid *N*-citraconyl-dioleoylphosphatidyl ethanolamine **16** and its mechanism of acid-triggered cleavage.

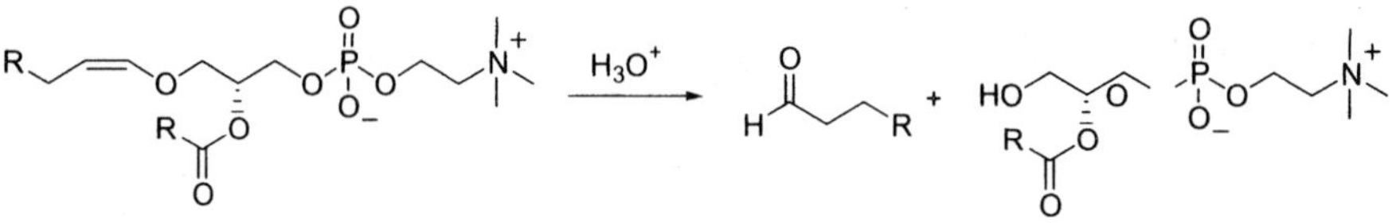

Figure 19.7. Hydrolysis of plasmalogen lipids.

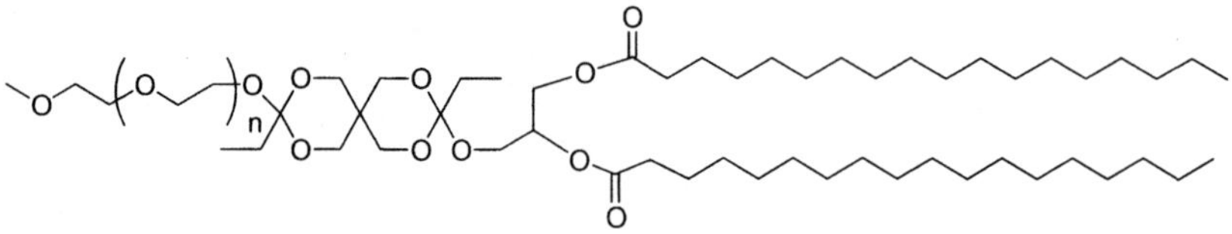

Figure 19.8. Structure of a PEGylated diorthoester distearoyl glycerol.

has been developed very recently.[127] The photosensitive group 3-nitroaminopyrine is linked to the head group region of a hydrophilic phosphocholine headgroup.

19.5.4. Redox-Triggered Liposomes

Other triggered mechanisms include redox-triggered PEG conjugate containing cysteine-cleavable dithiobenzyl urethane linkage (Figure 19.10).[128] The rationale for this approach is that the cytoplasm has a lower redox potential and substantially more molecules with free sulfhydryl groups. Liposomes containing 3 mol% of compound **23** and 97% DOPE were stable in plasma for over a day but released most of their contents within 1 hour in the presence of dithiotreitol. Disulfide-containing cationic lipids have been devised as improved carriers for gene transfection.

19.6. TARGETED DELIVERY

Targeted liposome drug delivery systems should be very useful in cancer therapy, because such liposomes should selectively localize anticancer drugs at the tumor site, thus reducing the toxicity of the drugs to normal cells and improving their therapeutic activity because of the higher drug levels delivered to the tumor. The targeting of liposomes to specific tissues has been used to increase drug delivery efficiency.[129–131] Various antibodies have been conjugated to the surface of

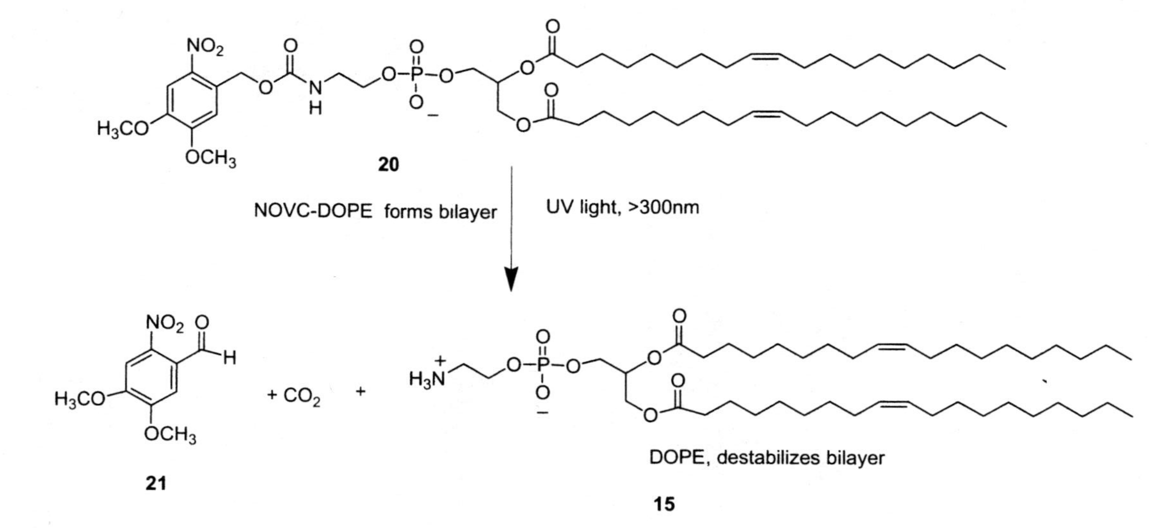

Figure 19.9. The photocleavage of NOVC-DOPE.

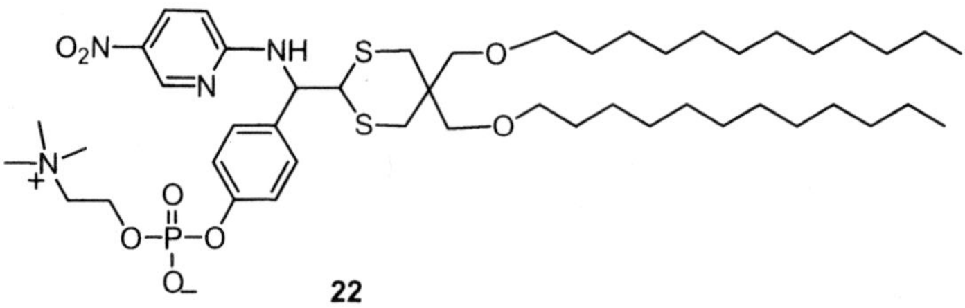

Figure 19.10. A new photosensitive lipid with a PC head group.

stealth liposomes to produce the so called immunoliposomes for active targeting. Extended circulation time is important for active targeting. Although the antibodies can improve the targeting of liposome delivery, studies have shown that accumulation of immunoliposome at the tumor site does not necessarily increase the tumoricidal activity compared to liposomes without antibodies. The main reason is that the drugs cannot escape from the carriers after reaching the tumor site. The immunoliposomes also have the drawback of complexity. Each disease-antibody combination requires a special carrier and manufacturing protocol.

Tumor-specific targeting is a critical goal in research on liposomal drug delivery.[130] The problems encountered when using immunoliposome-mediated targeting of anticancer agents and potential solutions have been reviewed recently.[131–136] Tumor cells aberrantly express tumor-associated antigens that can be utilized as suitable target molecules. Monoclonal antibodies against tumor-associated antigens have been successfully adopted for targeting cancer cells. One example of coupling antibody-Fab′ is shown in Figure 19.12. Tumor-specific ligands other than monoclonal antibodies have also been investigated as *in vivo* tumor-directing molecules. However, this strategy has achieved only limited success. Further studies of tumor-specific interactions and liposomal formulations are necessary for the application of tumor-specific drug delivery for anticancer chemotherapy or gene therapy. Other antibodies that have been investigated include anti-HER2 antibodies for the preparation of immunoliposomes.[135,136]

Besides antibodies, ligands for cell surface receptors are also used as active targeting groups.[137] Chapter 9 describes receptor-mediated drug delivery in detail. Therefore, only a brief description is presented here. Tumor cells contain increased numbers of transferrin receptors compared to normal cells. Transferrin has been used as a ligand for delivering anticancer drugs or drug-containing liposomes.[138–140] Folate receptors[141–144] have been found to be overexpressed by ovarian carcinoma. Folate-containing liposome delivery systems should benefit ovarian cancer treatment. A method of coupling folate to DSPE through a PEG linker is shown in Figure 19.13. Other ligands such as carbohydrates have been used to target selectins[145–149] and other cell surface receptors. Specific targeting to internalizing cell surface receptors has increased both *in vitro* cytotoxicity and *in vivo* efficacy of liposomal drugs.

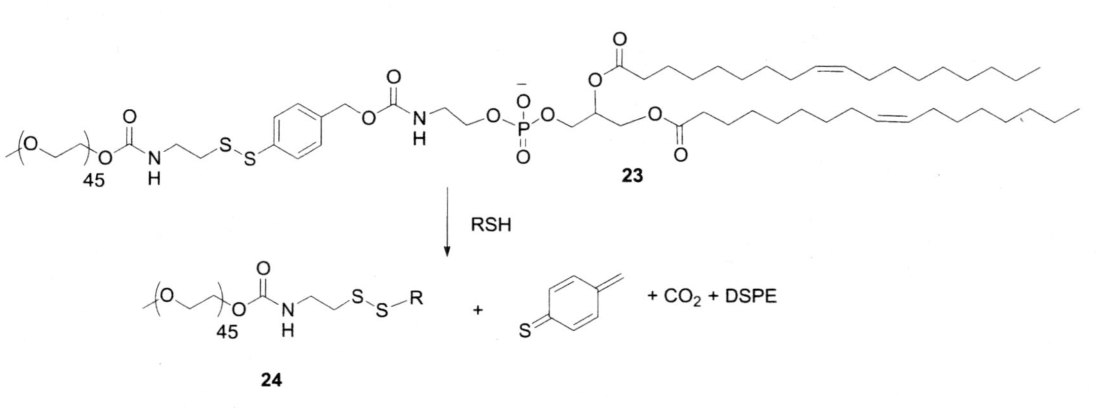

Figure 19.11. A redox-triggered PEG containing lipid (**24**) and its thiolytic cleavage.

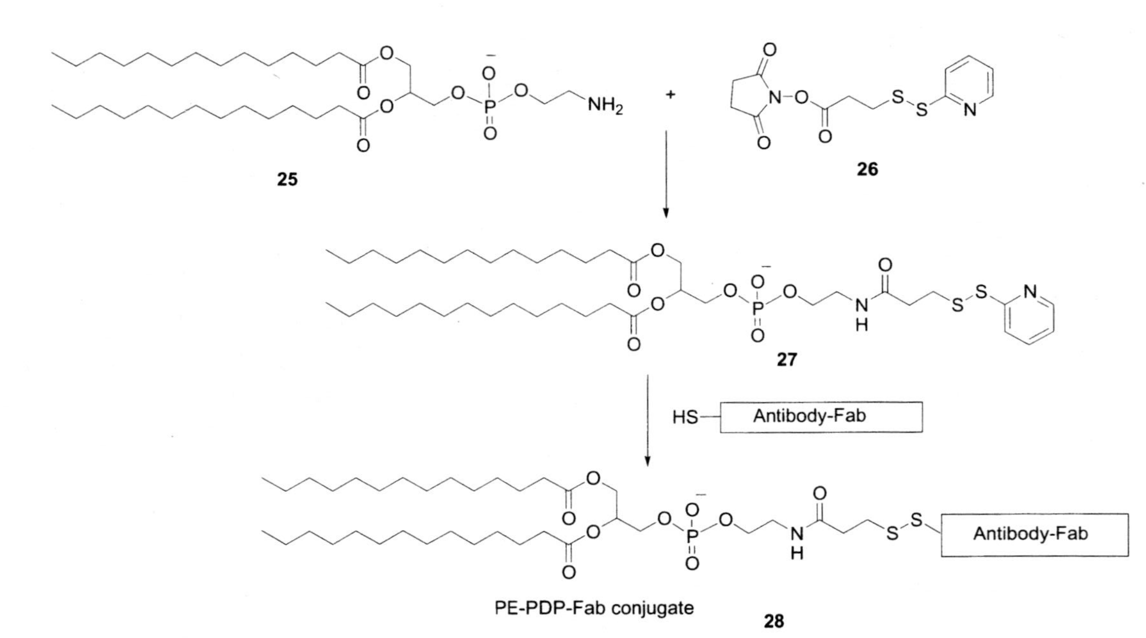

Figure 19.12. A method of coupling antibody Fab′ to PE using *N*-succinimidyl-pyridyl-dithioproprionate.

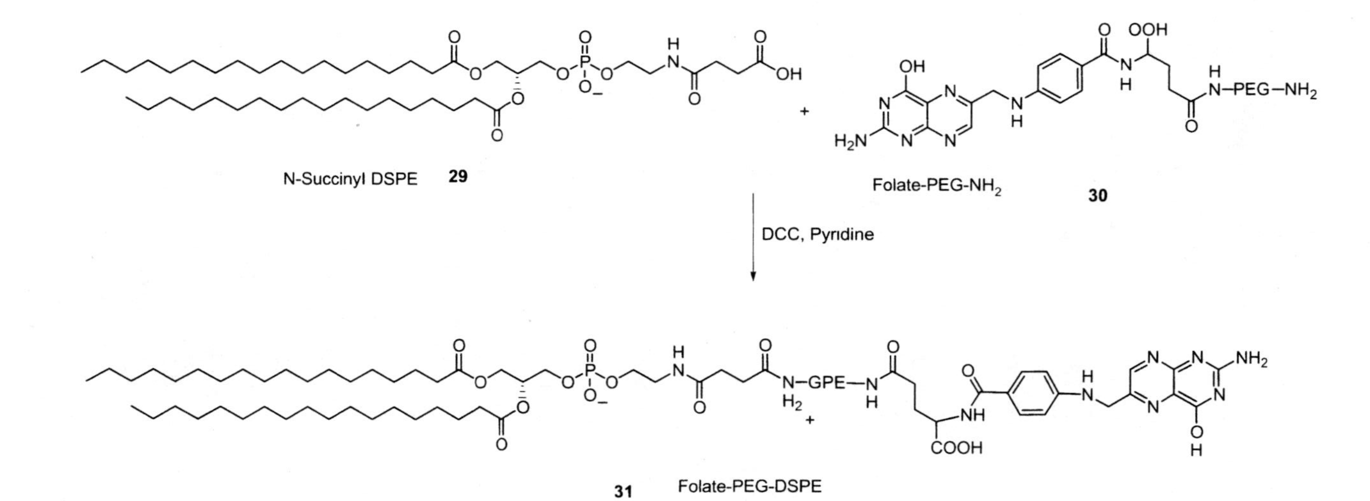

Figure 19.13. Synthesis of folate-PEG-DSPE for targeted drug delivery.

19.7. CONCLUSIONS

Liposomes are the most widely studied colloidal drug carriers in medicines, especially for anticancer agents, antibacterial agents, and gene transfection. The problems involved in developing good drug or gene carriers have been gradually resolved with the advent of improved liposomes and new lipids. These include long circulating stabilized liposomes, numerous cationic lipid systems for gene delivery, new generations of liposomes with triggered release mechanisms, and antibody- or receptor-mediated targeting delivery. Although there are still many hurdles to overcome before the full therapeutic potential of liposomes can be reached, there is great promise that smart liposomal formulations will greatly benefit anticancer therapy and the delivery of vaccines, proteins, and DNA.

REFERENCES

1. Bangham, A. D; Horne, R. W. *Nature* **1962**, *196*, 952–953.
2. Bangham, A. D.; Horne, R. W. *J. Mol. Biol.* **1964**, *8*, 660–668.
3. Drummond, D. C.; Meyer, O.; Hong, K.; Kirpotin, D. B.; Papahadjopoulos, D. *Pharmacol. Rev.* **1999**, *51*, 691–743.
4. Pinto-Alphandary, H.; Andremont, A.; Couvreur, P. *Int. J. Antimicrob. Ag.* **2000**, *13*, 155–168.
5. Nishikawa, M.; Hashida, M. *Biol. Pharm. Bull.* **2002**, *25*, 275–283.
6. Rao, M.; Alving, C. R. *Adv. Drug Deliv. Rev.* **2000**, *41*, 171–188.
7. Gregoriadis, G. *Liposomes as Drug Carriers, Recent Trends and Progress.* John Wiley & Sons: New York, **1988**.
8. Lopez-Berestein, G.; Fidler, I. J. *Liposomes: The Therapy of Infectious Diseases and Cancer.* Alan R. Liss: New York, **1989**.
9. Fielding, R. M. *Clin. Pharmacokinet.* **1991**, *21*, 155–164.
10. Maesaki, S. *Curr. Pharm. Design* **2002**, *8*, 433–440.
11. Barenholz, Y. *Curr. Opin. Colloid In.* **2001**, *6*, 66–77.
12. Barratt, G. *Cell Mol. Life Sci.* **2003**, *60*, 21–37.
13. Lis, L. J; McAlister, M.; Fuler, N.; Rand, R. P.; Parsegian, V. A. *Biophys. J.* **1982**, *37*, 657–665.
14. Wong, M.; Thompson, T. E. *Biochemistry* **1982**, *21*, 4133–4139.
15. Larrabee, A. L. *Biochemistry* **1979**, *15*, 3321–3326.
16. Viguera, A. R.; Alonso, A.; Goni, F. M. *Biochemistry* **1993**, *32*, 3708.
17. Leyland-Jones, B. *Semin. Oncol.* **1993**, *20*, 12–17.
18. Codde, J. P.; Lumsden, A. J.; Napoli, S; Burton, M. A.; Gray, B. N. *Anticancer Res.* **1993**, *13*, 539–543.
19. Jones, M. N.; Francis, S. E.; Hutchinson, F. J.; Handley, P. S.; Lyle, I. G. *Biochim. Biophys. Acta* **1993**, *1147*, 251–161.
20. Strauss, G. *J. Soc. Cosmet. Chem.* **1989**, *40*, 51–60.
21. Rädler, J. O.; Koltover, I.; Salditt, T.; Safinya, C. R. *Science* **1997**, *275*, 810–814.

22. Lasic, D. D.; Papahadjopoulos, D. *Science* **1995**, *267*, 1275–1276.
23. Behr, J.-P.; Demeneix, B.; Loeffler, J.-P.; Perez-Mutul, J. *Proc. Natl. Acad. Sci. USA* **1989**, *86*, 6982–6986.
24. Radler, J. O.; Koltover, I.; Jamieson, A.; Salditt, T.; Safinya, C. R. *Langmuir* **1998**, *14*, 4272–4283.
25. de Menezes, D. E. L.; Pilarski, L. M.; Allen, T. M. *Cancer Res.* **1998**, *58*, 3320–3330.
26. Allen, T. M. *Drugs* **1997**, *54*, 8–14.
27. Allen, T. M.; Moase, E. H. *Adv. Drug Deliv. Rev.* **1996**, *21*, 117–133.
28. Johnston, D. S.; Sanghera, S.; Manjon-Rubio, A.; Chapman, D. *Biochim. Biophys. Acta* **1980**, *602*, 213–216.
29. Johnston, D. S.; Sanghera, S.; Pons, M.; Chapman, D. *Biochim. Biophys. Acta* **1980**, *602*, 57–59.
30. Cav-Con Oncology, Comparison of "sterically stabilized" liposome technology versus LCM technology (Filmix®) for anti-tumor drug delivery; available at http://www.ntplx.net/~cavcon/da1-5.html
31. Patel, G. B.; Agnew, B. J.; Deschatelets, L.; Fleming, L. P.; Sprott, G. D. *Int. J. Pharm.* **2000**, *194*, 39–49.
32. Allen, T. M.; Hansesn, C. B.; Demenezes, D. E. L. *Adv. Drug Deliv. Rev.* **1995**, *16*, 267–284.
33. Gregoriadis, G. *Trends Biotech.* **1995**, *13*, 527–537.
34. Torchilin, V. P. *J. Microencapsulation* **1998**, *15*, 1–19.
35. Woodle, M. C.; Lasic, D. D. *Biochim. Biophys. Acta* **1992**, *1113*, 171–199.
36. Allen, T. M. *Adv. Drug Deliv. Rev.* **1994**, *13*, 285–309.
37. Woodle, M. C. *Adv. Drug Deliv. Rev.* **1998**, *32*, 139–152.
38. Ishida, T.; Harashima, H.; Kiwada, H. L. *Biosci. Rep.* **2002**, *22*, 197–224.
39. Allen, T. M. *Trends Pharmacol. Sci.* **1994**, *15*, 215–220.
40. Oku, N.; Namba, Y. *Crit. Rev. Ther. Drug* **1994**, *11*, 231–270.
41. Yang, L; Alexandridis, P. *Curr. Opin. Colloid. In.* **2000**, *5*, 132–143.
42. Laverman, P.; Boerman, O. C.; Oyen, W. J. G.; Corstens, F. H. M.; Storm, G. *Crit. Rev. Ther. Drug* **2001**, *18*, 551–566.
43. Cattel, L.; Ceruti, M.; Dosio, F. *Tumori* **2003**, *89*, 237–249.
44. Allen, C.; Dos Santos, N.; Gallagher, R.; Chiu, G. N. C.; Shu, Y.; Li, W. M.; Johnstone, S. A.; Janoff, A. S.; Mayer, L. D.; Webb, M. S.; Bally, M. B. *Biosci. Rep.* **2002**, *22*, 225–250.
45. Mueller, A.; O'Brien, D. F. *Chem. Rev.* **2002**, *102*, 727–758.
46. Rudolph, A. S.; Calvert, J. M.; Schoen, P. E.; Schnur, J. M. In *Biotechnological Applications of Lipid Microstructures*; Perseus Publishing: **1988**, pp.
47. Okada, S.; Peng, S.; Spevak, W.; Charych, D. *Acc. Chem. Res.* **1998**, *31*, 229–239.
48. Tieke, B.; Bloor, D. *Makromol. Chem.*, **1979**, *180*, 2275.
49. Day, D.; Ringsdorf, H. *J. Polym. Sci. Polym. Lett. Edn.* **1978**, *16*, 205.
50. Regen, S. L.; Singh, A.; Oehme, G.; Singh M. *J. Am. Chem. Soc.* **1982**, *104*, 791–795.
51. Wagner, N.; Dose, K.; Joch, H.; Ringsdorf, H. *FEBS Lett.* **1981**, 313–318.
52. Johnston, D. S.; Mclean, L. R.; Whitam, M. A.; Clark, A. D.; Chapman, D. *Biochemistry* **1983**, *22*, 3194–3202.

53. O'Brien, D. F.; Whitsides, T. H.; Klingbiel, R. T. *J. Polym. Sci. Polym. Lett. Ed.* **1981**, *19*, 95–101.
54. Wang, G.; Hollingsworth, R. I. *Langmuir* **1999**, *15*, 3062–3069.
55. Wang, G.; Hollingsworth, R. I. *Adv. Materials.* **2000**, *12*, 871–874.
56. Nakashima, N.; Asakuma, S.; Kunitake, T. *J. Am. Chem. Soc.* **1985**, *107*, 509–510.
57. Dorn, K.; Klinbiel, Specht, D. P.; Tyminski, P. N.; Ringsdorf, H.; O'Brien, *J. Am. Chem. Soc.* **1984**, *106*, 1627–1633.
58. Singh, A; Price, R.; Schnur, J. M.; Schoen, P.; Yager, P. *Polym. Prepr.* **1986**, *27*, 393–394.
59. Asakuma, S.; Okada, H.; Kunitake, T. *J. Am. Chem. Soc.* **1991**, *113*, 1749–1755.
60. Marra, K. G.; Kidani, D. D. A.; Chaikof, E. L. *Langmuir* **1997**, *13*, 5697–5701.
61. Dufrêne, Y. F.; Barger, W. R.; Green, J.-B. D.; Lee, G. U. *Langmuir* **1997**, *13*, 4779–4784.
62. Discher, B. M.; Hammer, D. A.; Bates, F. S.; Discher, D. E. *Curr. Opin. Colloid Interface Sci.* **2000**, *5*, 125–131.
63. Kwon, G. S.; Okano, T. *Adv. Drug Deliv. Rev.* **1996**, *21*, 107–116.
64. Lowe, S. E.; Zeikus, J. G. *Arch. Microbiol.* **1991**, *155*, 325–329.
65. Jung, S.; Lowe, S. E.; Hollingsworth, R. I.; Zeikus, J. G. *J. Biol. Chem.* **1993**, *268*, 2828–2835.
66. Berube, L. R.; Hollingsworth, R. I. *Biochemistry* **1995**, *34*, 12005–12011.
67. Jung, S.; Hollingsworth, R. I. *J. Lipid Res.* **1994**, *35*, 1932–1945.
68. Lee, J.; Jung, S.; Lowe, S.; Zeikus, J. G.; Hollingsworth, R. I. *J. Am. Chem. Soc.* **1998**, *120*, 5855–5863.
69. Jung, S.; Zeikus, J. G.; Hollingsworth, R. I. *J. Lipid Res.* **1994**, *35*, 1057–1065.
70. Chang, E. L.; Rudolph, A.; Lo, S. L. *Biophys. J.* **1990**, *57*, 220a-220a.
71. Chang, E. L. *Biochem. Biophys. Res. Commun.* **1994**, *202*, 673–679.
72. Lo, S. L.; Chang, E. L. *Biochem. Biophys. Res. Commun.* **1990**, *167*, 238–243.
73. Komatsu, H.; Chong, P. L.-G. *Biochemistry* **1998**, *37*, 107–115.
74. Elferink, M. G. L.; de Wit, J. G.; Driessen, A. J. M.; Knonings, W. N. *Biochim. Biophys. Acta* **1994**, *1193*, 247–254.
75. Svenson, S.; Thompson, D. H. *J. Org. Chem.* **1998**, *63*, 7180–7182.
76. Arakawa, K.; Eguchi, T.; Kakinuma, K. *J. Org. Chem.* **1998**, *63*, 4741–4745.
77. Eguchi, T.; Ibaragi, K.; Kakinuma, K. *J. Org. Chem.* **1998**, *63*, 2689–2698.
78. Eguchi, T.; Arakawa, K.; Terachi, T.; Kakinuma, K. *J. Org. Chem.* **1997**, *62*, 1924–1933.
79. Thompson, D. H.; Wong, K. F.; Humphry-Baker, R.; Wheeler, J. J.; Kim, J.-M.; Rananavare, S. B. *J. Am. Chem. Soc.* **1992**, *114*, 9035–9042.
80. Yamauchi, K.; Sakamoto, Y.; Moriya, A.; Yamada, K.; Hosokawa, T.; Higuchi, T.; Kinoshita, M. *J. Am. Chem. Soc.* **1990**, *112*, 3188–3191.
81. Menger, F. M.; Chen, X. Y.; Brocchini, S.; Hopkins, H. P.; Hamilton, D. *J. Am. Chem. Soc.* **1993**, *115*, 6600–6608.
82. Moss. R. A.; Li, J.-M. *J. Am. Chem. Soc.* **1992**, *114*, 9227–9229.
83. Moss. R. A.; Fujita, T.; Okumura, Y. *Langmuir* **1991**, *7*, 2415–2418.
84. Führhop, J.-H.; Liman, U.; Koesling, V. *J. Am. Chem. Soc.* **1988**, *110*, 6840–6845.
85. Kim, J-M.; Thompson, D. H. *Langmuir* **1992**, *8*, 637–644.
86. Führhop, J.-H.; Hungerbühler, H.; Siggel, U. *Langmuir* **1990**, *6*, 1295–1300.

87. Wang, G.; Hollingsworth R. I. *J. Org. Chem.* **1999**, *64*, 4140–4147.
88. Patel, G. B.; Sprott, G. D. *Crit. Rev. Biotech.* **1999**, *19*, 317–357.
89. Hanford, M.; Peeples, T. L. *Appl. Biochem. Biotech.* **2002**, *97*, 45–62.
90. Gliozzi, A.; Relini, A.; Chong, P. L. G. *J. Membrane Sci.* **2002**, *206*, 131–147.
91. Uchegbu, I. F. *Pharm. J.* **1999**, *263*, 309–318.
92. Lasic, D. D.; Templeton, N. S. *Adv. Drug Deliv. Rev.* **1996**, *20*, 221–266.
93. Gao, X; Huang, L. *Gene Ther.* **1995**, *2*, 710–722.
94. Templeton, N. S.; Lasic, D. D. *Mol. Biotech.* **1999**, *11*, 175–180.
95. Safinya, C. R. *Curr. Opin. Struct. Biol.* **2001**, *11*, 440–448.
96. Oku, N.; Yamazaki, Y.; Matsuura, M.; Sugiyama, M.; Hasegawa, M.; Nango, M. *Adv. Drug Deliv. Rev.* **2001**, *52*, 209–218.
97. Ilies, M. A.; Seitz, W. A; Balaban, A. T. *Curr. Pharm. Design* **2002**, *8*, 2441–2473.
98. Audouy, S. A. L ; de Leij, L. F. M. H.; Hoekstra, D.; Molema, G. *Pharm. Res.* **2002**, *19*, 1599–1605.
99. Templeton, N. S. *Curr. Med. Chem.* **2003**, *10*, 1279–1287.
100. Niculescu-Duvaz, D.; Heyes, J.; Springer, C. J. *Curr. Med. Chem.* **2003**, *10*, 1233–1261.
101. Hirko, A.; Tang, F. X.; Hughes, J. A. *Curr. Med. Chem.* **2003**, *10*, 1185–1193.
102. de Lima, M. C. P.; Neves, S.; Filipe, A.; Duzgunes, N.; Simoes, S. *Curr. Med. Chem.* **2003**, *10*, 1221–1231.
103. Ewert, K.; Slack, N. L.; Ahmad, A.; Evans, H. M.; Lin, A. J.; Samuel, C. E.; Safinya, C. R. *Curr. Med. Chem.* **2004**, *11*, 133–149.
104. Miller, A. D. *Curr. Med. Chem.* **2003**, *10*, 1195–1211.
105. Duzgunes, N.; de Ilarduya, C. T; Simoes, S.; Zhdanov, R. I.; Konopka, K.; de Lima, M. C. P. *Curr. Med. Chem.* **2003**, *10*, 1213–1220.
106. Nakanishi, M. *Curr. Med. Chem.* **2003**, *10*, 1289–1296.
107. Kumar, V. V.; Singh, R. S.; Chaudhuri, A. *Curr. Med. Chem.* **2003**, *10*, 1297–1306.
108. Liu, D. X.; Ren, T; Gao, X. *Curr. Med. Chem.* **2003**, *10*, 1307–1315.
109. Ozer, A. Y.; Farivar, M.; Hincal, A. A. *Eur. J. Pharm. Biopharm.* **1993**, *39*, 97–101.
110. Needham, D.; Dewhirst, M. W. *Adv. Drug Deliv. Rev.* **2001**, *53*, 285–305.
111. Kono, K. *Adv. Drug Deliv. Rev.* **2001**, *53*, 307–319.
112. Drummond, D. C.; Zignani, M.; Leroux, J.-C. *Prog. Lipid Res.* **2000**, *39*, 409–460.
113. Drummond, D. C.; Daleke, D. L. *Chem. Phys. Lipids* **1995**, *75*, 27–41.
114. Gerasimov, O. V.; Boomer, J. A.; Qualls, M. M.; Thompson, D. H. *Adv. Drug Deliv. Rev.* **1999**, *38*, 317–338.
115. Guo, X.; Szoka, F. C. *Acc. Chem. Res.* **2003**, *36*, 335–341.
116. Guo, X.; Szoka, F. C., Jr. *Bioconjug. Chem.* **2001**, *12*, 291–300.
117. Rui, Y.; Wang, S.; Low, P. S.; Thompson, D. H. *J. Am. Chem. Soc.* **2000**, *64*, 27–37.
118. Bondurant, B; O'Brien, D. F. *J. Am. Chem. Soc.* **1998**, *120*, 13541–13542.
119. Chesnoy, S.; Huang, L. *Annu. Rev. Biophys. Biomol. Struct.* **2000**, *29*, 27–47.
120. Bisby, R. H; Mead, C.; Morgan, C. C. *Biochem. Biophys. Res. Commun.* **2000**, *276*, 169–173.

121. Gerasimov, O. V.; Boomer, J. A.; Qualls, M. M.; Thompson, D. H. *Adv. Drug Deliv. Rev.* **1999**, *38*, 317–338.
122. Shum, P.; Kim, J.-M.; Thompson, D. H. *Adv. Drug Deliv. Rev.* **2001**, *54*, 273–284.
123. Zhang, Z. Y.; Smith, B. *Bioconjug. Chem.* **1999**, *10*, 1150–1152.
124. Bochet, C. G. *J. Chem. Soc., Perkin Trans.* **2002**, *1*, 125–142.
125. Adams, S. R.; Tsien, R. Y. *Annu. Rev. Physiol.* **1993**, *55*, 755–784.
126. Holmes, C. P. *J. Org. Chem.*, **1997**, *62*, 2370–2380.
127. Wan, Y.; Angleson, J. K.; Kutateladze, A. G. *J. Am. Chem. Soc.* **2002**, *124*, 5610–5611.
128. Zalipsky, S.; Qazen, M.; Walker, J. A.; II; Mullah, N.; Quinn, Y. P.; Huang, S. K. *Bioconjug. Chem.* **1999**, *10*, 703–707.
129. Moghimi, S. M.; Rajabi-Siahboomi, A. R. *Adv. Drug Deliv. Rev.* **2000**, *41*, 129–133.
130. Park, Y. S. *Biosci. Rep.* **2002**, *22*, 267–281.
131. Maruyama, K. *Biosci. Rep.* **2002**, *22*, 251–266.
132. Pinto-Alphandary, H.; Andremont, A.; Couvreur, P. *Int. J. Antimicrob. Agents* **2000**, *13*, 155–168.
133. Maruyama, K.; Ishida, O.; Takizawa, T.; Moribe, K. *Adv. Drug Deliv. Rev.* **1999**, *40*, 89–102.
134. Harasym, T. O.; Bally, M. B.; Tardi, P. *Adv. Drug Deliv. Rev.* **1998**, *32*, 99–118.
135. Papahadjopoulos, D.; Kirpotin, D. B.; Park, J. W.; Hong, K. L.; Shao, Y.; Shalaby, R.; Colbern, G.; Benz, C. C. *J. Liposome Res.* **1998**, *8*, 425–442.
136. Hung, M. C.; Chang, J. Y. J.; Xing, X. M. *Adv. Drug Deliv. Rev.* **1998**, *30*, 219–227.
137. Forssen, E.; Willis, M. *Adv. Drug Deliv. Rev.* **1998**, *29*, 249–271.
138. Singh, M. *Curr. Pharm. Design* **1999**, *5*, 443–451.
139. Wagner, E.; Curiel, D.; Cotton, M. *Adv. Drug Deliv. Rev.*, **1994**, *14*, 113–135.
140. Singh, M. *Curr. Pharm. Design.* **1999**, *5*, 443–451.
141. Lee, R. J.; Huang, L. *J. Biol. Chem.* **1996**, *271*, 8481–8487.
142. Lee, R. J.; Low, P. S. *Biochim. Biophys. Acta* **1995**, *1233*, 134–144.
143. Shudimack, J.; Lee, R. J. *Adv. Drug Deliv. Rev.* **2000**, *41*, 147–162.
144. Reddy, J. A.; Low, P. S. *Crit. Rev. Ther. Drug* **1998**, *15*, 587–627.
145. Diebold, S. S.; Kursa, M.; Wagner, E.; Cotten, M.; Zenke, M. *J. Biol. Chem.* **1999**, *274*, 19087–19094.
146. Remy, J. S.; Kichler, A.; Mordvinov, V.; Schuber, F.; Behr, J. P. *Proc. Natl. Acad. Sci. USA* **1995**, *92*, 1744–1748.
147. Wagner, E.; Zenke, M.; Cotton, M.; Beug, H.; Birnstiel, M. L. *Proc. Natl. Acad. Sci. USA* **1990**, *87*, 3410–3414.
148. Luo, Y.; Ziebell, M. R.; Prestwich, G. D. *Biomacromolecules* **2000**, *1*, 208–218.
149. Hashida, M.; Nishikawa, M.; Yamashita, F.; Takakura, Y. *Adv. Drug Deliv. Rev.* **2001**, *52*, 187–196.

20

REGULATORY AND INTELLECTUAL PROPERTY ISSUES IN DRUG DELIVERY RESEARCH

SHIHONG NICOLAOU

Technology Transfer and Intellectual Property Services, University of California, San Diego, 9500 Gilman Drive, La Jolla, CA 92093

20.1. INTRODUCTION

The U.S. Patent Office gives the inventor or assignee the right to exclude others, for a limited time, from making, using, or selling an invention by granting a patent. What is not well understood by the public is that the grant of a patent does not confer on the inventor the right to make, use, or sell the invention. Rather, it permits the

Drug Delivery: Principles and Applications Edited by Binghe Wang, Teruna Siahaan, and Richard Soltero
ISBN 0-471-47489-4

inventor to exclude others from such acts. For pharmaceuticals, the right comes only with an approval from the U.S. Food and Drug Administration (FDA). In the United States, patents can be obtained on processes, machines, articles of manufacture, and compositions of matters. The inventions are patentable as long as they are new, useful, and unobvious and as long as certain events, commonly referred to as "statutory bars," have not occurred before a patent application is filed.[1]

Among all the patents awarded each year by patent offices worldwide, a large proportion concerns pharmaceutical inventions. Research-based pharmaceutical companies rely on patents to protect their investments in researching and developing new drugs. Unlike other industries, the pharmaceutical industry is regulated by government agencies in order to assure the safety and efficacy of the invented therapeutic products. Therefore, there is a delay between filing a patent and placing a product in the market. It is then conceivable that the pharmaceutical companies receive a more limited patent exclusivity. Although pharmaceutical companies have developed various strategies to protect their intellectual property and to extend the patent terms by any available means, generic companies have been aggressively attacking the validity of some patents in order to obtain an early market entry. One successful strategy for many pharmaceutical companies is the development of products with embedded drug delivery designs. These approaches target existing drugs, which may potentially delay the expiration of the patent term of the product and consequently hinder the generic competitors.

20.2. TRENDS IN DRUG DELIVERY RESEARCH

A wave of drug patent expirations is providing generic companies with many opportunities to develop and market less expensive generic versions of prescription drugs. In recent years, tremendous efforts have been made to provide improved versions of these drugs to address issues related to patient compliance, ease of drug administration, and reduced side effects. Drug delivery technologies are developed to address the drawbacks of conventional dosage forms and administration. Patents for a large number of blockbuster drugs will soon expire. Many of these prescription drugs have unfavorable physicochemical and/or pharmacokinetic properties; thus, there are various limitations on the dosing regimen and undesirable side effects in the conventional delivery systems. There can be increased therapeutic value when the drugs are delivered through the appropriate delivery systems or formulations.

For example, oral controlled-release systems can better regulate blood level concentrations of the active drug and diminish side effects caused by spikes in blood concentrations. Nasal and pulmonary delivery systems may permit delivery of biopharmaceuticals such as proteins and peptides that would otherwise need to be administered by injection. Transdermal delivery systems can be effective for certain drugs because they can help to avoid first-pass metabolism and to better regulate systemic delivery of the active drug. Due to the need for more effective drug delivery technologies, there is growing competition in the area of drug delivery research.

20.3. CURRENT PATENT SYSTEM FOR PHARMACEUTICALS

The patent system allows inventors or patent owners an exclusive right to market their invention for a permitted limited time. Prior to June 8, 1995, the term of a U.S. patent was 17 years from the date of issue. After that date, the patent term was changed to 20 years from the earliest filing date. Thus, the patent term is being consumed prior to grant while a patent application is being examined. This new patent rule has led to a serious decrease in the term of patents. Pharmaceutical and biotechnology-based patent applications are most affected because of delays in processing and the frequent need to submit supplemental data to the Patent Office. As a result of recent changes in U.S. patent law, patent term can now be adjusted if the Patent Office unreasonably delays processing of the application and if patent issuance is delayed due to patent appeals and interferences.[2] In addition, the patent term can be adjusted for pharmaceutical inventions as a result of the Hatch-Waxman Act. Illustrated in Figure 20.1 is a general timeline of the patent prosecution activity relative to the drug development and approval process.

The Hatch-Waxman Act, passsed in 1984, has substantially shaped the legal environment governing FDA approval of generic drug products and U.S. patent terms.[3] It set up a regulatory framework that sought to balance incentives for continued innovation by research-based pharmaceutical companies and opportunities for market entry by generic drug manufacturers. The law now allows generic drug makers to file abbreviated new drug applications (ANDAs) for products covered by patents that are not scheduled to expire for years. Generic drug companies do not have to endure costly and time-consuming safety and efficacy clinical trials by the FDA to bring their products to market. All they need to do is to file an ANDA. As long as a generic product is equivalent to its previously patent-protected predecessor, the generic drug company can sell its generic version of the product after a relatively short FDA review process. The Hatch-Waxman Act was also designed to facilitate drug discovery and development by restoring effective patent life to innovators. It created a system whereby any unexpired patent that covers a pharmaceutical product subject to regulatory review prior to its first commercial marketing can obtain a patent term extension to compensate for regulatory delays.

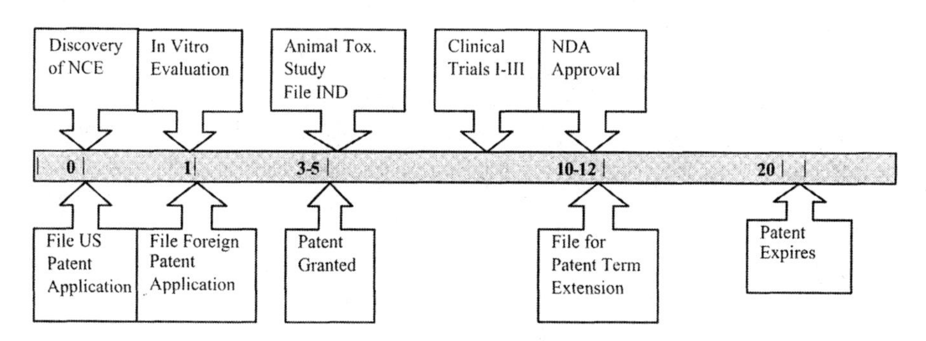

Figure 20.1. Patent Prosecution and Regulatory Timelines.

20.4. STRATEGIES FOR PHARMACEUTICAL PATENTING

Pharmaceutical patents are generally drafted in a form in which the claimed drug is described according to its chemical composition. Additionally pharmaceutical patents are drafted for method-of-use claims that describe the function and the therapeutic use of the drug; for formulation claims that describe the inactive ingredients used in the final dosage form or the delivery device; and for process claims that describe the processes used for the drug's production. The latter three types of claims generally broaden the protection that is supported by the structurally defined claims. Drug patents which claim therapeutic use, formulations, and processes may allow exclusive rights well beyond the expiration of the original composition patent.

There is a hierarchy of protection in pharmaceutical patents. At the apex are composition of matter patents, followed by method-of-use patents, formulation patents, and manufacturing process patents. For example, claiming the chemical structure of acetylsalicylic acid is the composition patent for aspirin. A method-of-use patent would be directed to the relief of pain and inflammation using aspirin. Formulation patents could cover the tablet, the capsule, or other dosage delivery forms of aspirin in commercial use. A synthetic method of producing acetylsalicylic acid would be covered in the process patent. In general, composition patents provide the strongest exclusivity and therefore are the most desirable. Method-of-use patents are the second most important. Formulation and process patents can help with complex and hard-to-manufacture drugs.

Various strategies are pursued by pharmaceutical companies to extend the product life cycle. Under the Hatch-Waxman Act, patent term extensions are available for patented drug substances, drug products, method of use, and method of manufacturing to compensate for the time lost in the FDA's regulatory approval process. However, the limitations of the statute of the following aspects have to be considered:

1. A patent term can be extended only once;
2. Only one patent can be extended for a given regulatory review period;
3. FDA authorization of the drug product must be for the first commercial marketing or use of the drug product's active ingredient; and
4. Extensions are limited to what is covered by FDA authorization.

An optimal patent strategy has to be designed to maximize the term extension under all these limitations. For example, several related chemical entities can be discovered simultaneously, including the active ingredient and its stereoisomers or different crystalline forms. For patent purposes, each form may be considered to be a new active ingredient, for which a patent term extension is potentially available under condition 3 above. However, if multiple forms of the active ingredient were claimed in the same patent, the patentee's term extension would be limited only to the first of these approved by the FDA (condition 1). It would then be advantageous to claim the different ingredients in separate patents in order to preserve the possibility of extending the patent life for each ingredient.

Most often, separate patents which cover composition claims of new active ingredients and method-of-use claims are obtained. Because both patents will presumably have a common FDA regulatory review period, condition 2 limits term extension to only one of the patents. Thus, the composition claims and method-of-use claims for the single drug product should be combined, if possible, into a single patent so that all the claims are eligible for extension. Sometimes the pharmaceutical company must decide which patent to extend when two or more patents that cover the same approved drug product are issued. Although a composition patent is typically the preferred choice for the extension, this may not be the best strategy when a formulation patent or a new drug delivery patent has a longer remaining term.

20.5. ISSUES CONCERNING DRUG DELIVERY PATENTS

Drug delivery patents presumably create new opportunities for product exclusivity. For instance, delivery device patents which cover nasal sprays, metered dose inhalers (MDIs), and dry powder inhalers (DPIs) can protect a drug long after their composition patent expires. Sustained-release delivery systems can also provide extended protection for approved drugs. Special drug delivery technologies that can potentially increase the therapeutic benefit and extend the exclusivity of the product include prodrugs and active metabolites. Some of these delivery technologies, however, present complicated issues in terms of intellectual property protection. Litigation relating to drug delivery patents is abundant. A few examples are provided in the following sections to demonstrate the complexities of the issue.

20.5.1. Patents on Oral Drug Delivery

Oral drug delivery inventions represent the most successful of all delivery technologies. Specifically, the oral controlled-release technology is the most adaptive technology, giving the research-based pharmaceutical companies a proprietary position.

Although Paxil's patent expires in 2006, a number of generic companies have already filed ANDAs to market generic versions of the drug (paroxetine hydrochloride). Paxil's patent holder, GlaxoSmithKline, has been fending off patent challenges by filing patent-infringement suits against these generic companies. At the same time, GlaxoSmithKline had extended the Paxil product line by launching a new controlled-release formulation and by gaining approval for new indications. Paxil has been approved in 28 countries for treating posttraumatic stress disorder. Paxil CR is a controlled-release formulation of paroxetine which was launched in the United States in April 2002. It is intended for the treatment of major depressive disorder and panic disorder. Paxil CR combines the efficacy of paroxetine with an advanced technology that controls dissolution and absorption of the drug in the body. In addition to demonstrating efficacy, clinical studies of Paxil CR indicated a favorable tolerability profile with a lower incidence of patient noncompliance due to adverse events. GlaxoSmithKline's efforts were successful in extending the product life cycle by applying the controlled-release technology and by obtaining drug delivery patents.

Pfizer's Procardia XL is a once daily calcium channel blocker (CCB) for hypertension and angina. It delivers nifedipine via Alza's osmotic pump, a patented high-tech sustained-release (SR) formulation. Several generic companies are currently seeking FDA approval for an "A/B" substitute using low-tech SR formulations. The generic companies claim that they are not infringing Pfizer's patent estate, which includes:

1. Nifedipine in combination with polymeric materials (November 2000);
2. Alza's GITS sustained-release system (September 2003); and
3. Nifedipine small crystals (November 2010).

Penwest and its partner, Mylan, were the first generic drug companies to file an ANDA (June 1997); the FDA awarded final approval for the generic Procardia XL 30 mg in December 1999. However, Mylan was not expected to launch their version of the product until Pfizer's lawsuit against Mylan was resolved. In March 2000, Pfizer settles its litigation with Mylan. Mylan agreed not to launch its generic version of Procardia XL, but instead will sell Pfizer's Procardia XL under a different brand name. It appears that Pfizer has managed to escape rapid generic competition for Procardia XL, which should now decrease managed to escape in a steady manner.

20.5.2. Patents on Prodrugs

In recent years, there have been several litigated cases focusing on prodrug patents. The issues are related to whether infringement occurs when a prodrug is metabolized into a patented compound in the body of a patient.

The prodrug issue was discussed in a lawsuit of *Hoechst-Roussel Pharmaceuticals, Inc. v. Lehman* in 1997.[4] Hoechst-Roussel owned a U.S. patent which contained claims covering 1-hydroxy-tacrine, a drug for treatment of Alzheimer's disease. Hoechst sued Warner-Lambert for patent infringement, asserting that tacrine hydrochloride, marketed by Warner-Lambert under the brand name COGNEX, infringed their patent. COGNEX is metabolized into 1-hydroxy-tacrine in the body of a user. In a consent decree, Warner-Lambert admitted that COGNEX infringed Hoechst's patent.[4] However, the Patent Office determined that Hoechst was not eligible for term extension based on the regulatory review period for Warner-Lambert's COGNEX. The ruling was based on the fact that Hoechst was not involved directly or indirectly in the regulatory approval for tacrine hydrochloride, and their patent did not claim tacrine hydrochloride, as required by the statute. In the appeals, both the district court and the federal circuit court ruled in favor of the Patent Office, stating that Hoechst was not entitled to statutory extension.

In a lawsuit involving the Ortho Pharmaceutical Corporation, the issue was whether Ortho's norgestimate infringed the patentee's claims to norgestrel under the doctrine of equivalence.[5] Both compounds are progestin-type steroid hormones used as oral contraceptives. The two compounds share the same fused ring core structure but have different substituents at two positions on the ring system. At the trial, there seemed to be no doubt that norgestimate is converted to norgestrel by metabolism in the human body. The court found that both drugs act by the same biochemical mechanism, i.e., induction of a progestational response through binding to progestational receptors. On this basis, the court found infringement under the doctrine of equivalence.[5] The court specifically stressed the fact that the Ortho scientists who originally synthesized norgestimate did so by starting with norgestrel and using well-known basic laboratory techniques. Thus, Ortho's norgestimate was found to infringe the patentee's claims to norgestrel in the trial.

In 2002, another judicial trial in claim construction prevented a patentee from covering *in vivo* conversion. Bristol-Myers Squibb has, upon expiration of its basic patent to the use of buspirone to treat anxiety, obtained a patent covering the systemic administration of an effective anxiolytic (anxiety-treating) dose of the 6-hydroxy metabolite of buspirone. The district court construed the claim as directed to the administration of an "externally-measured quantity of the metabolite into the body, and not to the administration of a dose of buspirone into the body, which, in turn, produces variable and changing levels (not doses) of the metabolite in the bloodstream."[6] Therefore, this patent could not be asserted against the generic manufacturers and sellers of generic buspirone.

Although the prodrug issue has arisen on several occasions, it is unclear if any U.S. court has found it necessary to rule on whether the *in vivo* conversion to a patented product constitutes patent infringement.

20.6. CONCLUSIONS

As drug delivery strategies become increasingly important in extending the product life cycle, there has been fierce competition in the area of drug delivery research. The legal system may see a proportional increase in litigation centering on these types of patents. Companies have traditionally used patent litigation as means of protecting their investment in drug development. A typical model of patent litigation is to assert ownership of one or more patents to prevent competitors from marketing the same drug.

Patent law in the area of pharmaceuticals in general has become more complex, both from the government regulation policy side and in terms of the details of patent term adjustment between the Patent Office and the FDA's approval process. A better understanding of drug patent and regulatory issues will help to maximize protection of an approved drug product.

REFERENCES

1. 35 U.S.C. §102.
2. The American Inventors Protection Act, **1999**.

3. Drug Price Competition and Patent Term Restoration Act, Pub. L. No. 98-417, 98 Stat. 1585, **1984**.
4. *Hoechst-Roussel Pharmaceuticals, Inc. v. Lehman*, 109 F. 3d 756, 42 USPQ2d 1220 (Fed. Cir. **1997**).
5. *Ortho Pharmaceutical Corp. v. Smith*, 18 USPQ2d 1977 (E. D. PA. **1990**).
6. *In re. Buspirone*, 185 F. Supp. 2d 340 (S.D.N.Y. **2002**).

INDEX

Drug Delivery: Principles and Applications Edited by Binghe Wang, Teruna Siahaan, and Richard Soltero
ISBN 0-471-47489-4